AF708863

Media and Radio Signal Processing for Mobile Communications

Get to grips with the principles and practise of signal processing used in real mobile communications systems. Focusing particularly on speech and video processing, pioneering experts employ a detailed, top-down analytical approach to outline the network architectures and protocol structures of multiple generations of mobile communications systems, identify the logical ranges where media and radio signal processing occur, and analyze the procedures for capturing, compressing, transmitting and presenting media. Chapters are uniquely structured to show the evolution of network architectures and technical elements between generations up to and including 5G, with an emphasis on maximizing service quality and network capacity through reusing existing infrastructure and technologies. Examples and data taken from commercial networks provide an in-depth insight into the operation of a number of different systems, including GSM, cdma2000, W-CDMA, LTE, and LTE-A, making this a practical, hands-on guide for both practicing engineers and graduate students in wireless communications.

Kyunghun Jung is a Principal Engineer at Samsung Electronics, where he leads the research and standardization for bringing immersive media services and vehicular applications to 5G systems.

Russell M. Mersereau is Regents Professor Emeritus in the School of Electrical and Computer Engineering at the Georgia Institute of Technology, and a Fellow of the IEEE.

This impressive book provides an excellent comprehensive explanation of the principles and practices of media and radio signal processing in real mobile communications systems. It also wonderfully explains the evolution of signal processing operations and thereby gives the reader a deep insight into the challenges and how they were overcome.

Kari Järvinen, Nokia Technologies

With today's mobile user experience so influenced by multimedia services, providing a clear background on the fundamentals of the entire protocol stack, from the physical layer to the multimedia codecs, media handling, and immersive media, is an invaluable book for understanding today's mobile cellular systems. The authors' experience with the development of the protocols and standards of these systems provides unknown insights into the reason for their development that allows the reader to better understand these technologies.

Nikolai Leung, Qualcomm

Media and Radio Signal Processing for Mobile Communications

KYUNGHUN JUNG
Samsung Electronics

RUSSELL M. MERSEREAU
Georgia Institute of Technology

CAMBRIDGE
UNIVERSITY PRESS

CAMBRIDGE
UNIVERSITY PRESS

University Printing House, Cambridge CB2 8BS, United Kingdom

One Liberty Plaza, 20th Floor, New York, NY 10006, USA

477 Williamstown Road, Port Melbourne, VIC 3207, Australia

314–321, 3rd Floor, Plot 3, Splendor Forum, Jasola District Centre, New Delhi – 110025, India

79 Anson Road, #06–04/06, Singapore 079906

Cambridge University Press is part of the University of Cambridge.

It furthers the University's mission by disseminating knowledge in the pursuit of education, learning, and research at the highest international levels of excellence.

www.cambridge.org
Information on this title: www.cambridge.org/9781108421034
DOI: 10.1017/9781108363204

© Cambridge University Press 2018

This publication is in copyright. Subject to statutory exception and to the provisions of relevant collective licensing agreements, no reproduction of any part may take place without the written permission of Cambridge University Press.

First published 2018

Printed in the United Kingdom by TJ International Ltd. Padstow Cornwall

A catalogue record for this publication is available from the British Library.

Library of Congress Cataloging-in-Publication Data
Names: Jung, Kyunghun, 1970– author. | Mersereau, Russell M., author.
Title: Media and radio signal processing for mobile communications /
Kyunghun Jung, Samsung Electronics, Russell M. Mersereau, Georgia Institute of Technology.
Description: New York, NY : Cambridge University Press, 2018. |
Includes bibliographical references and index.
Identifiers: LCCN 2017054695 | ISBN 9781108421034 (alk. paper)
Subjects: LCSH: Multimedia communications. | Mobile communication systems. |
Signal processing–Digital techniques.
Classification: LCC TK5105.15 .J86 2018 | DDC 621.39/167–dc23
LC record available at https://lccn.loc.gov/2017054695

ISBN 978-1-108-42103-4 Hardback

DAMAGED

Cambridge University Press has no responsibility for the persistence or accuracy of URLs for external or third-party internet websites referred to in this publication and does not guarantee that any content on such websites is, or will remain, accurate or appropriate.

© 2001. 3GPP™ TSs and TRs are the property of ARIB, ATIS, CCSA, ETSI, TTA and TTC who jointly own the copyright in them. They are subject to further modifications and are therefore provided to you "as is" for information purposes only. Further use is strictly prohibited.

To Bongho, Hyesook, and Hoonjung, and to Martha

Contents

Preface

Advances in media and radio signal processing have been the driving forces behind the industrial and social changes enabled by the widespread use of smartphones and mobile multimedia communications. We started our research on these exciting topics in January 1999, as the expectations for 3G mobile communications systems and their multimedia services were generating great excitement. Our research, initially from an academic viewpoint for a doctoral dissertation, shifted to more practical concerns when Dr. Jung joined Samsung Electronics and began working to design 3G and 4G mobile communications systems. We realized that many of the approaches and assumptions made in the literature were not realistic in actual systems and we identified new opportunities for improvements.

Some of these approaches, which were based on extensions of conventional joint source-channel coding, were inadequate to reflect real situations, such as the high cost of frequency spectrum or the need for a network entity to be responsible for controlling the tradeoff between media quality and network capacity. Books that analyzed real mobile communications systems, on the other hand, focused on the radio signal processing and network architectures, while providing limited guidance on the needs of the media signal processing. In light of the significant discrepancy between the work in academia and industry that we observed, we prepared this book to explain the principles and practices of both media and radio signal processing used in actual mobile communications systems.

We examine multiple generations of commercially deployed or standardized mobile communications systems and analyze in detail the areas where the media and radio signal processing take place and interact. We trace the evolution of the signal processing operations, as new technical elements were introduced to meet the challenges. We identify where elements were inherited from earlier systems for compatibility, and explain how the media codecs, network architectures, and radio access technologies interact to maximize quality and capacity in a consistent, top-down fashion. From Chapter 2 to Chapter 9, each pair of chapters covers the basic construction and operating principles of a mobile communications system and its evolved version in which the initial limitations are partially solved. Each pair is self-contained and can be read independently. Proceeding to the next pair shows more radical approaches made when evolutionary enhancements were not sufficient and completely new elements were required. We would like to point out that the signal processing techniques in the early chapters are no less important than those in the later chapters on more state-of-the-art systems, as they often become critical design constraints when new systems are designed.

Several media compression and wireless transmission techniques that looked promising from their theoretical analysis and even made it into standardization and implementation, ultimately proved to be unsuccessful in attaining the envisioned performance in real environments. Since managing complexity and stability is a key requirement in the design of complex systems such as mobile communications, many procedures designed for previous systems are re-used. We discuss examples of borrowed technical concepts from earlier systems that are applied to different areas successfully. The simulations of communications systems often produce varying results depending on the complexity of the system models or configuration of their parameters. Moreover, evaluations of media quality often require subjective testing. In this book, we present the highest-quality simulation results recognized by the standardization organizations, official results of subjective testing administered by expert agencies contracted for those services, and field measurements from commercially-operational GSM, cdma2000, W-CDMA, and LTE and LTE-A handsets and networks that show the variation of key media and radio parameters during compression and transmission in the time domain.

The trajectory of the technical evolution covered throughout the chapters shows that each generation has introduced new technical elements or absorbed elements previously not included in mobile communications systems, as video in 3G and IP in 4G. These require new types of signal processing to meet the harsh mobile environment. Historically, compression and transmission of media have been the focus of mobile communications, but it is envisioned that other areas of signal processing, e.g., recognition and synthesis of media, will play key roles in next generation systems providing immersive media and vehicular applications. We expect that this book will bridge the gap between academia and industry, and provide its readers with insight for the design, analysis, and implementation of mobile multimedia communications systems.

Acknowledgments

We started our signal processing careers through the books, teaching, and collaborations of A. V. Oppenheim, R. W. Schafer, T. P. Barnwell III, J. H. McClellan, and L. R. Rabiner, whose influence can be seen in the early chapters of this book. It was the DSP Leadership Universities Program of Texas Instruments, granted in April 1999 with the consideration of Gene Franz, Bob Hewes, Panos Papamichalis, and Raj Talluri, that enabled us to initiate our long-term research on the handling and interaction of media over mobile communications. In the systems from GSM to 5G, we were advised by many designers and developers of those systems. For GSM, Paolo Usai, Karl Hellwig, Stefan Bruhn, and Jongsoo Choi shared with us their experience and expertise on this fundamental and still dominant mobile communications system. For IS-95 and cdma2000, we were deeply influenced by the work of Jhongsam Lee, Vieri Vanghi, and Yucheun Jou. For W-CDMA and 3G-324M, Kwangcheol Choi, Yonghyun Lim, and Youngmin Jeong helped us with the real-time transmission of media over the system. We enjoyed the development and deployment of EVS over 4G systems with Hosang Sung, Kihyun Choo, Jonghoon Jeong, and Woojung Park. Thomas Belling contributed advice and suggestions we learned about core network issues. Terence Betlehem helped us understand a new signal processing area, ambisonic audio, and write the audio section of virtual reality. We also appreciate the ongoing efforts of Kyungmo Park and Imed Bouazizi for the realization of 5G systems, as outlined in the last chapter. We would like to thank especially Kari Järvinen, Tomas Frankkila, and Nikolai Leung for their decade-long services at the MTSI SWG during the historical transitions from circuit-switched to packet-switched mobile multimedia communications systems and the introduction of IMS. This small group of experienced and versatile experts adroitly handled complex engineering problems in the last stage of standardization and development where many technical issues are interwoven, and shared the thrill of stabilizing those systems just before their worldwide launches. With Ingemar Johansson, we introduced the negotiation of video resolution to the Internet community, via RFC 6236. We would also like to thank Byungoh Kim who managed the hosting of standardization meetings, in which many important technical decisions were made, at exotic venues in Korea. Finally, we appreciate the generous permission of NTT DOCOMO, Innowireless, Accuver, 3GPP, 3GPP2, and Samsung Electronics for the use of their images, experimental data, and other precious information that constitute key features of this book.

Glossary

4GV 4th Generation Vocoder. 164

ACELP Algebraic Code Excited Liner Prediction. 64
ACI Adjacent Channel Interference. 91
ACK Acknowledgment. 328
ACS Active Codec-mode Set. 75
ADPCM Adaptive Differential Pulse Coded Modulation. 19
AES Advanced Encryption Standard. 313
AL Adaptation Layer. 263
AL2 Adaptation Layer Type 2. 198
AL3 Adaptation Layer Type 3. 263
AM Acknowledged Mode. 201
AMC Adaptive Modulation and Coding. 302
AMPS Advanced Mobile Phone System. 6
AMR Adaptive Multi-Rate. 71
AMR-WB Adaptive Multi-Rate Wideband. 354
AOP Anchor Operating Point. 169
APCM Adaptive Pulse Coded Modulation. 48
APS Adaptive Pulse Shaping. 96
AQPSK Adaptive Quadrature Phase Shift Keying. 92
ARFCN Absolute Radio Frequency Channel Number. 44
AS Application Specific. 366
AS Application Server. 364
ASN.1 Abstract Syntax Notation One. 278
ATM Asynchronous Transfer Mode. 198
AWGN Additive White Gaussian Noise. 86

BCH Bose–Chaudhuri–Hocquenghem. 12
BER Bit Error Rate. 34
BIC Blind Interference Cancellation. 87
BLER Block Error Rate. 228
BLP Bitmask of following Lost Packets. 408
BMC Broacasting and Multicasting Control. 202
BPSK Binary Phase Shift Keying. 196

BS Base Station. 7
BSC Base Station Controller. 2
BSR Buffer Status Report. 316
BSS Base Station Subsystem. 58
BTS Base Transceiver Station. 2
BWE Bandwidth Extension. 389
BWM Bandwidth Multiplier. 427

CA Carrier Aggregation. 319
CABAC Context Adaptive Binary Arithmetic Coding. 360
CAZAC Constant Amplitude Zero Auto-Correlation. 337
CCI Co-Channel Interference. 91
CCSRL Control Channel Segmentation and Reassembly Layer. 259
CCTrCH Coded Composite Transport Channel. 206
CDMA Code Division Multiple Access. 116
CDVCC Coded Digital Verification Color Code. 34
CELP Code Excited Linear Prediction. 130
CFN Connection Frame Number. 295
CID Context Identifier. 311
CIR Carrier-to-Interference Ratio. 49
CLDFB Complex Modulated Low Delay Filter Bank. 386
CLTD Closed-Loop Transmit Diversity. 223
CMC Codec Mode Command. 74
CMI Codec Mode Indication. 74
CMOS Complementary Metal Oxide Semiconductor. 235
CMR Codec Mode Request. 74
CNG Comfort Noise Generation. 384
CoID Codec Identifier. 426
CP Control Plane. 303
CP Cyclic Prefix. 321
CPICH Common Pilot Channel. 230
CQI Channel Quality Indicator. 328
CRC Cyclic Redundancy Check. 39
CRS Cell-specific Reference Signal. 338
CSMA Carrier Sense Multiple Access. 13
CSoHS Circuit-Switched Voice Services over HSPA. 295
CT Channel Type. 204
CTU Coding Tree Unit. 393
CVO Coordination of Video Orientation. 410
CVSD Continuously Variable Slope Delta Modulation. 416
CVT Continuously Variable Transmission. 175

D-AMPS Digital Advanced Mobile Phone System. 32
DARP Downlink Advanced Receiver Performance. 85

DC Direct Conversion. 216
DCI Downlink Control Information. 333
DCT Discrete Cosine Transform. 240
DFT Discrete Fourier Transform. 21
DL-SCH Downlink Shared Channel. 316
DM Device Management. 427
DMRS Demodulation Reference Signal. 328
DN Data Network. 453
DP Data Partitioning. 257
DPCCH Dedicated Physical Control Channel. 208
DPDCH Dedicated Physical Data Channel. 208
DRX Discontinuous Reception. 274
DS Dynamic Scheduling. 341
DS Direct Source. 435
DS Direct Spread. 117
DSP Digital Signal Processor. 183
DST Discrete Sine Transform. 394
DTCH Dedicated Traffic Channel. 204
DTMF Dual-Tone Multi-Frequency. 366
DTS DARP Test Scenario. 85
DTX Discontinuous Transmission. 39
DU Digital Unit. 297

E-UTRAN Evolved Universal Terrestrial Radio Access Network. 297
ECN Explicit Congestion Notification. 302
EDGE Enhanced Data Rates for GSM Evolution. 90
EEP Equal Error Protection. 38
EFR Enhanced Full Rate. 64
EIB Erasure Indicator Bit. 151
EMR Enhanced Measurement Report. 59
Enhanced aacPlus Enhanced Advanced Audio Coding Plus. 385
EO End Office. 8
EPC Evolved Packet Core. 298
ERT Error Resilience Tool. 256
ESN Electronic Serial Number. 141
ESP Encapsulating Security Payload. 311
EV-DO Evolution Data Only. 294
EVRC Enhanced Variable Rate Codec. 155
EVS Enhanced Voice Services. 303

F-FCH Forward Fundamental Channel. 149
FBI Feedback Information. 212
FBR Fixed Bit-Rate. 30
FC Full Context. 308

FCELP Full-Rate Code Excited Linear Prediction. 169
FDD Frequency Division Duplex. 7
FDMA Frequency Division Multiple Access. 33
FET Frame Early Termination. 180
FFT Fast Fourier Transform. 240
FI Framing Information. 281
FIR Full Intra Request. 407
FM Frequency Modulation. 6
FO First-Order. 307
FOV Field of View. 461
FPPP Full-Rate Prototype Pitch Period. 169
FR Full Rate. 47
FSK Frequency Shift Keying. 11

GBR Guaranteed Bit-Rate. 370
GMSK Gaussian Minimum Shift Keying. 54
GOB Groups of Block. 239
GP Guard Period. 45
GPRS General Packet Radio Service. 43
GSC Generic Signal Audio Coder. 384
GSM Global System for Mobile Communications. 18

HARQ Hybrid Adaptive Repeat and Request. 214
HCELP Half-Rate Code Excited Linear Prediction. 169
HEC Header Extension Code. 257
HEC Header Error Control. 261
HEVC High Efficiency Video Coding. 303
HMD Head Mounted Display. 453
HNELP Half-Rate Noise Excited Linear Prediction. 173
HR Half Rate. 66
HRM Half-Rate Max. 173
HRTF Head Related Transfer Function. 461
HSDPA High Speed Downlink Packet Access. 295
HSPA High Speed Packet Access. 294
HSS Home Subscriber Server. 364
HSUPA High Speed Uplink Packet Access. 296

I-CSCF Interrogation Call Session Control Function. 364
IC Interference Cancelation. 84
ICI Inter-Channel Interference. 92
ICM Initial Codec Mode. 81
IDR Instantaneous Decoding Refresh. 234
IE Information Element. 369
IETF Internet Engineering Task Force. 305
IF Intermediate Frequency. 216

IMS IP Multimedia Subsystem. 298
IOT Internet of Things. 452
IP Internet Protocol. 292
IR Initialization and Refresh. 307
IS-54 Interim Standard 54. 32
IS-95 Interim Standard 95. 114
ISDN Integrated Services Digital Network. 195
ISF Immittance Spectral Frequency. 357
ISI Inter-Symbol Interference. 144
ISO International Organization for Standardization. 253
ITU-T International Telecommunication Union Telecommunication Standardization Sector. 253

JBM Jitter Buffer Management. 363
JD Joint Demodulation. 87
JPEG Joint Photographic Experts Group. 258

LAR Log Area Ratio. 28
LCD Liquid Crystal Display. 242
LCG ID Logical Channel Group Identifier. 316
LCID Logical Channel Identifier. 315
LDPC Low Density Parity Check. 455
LEC Local Exchange Carrier. 8
LGP Linearized GMSK Pulse. 90
LOS Line-of-Sight. 44
LPC Linear Predictive Coding. 22
LS Last Segment. 281
LSB Least Significant Bit. 232
LSF Line Spectral Frequency. 28
LTE-A Long Term Evolution Advanced. 435

MAC Media Access Control. 201
MBM Motion Boundary Marker. 257
MBMS Multimedia Broadcast Multicast Service. 202
MBR Maximum Bit-Rate. 370
MC Multiplex Control. 261
MCPTT Mission Critical Push To Talk. 439
MCS Modulation and Coding Scheme. 330
MD Music Detector. 165
MDCT Modified Discrete Cosine Transform. 383
MIB Master Information Block. 342
MIMO Multiple-Input and Multiple-Output. 330
MIPS Million Instructions Per Second. 96
MM Mixed Mode. 135
MME Mobility Management Entity. 298

MO Management Object. 428
MONA Media Oriented Negotiation Acceleration. 281
MOS Media Oriented Setup. 282
MOS Mean Opinion Score. 6
MPC Media Preconfigured Channels. 282
MPEG Motion Picture Expert Group. 253
MPL Multiplex Payload Length. 264
MRC Maximal Ratio Combining. 127
MS Mobile Station. 2
MSC Mobile Switching Center. 2
MSE Mean Square Error. 237
MSRG Modular Shift Register Generation. 120
MTSI Multimedia Telephony Service for IMS. 299
MTSIMA MTSI Media Adaptation. 430
MTSINP MTSI Network Preference. 428
MTSO Mobile Telephone Switching Office. 7
MTU Maximum Transfer Unit. 301
MUD Multi-User Detector. 215
MuMe Multi-Media. 426
MUROS Multi-User Re-using One Slot. 109
MUX Multiplexer. 260

N-AMPS Narrowband Advanced Mobile Phone System. 32
NACK Negative Acknowledgment. 407
NAL Network Adaptation Layer. 359
NAS Non-Access Stratum. 303
NB Narrow Band. 19
NC No Context. 308
NELP Noise Excited Linear Prediction. 169
NFV Network Function Virtualization. 454
NMT Nordic Mobile Telephone. 43
NR New Radio. 454
NRZ Non-Return-to-Zero. 12
NSRP Numbered Simple Re-transmission Protocol. 259

O-mode Bi-directional Optimist Mode. 305
O-TCH/AHS Adaptive Multi-Rate Speech Channel at 8-PSK Half Rate. 87
OFDM Orthogonal Frequency Division Multiplexing. 319
OID Organization Identifier. 426
OLED Organic Light Emitting Diode. 242
OLTD Open-Loop Transmit Diversity. 223
OoBTC Out-of-Band Transcoder Control. 231
OQPSK Offset QPSK. 144
OSC Orthogonal Sub-Channel. 95

OSI Open Systems Interconnection. 202
OTD Orthogonal Transmit Diversity. 162
OTT Over The Top. 457
OVSF Orthogonal Variable Spreading Factor. 213

P-CSCF Proxy Call Session Control Function. 364
P-GW Packet Data Network Gateway. 298
PCC Primary Component Carrier. 422
PCCC Parallel Concatenated Convolutional Code. 210
PCEF Policy and Charging Enforcement Functionality. 365
PCell Primary Cell. 422
PCG Power Control Group. 140
PCM Pulse Coded Modulation. 2
PCRF Policy and Charging Rules Function. 364
PCS Personal Communications Service. 151
PDB Packet Delay Budget. 373
PDC Personal Digital Cellular. 74
PDCCH Physical Downlink Control Channel. 342
PDCP Packet Data Convergence Protocol. 201
PDSCH Physical Downlink Shared Channel. 332
PDU Protocol Data Unit. 200
pDVD Percentage Degraded Video Duration. 249
PELR Packet-Error Loss Rate. 373
PEMR Packet Enhanced Measurement Report. 59
PHICH Physical Hybrid-ARQ Indicator Channel. 327
PHR Power Headroom Report. 316
PHS Personal Handy-Phone System. 258
PHY Physical Layer. 201
PID Packet ID. 408
PIP Picture In Picture. 269
PLI Picture Loss Indication. 407
PLR Packet Loss Ratio. 405
PM Packet Marker. 261
PMI Precoding Matrix Indicator. 328
PMRM Power Measured Report Message. 152
PN Pseudo Noise. 117
PPI Pixels Per Inch. 452
PPP Prototype Pitch Period. 169
PRACK Provisional Response Acknowledgment. 375
PSD Power Spectral Density. 117
PSNR Peak Signal to Noise Ratio. 243
PSTN Public Switched Telephone Network. 2
PSVT Packet Switched Video Telephony. 404
PT Payload Type. 263

PUCCH Physical Uplink Control Channel. 316
PUSCH Physical Uplink Shared Channel. 323

QCELP Qualcomm Code Excited Linear Prediction. 130
QCELP-13 Qualcomm Code Excited Linear Prediction 13 kbps. 151
QCI QoS Class Identifier. 372
QNELP Quarter-rate Noise Excited Linear Prediction. 169
QOF Quasi-Orthogonal Function. 190
QoS Quality of Service. 199
QPP Quadrature Permutation Polynomial. 324
QPPP Quarter-rate Prototype Pitch Period. 169
QPSK Quadrature Phase Shift Keying. 196

R-FCH Reverse Fundamental Channel. 145
R-mode Bi-directional Reliable Mode. 305
RAB Radio Access Bearer. 201
RAT Radio Access Technology. 195
RATSCCH Robust AMR Traffic Synchronized Control Channel. 79
RB Resource Block. 320
RC Repeat Count. 262
RC Radio Configuration. 144
RCELP Relaxed Code Excited Linear Prediction. 155
Rev. E Revision E. 168
RF Radio Frequency. 3
RI Rank Indication. 328
RIV Resource Indication Value. 330
RLC Radio Link Control. 201
RM Resynchronization Marker. 256
RM Rate Matching. 221
RNC Radio Network Control. 198
ROHC Robust Header Compression. 201
RoT Rise over Thermal. 191
RPE-LTP Regular Pulse Excitation-Long Term Prediction. 47
RRC Radio Resource Control. 303
RRC Root-Raised Cosine. 96
RRH Remote Radio Head. 297
RS Rate Set. 154
RSCP Received Signal Code Power. 288
RSRP Reference Signal Received Power. 423
RSRQ Reference Signal Received Quality. 423
RSSI Received Signal Strength Indicator. 34
RTCP Real-time Transport Control Protocol. 299
RTP Real-time Transport Protocol. 299
RTT Round Trip Time. 441

RV Redundancy Version. 325
RVLC Reversible Variable Length Code. 257
RXLEV Received Signal Level. 55
RXQUAL Received Signal Quality. 55

S-CSCF Session Call Session Control Function. 364
S-GW Serving Gateway. 298
SACCH Slow Associated Control Channel. 34
SAD Sum of Absolute Difference. 237
SAIC Single Antenna Interference Cancellation. 86
SAO Sample Adaptive Offset. 394
SAT Supervisory Audio Tone. 14
SBC Sub-Band Codec. 416
SBR Spectral Band Replication. 390
SC Static Context. 308
SC-VBR Source Controlled Variable Bit-Rate. 379
SCC Secondary Component Carrier. 422
SCell Secondary Cell. 422
SCH Synchronization Channel. 219
SCPIR Sub-Channel Power Imbalance Ratio. 92
SCS Supported Codec-mode Set. 426
SDP Session Description Protocol. 365
SDU Service Data Unit. 200
SF Signaling Flag. 45
SFH Slow Frequency Hopping. 57
SFN System Frame Number. 342
SIB2 System Information Block 2. 348
SID Silence Descriptor Frame. 41
SIGCOMP Signaling Compression. 301
SIN System Identification Number. 12
SIP Session Initiation Protocol. 299
SIR Signal-to-Interference Ratio. 11
SLF Subscription Locator Function. 364
SMS Short Message Service. 202
SMV Selectable Mode Vocoder. 164
SNR Signal-to-Noise Ratio. 9
SO Second-Order. 307
SPC Signaling of Preconfigured Channels. 282
SPS Semi Persistent Scheduling. 341
SR Spreading Rate. 190
SR Scheduling Request. 316
SRB Signalling Radio Bearers. 201
SRP Simple Re-transmission Protocol. 259
SRS Sounding Reference Signal. 338

SRVCC Single Radio Voice Call Continuity. 391
SSAC Service Specific Access Control. 348
SSN Segment Sequence Number. 281
SSRG Simple Shift Register Generation. 121
ST Signaling Tone. 16
STS Space Time Spreading. 162
STTD Space-Time block coding based Transmit Diversity. 223

TACS Total Access Communications System. 43
TB Tail Bits. 45
TBS Transport Block Set. 204
TBS Transport Block Size. 316
TCH/AFS Full Rate Speech Traffic Channel for AMR. 74
TCH/AHS Half Rate Speech Traffic Channel for AMR. 74
TCH/EFS Full Rate Speech Traffic Channel for EFR. 65
TCH/FS Full Rate Speech Traffic Channel. 44
TCP Transport Control Protocol. 301
TCTF Target Channel Type Field. 204
TCX Transform Codec Excitation. 384
TDMA Time Division Multiple Access. 7
TF Transport Format. 206
TFCI Transport-Format Combination Indicator. 208
TFI Transport-Format Indicator. 206
TFO Tandem Free Operation. 232
TM Traffic Mode. 136
TMMBR Temporary Maximum Media Bit-rate Request. 402
ToC Table of Contents. 361
TPC Transmit Power Control. 212
TRAU Transcoder and Rate Adaptation Unit. 3
TrFO Transcoder Free Operation. 231
TSC Training Sequence Code. 45
TT Traffic Type. 136
TTI Transmission Time Interval. 204

U-mode Uni-directional Mode. 305
UCF Until Closing Flag. 262
UDP User Datagram Protocol. 299
UE User Equipment. 198
UEP Unequal Error Protection. 37
UI User Interface. 268
UICC Universal Integrated Circuit Card. 378
UL-SCH Uplink Shared Channel. 316
UM Unacknowledged Mode. 201
UMB Ultra Mobile Broadband. 451

UMTS Universal Mobile Telecommunications System. 195
UP User Plane. 303
UPF User Plane Function. 453
USAC Unified Speech and Audio Coding. 386

VAD Voice Activity Detector. 49
VAMOS Voice Services over Adaptive Multi-User Channels on One Slot. 91
VBR Variable Bit-Rate. 30
VLC Variable Length Code. 235
VLSI Very Large Scale Integration. 18
VoIP Voice over Internet Protocol. 3
VoLTE Voice over Long Term Evolution. 297
VR Virtual Reality. 457
VSELP Vector Sum Excited Linear Predictive. 32

W-CDMA Wideband Code Division Multiple Access. 195
W-CDMA+ Wideband Code Division Multiple Access Plus. 272
WiMAX Worldwide Interoperability for Microwave Access. 451
WMOPS Weighted Million Operations Per Second. 183

1 Introduction

1.1 Historical Background

Mobile communications systems require a significant financial investment to obtain radio spectrum, which consists of small, but expensive, frequency bands that are used to extend the networks over wide geographic areas. There are additional costs to operate and maintain those networks. Roaming agreements made between service providers can complement insufficient network coverage, but financial constraints still dictate that existing assets such as the backbone networks be re-used whenever possible. As a result, new mobile communications systems are rarely designed without incorporating some elements of earlier systems.

Before discussing the signal processing procedures used by the second and later generations of digital mobile communications systems, it is appropriate to describe the goals and define the performance criteria that were used to construct those procedures. Then we outline the key features of their precursors, the first generation analog mobile communications systems, which introduced the cellular concept. It will become apparent that the design of these early analog systems and the experience gained from operating them had a profound impact on the design of later systems. A more detailed discussion of the technical and social background that drove the development of early mobile communications systems can be found in [Lee (1995), Rappaport (2002)].

1.1.1 Problem Description

In the discussions in this text we have divided the signal processing operations in the mobile communications system into two subsystems: the speech signal processing subsystem and the radio signal processing subsystem. The former incorporates bandwidth-limiting, sampling, and encoding the speech waveform into as few bits as possible while maintaining acceptable speech quality. The latter is concerned with protecting those bits, packaging them, and transmitting them through the network. In some sense the distinction is artificial since the two subsystems interact and are typically implemented on the same processors. On the other hand, they were typically developed by researchers with different technical backgrounds and in most cases are defined by different standards or different parts of the same standard. Whenever we use the term signal processing operations, this should be understood to mean both subsystems taken together.

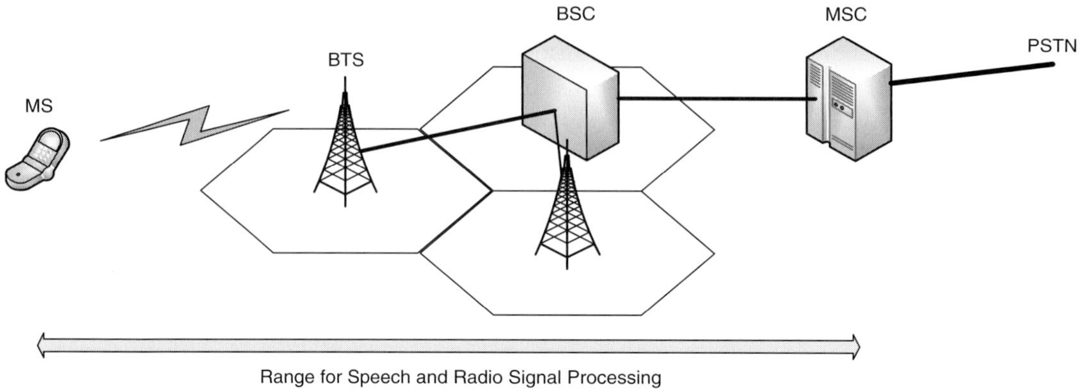

Fig. 1.1 Network architecture of circuit-switched mobile communications systems.

Figure 1.1 shows a generic network architecture that represents the signal processing operations employed by the second generation digital circuit-switched mobile communications systems. In this architecture, once a call is established, the Mobile Station (MS) transforms a short, typically 20 ms, segment of speech into an appropriate digital format, and then transmits it to one or more Base Transceiver Stations (BTS). During the end-to-end transmission from the MS to the far-end device, which may be another MS or a fixed telephone, the speech is represented in several digital formats at different bit-rates, depending on the communications links over which the speech is transported.

A set of BTSs is controlled by a Base Station Controller (BSC). The BSC sets up and terminates calls to and from the BTSs and hands over ongoing calls among the BTSs based on the quality of the wireless links between the MS and the BTSs or the level of cell loading. The Mobile Switching Center (MSC) manages the operation of the controllers and connects them with either the Public Switched Telephone Network (PSTN) or other circuit-switched mobile communications networks. The 64 kbps Pulse Coded Modulation (PCM) format is typically used from the BSC and upward, i.e., in the direction of the MSC. The speech delivered to the MS undergoes the reverse signal processing operations.

The link between the MS and the BTS is not the only wireless link in the end-to-end speech transmission paths. In addition, microwave links, consisting of one or more T1 (1.544 Mbps) or E1 (2.024 Mbps) lines modulated in the high-frequency carriers, are often used as *backhaul* between the base station and the switching center, or in locations where fixed networks are not available or economical. Since the microwave is operated as a high-powered, line-of-sight wireless link over a dedicated, low-cost frequency spectrum, it does not suffer from many of the limitations inherent in mobile communications. In this chapter, we confine our interest to the dynamic interactions of the MS, BTS, BSC, and MSC, which must be carefully coordinated to maximize both speech quality and network capacity. Similar approaches will be taken in the following chapters but the network architectures or the node names will change as the mobile communications systems evolve.

The term *circuit-switched* refers to the nature of communications links in which the information, such as digitally formatted speech, is transmitted with negligible variation in speed or delay, regardless of the link quality or network load. This definition does not necessarily imply that the level of data loss or the bit-rate is uniformly maintained over the end-to-end transmission paths, however. A circuit-switched network consists of a series of such communications links, each of which transports the speech or data of one or more users at a fixed bit-rate. The end-to-end paths meeting the required transport capabilities and and channel conditions must be established before the transmission begins.

The interface between two communications links where the bit-rate or the speech format needs to be changed may require an additional processing delay but such a delay is generally lower than that associated with *packet-switched* networks such as the Internet or Ethernet, where the data packets can be transmitted without establishing an end-to-end transmission path. Without an established path, data packets can be lost or delivered in an order that is different from the order in which they were initially transmitted. The maximum allowed total delay, i.e., the mouth-to-ear delay, in commercial voice telephony systems is required to be equal to or less than 280 ms for a satisfactory call quality [ITU-T (2003)]. The wired telephone networks and contemporary circuit-switched mobile communications networks often complete the entire procedures in less than 200 ms.

In circuit-switched networks, the coded received speech first encounters error correction decoding, which is followed by error concealment when uncorrected errors corrupt the speech. The decoded speech is then converted to an analog representation for play out. Re-transmission of missing or corrupted frames, which would increase the delay and its variability, is generally not used. In packet-switched networks, each network node is allowed to retransmit lost data packets reported by a neighboring node, provided that the total delay budget is met. As interim solutions that bridged the gap between these two fundamentally different transmission techniques, hybrid approaches that combined the benefits of circuit-switched wired networks and packet-switched wireless networks were proposed and standardized [Ozturk *et al.* (2010)]. With these approaches, speech handling in the wired portions of the network is identical to that in conventional circuit-switched networks while re-transmission of lost speech data and scheduling of shared channels are allowed in the wireless links between the MS and the BTS.

Figure 1.2 shows the signal processing operations employed when the speech is transmitted between two second generation digital circuit-switched networks, from GSM to cdma2000. The digitized and compressed speech is wirelessly transmitted by the MS and recovered from the Radio Frequency (RF) signal by the BTS. The compressed speech is then reconstructed at the Transcoder and Rate Adaptation Unit (TRAU), which can be located at either the BTS, BSC, or MSC. The farther the TRAU is separated from the BTS, the farther the speech is transported at its lowest bit-rate. This saves the infrastructure cost since a 64 kbps channel can transport four speech channels encoded at bit-rates lower than 16 kbps. Therefore it is advantageous to extend the distance between the speech encoder and the speech decoder as far as possible, in some cases covering the entire transmission path. Voice over IP (VoIP) is an example of such an extreme case.

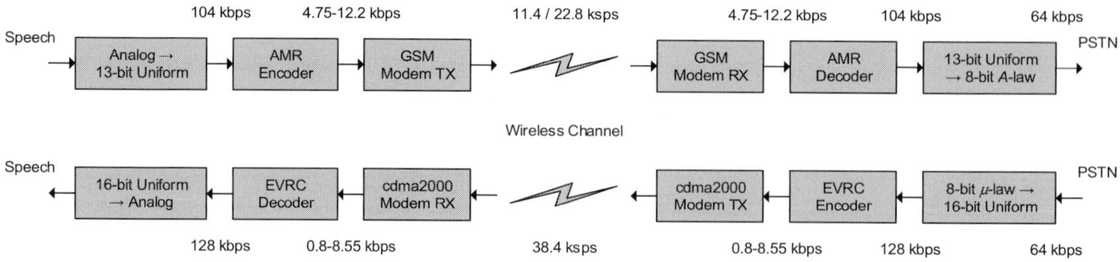

Fig. 1.2 Speech and radio signal processing operations from GSM to cdma2000.

The wireless link between the MS and the BTS is unique in that the bit-rate of the speech, which can change even during a call depending on the voice activity or the network control, is the lowest in the transmission path. Furthermore, because of the harsh nature of the wireless channel and the limited signal processing and transmit power of the MS, the speech is more likely to be damaged or lost in this short link than in any other. The transmission cost is also the highest in this link, because of the large investment for radio spectrum and network infrastructure.

The roaming capability, when extended globally, greatly increases the value of the radio spectrum that is shared by many countries. As a result, each generation of mobile communications systems has made more efficient use of the radio spectrum than its predecessors while simultaneously improving the speech quality. The main objectives of the signal processing operations in circuit-switched mobile communications networks can be simply summarized as the maximization of the number of satisfied users through efficient design of the network architectures and the procedures for all of the entities between the MS and the MSC. Beyond this point the existing PSTN infrastructure allows few opportunities for innovation.

Figure 1.3 illustrates the generic signal processing operations that occur between the MS and the BTS that are applicable to most digital circuit-switched networks. To counter the negative effects of the wireless channel including propagation loss and multipath fading of the transmitted signal, the MS and the BTS continuously control the bit-rate and transmit power, and report the channel status to the BSC so that the call can be transferred to a neighboring BTS with better link quality when the current BTS cannot support the necessary network services. The BTS, to which the call is handed over, may belong to the same or a different network type. During the transfer process, some of the speech signals en route to the destination BTS or MS can be lost, generating a small but audible loss of quality.

A number of metrics and criteria have been established to measure how well these performance objectives are met. This and the following chapters will show that these objectives can be achieved using a variety of approaches. These range from efficient speech compression algorithms that result in speech quality that is high enough for commercial services at low bit-rates to wireless communications techniques that use less bandwidth and/or less transmit power. In a restricted medium, such as the wireless channel, higher signal quality and higher network capacity are conflicting objectives. Thus, control mechanisms that trade one against the other play a key role in the overall system operation. New techniques for speech compression or wireless transmission need

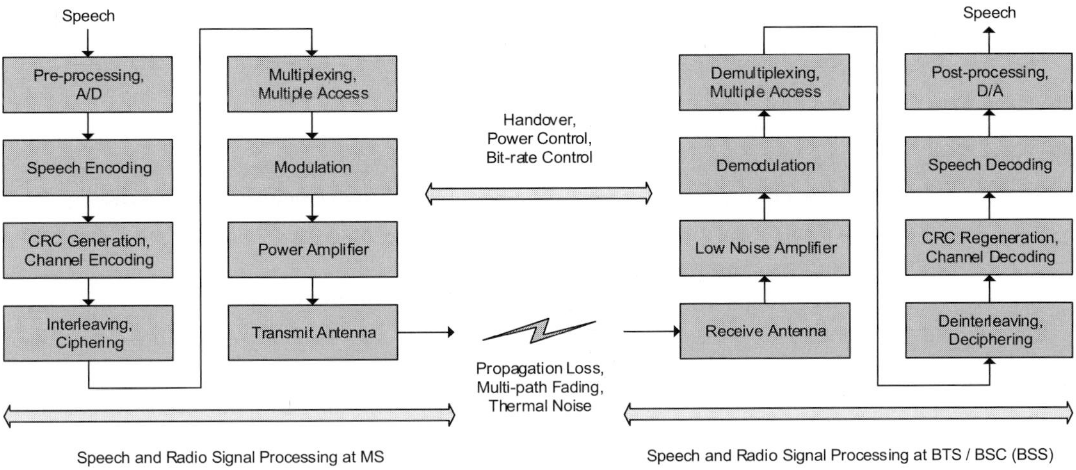

Fig. 1.3 Generic speech and radio signal processing operations in mobile communications systems.

to be incorporated carefully, however, to be compatible with the existing infrastructure. Changes can be made to operational networks but for MSs, once manufactured and activated, it can be very difficult, if not impossible, to make substantial changes other than software upgrades.

1.1.2 Performance Criteria

The speech and radio signal processing procedures used in mobile communications systems are designed to meet well-established criteria for maintaining speech quality. These fall into five types. The *blocked call rate* measures the capability of the network to handle incoming service requests. A request for setting up a call might be rejected because of insufficient radio resource or poor link quality. The blocked call rate does not differentiate among the possible sources of call blocking. This measure can be applied to a diversity of network types including fixed or mobile, analog or digital, and circuit-switched or packet-switched.

A second quality criterion is the *call drop rate*, which evaluates the capability of the network to maintain an established call. Conventional telephony systems such as the PSTN maintain a negligible call drop rate, but mobile communications systems are likely to exhibit rates as high as a few percent, regardless of the underlying radio access technologies or speech compression algorithms.

A third group of factors that affect the speech quality includes those that measure the reliability or link quality of the connection. These include the *bit error rate*, *frame error rate*, or *frame erasure rate*, all of which measure the probability that encoded speech frames are corrupted or lost in the channel during transmission. In the PSTN, the bit error rate is typically as low as 10^{-6} whereas a 1–3% corruption rate for transmitted speech frames is considered acceptable in mobile communications networks. When the received speech frames contain bit errors, the error control coding may identify the location of those errors and recover the speech. Error concealment methods, such as the methods that replace corrupted frames by interpolating or extrapolating nearby correctly decoded speech frames, can also be used to maintain an acceptable speech quality.

A fourth group of quality criteria relates to the operation of packet-switched networks, especially those built with the Internet Protocol (IP). This group includes such measures as the *packet loss rate* and *jitter loss rate* to evaluate the effects of different error types. Finally, there are measures of network acceptability that quantify the *end-to-end delay* of the voice services. These are often the most stringent to meet but they have a profound influence on the overall design of the system.

These five groups of quality criteria are defined mainly to establish a set of minimum requirements for *toll quality* or *carrier grade* services. In many cases these are objectively measurable but they cannot completely replace the important subjective criteria, as measured by the Mean Opinion Score (MOS) derived from subjective evaluations with human listeners. From the point of view of service providers, all of these criteria are used to maximize the number of simultaneous calls whose quality exceeds a set of minimum requirements, rather than to maximize the quality of each call for a fixed number of simultaneous calls.

1.2 Analog Mobile Communications Systems

When it was developed in the 1970s and commercially launched in 1982, the Advanced Mobile Phone System (AMPS) introduced many fundamental aspects of mobile communications systems, such as the frequency re-use to increase network capacity and the handover of ongoing calls between cells [Young (1979), MacDonald (1979)]. Some aspects were necessary to cope with the regulatory limitations. One of these was the restricted bandwidth that was allocated. The AMPS was initially assigned two 25 MHz bands located above 800 MHz for the forward (BTS to MS), and reverse (MS to BTS) channels. The AMPS was adopted by many countries and often operated in frequency bands slightly different from the original ones.

For the Frequency Modulation (FM) techniques that were used in AMPS, radio spectrum below 800 MHz would have been preferred but was not then available. The 800 MHz bands were from a part of the radio spectrum that had previously been occupied by television channels. This bandwidth had been freed after the channels were relocated to cable. When the number of people using mobile communications continued to increase, the need to accommodate additional customers, coupled with the difficulty of obtaining additional spectrum, resulted in technical decisions made during the redesign of AMPS that influenced many key aspects of the next generation digital mobile communications systems.

Before proceeding to more detailed descriptions of AMPS, it is important to distinguish between a *band* and a *channel* as these terms are used in this book. We follow the definitions of [Razavi (2011)], in which a band refers to the entire radio spectrum in which the MSs of a mobile communications system are allowed to communicate, while a channel refers to the smaller bandwidth assigned to one or more MSs for services. These definitions match well with the spectrum allocation practices of both the

Table 1.1 Channel numbering system.

Channel number n	Reverse channel frequency (MHz)	Forward channel frequency (MHz)
1–799	$825 + 0.03n$	$870 + 0.03n$
991–1023	$825 + 0.03(n - 1023)$	$870 + 0.03(n - 1023)$

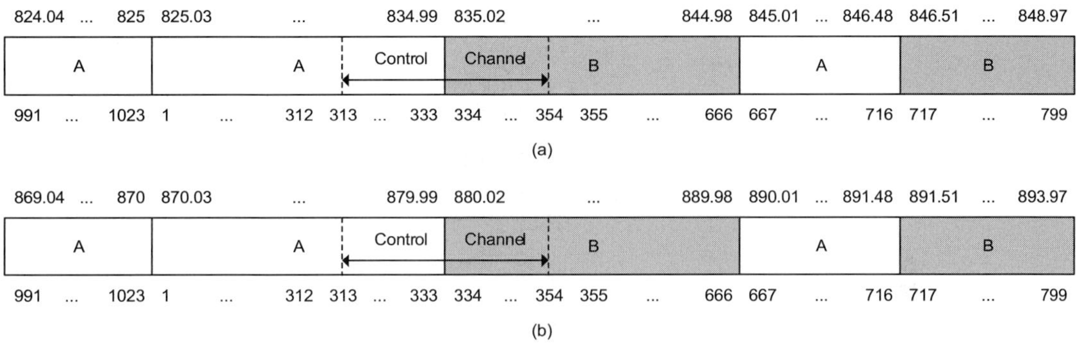

Fig. 1.4 Spectrum allocation. (a) Reverse channels. (b) Forward channels.

first generation analog mobile communications systems and the second generation Time Division Multiple Access (TDMA) systems. In these systems, signals from one or more MSs are transmitted over a narrowly confined channel of 30–200 kHz. Each channel has the center frequency of an RF carrier and an integer is typically assigned to label each channel. Then, a set of contiguous channels constitutes a band. An MS in a mobile communications system requires at least one band for the reverse channels and another for the forward channels, if it is operated in the Frequency Division Duplex (FDD) mode.

1.2.1 Network Architecture

With a 25 MHz band and a channel spacing of 30 kHz, AMPS provides 832 channels that can be divided between one or more service providers in each area. A typical configuration might be that a half of the total capacity, i.e., a combination of 395 channels for voice service and 21 channels for call control, would be assigned to each service provider in a market where two providers compete. Figure 1.4 shows the spectrum allocation of AMPS in the US in the 1980s when two types of service providers, a non-wireline (A) operator and a wireline (B) operator, shared the band. Channels 313–333 and 334–354 are the control channels assigned for each operator. The channel numbers and carrier frequencies of AMPS are related as shown in Table 1.1.

The earlier architecture of a generic circuit-switched mobile communications network, shown in Fig. 1.1, is directly applicable to AMPS whose network architecture is shown in Fig. 1.5; only the terminology is different. In AMPS, the Base Station (BS) performs similar network operations as the BTS, and the Mobile Telephone Switching Office (MTSO) performs management tasks similar to those of the BSC and the MSC. It controls the call processing and manages the cellular operation.

Table 1.2 Control partitioning of AMPS.

MS	BS	MTSO
Setup channel selection	Radio control	Standard local switching
Channel tuning	Location data collection	Radio channel management
Message reception and	Component calibration	Remote unit maintenance
transmission	MS control	BS and MS control
Failure sequence timing	Message relaying and	Message administration MS
Tone switch-hook	reformatting	location tracking
supervision	Switch-hook and fade	Handover synchronization
Pre-origination dialing	supervision	

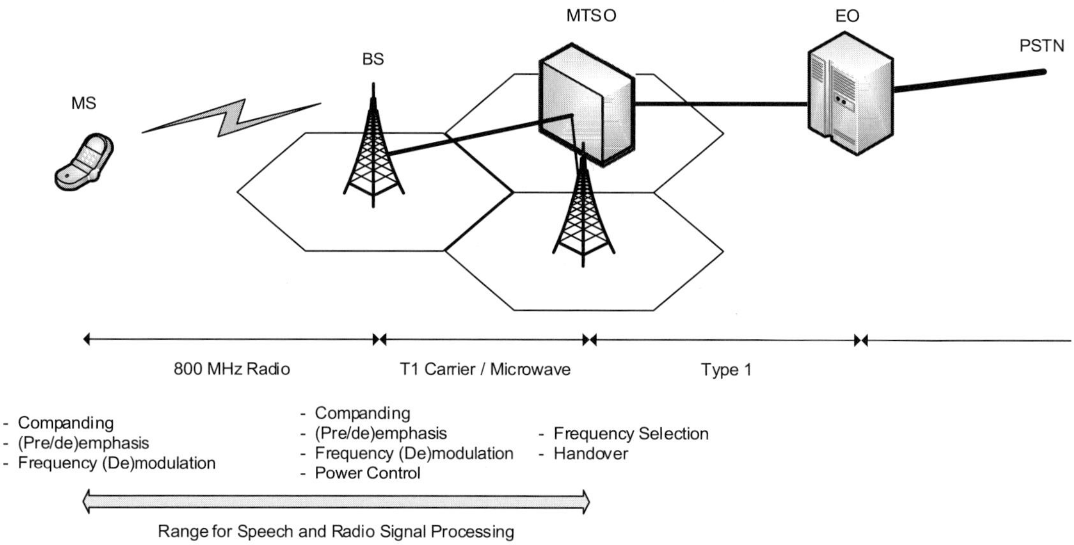

Fig. 1.5 Network architecture of AMPS.

MTSO can be interconnected to a Local Exchange Carrier (LEC) End Office (EO) with Type 1 interconnection link. Table 1.2 outlines the technical responsibilities of the MS, BS, and MTSO partitioned among these three entities [Fluhr and Porter (1979)]. Although AMPS is often classified as an analog mobile communications system, many signal processing and control channel operations are represented in digital formats.

Figure 1.5 also shows the interface types used between the network entities of AMPS. We focus on the link between the MS and the BS, and the link between the BS and the MTSO. The first link is of crucial importance since no further signal processing is performed after the analog speech waveform from the MS is digitized and encoded into a 64 kbps PCM format at the BS. This basic format for speech is maintained throughout the transmission paths until the signal reaches either another BS or a fixed telephone.

Conversion between two PCM formats, μ-law and A-law, may occur at intermediate locations but this would have little impact on the speech quality or total delay since the two formats are similarly defined and the conversion requires a negligible amount of computation. A T1 carrier, microwave, or Type 1 interconnection link carries large numbers of 64 kbps PCM channels. The second link, between the BS and the MTSO, is also important since the MTSO is responsible for controlling speech quality and network capacity by indirectly controlling the MS through the BS. The measures available to the MTSO include the handover to another BS or channel in the same cell, and power control.

1.2.2 Speech and Radio Signal Processing Operations

Figure 1.6 shows the speech and radio signal processing operations in AMPS for the transmit and receive sides [Arredondo *et al.* (1979)]. In the first step, the sound pressure level of speech is converted to voltage variations by the microphone, and then band-pass filtering limits the bandwidth of the signal to 300–3000 Hz. Because the waveform will be frequency modulated, the signal amplitude is also limited to control the amount of energy that would be leaked to adjacent channels. This is done by companding, i.e., non-linearly compressing the amplitude at the transmitter and expanding it at the receiver. AMPS uses a 2:1 compander, through which a 2 dB change in the input voltage level is compressed to a 1 dB change. The compander also has the effect of improving the subjective speech quality in poor channel conditions. Figure 1.7 illustrates the companding and modulation procedures used.

The energy of speech signal, after filtering and compression, is concentrated in the low frequency bands and the Signal-to-Noise Ratio (SNR) in the high frequency bands is reduced. It is further degraded in the FM and carrier modulation process. Figures 1.7(a) and 1.7(b) show the input-output characteristics of the compander and the deviation limiter, respectively. With a channel width of 30 kHz, the frequency deviation of speech signal is confined to approximately 24 kHz, around a center frequency f_c,

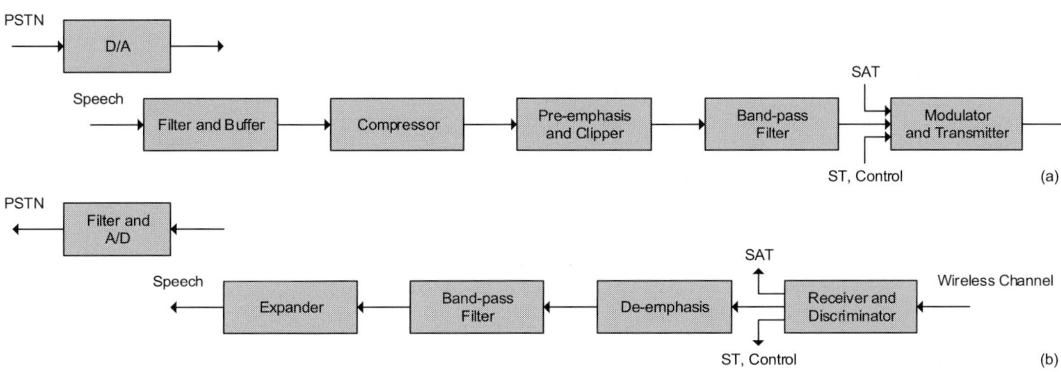

Fig. 1.6 Speech and radio signal processing operations of AMPS. (a) Transmitter side. (b) Receiver side.

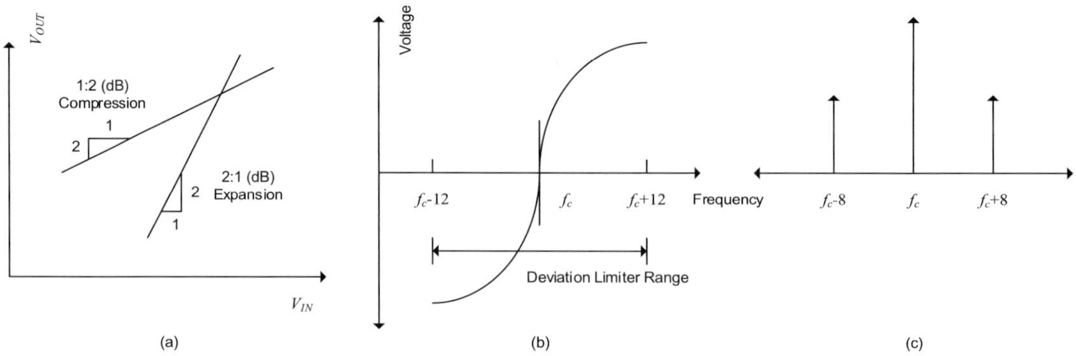

Fig. 1.7 (a) Compander input-output characteristics. (b) Frequency deviation limiting. (c) ±8 kHz binary FSK.

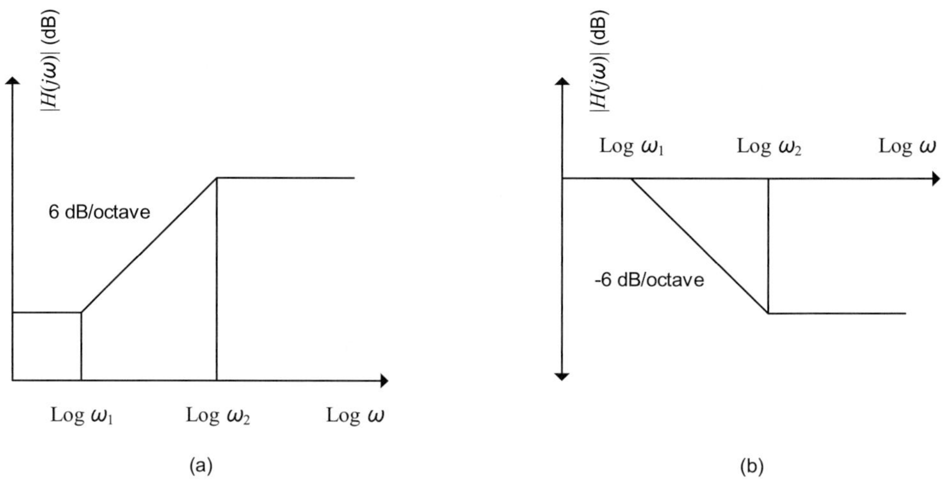

Fig. 1.8 Frequency response. (a) Pre-emphasis. (b) De-emphasis.

to reduce the interference to and from the adjacent channels. Pre-emphasis boosts the high-frequency components of speech signal at the transmitter while de-emphasis compensates for this at the receiver. Figures 1.8(a) and 1.8(b) show the frequency response of the pre-emphasis and de-emphasis filters, respectively, where the angular frequency $\omega = 2\pi f$. These analog signal processing operations are common to both the forward and reverse channels of AMPS.

There are no measures that prevent unauthorized eavesdropping of ongoing calls in AMPS, since providing reliable security with analog signal processing is very difficult. Anyone with the intention and capability of scanning channels can listen to or record the conversations. This fundamental limitation of AMPS was overcome in the next generation digital systems in which ciphering of digitally compressed speech became a basic feature.

It is not easy to track a speaker when speech signals to and from multiple MSs are transmitted over the same channel. Therefore it is important to allocate the channels carefully over the cells to insure that this collision does not happen. For frequency-modulated speech processed as in Fig. 1.6 to be of acceptable quality, a Signal-to-Interference Ratio (SIR) of at least 18 dB is required over 90% of the network coverage. The 7-cell ($K = 7$) frequency re-use pattern shown in Fig. 1.9 has been found to be the smallest re-use factor that meets the requirements for channel efficiency with 120-degree directional antennas. Two MSs that use the same channel should be sufficiently separated to avoid the mutual interference. With the 7-cell re-use, two layers of cells provide enough propagation loss to insulate two cells that share the same set of channels. In Fig. 1.9, those cells marked with the same character can share the same set of channels. Note that in practice the cells are rarely regularly hexagonal in shape and may differ widely in size. The channels are assigned to each cell based on the amount of expected voice traffic.

Referring to Fig. 1.4, if there are 395 voice channels and 21 control channels assigned to each service provider, with the 7-cell re-use pattern, approximately 56 voice channels can be assigned to each cell. One-third of the channels assigned to each cell, i.e., 17 or 18, can be allocated to each sector of a three-sector antenna. Note that using directional antennas reduces the interference but cannot increase the number of channels. The 42 control channels, channels 313–354, are located in the middle of the 25 MHz band to facilitate the operation of a channel-scanning frequency synthesizer, especially for those in the MS whose tuning capability is limited. Unlike the voice channels, the control channels apply a form of digital modulation, binary Frequency Shift Keying (FSK), to modulate the Manchester-coded data. Figures 1.10(a) and 1.10(b) show the formats of the forward and reverse control channels, which transmit control data at 10 kbps.

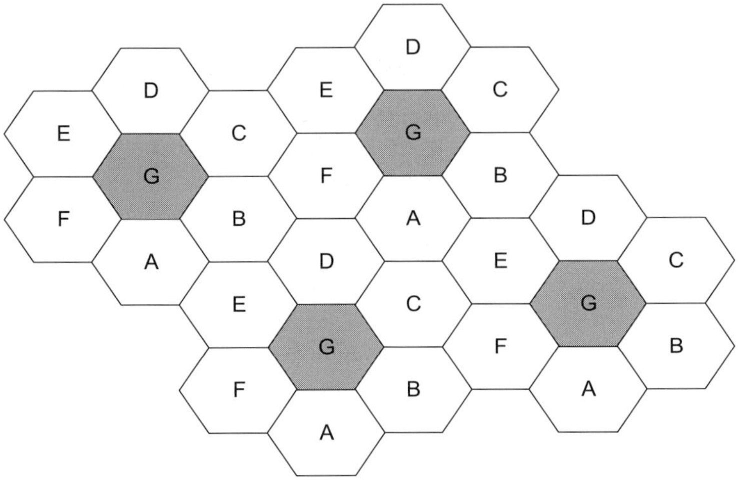

Fig. 1.9 7-cell re-use pattern.

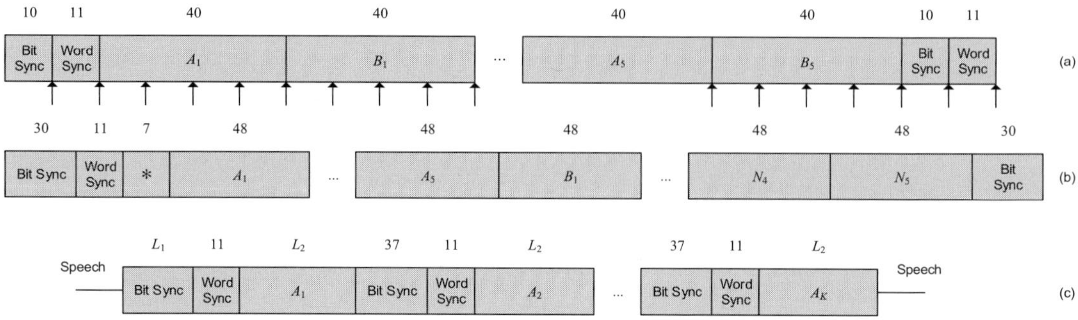

Fig. 1.10 Control channels. (a) Forward control channel. (b) Reverse control channel. (c) Voice control channel.

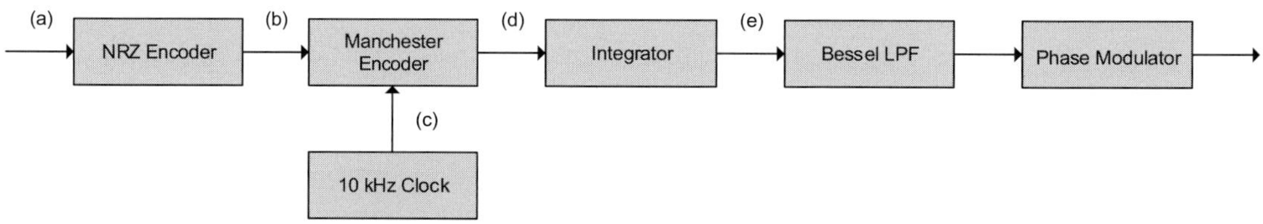

Fig. 1.11 Control signal processing procedures.

The binary control data is first converted to a Non-Return-to-Zero (NRZ) format, and further encoded to the Manchester (bi-phase) code, as shown in Figures 1.11 and 1.12. The benefit of Manchester coding is that it concentrates signal energy within a 10 kHz band, enabling an easy detection of signal at the receiver. The Manchester-coded data is integrated and low-pass filtered. Finally each symbol is represented with one of two possible frequency deviations and phase modulated with an RF carrier, as shown in Fig. 1.7(c).

Figure 1.10 shows that the formats of the forward and reverse control channels are not identical. Each control channel is separated into the A and B messages so that MSs with even phone numbers read and write the A messages, and MSs with odd numbers use the B messages. Throughout the forward control channels, key information to be used by the MS is periodically broadcast by the BS. This includes the System Identification Number (SIN) of the network and the power level for initial transmission. A burst-idle bit, as shown in Fig. 1.10 with an $\uparrow$, is inserted after each ten message symbols, after a Bit Sync, and after a Word Sync, to turn the receiver off during an idle period.

In the forward control channels, ten messages follow the Bit Sync and Word Sync, and the A and B messages alternate. After a (40,28) Bose–Chaudhuri–Hocquenghem (BCH) coding is applied, each message is repeated five times. Thus, in Fig. 1.10(a), $A_1 = A_2 = \cdots = A_5$, and $B_1 = B_2 = \cdots = B_5$. The receiver applies the majority logic to recover the correct message. With a minimum distance of 5, the BCH code used in the forward control channel can correct one bit error or detect up to two bit errors in each message. The 10-bit Bit Sync, 1010101010, and the 11-bit Word Sync, 11100010010, are unique

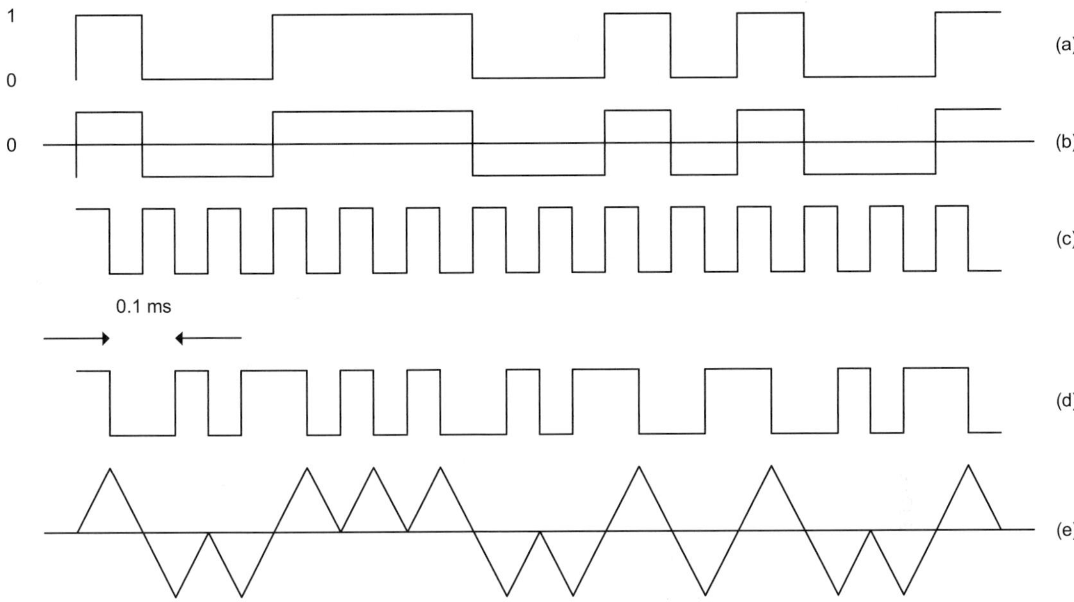

Fig. 1.12 Waveforms of control signal processing. (a) Binary data. (b) NRZ-coded. (c) 10 kHz clock. (d) Manchester-coded. (e) Integrator output.

bit patterns used by the receiver to facilitate detecting the message boundaries. The messages are not allowed to have these bit patterns. With a clock frequency of 10 kHz, the duration of a message and four burst-idle bits is 4.4 ms. Therefore, although the bit-rate at the phase (binary FSK) modulator is 10 kbps, with two MSs sharing a forward control channel, the actual bit-rate for an MS is $\frac{28\times 10}{(10+11+40\times 5\times 2+42)} = 0.6$ kbps.

The use of the forward control channels is managed by the network in a centralized fashion, but the use of the reverse control channels is left to the discretion of the MS. To avoid the collisions of messages sent from multiple MSs, the Carrier Sense Multiple Access (CSMA) is employed. An MS tries to detect the presence of transmissions by other MSs before attempting to transmit by checking the burst-idle bits. If one or more carriers are sensed, the MS waits until the end of any ongoing transmissions and then initiates its own transmission. Each message, consisting of one–five 36-bit words, is (48,36) BCH-coded, and repeated five times. This BCH code can correct up to one bit error. Bit Sync is made up of 30 bits of alternating ones and zeros, i.e., 1010, ... ,1010, and Word Sync is 11100010010. Digital Color Code, as shown in Fig. 1.10(b) with $*$, can be one of four 7-bit sequences that identify the target BS. Following a similar development as in the forward control channel, assuming $N \leq 5$ distinct messages, the actual bit-rate of the reverse control channel is $\frac{36\times N\times 10}{48+48\times 5\times N} = \frac{7.5N}{1+5N}$ kbps, which corresponds to 1.25 kbps for $N = 1$ and 1.44 kbps for $N = 5$, for an MS.

The 21 pre-defined control channels may not be sufficient to manage the operation of AMPS when the network is overloaded, but defining additional control channels from the spectrum would reduce the gain in network capacity gained from frequency re-use. Since voice activity is typically absent for more than half of a call duration, the gaps in the signals in the voice channels can also be used to transmit control information.

Table 1.3 Voice control channel parameters.

	Forward control channel	Reverse control channel
L_1	100	101
L_2	40	48
K	11	5

Figure 1.10(c) shows the format of a voice control channel whose key parameters are outlined in Table 1.3. Naturally the bit-rate of a voice control channel is lower than that of the dedicated control channels.

From channel parameters, the actual bit-rate for an MS is $\frac{28\times10}{100+11+40+(48+40)\times10} = 0.27$ kbps in the forward voice control channel. In the reverse voice control channel, the actual bit-rate is $\frac{36\times10}{101+11+48+(48+48)\times4} = 0.66$ kbps. When there is an urgent need for control signaling, the voice can be interrupted for a period short enough not to be perceived, and the control information can be transmitted at 10 kbps during this interval. The majority logic is used as in the control channel to assist the reception of messages. The messages are repeated 11 times in the forward voice control channels while in the reverse voice control channels they are repeated five times.

The main use of voice channels for control purposes is for signaling the handover messages, which are usually transmitted at low SNR by MSs located at the cell edges. This strategy, called *blank and burst*, of temporarily transmitting control information instead of voice is also employed in many other digital mobile communications systems. Table 1.4 shows the control information exchanged between the MS and the BS. In the call setup using dedicated control channels, more bits are exchanged than during the call using the voice control channels. Note that the 64-bit dialed digits do not have tight delay requirements as they are typically input manually. Note also that the transmit power is directly controlled only in the forward channels but mutual, and faster, control of the power would reduce the interference and increase network capacity.

1.2.3 Cellular Operation

In AMPS, speech quality of each MS needs to be constantly monitored during conversations so that an acceptable quality, the toll quality, is maintained. The forward and reverse control channels, because of their limited capacity and the delay required to obtain an access opportunity, are not appropriate for this type of persistent measurement and immediate signaling. With the voice control channel, it is possible to continuously monitor the speech quality but the amount of information required for informing such measurement is still excessive for its negligible bit-rates. As a compromise that meets the two conflicting requirements, a type of analog signaling that spectrally overlaps with the voice signal can be used, which does not noticeably interrupt the conversation.

To indicate that a channel of each BS is currently alive, Supervisory Audio Tone (SAT), a single-frequency signal at either 5970, 6000, or 6030 Hz above the center frequency, is continuously transmitted with the speech signal. Only one of the SATs is

Table 1.4 Control information on dedicated control and voice control channels.

Channel	Control information	Bits
Forward control	MS page	24, 34
	Channel designation	11
	MS transmit power	2
	Overhead (local parameters)	22–30
	System control	4
Reverse control	Identification	56, 66
	Dialed digits	64
	System control	4
Forward voice control	Orders	5
	Channel designation	11
	MS transmit power	2
	System control	4
Reverse voice control	Order confirmation	5
	Dialed digits	64
	System control	4

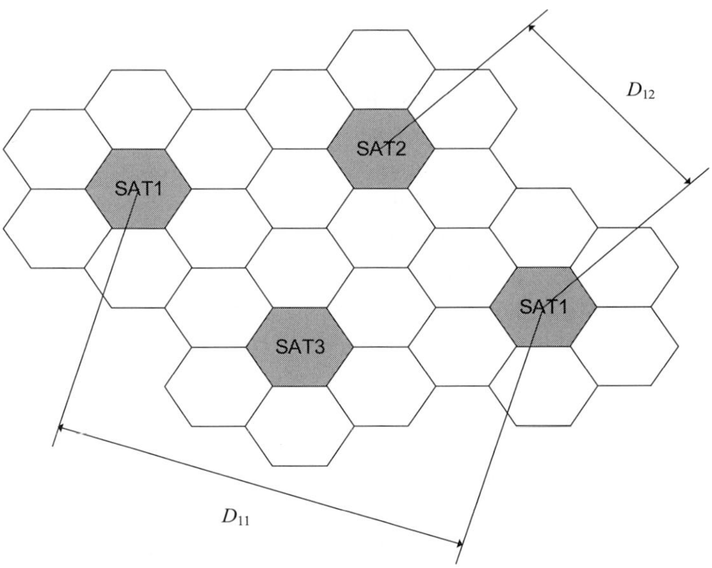

Fig. 1.13 Spatial allocation of SATs.

used by each BS. If an SAT is not detected for more than a pre-defined time, the call is disconnected by either the MS or the BS. When the MS does not detect the SAT from the serving BS but returns an SAT at other frequency, which may happen, for example, when the SAT from another BS using the same channel is stronger, the call is also disconnected. Therefore, like voice channels, the SAT frequencies have to be carefully allocated over the cells. Figure 1.13 shows the allocation of SATs over neighboring cells,

Table 1.5 Supervision decisions for SAT and ST.

	ST on	ST off
SAT received	MS on-hook	MS off-hook
SAT not received	MS in fade or transmitter turned-off	

Table 1.6 System parameters of analog mobile communications systems.

	NTT	TACS	NMT
Region	Japan	UK	Scandinavia
Reverse channel frequency (MHz)	870–885	917–950	463–467.5
Forward channel frequency (MHz)	925–940	872–905	453–457.5
Reverse/forward channel spacing (MHz)	55	45	10
Channel bandwidth (kHz)	25/12.5	25	25
Number of channels	600	1320	180
Modulation (voice)	FM	FM	FM
Modulation (control)	FSK	FSK	FSK

in which D_{11} should be sufficiently larger than D_{12} to re-use the SAT. Since the SAT is located above the spectrum of the speech, the SAT can be combined with the speech after band-pass filtering at the transmitter and removed from the speech by band-pass filtering at the receiver, as shown in Fig. 1.6.

In addition to detecting the presence of an SAT, signal amplitude is used to monitor the health of a channel. If the power level of SAT measured by the BS falls below a threshold, the MTSO first signals the MS to increase its transmit power, which can have one of nine pre-defined levels. If this is not effective, or not possible because the MS is already transmitting at its maximum power, the MTSO asks the neighboring BSs to measure the signal strength of the MS. If stronger measurements are reported, the MTSO initiates the handover to a new BS with a stronger signal. A 10 kHz Signaling Tone (ST) is sometimes transmitted with the speech signal for control purposes. Table 1.5 outlines the decisions to be made by the BS based on the combinations of the SAT and the ST.

Until now we have summarized the signal processing and quality control procedures of AMPS, a first generation analog mobile communications system. Table 1.6 outlines the key parameters of some mobile communications systems that were contemporary with AMPS, among which the NTT system was the first to be commercially deployed. These use similar analog signal processing operations. Many essential features that later digital mobile systems inherited were introduced in AMPS, including the handover, power control, and in-band signaling. Although AMPS became a dominant mobile communications system in the first generation, technical opportunities for improving speech quality or network capacity can be easily identified. For example, when there is no voice activity, the transmitter can be completely turned off to reduce power consumption and the interference to other cells. Because of the fundamental limitations of analog signal processing, several of the basic procedures shown in Fig. 1.6 have to be maintained, regardless of the cell loading level or link quality. This separation of the speech signal processing from the radio signal processing gradually disappeared in the next generation of digital mobile communications systems. Their interaction was exploited for higher speech quality and network capacity.

1.3 References

Arredondo, G. A., Feggeler, J. C., and Smith, J. I. 1979. Advanced Mobile Phone Service: Voice and Data Transmission. *The Bell System Technical Journal*, **58**(1).

Fluhr, Z. C., and Porter, P. T. 1979. Advanced Mobile Phone Service: Control Architecture. *The Bell System Technical Journal*, **58**(1).

ITU-T. 2003. G.114 International Telephone Connections and Circuits – General Recommendations on the Transmission Quality for an Entire International Telephone Connection; One-Way Transmission Time. May.

Lee, W. C. Y. 1995. *Mobile Cellular Telecommunications: Analog and Digital Systems*. 2nd edn. McGraw-Hill Professional.

MacDonald, V. H. 1979. Advanced Mobile Phone Service: The Cellular Concept. *The Bell System Technical Journal*, **58**(1).

Ozturk, O., Kapoor, R., Chande, V., Hou, J., and Mohanty, B. 2010. Circuit-Switched Voice Services over HSPA. *IEEE Vehicular Technology Conference*, May.

Rappaport, T. S. 2002. *Wireless Communications: Principles and Practice*. 2nd edn. Prentice Hall.

Razavi, B. 2011. *RF Microelectronics*. 2nd edn. Prentice Hall.

Young, W. R. 1979. Advanced Mobile Phone Service: Introduction, Background, and Objectives. *The Bell System Technical Journal*, **58**(1).

2 Signal Processing in TDMA Systems

We begin this chapter with a short discussion describing the digital compression of speech signals in mobile communications systems. In the fixed networks used by analog mobile communications systems, the speech is already represented in simple, but inefficient, digital formats such as 64 kbps PCM but analog waveforms are still employed to represent the speech over the wireless channel itself. We outline the basic idea behind Linear Predictive Coding (LPC), which increases the compression efficiency to the levels required for the limited frequency spectrum of mobile communications, by focusing on the types of input signals generated by the human vocal tract. Then we introduce approaches used to improve the analog mobile systems when their network capacity was challenged by a rapidly increasing number of users. After analyzing analog and digital extensions of the AMPS, we discuss the speech and radio signal processing operations used by a new Time Division Multiple Access (TDMA) mobile communications system, the Global System for Mobile Communications (GSM).

2.1 Speech Signal Processing

In AMPS, once an FM channel has been assigned, it is monopolized by a single MS for the duration of the call. To transport the speech signals of multiple MSs on a single channel, the signal of each MS has to be digitally processed and separated from those of the other MSs so that the BS can separate them if no error occurs during transmission. In the system, data transmission rate of the control channel is only 10 kbps. However, advancements in the RF and Very Large Scale Integration (VLSI) technologies since the AMPS was developed now enable more spectrally efficient digital modulation and robust error correction coding schemes that can boost the data transmission rate by a factor of four to five.

When the speech signal is represented using a conventional digital format such as 64 kbps PCM, the bit-rate is too large to be transmitted over a 30 kHz channel even with digital modulation and coding techniques. Furthermore, the bit-rate is likely to be increased further when the overhead to protect the digitally processed speech bit-stream and maintain the connection of each MS with the BS is added. To enable transmission over a 30 kHz channel, it is necessary to compress the speech so that its bit-rate is reduced significantly, to around 10 kbps. This must be done in such a way that the quality is not noticeably compromised by the compression.

Classical waveform-based coding techniques [Jayant and Noll (1984)], such as the PCM or the Adaptive Differential Pulse Coded Modulation (ADPCM), attempt to model the waveform shape of the speech signal as faithfully as possible. These techniques have low computational complexity but they cannot reach the required low bit-rates without significantly sacrificing the accuracy of the representation. As an alternative, model-based coding techniques, which simulate the speech generation process of the human vocal tract, can be used to compress and reconstruct speech signals that sound similar to the original but require lower bit-rates. The model-based speech compression techniques target perceptional similarity but do not attempt to match the original waveform shape.

The frequency at which the raw speech is sampled prior to compression also has a profound effect on the achievable quality of the reconstructed speech. Figure 2.1 shows a series of speech waveforms sampled at 8000 Hz (upper) and their associated spectrograms (lower). The spectrogram is a visual representation of the instantaneous frequency spectrum as the speech signal evolves with time. The waveforms, which consist of the speech signals in several languages, speech with music or noise, music, and the sound of a bugle, are low-pass filtered before and after the sampling to contain only those frequency components between 100 and 3500 Hz. Inspection of the spectrograms reveals that there are periodic structures in the frequency domain that may be exploited to design efficient speech compression algorithms. When the speech is represented at 8000 samples/s, as in PSTN, it is referred to as the Narrowband (NB) speech, to distinguish it from representations that use higher sampling rates. As the spectrum of the human voice is often extended to higher frequencies, representation of speech at higher sampling rates than 8000 Hz can improve the quality and resilience against acoustic noise while incurring additional bit-rate and complexity.

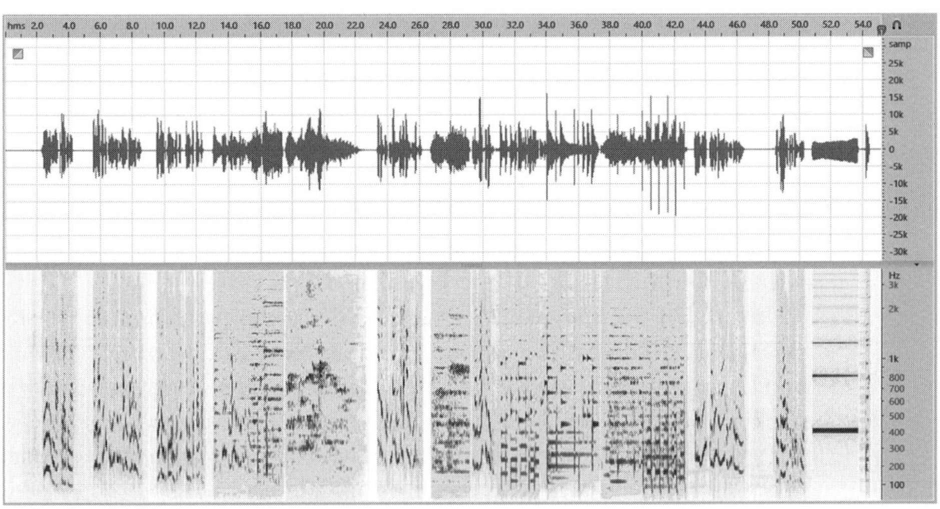

Fig. 2.1 Spectrogram of narrowband speech waveforms.

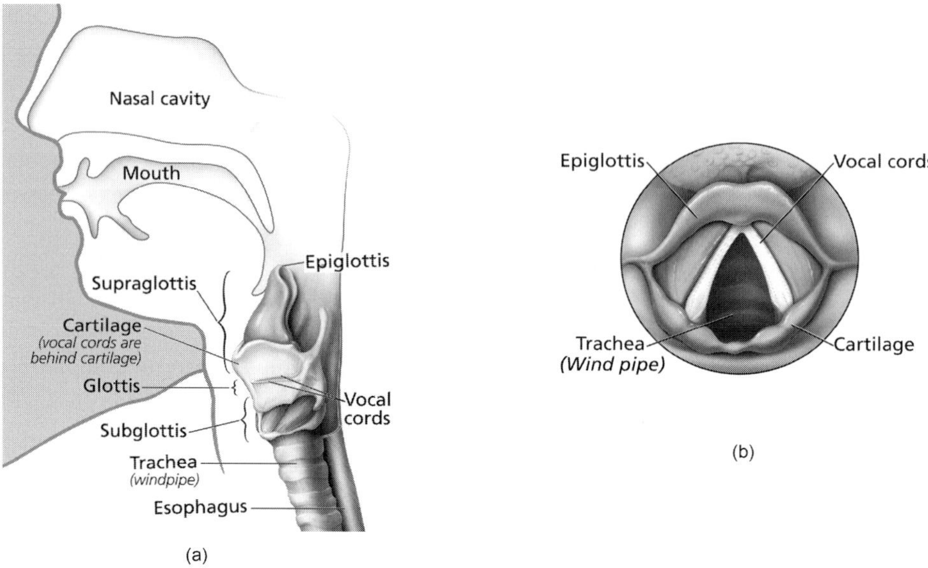

Fig. 2.2 Speech production system (courtesy of Alan Hoofring and the National Cancer Institute). (a) Larynx and nearby structures. (b) Larynx (top view).

In the model-based coding, the physical, biological human speech production anatomy is simulated to represent the speech at minimal bit-rates. Figure 2.2 depicts the anatomy of the speech production system in which the force driving production is provided by a flow of air from the glottis. The vocal tract is the acoustical path from the larynx to the mouth, nasal cavity, or both. The flow of air is modulated by the time-varying vocal tract shape. Detailed analysis of the interactions between the structures involved in speech production can be found in [Quatieri (2001), Rabiner and Schafer (2010)].

Figure 2.3(a) shows a 3.072 second speech waveform, which sounds the phrase *Add the sum to the product of these three.* The waveform consists of 24576 samples digitized at 8000 samples per second. The proportion of silence is comparable to that of actual speech, which suggests another opportunity to reduce the bit-rate, e.g., by using lower bit-rates or transmitting no signals at all in such periods. The waveform shown in Fig. 2.3(b) corresponds to the beginning of the speech segment, called its *onset*, which consists of the first 1400 samples of the segment. The first 500 samples contain no voice activity but the following 800 samples show a weak noisy signal. Figure 2.3(c) expands the portion of the waveform represented by samples 4000–4300.

This part of the speech waveform includes some periodic components, which are overlapped by non-periodic components. The periodic patterns originate from *voiced sound* where the vocal tract is excited by quasi-periodic pulses created by adjusting the tension in the vocal cords. The non-periodic waveforms are typically generated from background noise or *unvoiced sound*, created when the vocal tract is excited by turbulence which has a wide, flat spectrum. The vocal tract excitations corresponding to these two types of speech waveforms can be modelled reasonably well by a periodic impulse train and white noise, respectively.

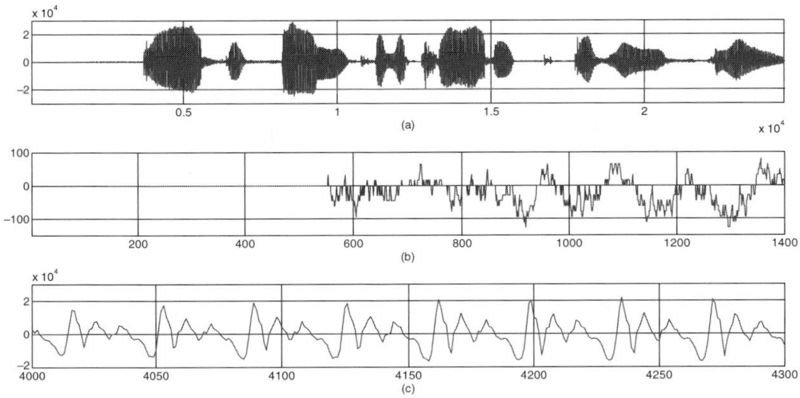

Fig. 2.3 (a) Speech waveforms. (b) Waveform for silence and unvoiced speech. (c) Waveform for voiced speech.

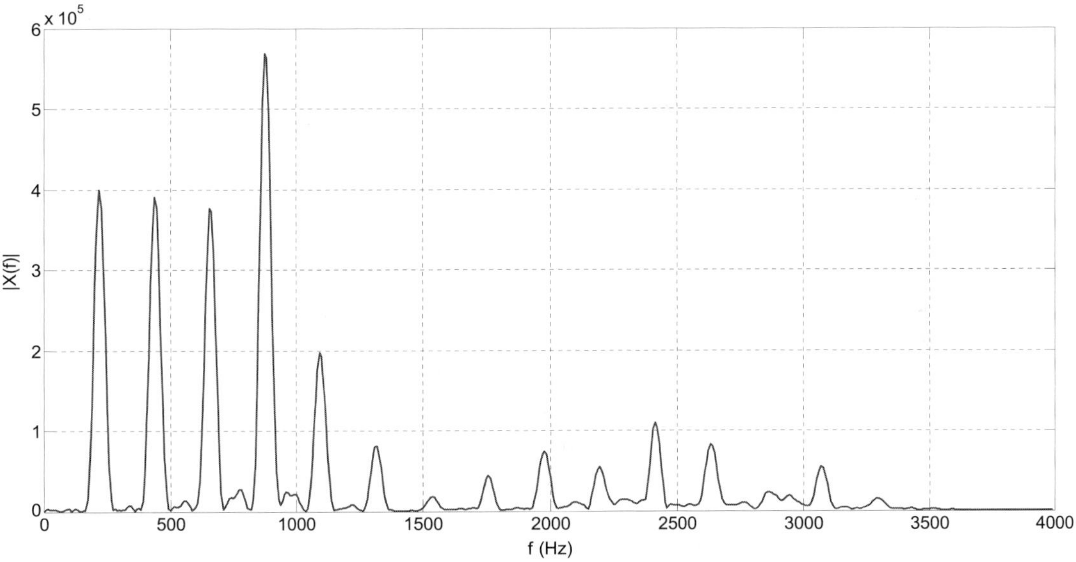

Fig. 2.4 Frequency domain representation of voiced speech waveform.

To picture the periodic structure of a voiced speech waveform in the frequency domain, samples 4000–4300 were transformed using a Discrete Fourier Transform (DFT), whose magnitude is shown in Fig. 2.4. The spectrum of this portion of the signal contains peaks that are equispaced along the frequency axis, reflecting the periodic structure of the underlying waveform. The duration of each cycle of voiced speech is called the pitch period length, τ, and the fundamental frequency, or pitch, is the reciprocal of the pitch period length, $F_0 = \frac{1}{\tau}$. F_0 is a measure for how high or low the voice sounds. From Fig. 2.4, it can be estimated that the fundamental frequency F_0 is 230 Hz. It should be noted that some of the spectral peaks have larger amplitudes than others. The local maxima in the envelope that connect the peaks are called *formants*.

They reflect the resonant frequencies of the vocal tract and are essential components for the intelligibility of speech. For this waveform the formants are located at $F_1 = 870$ and $F_2 = 2440$ Hz, where the peaks occur. These estimated parameters and acoustic characteristics of speech can be exploited in the reduction of bit-rate to represent the signal. For generic audio signals, however, such approaches specifically assuming the origin of the input signal would result in limited quality if other types of source signals are entered.

2.1.1 Linear Predictive Coding

Linear Predictive Coding (LPC) exploits the structure of the speech production system and the spectro-temporal characteristics of the speech waveform to a greater extent than the waveform-based coding techniques. By analyzing the acoustic nature of a short speech segment, during which the statistical characteristics of sound are assumed to remain stationary, LPC computes another segment of the same length in the time domain that can be represented at a lower bit-rate but sounds similar to the original. The bit-rate reduction is achieved by modeling the speech segment as the output of a linear time-varying digital filter whose coefficients and input signals are determined by the segment. The filter coefficients, and the gain and type of input signal constitute the bit-stream generated for each speech frame.

Figure 2.5* shows the modeling of human vocal tract as a series of N concatenated tubes. In Fig. 2.5(a), the glottis is assumed to be located to the left of tube 1, and the lips are connected to tube N. A set of wave equations relating the air pressure and the volume velocity are derived at the boundaries of the tubes, following the development of [Quatieri (2001), Rabiner and Schafer (2010)]. In tube k, the volume velocity at location x and time t, denoted as $u_k(x,t)$, is defined as the rate at which the air particles flow perpendicularly through an area A_k. In tube k, A_k is constant. $u_k(x,t)$ is constructed from the forward- and backward-traveling wave components. $p_k(x,t)$ represents the incremental pressure with respect to atmospheric pressure in tube k, and c represents the speed of

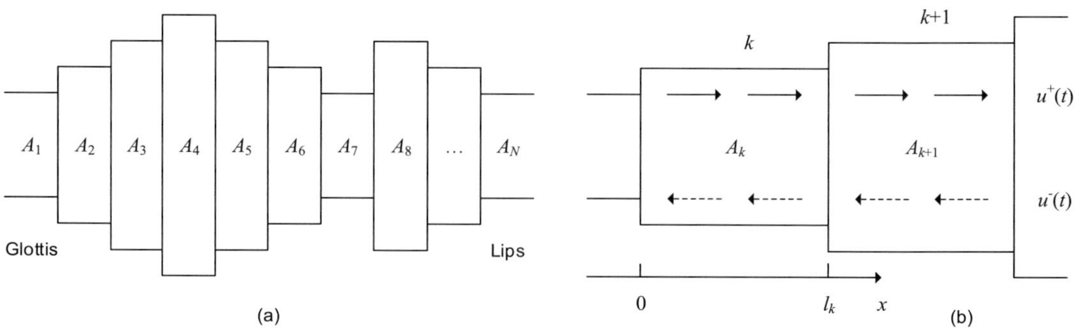

Fig. 2.5 (a) Concatenated tube model of vocal tract. (b) Forward and backward-traveling sound waves.

*Rabiner, Lawrence R.; Schafer, Ronald W., *Digital Processing of Speech Signals*, 1st ed., © 1979. Reprinted by permission of Pearson Education, Inc., New York, New York.

sound. It is assumed that the energy is lost only at the end of the tubes, the lips, when the sound waves propagate into free space.

The volume velocity and pressure for tube k are given by

$$u_k(x,t) = u_k^+\left(t - \frac{x}{c}\right) - u_k^-\left(t + \frac{x}{c}\right), \tag{2.1}$$

$$p_k(x,t) = \frac{\rho c}{A_k}\left[u_k^+\left(t - \frac{x}{c}\right) + u_k^-\left(t + \frac{x}{c}\right)\right], \tag{2.2}$$

where $0 \leq x \leq l_k$, and l_k is the length of tube k. ρ is the density of air particles, which is assumed to be constant in the atmosphere. $u_k^+(t)$ and $u_k^-(t)$ represent the velocity of forward- and backward-traveling waves, respectively. To solve the wave equations, the boundary conditions at the edges of the tubes can be exploited. For the volume velocity and the incremental pressure to be continuous in both time and space between tubes k and $k+1$, the following conditions must be met,

$$u_k(l_k, t) = u_{k+1}(0, t), \tag{2.3}$$

$$p_k(l_k, t) = p_{k+1}(0, t), \tag{2.4}$$

for $k = 1, 2, \ldots, N-1$.

Let $\tau_k = \frac{l_k}{c}$ be the time for the sound wave to propagate through tube k. Then, after some manipulation of the equations and the boundary conditions, the following relationships can be obtained for the volume velocity,

$$u_{k+1}^+(t) = \frac{2A_{k+1}}{A_{k+1} + A_k} u_k^+(t - \tau_k) + \frac{A_{k+1} - A_k}{A_{k+1} + A_k} u_{k+1}^-(t), \tag{2.5}$$

$$u_k^-(t + \tau_k) = -\frac{A_{k+1} - A_k}{A_{k+1} + A_k} u_k^+(t - \tau_k) + \frac{2A_k}{A_{k+1} + A_k} u_{k+1}^-(t), \tag{2.6}$$

which implies that in each tube, a part of the traveling wave in each direction propagates to the next tube while a part is reflected back to the current tube. Likewise, a part of the reflected wave continues propagating to the previous tube while another part is re-reflected. The reflection coefficient at the kth junction can be derived as

$$r_k = \frac{A_{k+1} - A_k}{A_{k+1} + A_k}, \tag{2.7}$$

which is the fraction of the wave at the junction between tubes k and $k+1$ that propagates backward into tube $k+1$.

In other words, r_k is the amount of $u_{k+1}^-(t)$ reflected at the junction. Since $A_k > 0$, $-1 \leq r_k \leq 1$. The magnitude of the reflection coefficient is identical between the two tubes, regardless of the direction in which the wave approaches the boundary. The signal flow graph in Fig. 2.6* can model the wave propagation inside the vocal tract. The forward- and backward-traveling sound waves are abstracted with the signals, gains on the branches, and the delay elements.

From the structure of signal flow graph, it is seen that the length of the impulse response is infinite but, in practice, the waves decay rapidly as the impact of the reflection coefficients accumulates exponentially. The interconnection of the identical

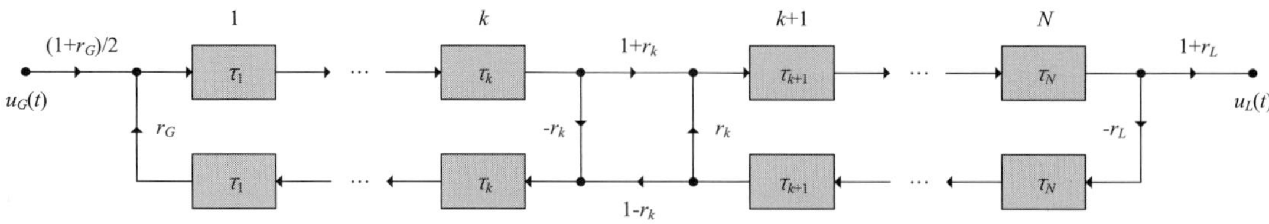

Fig. 2.6 Signal flow graph for lossless tube model of vocal tract.

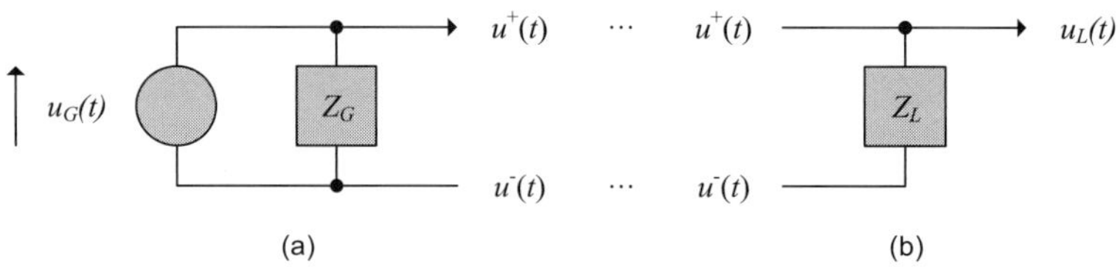

Fig. 2.7 Modeling boundary conditions. (a) Glottis. (b) Lips.

modular structures with different delays and reflection coefficients can be used to represent the wave propagation in tubes 2 through $N-1$. However, different structures may be required in tubes 1 and N, to represent the unique roles of the glottis and the lips in the generation of speech. To derive the modular structures of the first and last tubes, we assume that no part of the backward-traveling wave in tube 1 proceeds beyond the glottis. Likewise, it is assumed that no part of the backward-traveling wave enters tube N from free space. To describe the relationships of the volume velocity and the incremental pressure at the glottis and the lips, the boundary conditions are modeled as electrical circuits, as shown in Fig. 2.7*.

The input signal at the glottis, $x = 0$ in tube 1, can be visualized as an electrical circuit consisting of a source, $u_G(t)$, and an impedance, Z_G, as shown in Fig. 2.7(a). The relationship between pressure and velocity is conceptually similar to that between voltage and current. Using the wave equations and the acoustic assumptions, r_G in Fig. 2.6 is represented by Z_G, A_1, and the two constants. It is assumed that the backward-traveling wave in tube 1 does not proceed farther. Under this assumption the following relationships can be derived from the boundary conditions,

$$u_1(0,t) = u_G(t) - \frac{p_1(0,t)}{Z_G}, \tag{2.8}$$

$$u_1^+(0,t) - u_1^-(0,t) = u_G(t) - \frac{\rho c}{A_1}\frac{[u_1^+(0,t) + u_1^-(0,t)]}{Z_G}. \tag{2.9}$$

If we ignore the spatial parameter x, r_G is derived as

$$u_1^+(t) = \frac{(1+r_G)}{2}u_G(t) + r_G u_1^-(t), \tag{2.10}$$

$$r_G = \frac{Z_G - \frac{\rho c}{A_1}}{Z_G + \frac{\rho c}{A_1}}. \tag{2.11}$$

Likewise, the output signal at the lips, $x = l_N$ of tube N, can be visualized as an electrical circuit loaded with an impedance, Z_L. Using the wave equations and the acoustic assumptions, the relationship between Z_L and r_L can be similarly derived. Assuming that no traveling wave enters tube N from free space, with the boundary conditions, the following relationships must be met,

$$p_N(l_N, t) = Z_L u_N(l_N, t), \tag{2.12}$$

$$\frac{\rho c}{A_N}[u_N^+(t - \tau_N) + u_N^-(t + \tau_N)] = Z_L[u_N^+(t - \tau_N) - u_N^-(t + \tau_N)]. \tag{2.13}$$

Then the backward-traveling wave in tube N can be combined with the corresponding forward-traveling wave such that

$$u_N^-(t + \tau_N) = -r_L u_N^+(t - \tau_N), \tag{2.14}$$

$$r_L = \frac{\frac{\rho c}{A_N} - Z_L}{\frac{\rho c}{A_N} + Z_L}. \tag{2.15}$$

Let the lengths of N tubes be identical, $l = l_1 = l_2 = \cdots = l_N$, and let $\tau = \frac{l}{c}$ be the delay for propagating through a tube. Then the continuous signal flow graph of Fig. 2.6 can be converted into an equivalent discrete-time model as shown in Fig. 2.8*. Its impulse response is

$$h[n] = \sum_{k=0}^{\infty} b_k \delta[n - N - 2k] = b_0 \delta[n - N] + \sum_{k=1}^{\infty} b_k \delta[n - N - 2k], \tag{2.16}$$

which implies that after the input signal is applied at time $t = 0$, the earliest arrival of sound waves to the lips will be at $t = N\tau$, and that further arrivals will occur at multiples of 2τ after the first one.

The half-sample delays in the discrete-time model, which cannot be implemented exactly, can be removed in the lower backward branches, and replaced with one-sample delays in the upper forward branches. The gains in the branches remain the same, and a delay of $-\frac{N}{2}$ in the last branch toward the lips can compensate for the increased delay in the upper branches to make the two models equivalent.

Therefore, instead of a speech segment, N reflection coefficients and another value related to the amplitude of the segment can be transmitted, if they can be computed from the speech segment, and the bit-rate for these $N + 1$ values is lower. Alternatively, it can be shown that the transfer function that relates the volume velocity at the glottis and the lips is of the form,

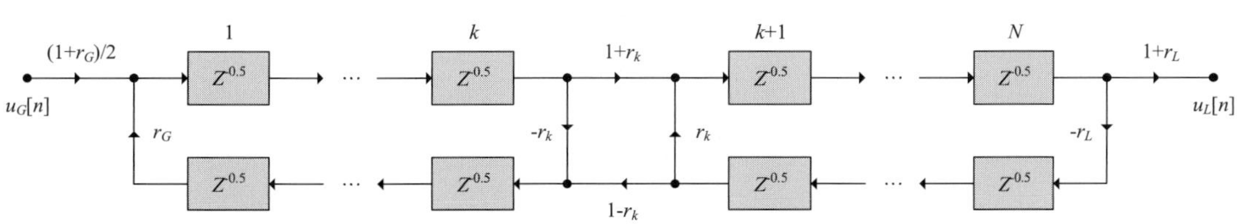

Fig. 2.8 Discrete-time lossless tube model of vocal tract.

$$H(z) = \frac{Az^{-\frac{N}{2}}}{1 - \sum_{k=1}^{N} a_k z^{-k}}. \tag{2.17}$$

Note that $z^{-\frac{N}{2}}$ in the numerator of $H(z)$ corresponds to a shift of $\frac{N}{2}$ samples in the time domain. Removing this factor does not influence the accuracy of modeling significantly. The N poles define the frequencies of the formants. More complex effects in the vocal tract can be reflected in the transfer function by adding additional poles and zeros. In the all-pole modeling of LPC, the process of human speech generation is simulated using a time-varying digital filter with a steady-state transfer function of the form,

$$H(z) = \frac{U_L(z)}{U_G(z)} = \frac{A}{1 - \sum_{k=1}^{N} a_k z^{-k}}, \tag{2.18}$$

where A is the gain and the a_k are the filter coefficients, which vary slowly with time. $U_G(z)$ and $U_L(z)$ are the z-transforms of the discrete-time volume velocity at the glottis and the lips, $u_G[n]$ and $u_L[n]$, respectively. The bit-rate reduction is achieved if fewer bits are required to represent $A, a_1, a_2, \ldots, a_N$ than are required for the original speech segment. These $N + 1$ parameters, N reflection coefficients and a gain, constitute the essential information for an encoded speech frame representing the input signal for a short period.

This all-pole model is a reasonable representation for non-nasal voiced speech. Although more detailed acoustic models for nasal and fricative sounds require more complex transfer functions, the all-pole model is adequate for most types of speech provided that the order, N, is sufficiently high. For the all-pole model to be valid, duration of input signal needs to be limited so that each speech sample can be linearly predicted, i.e., estimated from a linear combination of the previous samples. Therefore in LPC, the typical duration of a speech segment represented by the all-pole model is 10–30 ms, and the gain in accuracy from using a larger N diminishes when the order reaches about 10.

Let $s[n]$ be the output speech signal generated by applying $u_G[n]$ to a vocal tract modeled by $h[n]$. Then their z-transforms are related by

$$S(z) = H(z)U_G(z), \tag{2.19}$$

$$S(z) = AU_G(z) + \sum_{k=1}^{N} a_k S(z) z^{-k}. \tag{2.20}$$

When $u_G[n]$ is zero, e.g., after $n = 0$ when the input to the glottis is an impulse, the output speech signal can be estimated from the previous samples and the error can be expressed as

$$s[n] = Au_G[n] + \sum_{k=1}^{N} a_k s[n-k], \tag{2.21}$$

$$\approx \sum_{k=1}^{N} a_k s[n-k] = \tilde{s}[n], \tag{2.22}$$

$$e[n] = s[n] - \tilde{s}[n] = s[n] - \sum_{k=1}^{N} a_k s[n-k]. \tag{2.23}$$

These assumptions are also valid for the pitch period when the input signal is an impulse train. If Z_G and Z_L are real, all coefficients of the impulse response or the transfer function are also real. Then a_k can be found by minimizing the mean-square error,

$$\frac{\partial e^2[n]}{\partial a_k} = \frac{\partial}{\partial a_k}(s[n] - \tilde{s}[n])^2 = 0, \tag{2.24}$$

for $k = 1, 2, \ldots, N$. However, inverting the $N \times N$ matrix directly, for each speech frame, may incur excessive computation for the processors in the MS.

Define the autocorrelation $R(i)$ as $\sum_{n=0}^{\infty} s(n)s(n-i)$. Then the N equations can be rearranged into the form,

$$\sum_{k=1}^{N} a_k R(|i-k|) = R(i), \tag{2.25}$$

for $i = 1, 2, \ldots, N$, which in matrix form becomes

$$\begin{bmatrix} R(0) & R(1) & \cdots & R(N-1) \\ R(1) & R(0) & \cdots & R(N-2) \\ \vdots & \vdots & \vdots & \vdots \\ R(N-1) & R(N-2) & \cdots & R(0) \end{bmatrix} \begin{bmatrix} a_1 \\ a_2 \\ \vdots \\ a_N \end{bmatrix} = \begin{bmatrix} R(1) \\ R(2) \\ \vdots \\ R(N) \end{bmatrix}. \tag{2.26}$$

The a_k can be computed by inverting the matrix. The value of gain A can be computed from the following relationship [Hayes (1996)],

$$A^2 = R(0) - \sum_{k=1}^{N} a_k R(k) = \varepsilon_N, \tag{2.27}$$

where ε_N is the minimum prediction error. However, direct inversion of a matrix of this size is again a computationally intensive operation that should be avoided if there are alternatives. The complexity can be reduced significantly if the Toeplitz structure of the matrix is exploited, since the elements along each of the diagonals are equal. With this structure, it is possible to derive the values of a_k using recursion-based methods.

In the MS of a digital mobile communications system, GSM, which uses the Full-Rate (FR) speech codec, the reflection coefficients are computed and converted as follows. First, to remove the DC offset of the input signal sampled at 8000 samples/s, the speech samples, $s_o(n)$, are notch-filtered by the operation

$$s_{of}(n) = s_o(n) - s_o(n-1) + \alpha * s_{of}(n-1), \tag{2.28}$$

where α is $32735 * 2^{-15}$. The filtered signal is then pre-emphasized using

$$s(n) = s_{of}(n) - \beta * s_{of}(n-1), \tag{2.29}$$

with β set to $28180 * 2^{-15}$. For a 20 ms speech segment, the nine autocorrelation values when $N = 8$ are computed using

$$R(k) = \sum_{i=k}^{159} s(i)s(i-k), \tag{2.30}$$

for $k = 0, 1, \ldots, 8$.

From these values, the reflection coefficients can be computed using several recursions, which differ in the numbers of required multiplications and additions, and in the memory required to store the program and data. A straightforward approach for matrix inversion, such as the Gaussian elimination, requires a number of multiplications and divisions proportional to N^3. A recursion-based inversion technique, the Levinson–Durbin recursion, reduces the number to N^2. In addition, the required memory is reduced from N^2 to $2(N + 1)$. The Schur recursion is slightly more efficient than the Levinson–Durbin recursion as well as more friendly to implementations that use parallel processing. Figure 2.9 illustrates the flow of Schur recursion used in the FR speech encoder of GSM.

The Schur recursion computes the reflection coefficient r_i and the error ε_i using the ith-order filter for each i while the Levinson–Durbin recursion generates the filter coefficient a_i and the gain A. ε_i and A are related by $A = \sqrt{\varepsilon_N}$. Figure 2.10 shows that from a segment of speech waveform, three equivalent sets of $N + 1$ parameters, the autocorrelation values, the reflection coefficients and the error, and the filter coefficients and the gain, can be computed and converted to each other, using appropriate recursion techniques. The Inverse Levinson–Durbin, Step-Up, and Step-Down recursions complete the chain of parameter transformations. Notice that if the length of speech signal, e.g., 160 samples for 20 ms, is much larger than the filter order, e.g., $N = 8$, the computational cost for obtaining the autocorrelation values dominates the complexity but for the MS, the differences among the recursions can also be significant.

In Fig. 1.2, it is shown that the analog speech signal is digitized into an intermediate 13- or 16-bit format, which is compressed further to around 10 kbps using a speech encoder. If two bytes are assigned for each value of a_k and A in an all-pole model with $N = 8$, the bit-rate is reduced from 104 or 128 kbps to 7.2 kbps. Since $-1 \leq r_i \leq 1$ for all i, it can be shown using the Levinson–Durbin recursion that the system is stable, and all poles are located within the unit circle. However, stability of the transfer function might be lost if the values of reflection coefficients or filter coefficients are changed by errors occurring during transmission. A simple approach to preserve the stability is to convert the coefficients into another format that is more resilient to errors, such as the Line Spectral Frequencies (LSF), which are related to the pole locations [Itakura (1975)].

In the FR speech encoder, the ith reflection coefficients are converted to the Log Area Ratios (LAR), which are defined as

$$LAR(i) = Log_{10}\left(\frac{1 + r_i}{1 - r_i}\right), \tag{2.31}$$

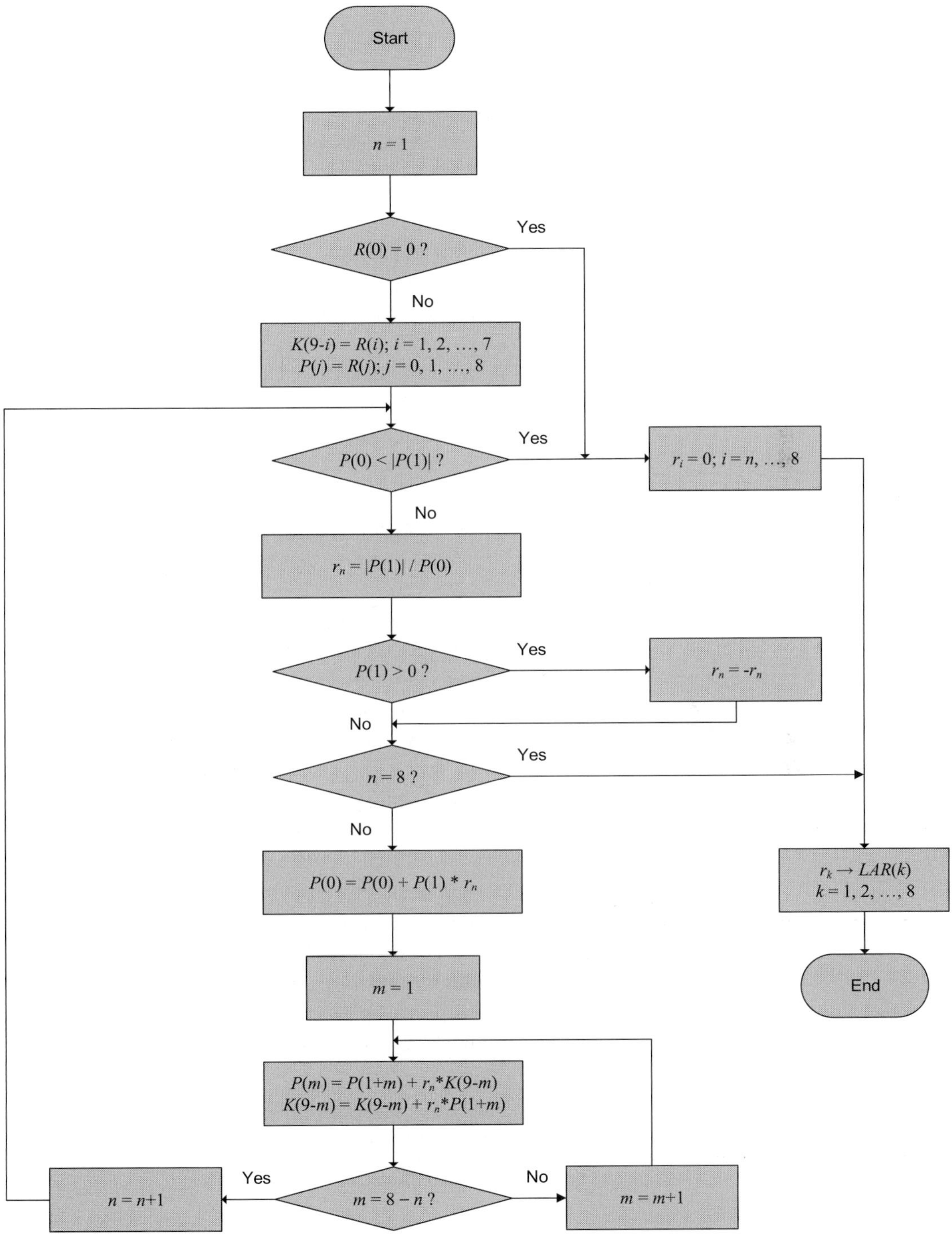

Fig. 2.9 LPC analysis using Schur recursion in GSM FR [3GPP (2000a)].

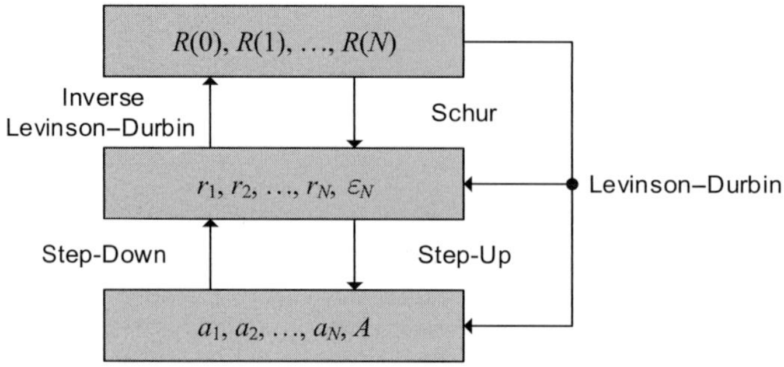

Fig. 2.10 Equivalence between autocorrelation sequence, reflection coefficients, and all-pole model parameters.

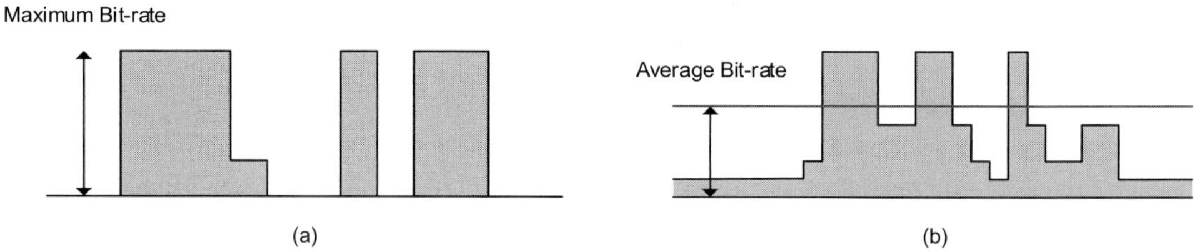

Fig. 2.11 Network control of speech bit-rate. (a) Fixed bit-rate. (b) Variable bit-rate.

for better quantization characteristics and higher error resilience. To save the complexity for implementing the logarithmic function, the following segmented piecewise approximation is used instead,

$$LAR(i) = \begin{cases} r_i & |r_i| < 0.675, \\ \mathrm{sign}[r_i] * [2|r_i| - 0.675] & 0.675 \le |r_i| < 0.950, \\ \mathrm{sign}[r_i] * [8|r_i| - 6.375] & 0.950 \le |r_i| \le 1.000, \end{cases} \tag{2.32}$$

which does not need the costly division and logarithm operations.

2.1.2 Fixed Bit-Rate versus Variable Bit-Rate Coding

Although the all-pole modeling can be used for most types of speech signals, different models can also be considered when the simple model fails. Typical speech compression suites include several models designed to meet the diverse acoustic nature of the input signals as closely as possible. Depending on whether or not the bit-rates required to represent the coefficients of the models are identical for each speech segment, the speech compression schemes can be classified as being either Fixed Bit-Rate (FBR) or Variable Bit-Rate (VBR).

Figure 2.11 illustrates the principles of FBR and VBR speech coding. Although drawn identically, the maximum bit-rates of two schemes are not necessarily equal. Figure 2.11(a) illustrates the output bit-rate of an FBR speech encoder where a fixed bit-rate is used whenever voice activity is present. Otherwise, the bit-rate falls to zero or to a low level that corresponds to a special frame type containing only the background noise.

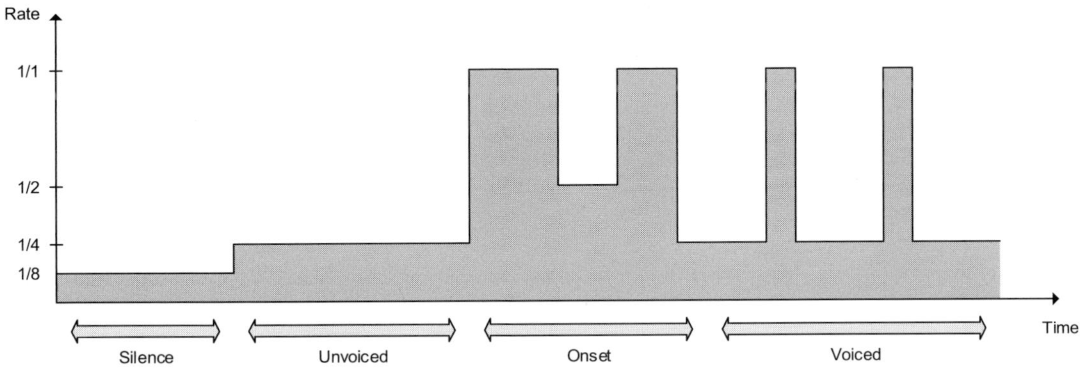

Fig. 2.12 4-level VBR speech coding.

An FBR speech encoder exploits the observations made in Fig. 2.3 that the duration of silence can be comparable to that of actual speech, to reduce the average bit-rate. In contrast, a VBR speech encoder generates multiple types of output speech frames, to match the time-varying nature of the input signal, as illustrated in Fig. 2.11(b). Depending on the algorithms, the lowest bit-rate of a VBR speech codec might be zero.

VBR can be considered a capacity-driven approach that is especially effective if the bit-rate saved by an MS can be used by other MSs. The more conventional FBR applies an identical bit-rate regardless of the nature of the input signal as long as voice activity is present. Therefore, in mobile communications systems using FBR, the network typically controls the maximum bit-rate of each MS to control the tradeoff between speech quality and network capacity. It is generally possible to apply different bit-rates to each MS, depending on the channel conditions to and from the MS. In addition to the bit-rate, the network also controls the transmit power of the MSs and base stations, as another resource available to control the quality–capacity tradeoff.

In contrast, in mobile communications systems that use VBR, the network typically controls the average bit-rate of each MS. Figure 2.12 illustrates the principles of a VBR speech encoder supporting four bit-rates, where different combinations of bit-rates are assigned for the silence, unvoiced speech, onset of speech, and voiced speech. Note that in the figure, the ratio of the four bit-rates is eight, four, two, and one but VBR itself does not limit the ratios. From the differences in the variation pattern of bit-rates, it can be seen that in VBR, it would be necessary to inform the receiver of the bit-rate used for speech somehow, periodically or when it changes. The LPC principles can be applied to either FBR or VBR as the bit-rate can be differentiated in the coefficient quantization stages.

2.2 AMPS Enhancements

2.2.1 Narrowband AMPS

The mobility and coverage of AMPS increased the number of mobile telephone users to an unexpected degree. The straightforward approach of meeting this challenge to

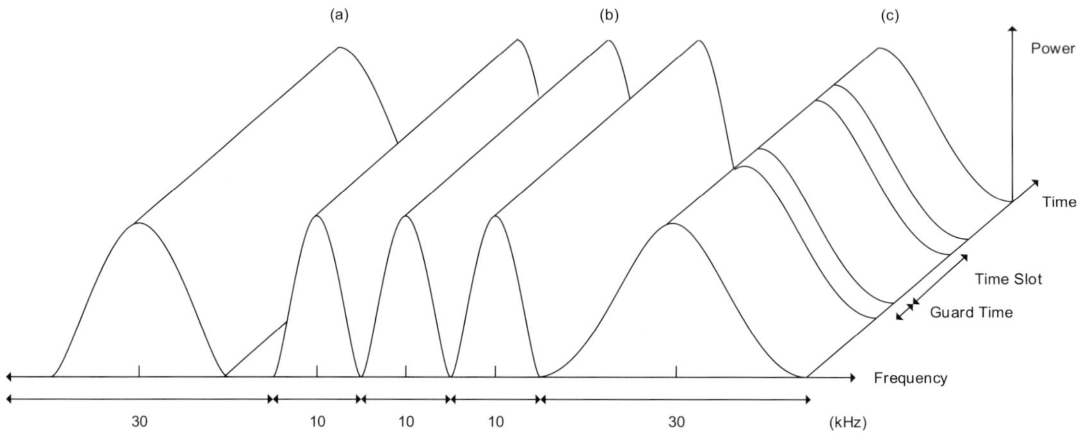

Fig. 2.13 Frequency-, time-, and power-domain representation. (a) AMPS. (b) N-AMPS. (c) D-AMPS.

increase network capacity, adding more frequency spectrum, was not particularly attractive as it would require the RF circuitry to operate over a wider bandwidth. This posed an implementation difficulty at that time. Another possible approach was to pack multiple RF carriers into a single 30 kHz channel. A new system designed along these lines, known as the Narrowband AMPS (N-AMPS), could theoretically provide three times the capacity of AMPS with a concomitant reduction of speech quality, by replacing a 30 kHz channel with three 10 kHz channels [Harte *et al.* (1998)].

The simultaneous transmission of speech and data could be supported by a 10 kHz channel augmented with 100 bps subband signaling. Figures 2.13(a) and 2.13(b) compare the AMPS and N-AMPS in the frequency-time-power domain. Although the N-AMPS was a conceptually simple approach, implementing the required analog signal processing shown in Fig. 1.6 for a third of the channel bandwidth presented a very serious challenge in the design of the analog filters. For this reason, the N-AMPS or any other analog extension of AMPS did not become a successful follow-on from the AMPS. Alternative approaches based on fully digital signal processing were considered to be more competitive.

2.2.2 Digital AMPS

The Interim Standard 54 (IS-54) defined a digital mobile communications system often called Digital AMPS (D-AMPS), which replaced the 30 kHz analog channel of AMPS with a digital channel of the same bandwidth. This digital channel can simultaneously support up to three MSs whose speech signals are encoded at 7.95 kbps using the Vector Sum Excited Linear Prediction (VSELP) speech codec. This codec is often referred to as the Full-Rate (FR), to distinguish it from some later codecs that operate at lower bit-rates. D-AMPS thereby achieves the same three-fold capacity increase that the N-AMPS promised. However, because of its digital modulation and error correction technologies, the SNR required to maintain an acceptable speech quality was reduced from 18 dB

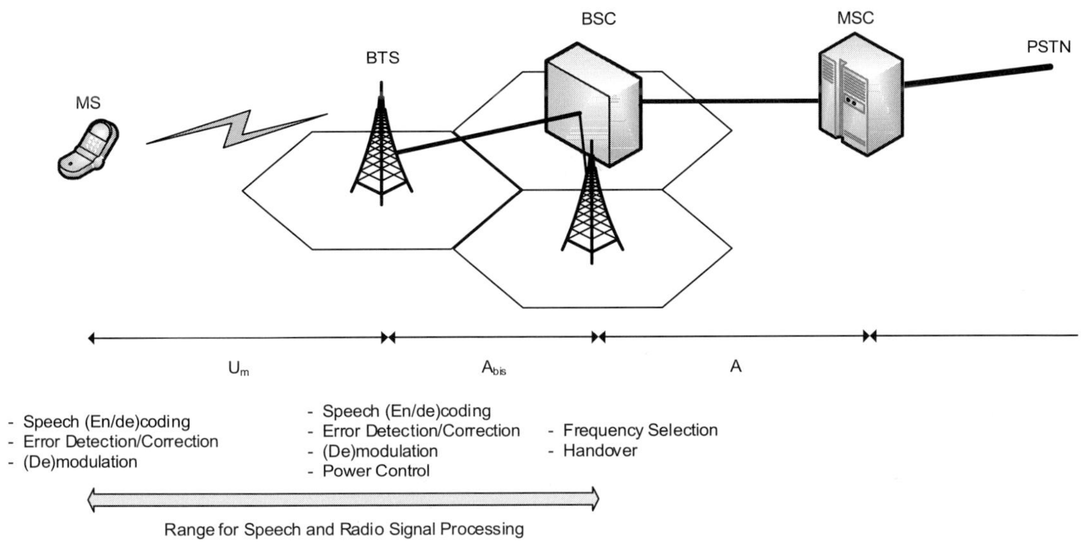

Fig. 2.14 Network architecture of D-AMPS.

required by the AMPS to 16 dB. This permitted a smaller frequency re-use factor and produced a correspondingly larger network capacity. In the case of AMPS, a cell radius of about 0.5 mile was considered to be the minimum allowable distance, but the cell radius of D-AMPS can fall below that, allowing channel re-use patterns that are smaller than $K = 7$. The network architecture of D-AMPS is shown in Fig 2.14.

Network Architecture

Processing the speech digitally means that the advancements in speech compression and signal processing capability could be incorporated to achieve higher quality and capacity. The Half-Rate (HR) is another speech codec developed for D-AMPS with which the speech is encoded at 3.975 kbps, thereby doubling the network capacity again albeit at a lower perceptual quality that is acceptable only when the level of voice traffic demands its use. Figure 2.13(c) illustrates D-AMPS in the frequency-time-power domain along with AMPS and N-AMPS. AMPS, N-AMPS, and D-AMPS are all classified as Frequency Division Multiple Access (FDMA) mobile communications systems, since the total spectrum is divided into smaller pieces in the frequency domain and assigned to different MSs for services. D-AMPS is further classified as a Time Division Multiple Access (TDMA) system, as each channel is again divided into smaller pieces in the time domain and assigned to each MS.

Channel Structure

In D-AMPS, a channel consists of a series of 40 ms TDMA frames, each of which is made up of six 6.667 ms time slots. Unlike the voice channels of AMPS, where the control information is either inserted when it is necessary or spectrally overlapped, in TDMA mobile communications systems such as D-AMPS, each time slot is structured so that speech data and control information are transported at pre-defined locations.

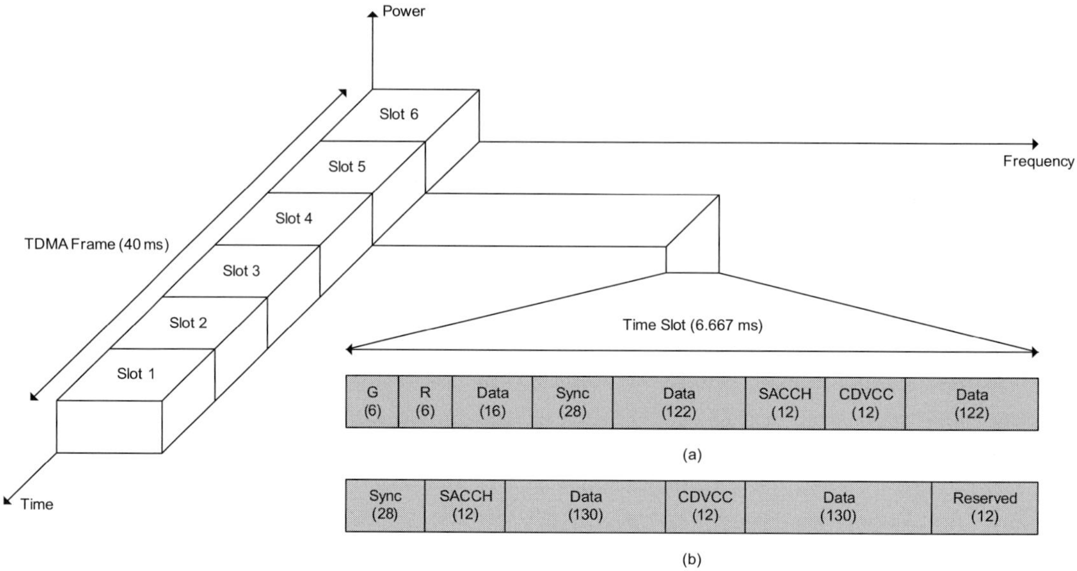

Fig. 2.15 TDMA frame structure. (a) Reverse channel. (b) Forward channel.

Such a structure, with dedicated fields for control information, may seem to be spectrally inefficient, but the fixed structure helps the receiver extract each type of information efficiently while enabling fast synchronization with 28 synchronization (Sync) bits. Figure 2.15 shows the structure of a TDMA frame in which the forward and reverse channels have different slot formats. A total of 324 bits, among which 260 bits are speech data, is transported in each slot; this corresponds to 48.6 kbps. In the full-rate mode, two slots of each TDMA frame separated by two slots are assigned to each MS. For example, slots 1 and 4, 2 and 5, and 3 and 6 can be assigned to three MSs respectively.

Coded Digital Verification Color Code (CDVCC) is a 12-bit field in each slot in the forward and reverse channels that plays a conceptually similar role to the Supervisory Audio Tone (SAT) of AMPS. While the SAT is represented by one of three frequencies, the CDVCC is an 8-bit integer protected with a 4-bit redundancy. In each slot, the Base Transceiver Station (BTS) transmits a CDVCC value which the MS should return exactly and immediately to continue communication. CDVCC can also be used to identify the BTSs sharing a channel. Slow Associated Control Channel (SACCH) replaces the voice control channel of AMPS, with which the BTS controls the transmit power of the MS or signals handover. With the SACCH, the MS reports the measured signal level, the Received Signal Strength Indicator (RSSI), and the Bit Error Rate (BER) of the current forward channels to assist the handover decisions of the BSC.

The presence of a data field in each time slot can also be used for signaling purposes. One important advantage of digital mobile communications systems is that since the speech is processed digitally, it is easier to transmit control information without noticeably reducing the speech quality by waiting for transmission opportunities when voice activity is absent, and using the empty slots. With AMPS, the analog speech signal is

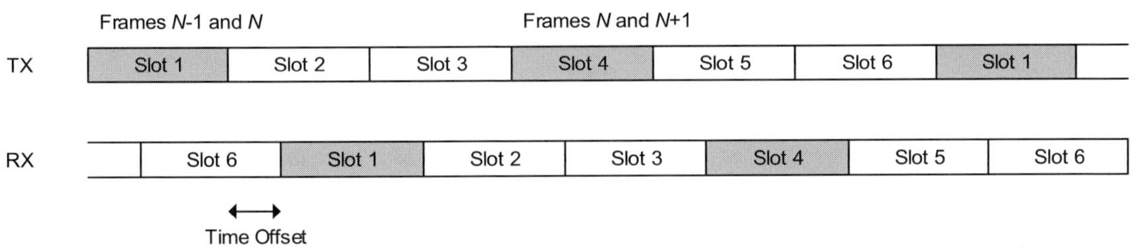

Fig. 2.16 Time offset between transmission and reception.

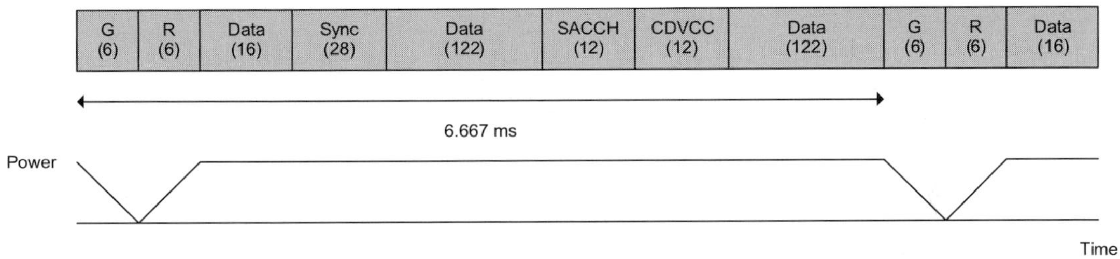

Fig. 2.17 Reverse channel transmission.

simply bypassed when the channel needs to be used for control signaling. Therefore to prevent the interruptions from being audible, the AMPS limits the duration of control data transmission to 34–54 ms. D-AMPS continues to use the 21 control channels of AMPS but it can temporarily assign additional control channels when they are necessary.

Although the forward and reverse channels are separated in the frequency domain, time slots of the channels are also separated in the time domain to reduce the mutual interference, i.e., the interference between the transmitter and the receiver. Thus, in a strict sense, conversations in D-AMPS are not full-duplex. The time offset between transmission and reception hides the limitations of using low-cost RF circuitry that can either transmit or receive at any one time. In D-AMPS, the transmit time of the BTS is synchronized in all forward channels but that of each MS must be continuously controlled to maintain the necessary separation in the time domain; the time slot of each MS is required to reach the BTS at close to its assigned temporal location so that the adjacent slots from two other MSs do not overlap. Figure 2.16 shows the time offset between the forward and reverse channels, defined as the time between the transmission of slot N and the reception of slot N. The default time offset of 44 bits can be reduced by the BTS, based on the estimated distance of each MS, by increments of between one-half and 15 bits. Notice that a term *symbol* is often used to represent the bits including speech data and control information before modulation.

During the guard time, represented in Fig. 2.15 as G, the transmitter of an MS is turned off for 0.123 ms, to avoid collision with a previous slot. To reduce the interference on neighboring channels that could arise from the instantaneous turn-on or turn-off of the transmitter, the transmit power is gradually increased or decreased during the ramp time, R, illustrated in Fig. 2.17 as slopes. In the forward channels, neither a guard

Table 2.1 System parameters of AMPS and D-AMPS.

	AMPS	D-AMPS
Multiple access	FDMA	FDMA/TDMA
Modulation	FM (voice), FSK (data)	$\pi/4$ DQPSK
Channel bandwidth (kHz)	30	30
Voice channels per carrier	1	3 (FR), 6 (HR)

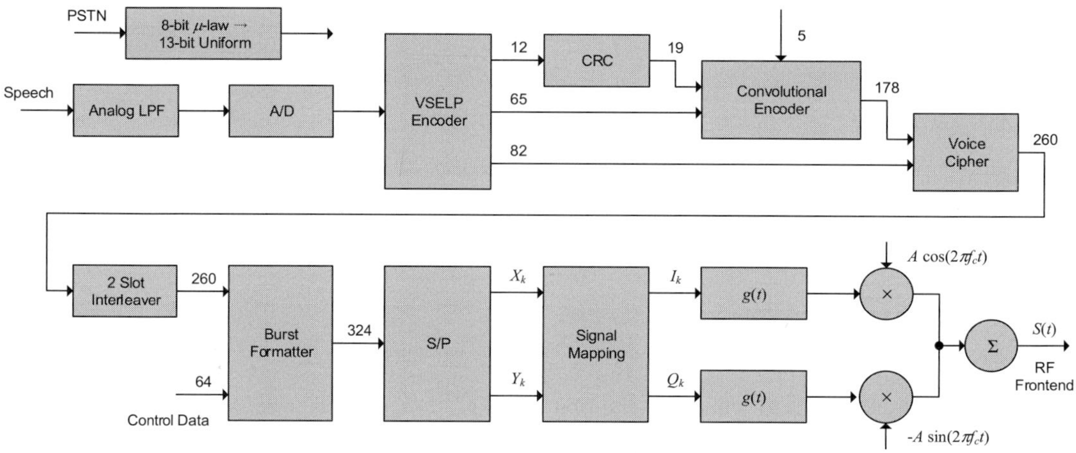

Fig. 2.18 Speech and radio signal processing operations of D-AMPS for VSELP.

time nor a ramp time is necessary since the BTS transmits over all channels and slots simultaneously. The 28-bit Sync field is a binary sequence with low autocorrelation values used by the receiver to locate the edge of each slot. The major parameters of AMPS and D-AMPS are compared in Table 2.1.

Speech and Radio Signal Processing Operations

Figure 2.18 shows the signal processing operations performed in D-AMPS for the forward and reverse channels. For each 20 ms interval, the speech signal from that interval is digitized and encoded in the VSELP speech encoder. Multiple levels of redundancy are applied to the encoded speech for error detection and correction. The speech data is then ciphered, i.e., shuffled in a complex way to make eavesdropping difficult. By contrast, in AMPS, as previously mentioned, no mechanisms to prevent eavesdropping are provided. Note that ciphering itself does not increase the bit-rate, although it does increase the processing delay at both the transmitter and the receiver. After ciphering, the speech data is mixed with the data of the previous or next speech frame, and is then split into two groups to fill two time slots in a TDMA frame. In addition to the speech data, the remaining fields of each slot are filled with appropriate control data. Finally the completed slots are modulated and transmitted over the wireless channel.

One of the important differences between the D-AMPS and the AMPS is the point at which digital representation of the speech signal begins and ends. In AMPS, digital representation of the speech begins and ends at the BS; in D-AMPS, once digitized, the speech remains in digital formats throughout the end-to-end transmission paths. In the forward channel, the speech signal in 64 kbps PCM is represented by eight bits per speech sample. This is then expanded to the 13-bit Uniform format that is sent to the VSELP speech encoder, which compresses 160 speech samples into a 159-bit frame. The same operations occur in the reverse channel. Detailed descriptions of the signal processing algorithms of the VSELP speech codec can be found in [EIA/TIA (1990)].

The 2:1 analog compander of AMPS applies a fixed amplitude compression ratio for any input within a range, whereas digital companding techniques, such as the A-law or μ-law, enable the use of different amplitude compression ratios matched to the input signal. A-law applies a linear companding characteristic near the origin but a logarithmic one when the input signal is large. The companded signal $c(x)$ of A-law is defined by

$$c(x) = \begin{cases} \frac{A|x|}{1+\ln A}\mathrm{sgn}(x) & 0 \leq \frac{|x|}{x_{\max}} \leq \frac{1}{A}, \\ x_{\max}\frac{1+\ln(A|x|/x_{\max})}{1+\ln A}\mathrm{sgn}(x) & \frac{1}{A} < \frac{|x|}{x_{\max}} \leq 1, \end{cases} \tag{2.33}$$

where x is the input signal, and sgn(x) is the signum function whose value is 0 when $x = 0$, and 1 when $x > 0$. A is a parameter that controls the shape of companding. Larger values of A compress the input signal more while $A = 1$ results in no companding. In the European PCM standard, A is set to 87.56. While companding in analog signal processing contributes only to a reduction of perceptual quality degradation from clipping or frequency deviation, companding in the digital case can increase the spectral efficiency. When the frequency content of a large input signal is low, the bit-rate can be statistically reduced even though the quantization error is large. A uniform quantizer would require about four more bits on average than an A-law quantizer for the same SNR at low input signal levels.

Figure 2.19 shows the block diagram of the VSELP speech encoder. The output of 159 bits per 20 ms consists of the reflection coefficients, which are mappings of the filter coefficients (38 bits) that have better quantization behavior, frame energy (5 bits), pitch lags (28 bits), codebook entries (56 bits) corresponding to the filter input, and the gain values (32 bits). Compared with the bit-stream of 64 kbps PCM where each speech sample is independently represented by the same number of bits, an encoded frame of a Linear Predictive Coding (LPC) speech encoder, such as VSELP, consists of different classes of bits whose contributions to the reconstructed speech quality when lost are not the same. Depending on which bits are lost or inverted, the speech decoder may be able to reconstruct a speech signal whose perceived quality is close to the original, or fail at decoding and be forced to replace the damaged frame by interpolating or extrapolating the segment from nearby successfully decoded frames.

The difference in the importance of the bit-stream elements may necessitate using Unequal Error Protection (UEP) to protect some of them more than others. With UEP, each bit or group of bits receives a different level of error detection or error protection capability, depending on its relative importance. This class of transmission techniques

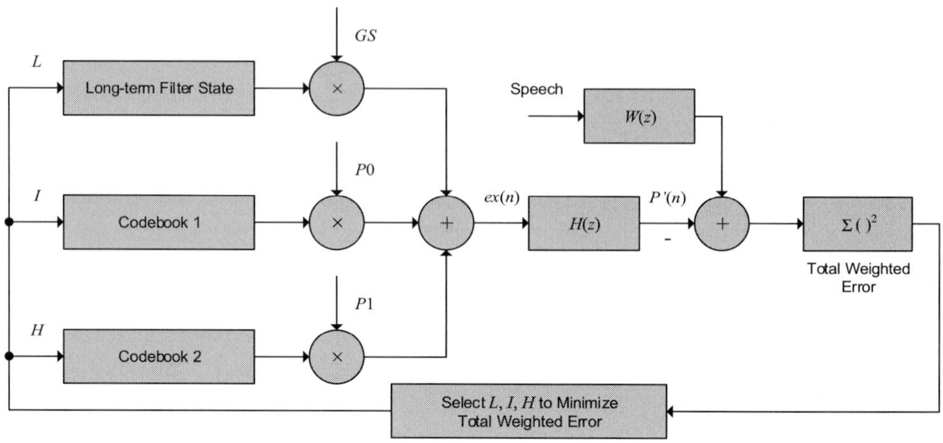

Fig. 2.19 Block diagram of VSELP speech encoder.

is effective when the error rate is limited. In circuit-switched networks such as the D-AMPS, such benign channel conditions can normally be maintained. However, if a significant proportion of the bits is corrupted, or if many speech frames do not arrive at the speech decoder, then UEP may be little more effective than Equal Error Protection (EEP).

This is often the case in packet-switched networks such as the Internet or in circuit-switched networks when the intermediate network nodes drop erroneous data units, as with the Radio Link Control (RLC) protocol of W-CDMA. For the UEP to be efficient, the encoded media frames should be short, typically fewer than a few hundred bits. The interdependence between successive media frames also reduces the effect of UEP as the corruption of a frame can propagate to the succeeding frames. For example, in many video codecs, decoding of the current frame often requires key parameters from the previous frames, which may have been compromised during transmission.

The relative importance, or subjective relevance, of encoded speech bits can be assessed by inverting a bit of each speech frame for a short period, e.g., 2 seconds. The bit-stream is then decoded and subjectively judged to construct a relevance table measuring the impact of inverting each bit [Hanzo *et al.* (2007)]. The sensitivity of the bit-stream can be improved by transforming key parameters into less sensitive formats, which contain equivalent information but have a higher level of resilience against errors, such as replacing the linear prediction coefficients by the reflection coefficients in LPC. Table 2.2 summarizes the bit-stream elements of the VSELP speech codec.

The 159 bits generated by the VSELP speech encoder for each 20 ms speech frame fall into two classes. Class 1 contains the 77 perceptually most important bits for the reconstruction of the speech; the remaining 82 form the Class 2 bits, which do not undergo any further processing for error detection or correction. The Class 1 bits include the reflection coefficients, gains, and the pitch lags. The 10th-order synthesis filter used in the codec requires ten reflection coefficients, of which the lower-order coefficients are

Table 2.2 Bit allocation of VSELP speech codec.

	Bits/subframe	Bits/frame
LPC coefficient		38
Frame energy $R(0)$		5
Pitch lag L	7	28
Codebook entry I, H	7, 7	56
Gain GS, $P0$, $P1$	8	32
Total		159

considered more important. Therefore to the ten reflection coefficients, 6, 5, 5, 4, 4, 3, 3, 3, 3, and 2 bits are assigned from the lowest- to highest-order coefficient, respectively.

The Class 1 bits are further re-ordered based on their subjective importance prior to the channel coding. A 7-bit Cyclic Redundancy Check (CRC) is computed and applied to the twelve most important Class 1 bits. Five zeros are appended to the end of the remaining 65 Class 1 bits to flush the shift registers of the convolutional encoder. Then rate $\frac{1}{2}$ convolutional encoding doubles the number of bits from 89 to 178. When the 82 Class 2 bits are appended, construction of a 260-bit frame of the VSELP speech codec is complete. After the ciphering and interleaving, which increases the delay but not the bit-rate, and adding 64 bits of control information made up of the G, R, Sync, SACCH, CDVCC, and Reserved fields, the 260 bits are transported on two time slots of a TDMA frame. Figure 2.18 shows the number of bits at each step in the speech and radio signal processing of D-AMPS. The procedures are applied to both the forward and reverse channels. The frame structures of the two channels are slightly different, as shown in Fig. 2.15, but their capacity for data transport is identical.

As can be seen in Fig. 2.3, it has already been observed that in a typical conversation, each speaker talks for less than a half of the call duration. The temporary absence of voice activity can be exploited to reduce the interference by turning off the transmitter. This affords two advantages: it increases the network capacity and it also extends the battery life. D-AMPS provides two states for the operation of a transmitter during silence, Discontinuous Transmission (DTX) High and Low, to classify the periods when voice activity is not present. In DTX High, no data is transmitted but information for control or about channel quality, such as the CDVCC and SACCH fields, is sent continuously; in DTX Low, the transmitter is periodically turned on to transmit the information on channel quality so that the BTS can update the channel status of each MS.

After ciphering, in which the 260 bits are shuffled in a very complex fashion such that only the serving BTS or the intended MS can reverse the process and recover the bits, the speech data is interleaved to spread the effects of bit errors as widely as possible over the bit-stream. One-half of the current speech frame is mixed with the bits of the previous speech frame and the other half is mixed with those of the following speech frame. Let the 130 bits of the current frame be $0X, 2X, \ldots$, and $258X$, and the 130 bits of the next frame be $1Y, 3Y, \ldots$, and $259Y$. Then the two bit-streams are entered into a 26×10 buffer column-wise and output row-wise, as illustrated in Fig. 2.20. An MS

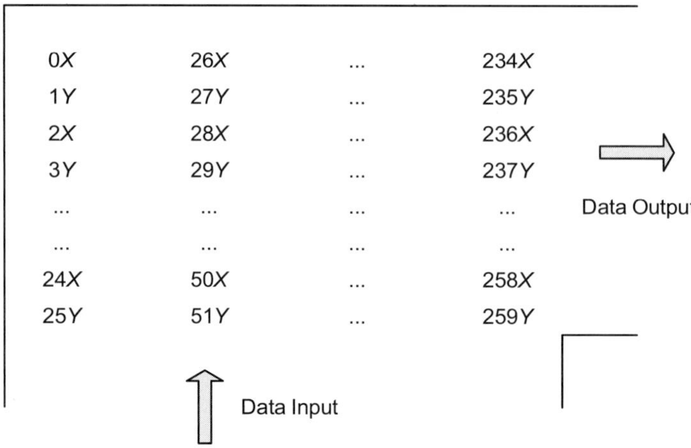

Fig. 2.20 Interleaving in full-rate mode.

operates in the full-rate mode when the VSELP speech codec is used. If a lower bit-rate codec that requires only one time slot per TDMA frame is used, then the MS operates in the half-rate mode.

The burst formatter constructs 324-bit slots with the interleaved speech data and control information. The assembled bits in each slot are separated into two streams, the odd stream, X_k, and the even stream, Y_k, which are mapped onto the in-phase component, I_k, and the quadrature component, Q_k, respectively. The two sequences are applied to a low-pass filter with an impulse response $g(t)$ respectively, generating two waveforms. The waveforms are modulated to RF carriers after an appropriate channel gain A is applied and summed to generate $S(t)$,

$$S(t) = \sum_k Ag(t - kT) \cos \varphi_k \cos 2\pi f_c t - \sum_k Ag(t - kT) \sin \varphi_k \sin 2\pi f_c t, \tag{2.34}$$

which is amplified and transmitted over the wireless channel. It is A that is adjusted when the transmit power at the MS or the BTS needs to be controlled.

At the receiver of the BTS or the MS, the speech is reconstructed by reversing the procedures. Many signal processing operations, such as CRC computation, ciphering, and interleaving, require comparable levels of complexity and delay at both the transmitter and the receiver. However, in general, the speech encoder has higher complexity than the speech decoder while the complexity of the convolutional encoder is almost negligible when compared to that of the Viterbi decoder, which is applicable to other channel codes. On the other hand, complexity of the signal processing procedures, which transform I_k and Q_k to $S(t)$, is less significant than that of the reverse procedures at the receiver, which tracks the time-varying phase difference and channel gain to equalize the impact of fading. This asymmetric partitioning of complexity between the source encoder and the source decoder, between the channel encoder and the channel decoder, and between the transmitter and the receiver can be found in most digital media compression technologies and in most digital mobile communications systems.

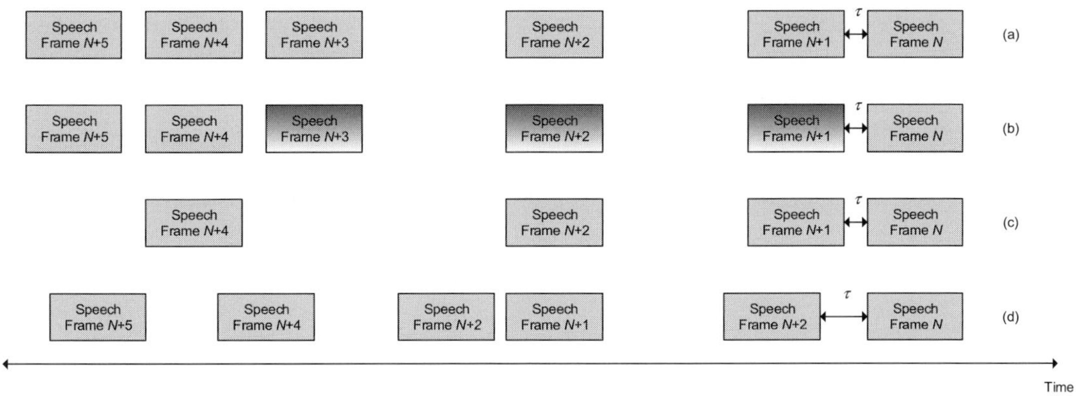

Fig. 2.21 Input types to speech decoder. (a) Clean channel. (b) Bit/frame error channel. (c) Erasure channel. (d) Packet-lossy channel with jitter.

The input to the speech decoder can be one of several types depending on the characteristics of the mobile communications system. In Fig. 2.21(a), the speech encoder generates speech frames that are transmitted over an error-free, clean channel. It is assumed that the speech frames are of the same type but a special frame type, Silence Descriptor Frame (SID), is generated when voice activity is absent and transmitted periodically. In this situation, all transmitted frames arrive without error. Figure 2.21(b) illustrates the influence of bit-level corruptions, where the received speech frames include bit errors that the channel decoder is unable to remove completely. However, even with the errors, the transmitted frames arrive at the speech decoder, and the interval between two consecutive speech frames, shown in Fig. 2.21 as τ, is almost constant. This type of channel impairment is observed in many circuit-switched digital mobile communications systems where the receiver can identify that some speech frames contain errors by CRC recomputation, but nonetheless forwards the frames containing those errors to the speech decoder, which may be able to salvage some bits from the corrupted frames by using knowledge of the frame structure. Alternatively, the decoder may apply error concealment techniques rather than recycle the corrupt frames.

In Fig. 2.21(c), the same type of channel impairment is encountered but the receiver does not forward speech frames containing errors to the speech decoder. This might happen when the receivers of intermediate network nodes drop frames containing errors. In the figure, frames $N + 3$ and $N + 5$ are missing. Therefore the speech decoder must use the successfully decoded frames to cover the loss, e.g., by interpolating the samples of successfully decoded speech frames. If a large number of speech frames is lost or dropped by the receiver, the gap between two successfully decoded speech frames can be quite large, which may exceed the capability of signal interpolation or extrapolation techniques to cover the loss, resulting in perceivable degradation of speech quality. However, τ remains almost constant.

Figure 2.21(d) illustrates the influence of transmission in very different types of networks, where speech frames containing errors are not fed to the speech decoder. The time between two speech frames can vary significantly even within a call, depending on the channel condition and the cell loading level. Moreover, in this case, the speech frames may be received multiple times or received in a different order than they were transmitted. In the figure, frame $N + 2$ is received twice and frame $N + 1$ is received after frame $N + 2$. In addition, frame $N + 3$ is absent. Although the size of each speech frame is shown as identical, frames containing speech data and background noise can be of different sizes. In many speech codecs, the latter frame type is likely to be smaller. The speech frames themselves can also be of different sizes when the variable bit-rate (VBR) coding is employed.

2.2.3 Further Opportunities

During the transition from AMPS to D-AMPS, the VSELP-encoded speech of D-AMPS was reported to be of a noticeably higher quality than that received by AMPS, especially at the cell edges. This may have been due to the effects of error detection and correction procedures. However, it was also observed that near the BS or in the absence of significant fading, AMPS often outperformed D-AMPS in terms of speech quality, perhaps due to its continuous transmission.

The second generation digital mobile communications systems, which were commercially more successful than the D-AMPS, were designed with fewer limitations inherited from the first generation systems. They achieved higher speech quality and larger network capacity by employing new wireless transmission techniques, such as advanced modulation and error control coding. However, they also exploited the statistical characteristics of speech signals in a more sophisticated way than in the simple two-state DTX mechanisms of D-AMPS. For example, the speech signal could be analyzed to apply different encoding methods and bit-rates to each speech segment of 20 ms. This can reduce the average bit-rate and transmit power without harming the speech quality. On the other hand, the bit-rate could be dynamically assigned to each MS by the network or between the encoded speech and the channel code to match time-varying channel conditions and traffic levels.

The *sunset* for AMPS and D-AMPS, shutting down the obsolete networks and re-assigning the frequency spectrum for new networks and services, began after the deployment of the third generation systems. The two early analog and digital systems established many of the essential concepts of cellular mobile communications and influenced the design of speech and radio signal processing procedures in many contemporary and next generation mobile communications systems. Some of the key constraints imposed on the design of AMPS and D-AMPS, such as the fixed channel bandwidths, were finally relaxed in the fourth generation systems.

2.3 Global System for Mobile Communications

The Global System for Mobile Communications (GSM), a second generation digital mobile system, was not designed as a migration path from the previous analog systems. With a channel bandwidth much wider than that of D-AMPS, each channel of GSM can transport the speech for eight MSs encoded at 13 kbps. Because fewer limitations were placed on the design of its signal processing procedures, GSM is able to reduce the operational SNR to as low as 9 dB, which enables a smaller frequency re-use factor than was possible in D-AMPS. The large channel bandwidth, 200 kHz, enables higher bit-rate services when multiple slots are assigned to an MS, as was provided for the General Packet Radio Service (GPRS).

2.3.1 Network Architecture

As outlined in Table 1.6, there were analog mobile communications systems that used technologies similar to those of AMPS, such as the Total Access Communications System (TACS) or the Nordic Mobile Telephone (NMT) [Cox (1999)], that were deployed locally in Europe, but GSM was designed with a far-sighted plan to provide pan-European coverage with sufficient flexibility for further enhancements. There was no obligation to re-use existing analog infrastructure and a contiguous bandwidth of 25 MHz in the 900 MHz spectrum was reserved in most European countries for GSM. These were luxuries that few designers of mobile communications systems have seen since.

As the number of countries deploying GSM increased, the economies of scale lowered the equipment and handset costs. In addition, the ability to roam over many countries became a key advantage of GSM in the competition with other digital mobile communications systems, some of which were considered technically superior. The network architecture of GSM illustrated in Fig. 2.22 is almost identical to that of D-AMPS, with the major differences in the link adaptation based on joint source-channel coding as an advanced operating mode.

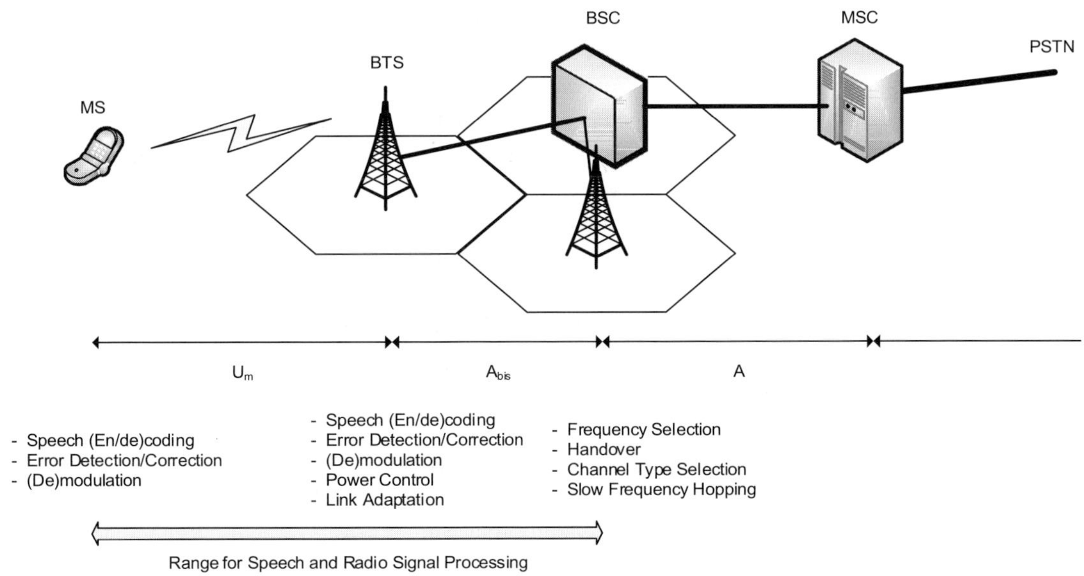

Fig. 2.22 Network architecture of GSM.

Table 2.3 Channel numbering system.

Channel number n	Uplink channel frequency (MHz)	Downlink channel frequency (MHz)
1–124	$890 + 0.2n$	$935 + 0.2n$

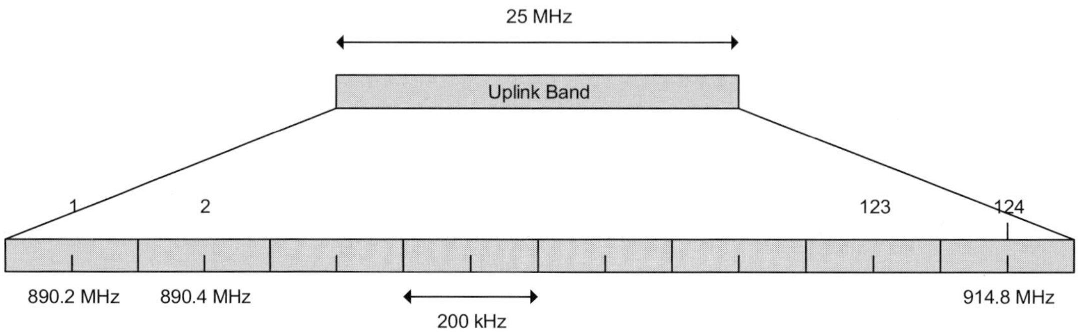

Fig. 2.23 Primary GSM-900 band.

2.3.2 Channel Structure

The original frequency spectrum for GSM was the Primary GSM-900 band, which includes 890–915 MHz for the uplink, as shown in Fig. 2.23, and 935–960 MHz for the downlink assuming a 45 MHz offset between channels with the same number. In GSM terminology, the terms uplink and downlink represent the reverse channel and the forward channel, respectively. In most mobile communications systems that use separate frequency bands for the uplink and downlink, i.e., systems that operate in the Frequency Division Duplex (FDD) mode, the higher frequency band is used for the downlink since in the direction from the MS to BTS, the probability of finding line-of-sight (LOS) paths is lower than it is in the other direction. Since the BTS is typically located at a higher elevation, it is easier to identify the MS with this choice. On the other hand, the lower frequency band is preferred for the uplink, where the transmit power is limited, because it has better refraction characteristics. Later, other frequency bands were defined for GSM to serve countries where the 900 MHz bands were already occupied by other services. The channel number, Absolute Radio Frequency Channel Number (ARFCN), and the carrier frequency of GSM are related as in Table 2.3. Just as in AMPS and D-AMPS, there is a 45 MHz difference between an uplink channel and a downlink channel of the same ARFCN.

GSM provides a total of 124 channels by using a 25 MHz band and a channel spacing of 200 kHz. However, it is customary not to use channels 1 and 124 to limit the interference into adjacent bands, which leaves 122 channels for voice, data, or control. Each channel of GSM transmits a series of TDMA frames, each of which, in turn, contains eight time slots. GSM offers many types of frames depending on the objectives. The basic frame type used to transport speech data, the Full-Rate Speech Traffic Channel (TCH/FS), has a structure in which speech data and control information have

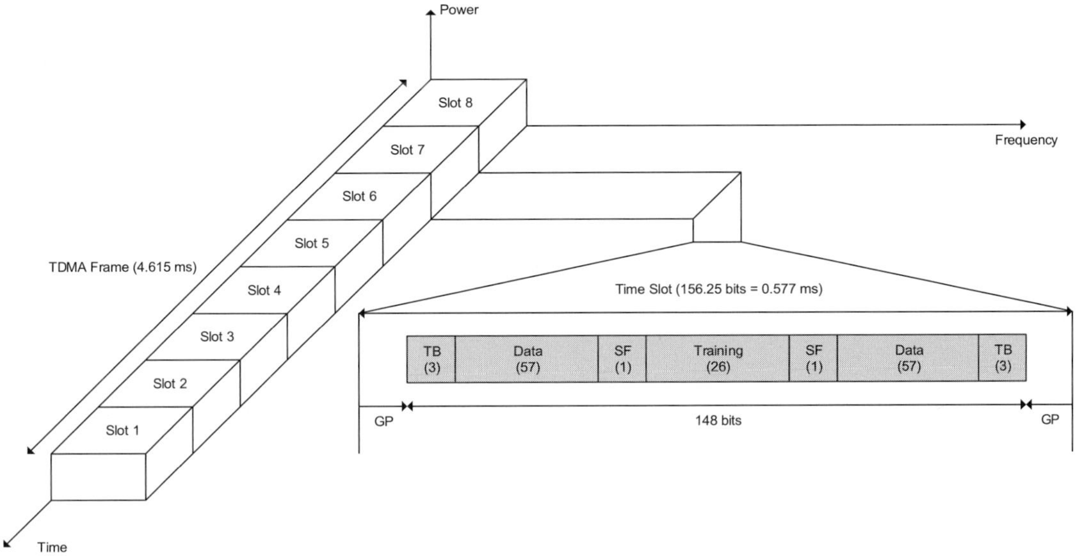

Fig. 2.24 Frame structure.

pre-defined locations, but unlike in D-AMPS, the structure of TCH/FS is symmetric, i.e., it is identical for the uplink and downlink. Figure 2.24 shows the detailed frame structure of TCH/FS in which 148 bits are transported in each time slot of 0.577 ms duration. In D-AMPS, the length of each TDMA frame is 40 ms, corresponding to the length of two speech frames. In GSM, the lengths of a TDMA frame and a speech frame have a more complex relationship; 26 consecutive TDMA frames, two of which do not carry speech data, are equal in length to six speech frames, 120 ms, as shown in the hierarchical frame structure of Fig. 2.26. If the guard bits are included, during which no bits are actually transmitted, the aggregate bit-rate of each channel is 270.83 kbps but the usable bit-rate assigned for each TCH/FS is $114 \times \frac{24}{120} = 22.8$ kbps. In this book, the unit symbols per second (sps) is interchangeably used with bits per second (bps) when it is necessary to differentiate the symbol transmission rate from that of information bits.

A Guard Period (GP) of 30.46 μs, corresponding to 8.25 bits, is inserted between two time slots to reduce the interference; a similar interval was inserted between two slots in D-AMPS. The Tail Bits (TB), consisting of three bits at the beginning and end of each slot, are set to logical 0; during this period the transmit power is increased or decreased gradually to reduce the interference on neighboring channels. After the tail bits at the beginning of a slot, 57 bits of speech data are transmitted, followed by a one-bit Signaling Flag (SF) that informs whether the payload is speech data or control information. The training sequence is a 26-bit sequence with a low autocorrelation that is used by the receiver to establish and maintain synchronization. As an example, one of the training sequences, Training Sequence Code (TSC) 7, is 11101111000100101110111100. If zero is mapped to +1 and one mapped to −1, the autocorrelation values decrease rapidly as the delay increases, as shown in Fig. 2.25(a). To establish synchronization, the MS correlates the received signal with a range of lag values until it reaches a peak value.

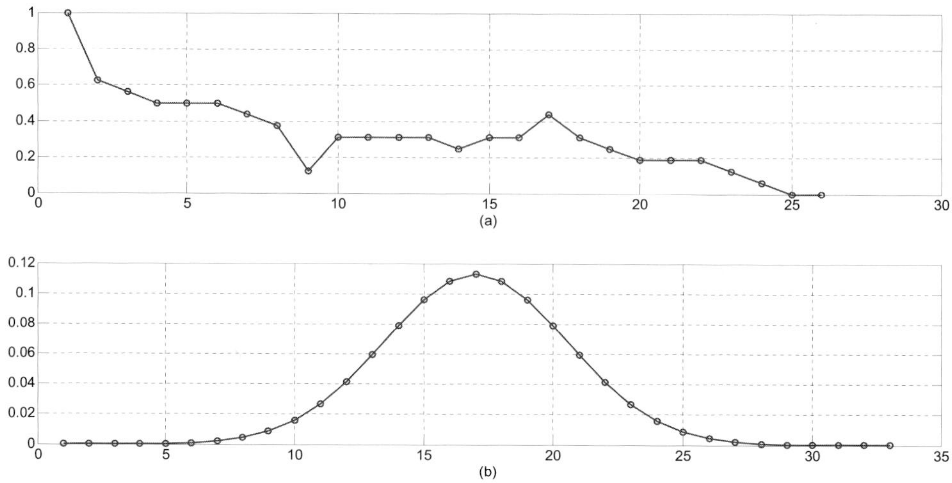

Fig. 2.25 (a) Estimated autocorrelation function of TSC 7. (b) Impulse response of a Gaussian filter.

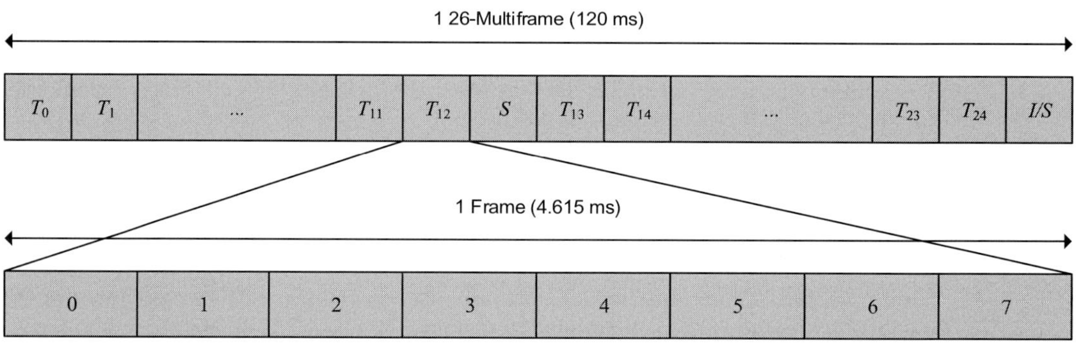

Fig. 2.26 Hierarchical frame structure of TCH/FS.

However, as the received training sequences are likely to be noisy, including errors, it is necessary to use robust techniques for estimating the autocorrelation. The synchronization process can be accelerated by applying the received signal to a set of correlators that operate in parallel.

A 26-multiframe consists of 24 TDMA frames containing speech data and 2 frames for control information. A TDMA frame for the Slow Associated Control Channel (SACCH) is inserted in the 13th TDMA frame, as shown in Fig. 2.26. The 26th frame is reserved for the SACCH in the Half-Rate Speech Traffic Channel (TCH/HS), but is not used in TCH/FS. Note that the Full-Rate and the Half-Rate refer to the speech codecs used by GSM but they are also used to represent the traffic channel types, TCH/FS and TCH/HS, respectively.

The control information can be transmitted either by *frame stealing*, i.e., replacing speech data with signaling messages, or by using some of the slots dedicated to SACCH. Since the former approach can influence the speech quality, it is used only in urgent

Table 2.4 System parameters of GSM.

Multiple access	FDMA/TDMA
Modulation	GMSK
Channel bandwidth (kHz)	200
Voice channels per carrier	8 (FR), 16 (HR)

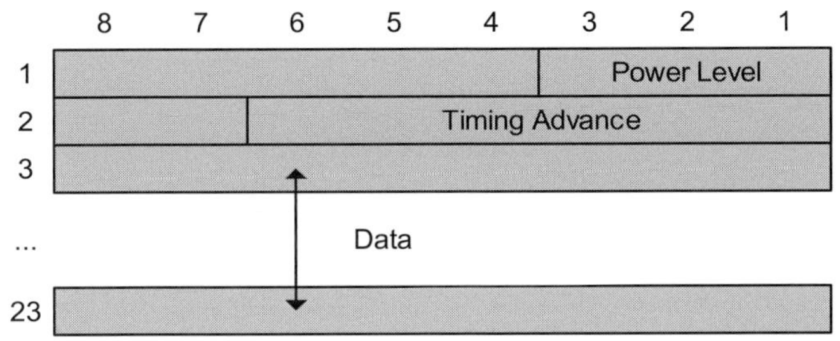

Fig. 2.27 SACCH format.

situations, such as handover, while the latter method is used to periodically transmit information essential to maintain link quality, such as the messages for power control or timing advance. The format of the SACCH is shown in Fig. 2.27. Notice that the number of bits assigned for each control operation determines the number of selectable levels.

The main system parameters of GSM, which are summarized in Table 2.4, are similar to those of D-AMPS since both belong to the second generation systems that share many TDMA features. Furthermore, the speech and radio signal processing operations of GSM shown in Fig. 2.32 are very similar to those of D-AMPS. However, GSM has a wider channel bandwidth, a modulation technique that maps fewer bits to a symbol, and shorter TDMA frames and time slots, which provide GSM with enough flexibility to allow further enhancements when higher signal processing capability and more efficient RF circuitry become available. In addition, GSM was equipped with new techniques including DTX or the Slow Frequency Hopping (SFH) that made it possible to exploit the unique nature of speech signals and reduce the impact of time-varying channel conditions in a more refined fashion than in D-AMPS, which suffered from numerous design constraints inherited from the AMPS.

2.3.3 Full-Rate Speech Codec

For each 20 ms period, the Full-Rate (FR) speech encoder transforms 160 speech samples originally in the 13-bit Uniform format into 260 bits, achieving a bit-rate of 13 kbps. The compression principle used is the Regular Pulse Excitation Long-Term Prediction (RPE-LTP), which provides acceptable quality with low complexity. Figure 2.28 shows a simplified block diagram of the FR speech encoder. First, the speech samples are pre-processed for offset compensation: the DC components are removed by notch filtering, and the samples are further filtered for pre-emphasis. Then the processed samples are

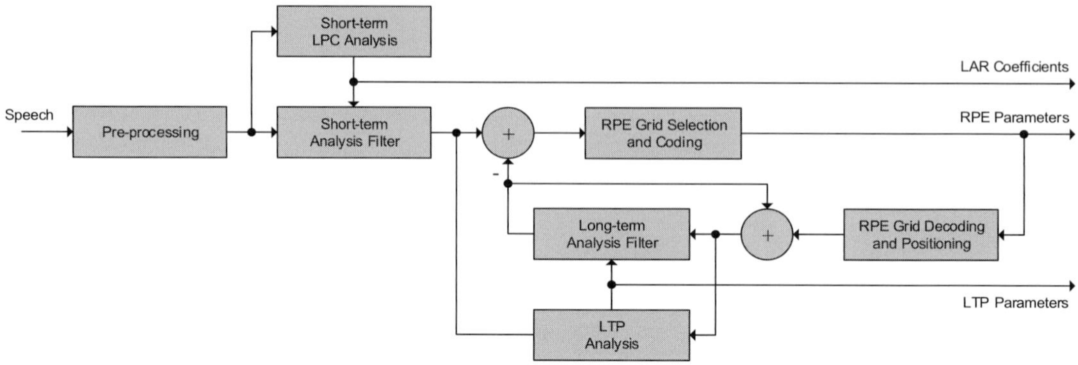

Fig. 2.28 Block diagram of FR speech encoder.

grouped into non-overlapping frames of 160 samples, each of which is analyzed using the LPC techniques to determine the coefficients of an 8th-order short-term analysis filter, as shown in Fig. 2.9.

The inverse of the analysis filter is used to filter the original 160 samples to obtain 160 samples of a short-term residual signal. The eight filter parameters, converted to reflection coefficients, are converted once again into 36-bit Log Area Ratios (LAR) for better quantization behavior, as in Eq. 2.32. In the quantization process, more bits are assigned to the lower-order coefficients, as they are more important for perception. Since the reflection coefficients affect the poles of the filter, errors from quantization or transmission could move the locations of the poles to result in an unstable filter. This can be avoided if the magnitude of the reflection coefficients after quantization is constrained to be strictly less than unity.

The short-term residual signal is broken into four subframes of 40 samples each. In each subframe a long-term correlation lag and an associated gain factor are computed by calculating the cross correlation of the subframe with 120 previously reconstructed short-term residual samples to find the highest correlation. The correlation lag typically has a value that lies between 40 and 120 to which seven bits are assigned; the gain is encoded with two bits. From the lag and the gain, an estimate of the short-term residual signal is computed, and 40 long-term residual signal samples are obtained by subtracting the estimate from the subframe containing the original short-term residual signal samples. The resulting block of 40 long-term residual signal samples is fed back for the regular pulse excitation (RPE) analysis.

After the RPE analysis, each block of 40 long-term residual samples is represented by one of four sub-sequences of length 13. Two bits are required to represent this sub-sequence, called the grid position in RPE terminology. The maximum of the absolute values of the sub-sequence is logarithmically quantized using six bits. The sub-sequence is encoded using Adaptive Pulse Coded Modulation (APCM) with three bits per sample. This thus requires 52 bits to represent each sub-sequence. For the last subframe, the information on the sub-sequence is not necessary since its long-term prediction (LTP) is computed within the boundary of the current speech frame, thereby requiring only the

Table 2.5 Bit allocation of FR speech codec.

	Bits/subframe	Bits/frame
LAR coefficient		36
LTP lag	7	28
LTP gain	2	8
RPE grid position	2	8
Block amplitude	6	24
RPE pulse index	39	156
Total		260

grid position and the block amplitude. Table 2.5 categorizes the 260 bits constituting each speech frame of the FR speech codec. It can be seen that identical bit allocation is applied to each subframe of the codec. Detailed information on the signal processing algorithms of the FR speech codec can be found in [3GPP (2000a)].

Discontinuous Transmission

Fig. 1.7 showed that in AMPS, the silence was translated into zero frequency-shift but did not otherwise affect the analog signal processing or transmit power. Although the power could be controlled and the channel could be re-assigned during a call, the absence of voice activity whose duration was shorter than the signaling delay between the MS and BS could not be exploited to increase the capacity. The two-state DTX mechanisms of D-AMPS left such opportunities to the discretion of the implementation [Delprat and Kumar (1999)], but GSM, by way of contrast, systematically exploits the silence to reduce the average interference, power consumption, and the required Carrier-to-Interference Ratio (CIR).

When DTX is enabled, the Voice Activity Detector (VAD) in the FR speech encoder is activated to determine whether or not the input includes speech signal. Figure 2.29 shows the VAD of the FR speech codec, whose operation is governed by three quantities calculated in the speech encoder [3GPP (2000c)]: ACF, N, and sof. Where ACF is the set of autocorrelation coefficients computed from the input speech frame to which adaptive filtering, energy computation, and averaging are applied, N is the LTP lag obtained for each subframe from which the pitch is extracted, and sof is the offset-compensated signal frame from which the information including tones is detected for the downlink.

VAD is the output parameter indicating the decision. VAD essentially operates as an energy detector by which the presence of voice activity in the input signal is declared, i.e., $VAD = 1$, if the decision variables computed from processing the internal and external metrics exceed certain thresholds. To ensure a reliable detection of voice activity, those thresholds should be set well above the noise level, but not so high as to miss the speech that becomes identified as noise. The thresholds in the VAD algorithms are continuously adapted to reflect the influences of time-varying local acoustic environments.

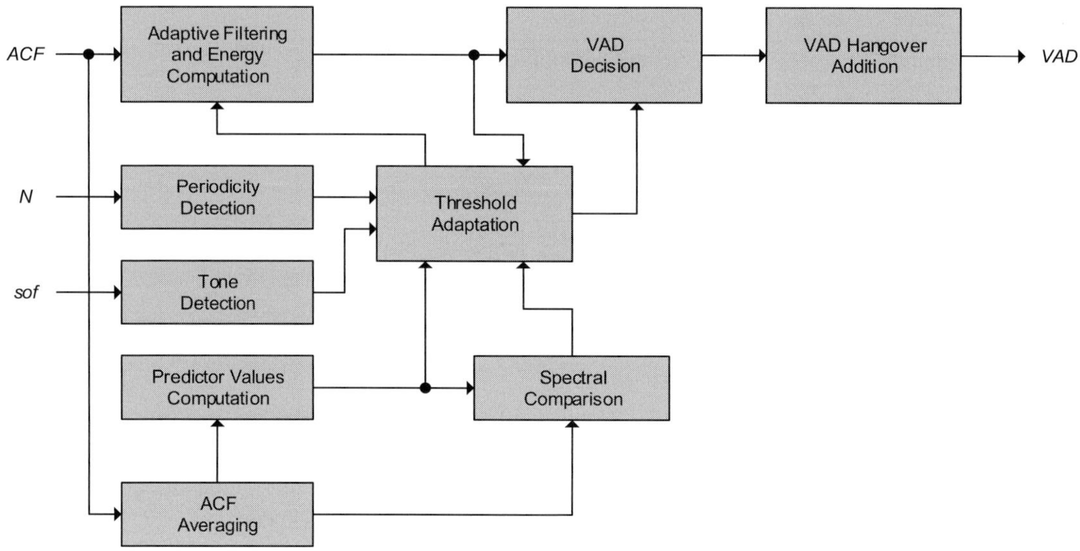

Fig. 2.29 Block diagram of FR VAD.

If the VAD declares that voice activity is present in a frame, the frame is encoded using the LPC algorithms; if not, the encoding methods designed to simulate the background acoustic noise are applied. These generate a special type of frame, Silence Descriptor Frame (SID). The primary objective in encoding SIDs is not to reconstruct the noise as faithfully as possible, but instead to deceive the listener into believing that the communications link is alive and that the sound at the far end is being continuously transmitted. The artificial noise generated for this objective is called *comfort noise*. SID contains only a minimum amount of information statistically related to the noise conditions at the transmitter. The speech decoder generates noise frames by combining information in the SID with locally generated parameters and setting the parameters of the speech decoder in a special fashion. During the silent period, an SID is transmitted when there is a need to update the key parameters used for noise generation.

The noise-related parameters in an SID of the FR speech codec are computed over four consecutive frames declared to be silent. In Fig. 2.30, an SID, SID_{k+1}, is created after the VAD identifies four consecutive speech frames as pause, i.e., noise. The SP flag, generated by the transmission DTX handler on the basis of the VAD flag, is set to 1 to indicate a speech frame and 0 to indicate an SID. During the four-frame *hangover period*, the noise is still encoded using the speech compression algorithms. This hangover period is necessary to prevent a loss of speech quality from mistakenly encoding speech as noise. Two constants, $burstconst = 3$ and $hangconst = 5$, and two integer variables, *burstcount* and *hangcount*, are used to control the hangover procedures. In the SID, eight mean log-area ratio parameters, computed by averaging the LAR coefficients of the current and previous frames, replace the 36 bits assigned for the LAR coefficients. A mean block amplitude coefficient, computed by averaging the block amplitudes of 16 subframes, is repeated four times, replacing the 24 bits assigned for

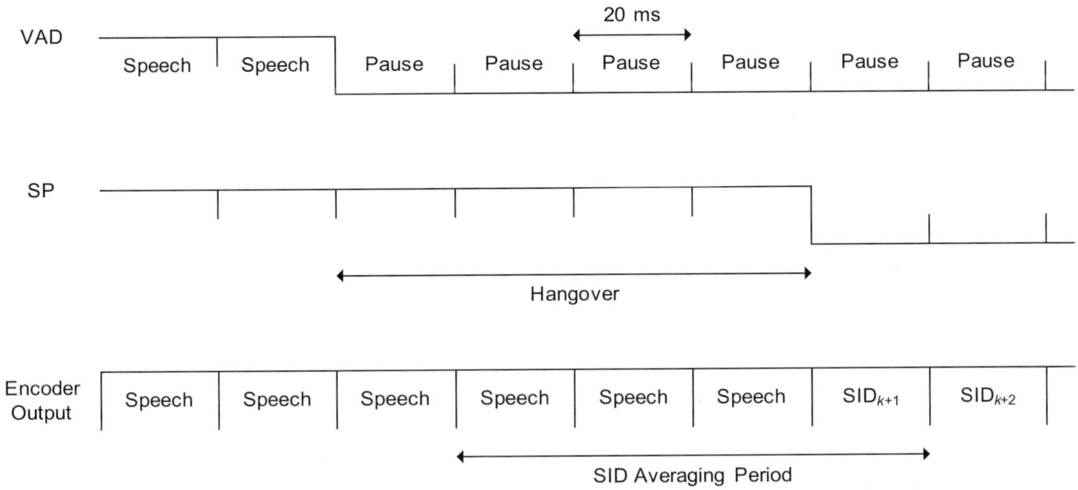

Fig. 2.30 DTX in full-rate mode.

the block amplitudes. The *SID averaging period* shows the scope of parameter averaging. The 60 bits of the comfort noise parameters fill in the same locations of the speech frame as the bits they replace. Finally 95 zero bits are used to indicate the presence of noise information. The remaining 105 bits in the SID are set to zero.

At the receiver, the reconstructed 260 bits are tested to identify whether they constitute a legitimate speech frame, an SID, or control information. A valid SID is declared only when fewer than two bits out of 95 zero bits are received in error. The 60 bits of noise parameters are combined with locally generated 150 bits to construct a complete 260-bit frame, which is applied to the speech decoder to synthesize the comfort noise. At the speech decoder, the RPE pulse amplitudes are replaced by a locally generated random integer sequence, each value of which is uniformly distributed between one and six. The RPE grid positions are also set to random integer values that are uniformly distributed between zero and three. The LTP gains are all set to zero, and the LTP lags are set to 40, 120, 40, and 120, respectively [3GPP (2001)].

Since an SID contains the same number of bits as an ordinary speech frame, the gain in network capacity from reduced interference is achieved only by infrequent transmission of SIDs. In principle, the speech encoder continues generating SIDs as long as there is no voice activity. However, most SIDs generated are not sent since the transmitter is turned off to reduce the interference and power consumption. In Fig. 2.30, the VAD continues declaring pause and the SIDs are generated. However, after SID$_{k+1}$ is transmitted, SID$_{k+2}$ and the later SIDs are not sent. Such a radio silence lasts for 24 frames.

If a short burst of speech frames is generated, within 24 frames after the last SID was transmitted, the last SID in the burst, as shown as SID$_k$ in Fig. 2.31, is sent to the radio subsystem, which includes the channel encoder and the interleaver, until an updated SID, shown as SID$_{k+1}$, is generated. This operating practice is intended to prevent a short spike of noise from influencing the noise pattern noticeably. Although the VAD is designed to exploit the structure of each speech codec, its algorithms

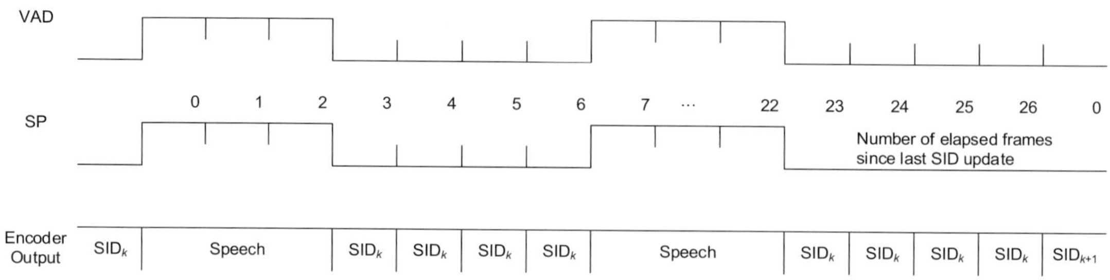

Fig. 2.31 DTX in short-burst mode.

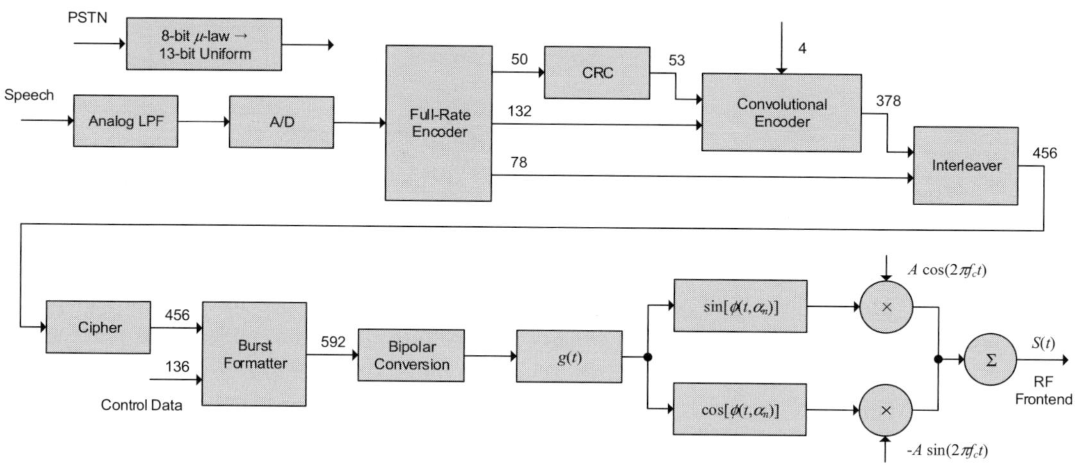

Fig. 2.32 Speech and radio signal processing operations of GSM for FR (TCH/FS).

can be continuously improved to enhance the accuracy of its decision, speech quality, and network capacity, with the main speech compression algorithms intact. Additional information on the VAD algorithms of the FR speech codec can be found in [3GPP (2000b)].

2.3.4 Uplink and Downlink Signal Processing

Figure 2.32 shows the signal processing operations of GSM. These are *symmetric* in that they are applicable to both the uplink and downlink. In the uplink, for each 20 ms period, the speech is sampled 160 times and digitized using the 13-bit Uniform quantizer. It is then encoded using the FR speech encoder. As in D-AMPS, the encoded speech bits are classified into three groups, depending on their relative perceptual importance. The redundancy is selectively applied to each group for a different level of error detection and correction capability. Then the resulting bits are interleaved and ciphered before being augmented with control information to construct each time slot. Finally the completed slots are modulated and transmitted over the wireless channel. For the downlink, the speech signal in the 8-bit 64 kbps PCM format is expanded to 13 bits and applied to the FR speech encoder, followed by the same procedures as the uplink.

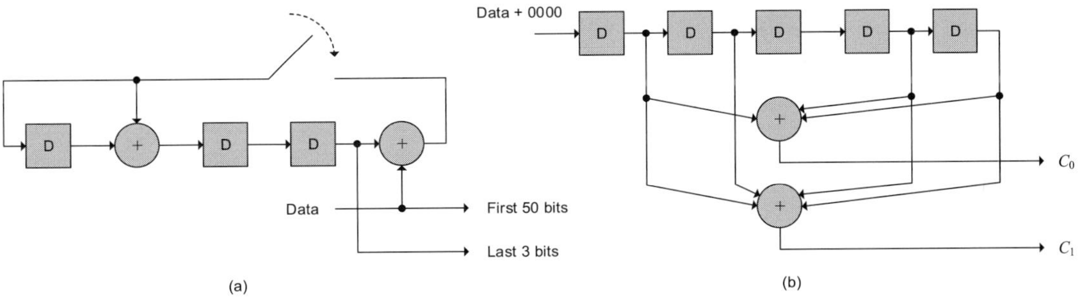

Fig. 2.33 (a) CRC generation. (b) Rate $\frac{1}{2}$ convolutional encoding.

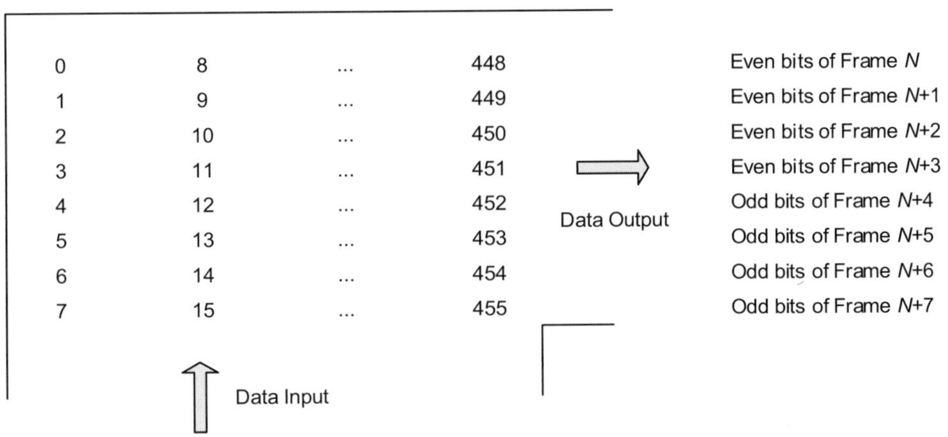

Fig. 2.34 Interleaving in full-rate mode.

For each 20 ms period the 260 bits from the FR speech encoder are grouped into three importance classes. The most important 50 bits are complemented with a CRC before they are channel coded with the next important 132 bits. In GSM nomenclature these protected 182 bits belong to Class 1 and the remaining unprotected 78 bits belong to Class 2. Figure 2.33 shows the shift register circuitry that generates the 3-bit CRC, and a $K = 5$, rate $\frac{1}{2}$ convolutional encoder. The convolutional encoder, whose input bits are padded with four zeros to flush the shift registers after encoding, can be described by two polynomials, $1 + D^3 + D^4$ and $1 + D + D^3 + D^4$. The CRC-augmented bits and the next important 132 bits are channel coded, generating 378 bits. The remaining 78 bits out of the original 260 bits are not protected at all [3GPP (2017c)].

Note that ten bits of the 60 noise parameters are not complemented with a CRC before they are channel coded as the CRC is computed from the more important 50 bits, and the 95 zero bits belong to Class 1. After all the procedures of speech and channel encoding, the 20 ms of speech produces 456 bits to be transmitted. These bits are interleaved, as illustrated in Fig. 2.34, ciphered, and distributed over eight time slots of eight TDMA frames, which may be shared with the data from other MSs. Once the 34 bits of control

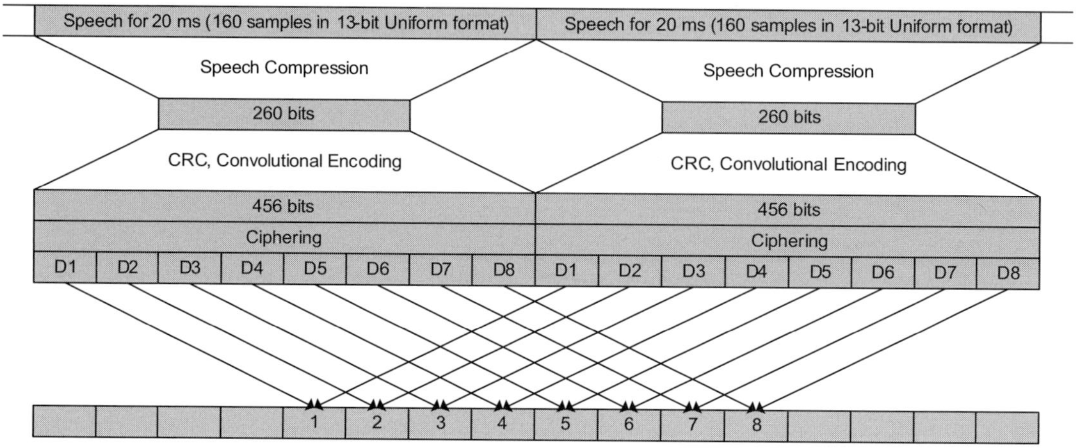

Fig. 2.35 Frame construction in full-rate mode.

information are added to each slot, the bits are ready to be modulated. Figure 2.35 shows the frame construction process when the FR speech codec is used.

GSM uses Gaussian Minimum Shift Keying (GMSK) in the modulation. The burst-formatted data of each slot, d_i, is differential coded using modulo-2 addition and transformed into a sequence of ± 1, α_i. This bipolar sequence is applied to a low-pass filter with an impulse response $g(t)$, formed as the convolution of a Gaussian filter, $h(t)$, with a rectangular pulse, $rect(\frac{t}{T})$. T is the duration of each information bit,

$$\widehat{d_i} = (d_i + d_{i-1}) \bmod 2, \tag{2.35}$$

$$\alpha_i = 1 - 2d_i, \tag{2.36}$$

$$g(t) = h(t) * rect\left(\frac{t}{T}\right), \tag{2.37}$$

$$h(t) = \frac{1}{\sqrt{2\pi}\sigma T} \exp\left(\frac{-t^2}{2\sigma T^2}\right). \tag{2.38}$$

Figure 2.25(b) shows the impulse response of $h(t)$, in which the duration of eight samples matches the duration of each information bit. The filter outputs are integrated to generate $\varphi(t, \alpha_n)$, the phase of the in-phase and quadrature components, which are then modulated to RF carriers after an appropriate channel gain A is applied. The carriers are then summed to generate $S(t)$, which is further amplified and finally transmitted over the wireless channel,

$$\varphi(t, \alpha_n) = \sum_i \alpha_i \pi \eta \int_{-\infty}^{t-iT} g(u)du, \tag{2.39}$$

$$S(t) = \cos[2\pi f_c t + \varphi(t, \alpha_n)]. \tag{2.40}$$

Here η is the modulation index used to control the maximum phase shift per bit, which is set to $\frac{1}{2}$ in GSM. Therefore the maximum phase change allowed is $\frac{\pi}{2}$ radians per bit. Further information on the GMSK modulation procedures can be found in [3GPP (2017d)].

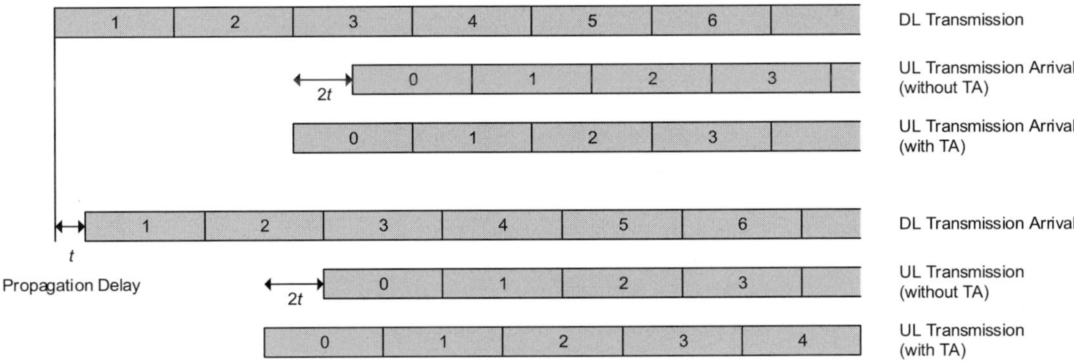

Fig. 2.36 Timing advance for frame synchronization.

Timing Advance

In D-AMPS, there are algorithms to control the transmit time of the reverse channels so that the data from each MS reaches the BTS within an acceptable timing interval. In GSM, similar methods are used to avoid the collision of data from multiple MSs at the BTS. As shown in the TDMA frame and time slot structures of Fig. 2.24, such an adjustment should be continuously made to align the time slots from the MSs. Figure 2.36 shows how the radio propagation delay necessitates adjusting the transmit time at the MS. In D-AMPS, the default offset between the forward and reverse channels is 44 bits, corresponding to 1.81 ms, but adjustments are made depending on the distance between the MS and the BTS.

In GSM, the BTS expects to receive the first bit of a time slot from an MS three slots after the BTS transmits the first bit of a slot for the MS. If the propagation delay from the BTS to the MS is t, and the MS transmits the first bit of a slot exactly three slots after it receives the first bit of a slot, then the BTS will receive the first bit delayed $2t$ from its expected temporal position. If $2t$ is larger than the guard period GP, the received signal from this MS can overlap with the signals of adjacent slots. By commanding the MS to advance the transmission time by $2t$, collisions of slots can be avoided. This allows the BTS to recover the transmitted data if no error of other origins occurs. In GSM, the BTS can control the timing advance in 64 increments: level 0 corresponds to no advance, and each level advances the transmit time by one bit period. Note that in Fig. 2.27, 6 bits are assigned in the SACCH for this operation. The timing advance also enables the BTS to estimate the distance to each MS. This information can be used as an important parameter for handover decisions.

Power Control

The transmit power of the uplink and downlink is nominally controlled in 2 dBm increments, based on the combination of the Received Signal Level (RXLEV) and the Received Signal Quality (RXQUAL), so that only minimal transmit power is used to maintain the target quality level. The transmit power of each MS is continuously controlled so that the BTS receives signals from the MSs in the cell at similar power levels

Table 2.6 Power control commands.

Code	Required action
0	Not used
1	Increase output power by 4 levels
2	Increase output power by 3 levels
3	Increase output power by 2 levels
4	Increase output power by 1 level
5	No output power level change
6	Decrease output power by 1 level
7	Decrease output power by 2 levels

to prevent stronger neighboring signals from overwhelming weaker ones and rendering them unidentifiable. Although the guard time and the guard band separate the signals to and from each MS in the time and frequency domain, it is not possible to completely avoid the spillover of signals. In addition to the measured power level, the results of channel decoding can also be taken into account, as the BTS commands the MS to increase the transmit power if more speech frames than acceptable fail in the CRC check.

The maximum transmit power of the MS depends on the frequency band. For example, in the 900 MHz band, the maximum transmit power is 39 dBm but in the 1800 MHz band, the MS is required to transmit at up to only 36 dBm [3GPP (2017e)]. Table 2.6 summarizes the codes that control the transmit power of an MS at eight levels, using 3 bits in the SACCH. If the MS cannot execute the power control command it receives, the MS should use a supported transmit power level as close as possible to the requested power level [3GPP (2017f)]. The maximum transmit power of the BTS also depends on the frequency band, as well as the power class of the amplifier. Although not specified strictly, to allow freedom in the implementation as long as the requirements are met, the transmit power is controlled somewhere in the signal processing chain between the bipolar conversion and the RF modulation, as shown in Fig. 2.32.

Power control is an important technique for achieving spectral efficiency in mobile communications systems. Historically it appeared as early as in AMPS to enable $K = 7$ frequency re-use. Power control techniques can be of two types. In *open-loop power control*, the transmit power is set based on the received signal level assuming the same amount of path loss in the uplink and downlink. Open-loop power control is typically used to roughly initialize the level of transmit power at the beginning of a call or after handover. In *closed-loop power control*, the transmit power is controlled in a more systematic fashion based on the signal level measurements or requests for adjustment. When closed-loop power control cannot track the variations of link quality fast enough, which may occur from fast fading or after handover, open-loop power control can intervene to help the transmitter converge to the target power level as quickly as possible.

The cost for a mismatch in the power control is generally greater in the uplink since downlink transmission is synchronized, which generates less co-channel interference. Therefore, in GSM, the transmit power is more frequently controlled in the uplink.

The cost for undershooting in power control is greater than that for overshooting. The transmit power is increased as fast as possible when necessary but decreased more conservatively by monitoring the call quality for a period after decreasing the transmit power one step at a time. The transmit power may be increased in a more careful fashion in other mobile communications systems if the signal is separated using measures other than guard time or guard band.

Slow Frequency Hopping

A major problem with narrow-band, multi-carrier mobile communications systems, such as AMPS, D-AMPS, and GSM, is frequency-selective fading, which can impair a channel significantly if the assigned channel is located in the middle of a deep fade. When the BSC identifies unacceptable link quality between the MS and the BTS, e.g., if the frame error rate exceeds a threshold but the situation does not improve even when the power control is active, the BSC may consider changing the channel to another frequency assigned to the same BTS (intra-cell handover) before handing over the call to a neighboring BTS (inter-cell handover), or disconnecting the call if a better channel is not available and the MS continues to be charged for unrecognizable communications.

Slow Frequency Hopping (SFH) is a key feature of the GSM in which the network assigns a set of channels to an MS but changes the carrier frequency for each TDMA frame, either in a cyclic or pseudo-random fashion. Details on the algorithms that compute the frequency hopping patterns, which enable an MS to change the carrier frequency without colliding with the transmission of other MSs, can be found in [3GPP (2017b)]. SFH reduces the influence of frequency-selective fading when the frequency band assigned for hopping is wider than the frequency band susceptible to fading. SFH can also reduce the required SNR for operation by 2–9 dB [Redl *et al.* (1998)], which can be used to increase the network capacity by shrinking cells or applying a smaller frequency re-use factor. SFH is an optional feature for the network but an MS must support either type of SFH operation. The *slow* in SFH implies that the rate of changing carrier frequency is slow when compared with the symbol rate. However, in practice, the MS has to change the carrier frequency faster than once per TDMA frame since it needs to tune to another channel for monitoring the transmission status of other cells, and then hop to the destination frequency within the time constraints.

Figure 2.37, drawn to ignore the influences of timing advance, shows the combination of SFH and neighboring-cell monitoring. It is assumed that this MS is assigned to the third slot of a TDMA frame for both the uplink and downlink. First, the MS tunes to a downlink frequency, C_i, receives a slot, tunes to an uplink frequency, $\underline{C_i}$, and transmits a slot there. Note that there is a time difference of three slots between the transmission (TX) and reception (RX) of the MS, as illustrated in Fig. 2.36. Then the MS tunes to a downlink frequency, D_i, belonging to a neighboring cell, and measures the signal strength. The MS tunes to another downlink frequency, C_k, belonging to the serving cell, and receives a slot there. This process continues during the call, and the signal quality of neighboring cells is monitored even when a voice call is not in progress. The signal strength of the serving and neighboring cells is periodically reported to the network for handover decisions. It is also assumed that the MS is

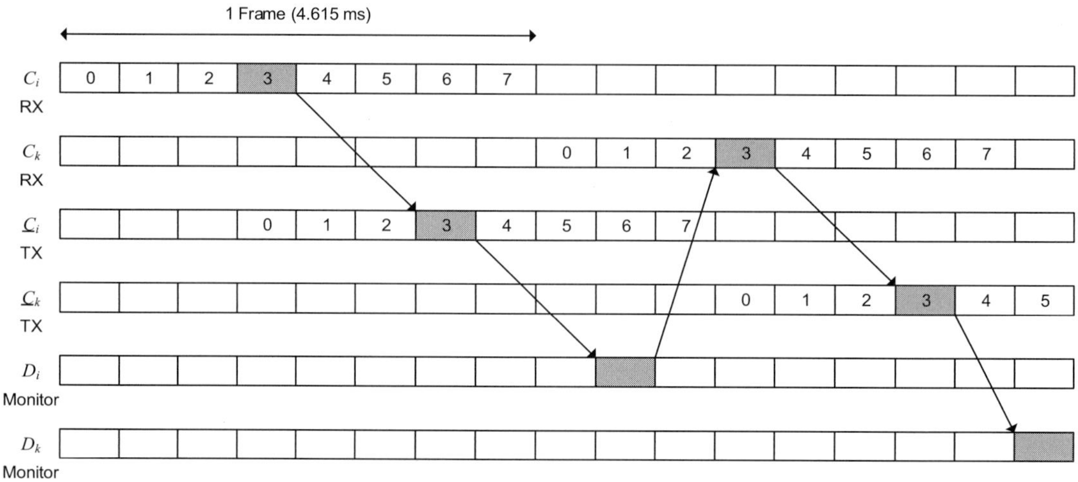

Fig. 2.37 Transmission, monitoring, and reception of time slots in SFH.

equipped with a single oscillator that generates sinusoidal signals at the channel frequency. The oscillator was considered to be a costly and power-consuming element in the RF circuitry when the GSM was designed. With enough time for frequency re-tuning between two time slots, using two oscillators brings little benefit. With a wide system bandwidth of 25 MHz reserved in many countries, SFH is an effective, low-complexity measure for unconsciously evading most types of frequency-selective fading. In some GSM systems that run on narrower bandwidth, e.g., the 450.6–457.6 MHz (uplink) band of GSM-450, the error-spreading effects of SFH are smaller but the lower frequencies provide better propagation characteristics and compensate for the reduction of system bandwidth.

Channel Information Reporting

Adaptive mechanisms to control the speech quality and network capacity of mobile communications systems require accurate information on the status of each traffic channel carrying speech data so that only a minimal number of bits or amount of transmit power is used. This information should be promptly and continuously supplied to the BTS or BSC to manage power control and handover. In GSM, two metrics measure the channel quality, RXLEV and RXQUAL, as defined in Tables 2.7 and 2.8. Notice that x dBm is an absolute unit for radio signal strength, related to the conventional unit (P) measured in Watts by $x = 10\log_{10} 1000P = 10\log_{10} P + 30$ dBm.

The RXLEV and RXQUAL values of the downlink, measured by the receiver of the MS, are periodically reported via SACCH to the BTS, which also measures the RXLEV and RXQUAL values of the uplink. The BTS and BSC, often collectively called the Base Station Subsystem (BSS), analyze the information to assess the channel status of each MS. When necessary the BSS initiates power control or handover to maintain speech quality or to balance network loading among the neighboring cells. If too many MSs are connected to a concentrated set of BTSs, the BSS can increase the relative

Table 2.7 Mapping of received signal level to RXLEV.

RXLEV	Lower bound (dBm)	Upper bound (dBm)
0		-110+SCALE
1	-110+SCALE	-109+SCALE
2	-109+SCALE	-108+SCALE
...	...	...
62	-49+SCALE	-48+SCALE
63	-48+SCALE	

Table 2.8 Mapping of received signal quality to RXQUAL.

RXQUAL	Lower bound (%)	Upper bound (%)	Assumed value (%)
0		0.2	0.14
1	0.2	0.4	0.28
2	0.4	0.8	0.57
3	0.8	1.6	1.13
4	1.6	3.2	2.26
5	3.2	6.4	4.53
6	6.4	12.8	9.05
7	12.8		18.10

signal strength of neighboring BTSs and induce the MSs to change their serving BTSs to less-loaded ones.

Although GSM allows only a pair of uplink and downlink channels separated by 45 MHz for voice services at any one time, each MS is required to continually measure and report the RXLEV values of the broadcasting channels from neighboring cells to the BSS, as illustrated in Fig. 2.37. Special downlink channels used to broadcast key information on the cell are measured since their transmit power is maintained at pre-defined levels. The measured channel quality is averaged and quantized into the 64 levels of RXLEV. The parameter SCALE is typically set to 0 but a non-zero value can be used in the Enhanced Measurement Report (EMR) and Packet Enhanced Measurement Report (PEMR) messages [3GPP (2017f)], when more detailed information is necessary for tighter control or packet data transmission. For RXQUAL, the bit error ratio is measured, averaged, and quantized to eight levels. Note that the period to observe these values also influences the accuracy of channel estimation: longer observation and averaging is likely to reduce noise but lose accuracy, and vice versa.

Delay Partitioning

For a toll quality, a total *mouth-to-ear* delay of less than 280 ms is typically required for the users to be satisfied. This can be distributed over the end-to-end transmission paths between the MS and the MSC [3GPP (2017a)], as shown in Fig. 2.38. The definitions and target values of the delay elements are set out in Tables 2.9, 2.10, and 2.11. The key delay element, Trftx, which is the time spent in the wireless channel between the MS and

Table 2.9 Delay elements.

	Definition
Tabisd	Time required to transmit a minimum number of encoded speech bits over downlink Abis interface to start encoding a speech frame
Tabisu	Time required to transmit a minimum number of encoded speech bits over uplink Abis interface to start decoding a speech frame
Ta/d	Delay in analogue to digital converter in uplink
Tbsc	Switching delay in BSC
Tbuff	Buffering time required for time alignment procedure for in-band control of remote transcoder
Td/a	Delay in digital to analogue converter in downlink
Techo	Delay induced by echo canceller
Tencode	Processing delay required to perform channel encoding
Tmsc	Switching delay in MSC
Tproc	Processing delay required to perform speech decoding
Trftx	Time required for transmission of a speech frame over air interface
Trxproc	Processing delay required to perform channel equalization (Trxproc eq.), channel decoding (Trxproc ch. dec.), and SID frame detection
Tsample	Duration of segment of PCM speech samples operated on by speech transcoder
Tsps	Worst case processing delay required by downlink speech encoder before an encoded speech bit can be sent over Abis interface
Ttransc	MS speech encoder processing delay, from input of last PCM sample to output of last encoded speech bit

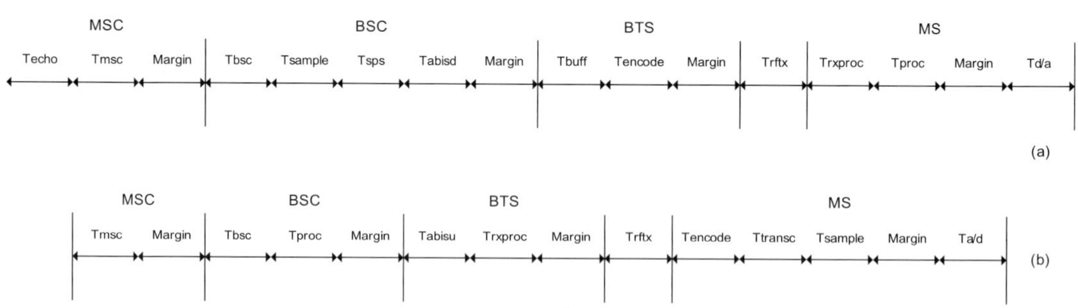

Fig. 2.38 Delay partitioning. (a) Uplink. (b) Downlink.

the BTS, is 37.5 ms in TCH/FS in the worst case. Trftx consumes a third of the delay budget assigned between the MS and the MSC. Notice that even in the case of calls between two MSs, the encoded speech spends only 70 ms over the air during its long journey from a mouth to an ear. The margin represents the implementation-dependent margin defined to compensate for the different system components.

Table 2.10 Uplink delay distribution in full-rate mode.

Node	Delay element	FR (ms)	EFR (ms)	AMR 12.2 (ms)
MSC	Tmsc	0.5	0.5	0.5
	Margin	0.5	0.5	0.5
BSC	Tbsc	0.5	0.5	0.5
	Tproc	1.5	1.27	1.816
	Margin	0.5	0.5	0.5
BTS	Tabisu	4	6.4375	6.625
	Trxproc eq.	8.8	6.84	6.84
	Trxproc ch. dec.	0	1.96	1.936
	Margin	3	3	3
MS	Trftx	37.5	37.5	37.5
	Tencode	1.6	0.32	0.272
	Ttransc	8	12.17	12.976
	Tsample	20	20	25
	Margin	2	2	2
	Ta/d	1	1	1
	Total	89.4	94.5	101.0

Table 2.11 Downlink delay distribution in full-rate mode.

Node	Delay element	FR (ms)	EFR (ms)	AMR 12.2 (ms)
MSC	Techo	1	1	1
	Tmsc	0.5	0.5	0.5
	Margin	0.5	0.5	0.5
BSC	Tbsc	0.5	0.5	0.5
	Tsample	20	20	25
	Tsps	1.6	2.3	2.28
	Tproc (Tsps)			1.816
	Tabisd	17.4	17.375	17.125
	Margin	0.5	0.5	0.5
BTS	Tbuff	1.25	1.25	1.25
	Tencode	1.6	1.6	0.272
	Margin	0.45	0.45	0.45
MS	Trftx	37.5	37.5	37.5
	Trxproc eq.	8.8	6.84	6.84
	Trxproc ch. dec.	0	1.96	1.936
	Tproc	1.5	1.27	1.816
	Margin	2	2	2
	Td/a	1	1	1
	Total	96.1	96.5	102.3

In general, improving speech quality by applying more complex compression algorithms adds to the processing delay. For example, the Enhanced Full-Rate (EFR) speech codec takes slightly longer to compress speech samples than the Full-Rate (FR), and the Adaptive Multi-Rate (AMR) speech codec requires an even larger encoding delay. However, the delay is very difficult to exploit in circuit-switched networks, since the entire signal processing operations are synchronized, leaving few opportunities to reduce or extend the delay of each path. In such networks, there are no measures to accelerate or decelerate the transmission of speech data. Interworking with less flexible networks such as the PSTN also limits the opportunity to control delay. In packet-switched networks, the delay can be exploited with packet scheduling, i.e., by controlling the timing and order of transmission, taking the relative importance or urgency of the packets into account. Thus, in packet-switched networks the end-to-end delay cannot be partitioned as precisely as implied in Tables 2.10 and 2.11. As SFH enables the transmission to evade fading in the frequency domain, scheduling can enable the transmission to evade unfavorable situations in the time domain, by waiting temporarily for the link quality to improve or until the priority for transmission is elevated, before looking for a vacancy in other, healthier channels. However, such a precise management of transmission opportunities would require channel information more accurate and frequent than can be provided by the simple methods illustrated in Fig. 2.37.

2.4 References

3GPP. 2000a. TS 06.01 V8.0.1 Full Rate Speech Processing Functions. November.

3GPP. 2000b. TS 06.31 V8.0.1 Discontinuous Transmission (DTX) for Full Rate Speech Traffic Channels. November.

3GPP. 2000c. TS 06.32 V8.0.1 Voice Activity Detection (VAD). November.

3GPP. 2001. TS 06.12 V8.1.0 Comfort Noise Aspects for Full Rate Speech Traffic Channels. June.

3GPP. 2017a. TR 26.975 V14.0.0 Performance Characterization of the Adaptive Multi-Rate (AMR) Speech Codec. March.

3GPP. 2017b. TS 45.002 V14.1.0 GSM/EDGE Multiplexing and Multiple Access on the Radio Path. March.

3GPP. 2017c. TS 45.003 V14.1.0 GSM/EDGE Channel Coding. March.

3GPP. 2017d. TS 45.004 V14.0.0 GSM/EDGE Modulation. March.

3GPP. 2017e. TS 45.005 V14.0.0 GSM/EDGE Radio Transmission and Reception. March.

3GPP. 2017f. TS 45.008 V14.0.0 GSM/EDGE Radio Subsystem Link Control. March.

Cox, D. C. 1999. Wireless Personal Communications: A Perspective. In: Gibson, J. D. (ed), *The Mobile Communications Handbook*, 2nd edn. CRC Press.

Delprat, M., and Kumar, V. 1999. Enhancements in Second Generation Systems. In: Gibson, J. D. (ed), *The Mobile Communications Handbook*, 2nd edn. CRC Press.

EIA/TIA. 1990. Interim Standard IS-54: Cellular System Dual-Mode Mobile Station – Base Station Compatibility Standard. May.

Hanzo, L., Somerville, F. C. A., and Woodard, J. P. 2007. *Voice and Audio Compression for Wireless Communications*. 2nd edn. Wiley-IEEE Press.

Harte, L. J., Smith, A. D., and Jacobs, C. A. 1998. *IS-136 TDMA Technology, Economics, and Services*. Artech House Publishers.

Hayes, M. H. 1996. *Statistical Digital Signal Processing and Modeling*. Wiley.

Itakura, F. 1975. Line Spectrum Representation of Linear Predictor Coefficients of Speech Signals. *The Journal of the Acoustical Society of America*, **57**(April).

Jayant, N. S., and Noll, P. 1984. *Digital Coding of Waveforms: Principles and Applications to Speech and Video*. Prentice Hall.

Quatieri, T. F. 2001. *Discrete-Time Speech Signal Processing: Principles and Practice*. Prentice Hall.

Rabiner, L. R., and Schafer, R. W. 2010. *Theory and Applications of Digital Speech Processing*. Prentice Hall.

Redl, S., Weber, M., and Oliphant, M. W. 1998. *GSM and Personal Communications Handbook*. Artech House Publishers.

3 Evolution of TDMA Systems

The success of GSM increased the number of its users to such an extent that the basic speech and radio signal processing operations employed when the system was initially launched could not meet the high demand. In this chapter, we discuss various approaches that were developed to improve the speech quality or network capacity, or both, of GSM. These approaches range from enhanced speech compression, to more robust wireless transmission, and dynamic coordination of compression and transmission. The effectiveness of each approach is evaluated with experimental results in a variety of situations.

3.1 Enhancements in Speech Compression

3.1.1 Enhanced Full-Rate Speech Codec

Although higher speech quality in mobile communications can be achieved by using either higher bit-rate or larger transmit power, a more fundamental opportunity lies in the use of more computationally intensive signal processing operations that become feasible because of VLSI technology advances. More sophisticated speech compression algorithms allowed higher quality to be achieved at similar or even lower transmission bit-rates. The Enhanced Full-Rate (EFR) speech codec uses the Algebraic Code Excited Linear Prediction (ACELP) in which a limited set of distributed pulses is used as the excitation to a linear prediction filter [3GPP (2000c)]. ACELP has the big advantage of implementing a large algebraic codebook for improved compression efficiency with fewer complexity issues such as the storage or the search speed, which helped the principles be adopted by other codecs.

Figure 3.1 shows a simplified block diagram of the EFR speech encoder. When voice activity is present, the encoder generates 244 bits every 20 ms. Table 3.1 shows how those 244 bits are distributed over the bit-stream. An 8-bit CRC is computed from the 65 most important bits of Class 1 and appended to them. The four most important bits of Class 2 are repeated three times before the three classes of bit-streams are re-ordered to increase the resilience against errors. At this point there are 260 bits, the same number as produced by the FR speech encoder. Therefore the remaining portions of the signal processing chain for the FR speech encoder can now be re-used. EFR achieves significantly higher quality than FR by using the ACELP algorithms and more advanced quantization

Table 3.1 Bit allocation of EFR speech codec.

	1st and 3rd subframes	2nd and 4th subframes	Bits per frame
2 LSP sets			38
Pitch delay	9	6	30
Pitch gain	4	4	16
Sign information for pulse	5	5	20
Pulse position	30	30	120
Fixed codebook gain	5	5	20
Total			244

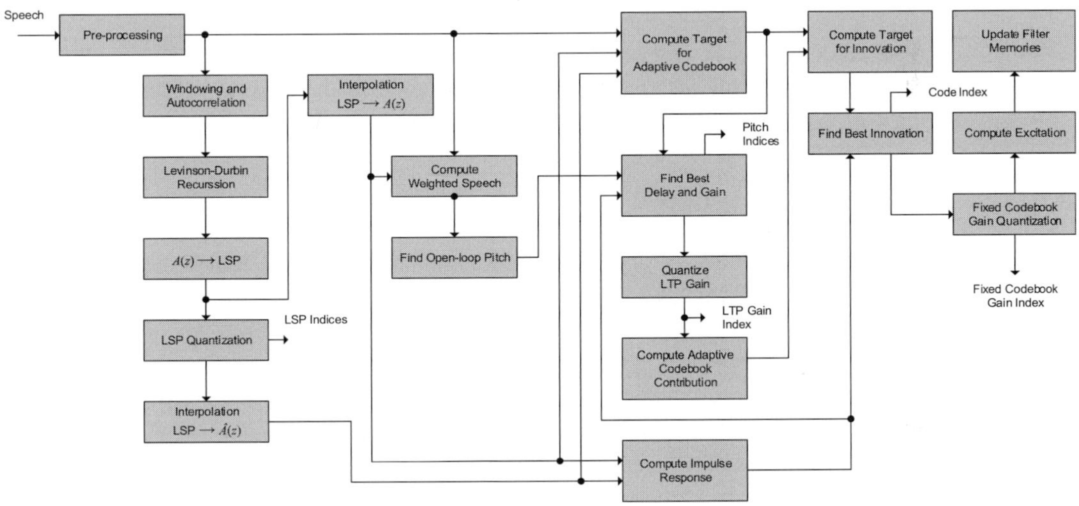

Fig. 3.1 Block diagram of EFR speech encoder.

techniques. EFR uses 16 fewer bits for the speech frame but spends these saved bits to reinforce the structure of its bit-stream.

EFR also supports discontinuous transmission. In the SID of EFR, the comfort noise parameters consist of averaged filter parameters and an averaged fixed codebook gain repeated four times; these are computed over eight consecutive frames not marked by the voice activity detector as speech. This is longer than the four-frame SID averaging period of FR. At the receiver, parameters not included in the SID are either randomly generated or set to some pre-defined values. Ninety-five ones are used to indicate the presence of noise information [3GPP (2000d)]. The EFR speech codec is also re-used by one of the most successful speech codecs, Adaptive Multi-Rate (AMR), which we shall encounter later as an operating mode. Figure 3.2 shows the number of bits at each step in the sequence of speech and radio signal processing operations of EFR. The channel used is the Full-Rate Speech Traffic Channel for EFR (TCH/EFS). With radio signal processing virtually intact, any performance improvement has to be made mainly from

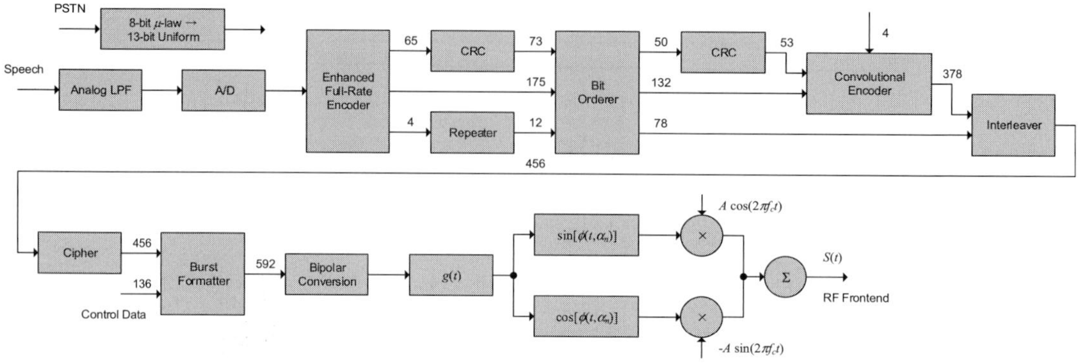

Fig. 3.2 Speech and radio signal processing operations of GSM for EFR (TCH/EFS).

enhanced compression and bit-stream packetization algorithms. Nevertheless EFR provides significantly higher quality than FR and performs better at the cell edge, thereby achieving both quality and capacity improvement.

3.1.2 Half-Rate Speech Codec

The three-fold increase in network capacity promised by the transition from AMPS to N-AMPS or D-AMPS was followed by an abrupt increase in the number of users of mobile phones, which meant that a shortage of network capacity quickly developed after each transition. One solution that was suggested was to use a speech codec, such as the Half-Rate (HR) speech codec, which allows more users to use the network by allocating a lower bit-rate to each user. Unfortunately, this lower bit-rate reduces the speech quality. An analysis of the traffic patterns of the mobile communications networks, however, revealed that the highest capacity was required only during limited periods in each day or week. The call arrivals could not be modeled by the simple Poisson process often assumed in the literature, however. Increased service demand is far more likely to exhibit periodicity during each work day with several sets of *busy hours*. Figure 3.3 gives an example of the distribution of voice service requests in a metropolitan area during weekdays where the cell loading threshold necessitating the deployment of a lower bit-rate speech codec is exceeded three times a day. Such distributions may exhibit different shapes during weekends, holidays, or at locations hosting special events.

One strategy for dealing with the increased demand while trying to maintain the speech quality without adding frequency spectrum or network infrastructure is to configure the network at a level that is sufficient to handle the call demand for all but the peak times and to incorporate adaptive mechanisms to boost network capacity temporarily while reducing speech quality to a controlled extent when the highest capacity is demanded. The HR speech codec, which operates at 5.6 kbps [3GPP (2000a)], accomplishes this by theoretically doubling the network capacity, albeit at a lower speech quality. Figure 3.4 shows a simplified block diagram of the HR speech encoder, which also uses the ACELP. $W(z)$ and $C(z)$ are the spectral weighting filter and a second weighting filter, respectively.

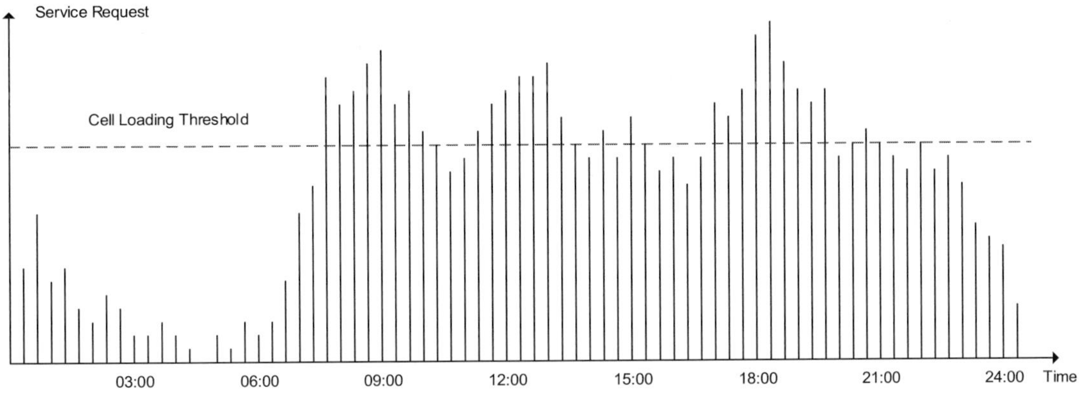

Fig. 3.3 Daily distribution of voice service requests.

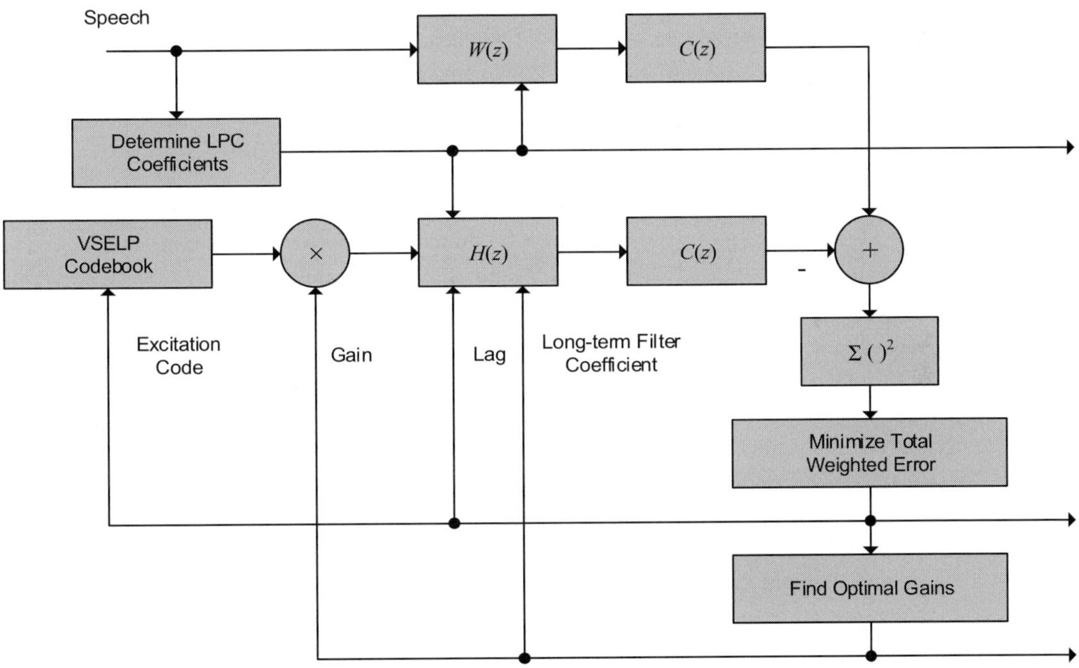

Fig. 3.4 Block diagram of HR speech encoder.

Table 3.2 shows the distribution of the 112 bits that constitute each encoded frame of the HR speech codec. To maximize the speech quality at the lower bit-rate, the HR speech codec operates in four modes depending on the level of voicing in the input speech, as outlined in Table 3.3. The mode selected is based on the LTP gain computed during speech encoding. The HR speech decoder applies identical reconstruction algorithms for Mode 1, 2, or 3. The mode for an unvoiced speech frame, Mode 0, necessitates a different bit allocation. Therefore two bits are required in the bit-stream of each frame to represent the voicing mode used. As the mode does not influence the channel bit-rate, the network neither checks nor exploits the mode being used.

Table 3.2 Bit allocation of HR speech codec.

	Mode 0	Mode 1, 2, or 3
Voicing mode	2	2
Frame energy	5	5
Reflection coefficient vector	28	28
Soft interpolation bit (INIT_LPC)	1	1
Codeword	56	36
Subframe lag		20
Subframe gain	20	20
Total	112	112

Table 3.3 Operating modes of HR speech codec.

Mode	Level of voicing
0	Unvoiced
1	Slightly voiced
2	Moderately voiced
3	Strongly voiced

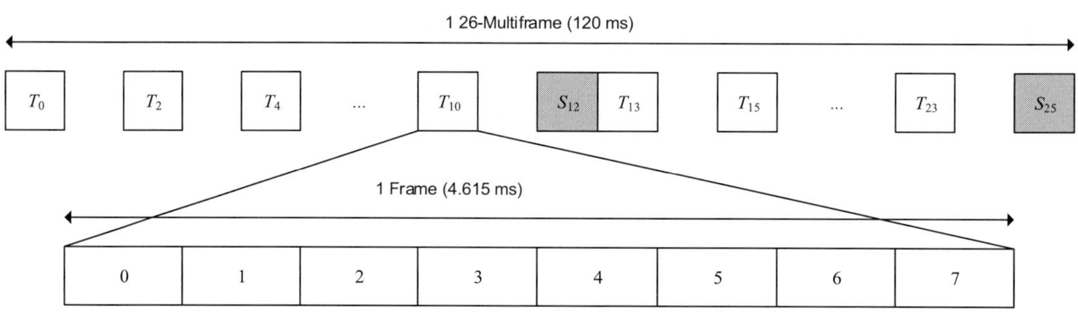

Fig. 3.5 Hierarchical frame structure of TCH/HS.

The encoded speech frames of HR use a time slot in every other TDMA frame, as shown in Fig. 3.5, in which the 13th and 26th frames of each 26-multiframe are used for SACCH in the Half-Rate Speech Traffic Channel (TCH/HS) to manage the link quality of two MSs. The HR speech encoder generates 95 Class 1 bits and 17 Class 2 bits per 20 ms. Applying the UED and UEP procedures to these 112 bits generates 228 bits.

A 3-bit CRC is computed and attached to the 22 most significant Class 1 bits for error detection. Then all the Class 1 bits together with the parity bits are encoded using a $K = 7$, rate $\frac{1}{3}$ convolutional code. At the convolutional encoder, three output bits are computed from each input bit using the polynomials, $1+D^2+D^3+D^5+D^6$, $1 + D + D^4 + D^6$, and $1 + D + D^2 + D^3 + D^4 + D^6$. The HR speech encoder employs a lower-rate channel code, which provides stronger protection by using more redundancy than the rate $\frac{1}{2}$ channel code of the FR speech encoder. In spite of the lower bit-rate budget, strong channel coding is necessary since each bit is now more perceptually important.

The convolutional encoding of $22 + 3 + 73 + 6 = 104$ bits generates 312 bits, 101 of which are punctured, leaving 211 bits. The input bits correspond to 95 Class 1 bits and six tail bits. The output bits from polynomial $1 + D + D^4 + D^6$ are punctured while for the three CRC bits, all of the output bits are retained. This puncturing process generates $3 \times \frac{2}{3}(95 + 6) + 3 \times \frac{3}{3}3 = 211$ bits. Puncturing is an efficient technique to match the number of bits to be transported to the available bit-rate, without hurting the error resilience excessively. Theoretical background on the puncturing of convolutional codes can be found in [Hagenauer (1988)].

The 17 Class 2 bits and the 211 bits from puncturing, exactly one-half of the bits TCH/FS carries, are interleaved and ciphered. As in the EFR speech codec, in the SID, indices of the frame energy value and the reflection coefficient vector are replaced by comfort noise parameters, which are calculated over eight consecutive frames not identified by the VAD as speech. At the receiver, other parameters not transmitted in the SID are either randomly generated or set to predefined values. To indicate the presence of noise information, 79 ones are sent [3GPP (2000b)].

In the frame construction process, each encoded frame of the HR speech encoder is interleaved over four TDMA frames, compared with eight in the full-rate mode. However, since HR uses a slot in every other TDMA frame, the total period over which a speech frame is interleaved remains the same, providing the same level of time-domain diversity. Figure 3.6 shows the number of bits at each step in the speech and radio signal processing of HR.

Due to its low bit-rate, the quality loss of using HR instead of FR or EFR is quite noticeable, even without errors, which increases the responsibility of error concealment when the channel condition becomes rough. The speech decoder can utilize various

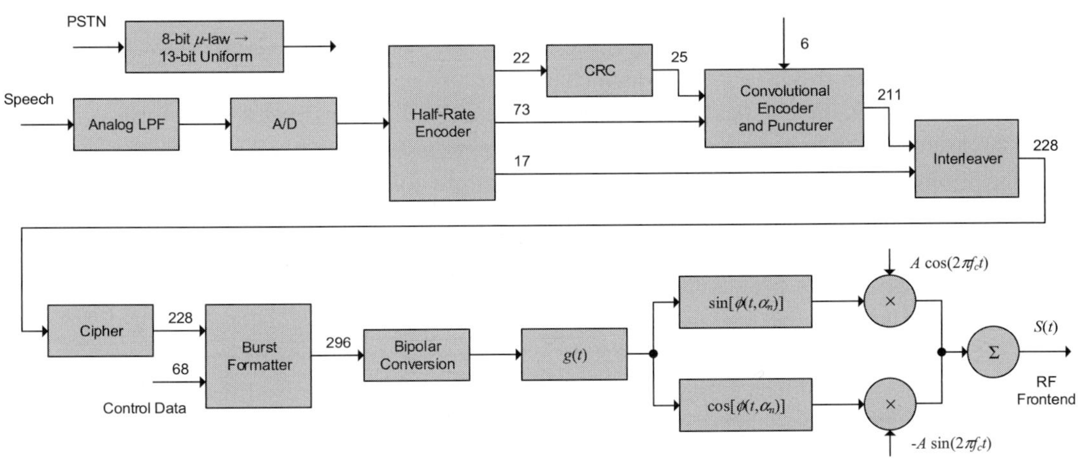

Fig. 3.6 Speech and radio signal processing operations of GSM for HR (TCH/HS).

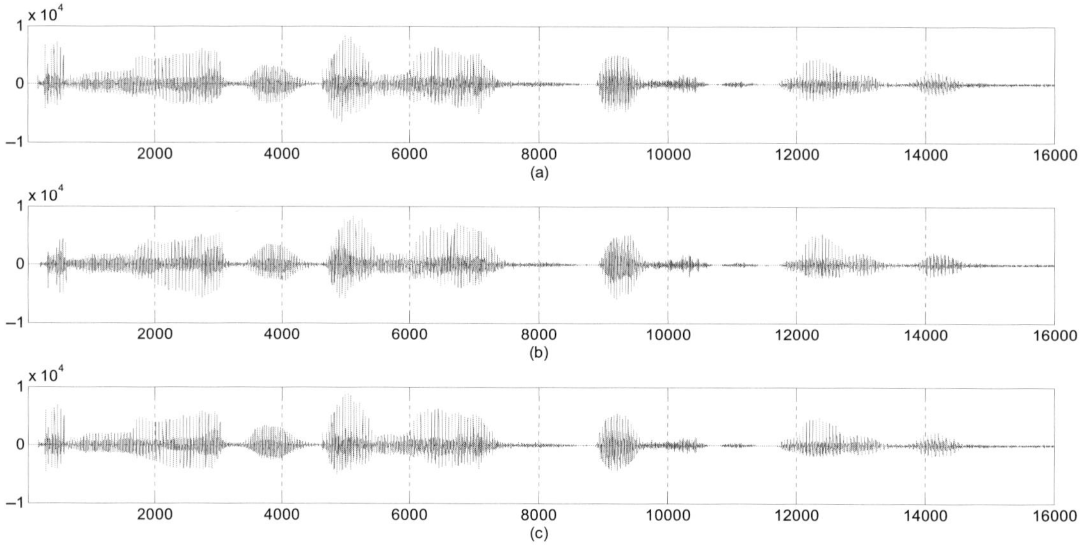

Fig. 3.7 Speech waveforms. (a) Uncompressed. (b) Compressed and reconstructed with HR. (c) Compressed and reconstructed with EFR.

strategies for covering the damage on speech quality, with the recent history of decoded data integrity and the information from soft decision metrics of the Viterbi decoder. The background noise or any deterioration of link quality can easily degrade the quality of voice call. However, when there is a large gap in the required network capacity during busy hours, deploying HR can be an economical approach to temporarily trade speech quality for network capacity.

Figure 3.7 compares the speech waveforms of the original with those of the encoded and reconstructed using HR and EFR, for two seconds or 100 frames. DTX is not active in both codecs, to align the waveforms. The difference in subjective quality between HR and EFR is in general significant, i.e., as high as 0.86 on the Mean Opinion Score (MOS) scale in clean channel according to the testing results of Table 3.23. However, in the figure, it can be seen that both reconstructed waveforms are very close to the original. EFR tends to suppress the impact of spikes in the input speech waveform, therefore containing the waveform to a smaller dynamic range than HR does.

HR can be deployed with several strategies. During busy hours, incoming service requests from MSs with good signal strength can be addressed with HR. Although it is possible to use call control procedures to convert ongoing calls from FR to HR, and vice versa, these practices are rarely used in network operation as the change of channel type can introduce a very noticeable degradation in the perceived speech quality. Therefore it is necessary to tune the deployment strategies for HR based on the observed or expected traffic patterns. HR and EFR are classified as the key features of GSM Phase 2, an evolutionary stage from the original GSM based on FR, which still uses GMSK.

3.2 Enhancements in Coordination of Compression and Transmission

3.2.1 Joint Source-Channel Coding Theory

The EFR speech codec uses a different allocation of the bit-rate between speech data and channel code while maintaining the same total bit-rate as the FR speech codec. Inasmuch as a fixed partitioning is unlikely to perform optimally under a wide range of channel conditions, it is reasonable to assume that the speech quality might be improved further if the relative numbers of bits for the encoded speech, CRC, and the channel codes are dynamically controlled depending on the time-varying channel conditions.

The Adaptive Multi-Rate (AMR) speech codec operates at eight different bit-rates, ranging from 4.75 to 12.2 kbps, based on the network control [3GPP (2017d)]. Upon receiving a service request from a local MS or from the PSTN, the network first decides the channel type, e.g., the full-rate or half-rate, and then assigns up to four bit-rates for the service request. A rough control of the tradeoff between speech quality and network capacity is made in this step.

AMR is based on classical joint source-channel coding theory whereby a fixed total bit-rate is optimally allocated between the source code and the channel code to minimize the combined distortion from data compression and transmission over error-prone channels [Shannon (1948), Massey (1978), Vembu *et al.* (1995)]. With a finite total bit-rate, the selection of R_s, the bit-rate for the encoded speech, automatically determines the value of R_c, the remaining bit-rate for the channel code. The channel code in this context includes the CRC, redundancy from convolutional coding, and any padding bits.

The objective of AMR operation is to determine an (R_s, R_c) pair from several candidates so that the end-to-end distortion between the uncompressed speech and the compressed, transmitted, and reconstructed speech, for a short-term period during which the statistical nature of speech waveform and channel condition does not change significantly, is minimized. Let D and $\frac{E_s}{N_0}$ denote the end-to-end distortion and the symbol-level SNR that specifies the channel condition, respectively. Then the problem of minimizing the distortion can be formulated as

$$\min_{R_s} D\left(R_s, \frac{E_s}{N_o}\right). \tag{3.1}$$

However, unlike in the classical joint source-channel coding theory in which the optimal bit-rate allocation between the source code and the channel code is searched iteratively over an infinite observation period to match stationary channel condition, AMR needs to track rapid variations of channel condition and select an allocation from several pre-defined candidates quickly. Thus, the problem can be re-formulated as

$$\min_{R_s} D\left(R_s, \frac{E_s}{N_o}(t)\right), \tag{3.2}$$

where the SNR changes with time. Since it requires a non-zero delay to compress and transmit the speech, any inaccuracy in the channel estimation or misguided adaptation will prevent the dynamic bit-rate allocation system from maintaining an optimal decision.

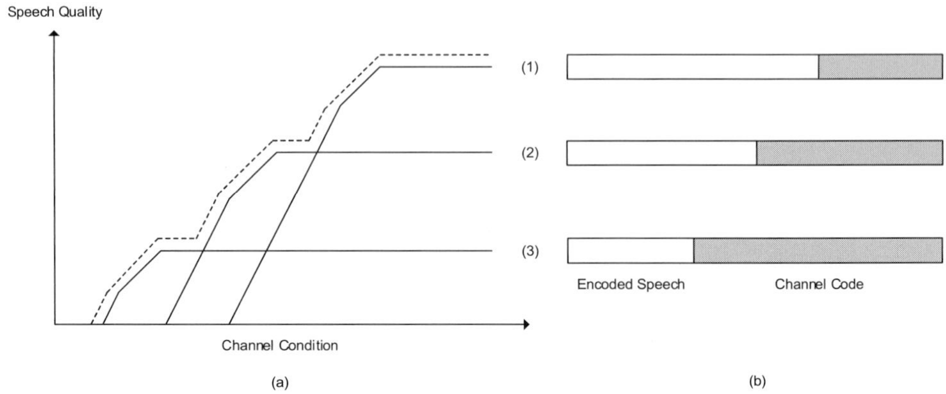

Fig. 3.8 Bit-rate allocations between encoded speech and channel code.

As an example, suppose that there are three bit-rate allocations between the encoded speech and the channel code. Figure 3.8(a) shows the three speech quality vs. channel condition curves for these allocations. As the channel condition deteriorates, i.e., as the SNR moves to the left, the quality of the compressed, transmitted, and reconstructed speech rapidly decreases and the overall distortion is dominated by the distortion from transmission errors. Once the SNR exceeds a certain level, the distortion from transmission errors becomes negligible. In this case, the overall distortion is dominated by the distortion caused by speech compression.

Assume that curve (2) corresponds to the current bit-rate allocation. If the speech encoder requires a higher bit-rate than that provided by (2), the bit-rate for the channel code will be correspondingly reduced to maintain the same total bit-rate as in allocation (1). The shape of the speech quality vs. channel condition curve for allocation (1) is similar to that of (2) but the curve for (1) is shifted up and to the right. In the clean or error-free channel conditions, bit-rate allocation (1) will provide higher speech quality than allocation (2) but allocation (1) is likely to suffer from a more abrupt loss of quality caused by transmission errors when the channel condition deteriorates. Therefore allocation (1) is better suited for good channel conditions than allocation (2).

On the other hand, if the speech is encoded at a lower bit-rate than that of allocation (2), the bit-rate for the channel code is automatically increased, as shown in allocation (3). The speech quality vs. channel condition curve is now shifted below and to the left from that of allocation (2). In the error-free channel conditions, allocation (3) will provide poorer speech quality than allocation (2) but the new allocation will be more resilient to transmission errors when the channel condition deteriorates. Consequently, allocation (3) will outperform the other two bit-rate allocations in poor channel conditions.

If a combined speech compression and transmission system supports a set of bit-rate allocations, and dynamically switches to the most appropriate allocation based on the channel conditions, its speech quality vs. channel condition curve will become the envelope of the curves of the three allocations, shown in the figure as dotted lines. Such an

AMR system should perform at least as well as or better than each individual allocation over a wide range of channel conditions [Corbun *et al.* (1998), Uvliden *et al.* (1998), Bruhn *et al.* (1999)].

The design of a joint source-channel coding system requires a set of bit-rate allocations that differ in the number of bits devoted to the encoded speech, CRC, and channel code while meeting the same total bit-rate budget. Each allocation should be chosen to serve a limited range of channel conditions in which it outperforms the other allocations. At the boundary between two ranges of channel conditions assigned for two neighboring allocations, either allocation may provide similar levels of performance. It is also necessary to select one or more metrics on which to base the switching. The metrics should be tracked continuously to trigger the switching between the bit-rate allocations.

Another important design issue is where the switching algorithms should be located. They can be at the sender or receiver of the encoded speech. The former policy, *sender-driven adaptation*, requires that the information on channel condition be available at the sender to base the decisions on. The benefit of this policy is that the switching algorithm analyzes the information and directly commands the speech encoder and the channel encoder based on the analysis. However, it has the disadvantage of requiring relatively higher bit-rate feedback channels for the information. Therefore this policy is appropriate for situations when the capacity of the feedback channel is enough or the bit-rate allocation does not need to be switched within strict timing constraints.

With the latter policy, *receiver-driven adaptation*, the receiver analyzes the status of media reception and asks the sender to switch to a preferred bit-rate allocation using a low bit-rate feedback channel that can be strongly protected. A major disadvantage of this policy is the latency between the analysis and the switching as the decision on allocation has to be made for channel condition in the future. In both policies, it is necessary to modify the transmitter, the receiver, and the frame/slot structure when the feedback information is introduced.

In GSM, room for the feedback channel needs to be created by reducing the bit-rate for either the encoded speech or the channel code. With a fixed total bit-rate and the need for continuously running the switching algorithms, a receiver-driven adaptation policy was adopted for AMR. The important information fed back to the sender does not require a high bit-rate but nevertheless needs to be transmitted continuously and safely. Therefore, a type of *in-band signaling*, in which the feedback information is attached to the encoded speech, is used. Other measures to carry the control information, such as SACCH or frame-stealing, cannot be used since neither signaling scheme can be used frequently or immediately enough.

The joint source-channel coding might be misunderstood as a technique for only improving speech quality that contributes little to network capacity, since the same total bit-rate is maintained over the wireless channel. However, this approach effectively replaces the quality loss caused by transmission with that caused by compression, which can be perceptually more acceptable. When the probability of losing TDMA frames during transmission is reduced by feedback-based bit-rate allocation, the transmit power can be reduced, resulting in lower interference and correspondingly higher network capacity. The transmit power can be increased if the combined error handling capability of

CRC and channel code is insufficient to recover the encoded speech and the probability that transmitted TDMA frames are erroneously decoded exceeds acceptable levels.

Under this design philosophy, the MS and the BTS need to monitor the quality of the downlink and uplink, respectively, and request changes in the bit-rate allocation if necessary. Since the BTS has a higher level of authority than the MS, a request from the MS is called a Codec Mode Request (CMR); it may be ignored by the BTS depending on the situation, while a request from the BTS is called a Codec Mode Command (CMC); it must be obeyed by the MS within a strict time constraint. During a call, the uplink and downlink may use different bit-rate allocations as their channel conditions are independent, but the same channel type and the same set of bit-rate allocations should be used by both links. With a fixed total bit-rate, designating the bit-rate of encoded speech is enough to specify the bit-rate of channel code or the allocation.

3.2.2 Adaptive Multi-Rate Speech Codec

The Adaptive Multi-Rate (AMR) speech codec was designed to meet these needs for joint source-channel coding with tight delay constraints. Eight bit-rates, ranging from 4.75 to 12.2 kbps, are defined in the AMR speech codec, any one of which can be used in the Full-Rate Speech Traffic Channel for AMR (TCH/AFS). The six low bit-rates can also be used in the Half-Rate Speech Traffic Channel (TCH/AHS). TCH/AFS and TCH/AHS are identical to TCH/FS and TCH/HS, respectively, in the TDMA frame level but differ in the lower levels. Therefore, even at the same speech bit-rate, the quality can differ significantly depending on the channel type and the channel condition. In AMR terminology, the bit-rates from 4.75–12.2 kbps are called codec modes 0–7.

The 6.70 and 7.95 kbps of AMR correspond to the Enhanced Full-Rate speech codec of the Japanese Personal Digital Cellular (PDC) System and the Full-Rate speech codec of D-AMPS, respectively. These are included in the AMR to ease the interworking of GSM with these contemporary TDMA networks by avoiding transcoding. In addition, the 12.2 kbps of AMR is basically the EFR speech codec of GSM but a new SID format is used in all of the codec modes. AMR shares the same architecture with EFR, but at bit-rates lower than 12.2 kbps it differs in areas such as the coefficient quantization used to generate the lower bit-rates.

Table 3.4 shows the bit allocation of AMR at each bit-rate or codec mode. Note that no bits are assigned to represent the codec mode, i.e., to signal which encoding method is used for the speech frame, although such information is necessary for decoding the frame. Note also that in Table 3.2, two bits are used in the bit-stream of the HR speech codec to signal the voicing mode, even if all modes use the same bit-rate. Therefore, at least three bits to signal the codec mode used are required to accompany each speech frame; these are transmitted using the in-band signaling. However, the bits for codec mode information also need to be protected with channel coding, which consumes additional bit-rate.

To save as many bits as possible for the encoded speech and the channel code, it is important to minimize the number of bits used to designate the codec mode for the current speech frame, Codec Mode Indication (CMI), and to designate the codec mode

Table 3.4 Bit allocation of AMR speech codec.

Codec mode	LSF submatrix index	Adaptive codebook index	Adaptive codebook gain	Sign information for pulse	Pulse position	Fixed codebook gain	Total
12.2	38	30	16	20	120	20	244
10.2	26	26		16	108	28	204
7.95	27	28	16	16	52	20	159
7.40	26	26		16	52	28	148
6.70	26	24		12	44	28	134
5.90	26	24		8	36	24	118
5.15	23	20		8	28	24	103
4.75	23	20		8	28	16	95

Table 3.5 Identification of codec modes for ACS.

Identifier	Legend
CODEC_MODE_1	Lowest codec mode (lowest bit-rate) of ACS
CODEC_MODE_2	Second lowest codec mode if ACS includes more than one mode
CODEC_MODE_3	Third lowest codec mode if ACS includes more than two modes
CODEC_MODE_4	Highest codec mode if ACS includes four modes

the far-end is requested/commanded to use for the next speech frame (CMR/CMC). The savings are made in two steps. First, the number of codec modes to be used in each call is limited to four or fewer. These should span as wide a bit-rate range as possible to adapt to a maximum range of channel conditions. For example, when four codec modes are used, the recommended set consists of 4.75, 5.90, 7.40, and 12.2 kbps. Since the same set of codec modes is always used for both directions, only two bits are required to signal any codec mode in the Active Codec-mode Set (ACS), the set of codec modes allowed for each call. Table 3.5 outlines the identifiers for four codec modes. If the result of channel decoding gives a codec mode not present in the ACS, the received CMI or CMR/CMC is assumed to be compromised during transmission. Second, the CMI and CMR/CMC are transmitted alternately. This does not significantly influence the performance of codec mode adaptation as the typical frequency of codec mode switching is less than a few times per minute. From the hierarchical frame structure, the receiver can always identify whether the in-band signaling is CMI or CMR/CMC.

As in EFR, in the AMR speech codec, the comfort noise parameters to be encoded in an SID are calculated over eight consecutive frames not identified as speech. The SID of AMR consists of 35 comfort noise bits which include a six-bit index representing the logarithmic frame energy, and four indicator bits representing the type of SID and the codec mode currently used by the AMR speech encoder. The frame energy is adjusted according to the codec mode for acoustically smoother transitions between the

Table 3.6 Bit allocation of AMR for TCH/AFS.

Codec mode	4.75	5.15	5.90	6.70	7.40	7.95	10.2	12.2
Class 1a bits	39	49	55	55	61	75	65	81
CRC	6	6	6	6	6	6	6	6
Class 1b bits	56	54	63	79	87	84	139	163
Convolutional encoder input	101	109	124	140	154	165	210	250
Convolutional coding rate	$\frac{1}{5}$	$\frac{1}{5}$	$\frac{1}{4}$	$\frac{1}{4}$	$\frac{1}{3}$	$\frac{1}{3}$	$\frac{1}{3}$	$\frac{1}{2}$
Padding bits	4	4	3	3	2	2	2	1
Convolutional encoder output	535	565	520	576	474	513	642	508
Punctured bits	87	117	72	128	26	65	194	60
Block-coded CMI/CMC/CMR	8	8	8	8	8	8	8	8
Total	456	456	456	456	456	456	456	456

comfort noise and the speech. The SID of AMR can be of two classes, SID_UPDATE, an ordinary SID containing the comfort noise parameters, and SID_FIRST, a marker indicating the end of a talkspurt in which all 35 bits are set to zero.

In the FR, HR, and EFR speech codecs, the speech is processed in one direction so that the signal processing procedures are influenced by mainly voice activity. When necessary, the MS or the BTS drops encoded speech frames and fills the time slots with control data, but does not change the operation of the speech encoder. AMR necessitates that the speech signal processing process, as well as the radio signal processing process, be directly controlled by the far-end, whether it is an MS or a BTS, while maintaining the same channel type.

Table 3.6 summarizes the bit allocation of AMR for TCH/AFS. One important difference from the previous speech codecs is the absence of Class 2 bits, i.e., the bits that are not protected by either the CRC or channel code. For the most important bits of each codec mode, the Class 1a bits, a six-bit CRC is computed for error detection. The bits of the lower codec modes are generally channel coded with lower-rate codes, which provide stronger protection. As in the HR speech codec, the channel-coded bits are punctured in all codec modes to meet the budget of 448 bits, reduced by in-band signaling from the 456 bits of TCH/FS or TCH/EFS. The 2-bit CMI or CMC/CMR is protected with a rate $\frac{1}{4}$ block code, as shown in Table 3.7, and added to the punctured bits from the convolutional encoder, filling the full-rate capacity of 456 bits for 20 ms of speech.

While some of the less privileged bits of the FR and EFR speech codecs are left unprotected, AMR in TCH/AFS provides at least a minimum level of protection to all bits while still meeting the bit-rate budget by puncturing the less important bits generated in the convolutional coding. The 39-bit SID, equal to the size of the smallest set of Class 1a bits in 4.75 kbps, enables error detection in any codec mode, improving the noise quality during discontinuous transmission. Figure 3.9 shows the signal processing operations of AMR in TCH/AFS, where an adaptation controller is introduced to the signal processing chain, forming another closed-loop control between the MS and the BTS, in addition to the power control loop that uses the SACCH.

Table 3.7 Block coding of in-band signal.

	TCH/AFS		TCH/AHS	
	Message	Coded message	Message	Coded message
CODEC_MODE_1	00	00000000	00	0000
CODEC_MODE_2	01	10111010	01	1001
CODEC_MODE_3	10	01011101	10	0111
CODEC_MODE_4	11	11100111	11	1110

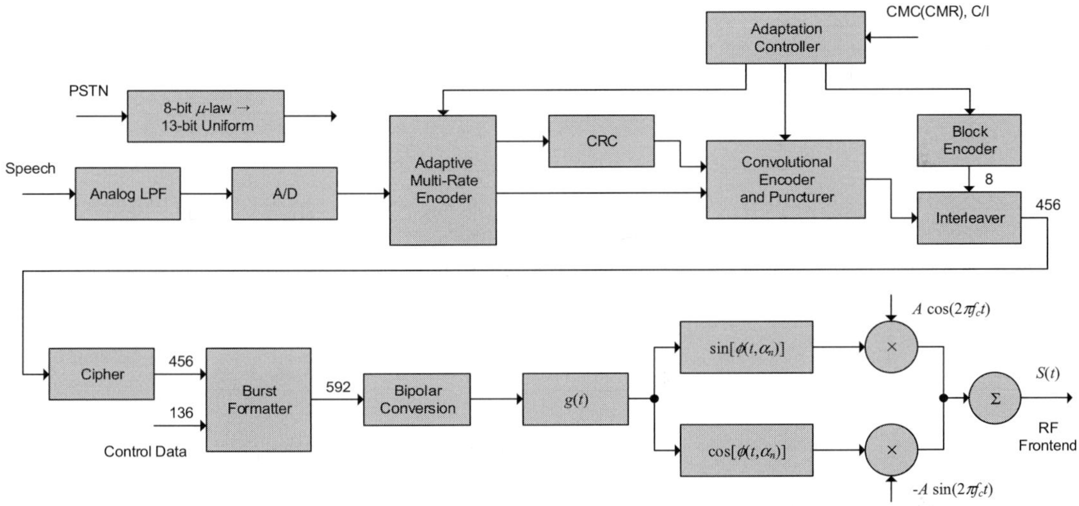

Fig. 3.9 Speech and radio signal processing operations of GSM for AMR (TCH/AFS).

Since the bit-rates of two codec modes are lower than that of HR, AMR can also be deployed in a half of the time slots. Table 3.8 summarizes the bit allocation of AMR for TCH/AHS. Because of the halved bit-rate budget, channel coding cannot be applied to all of the bits, leaving a set of from 12 to 36 bits in Class 2 of each codec mode vulnerable. The same six-bit CRC as in TCH/AFS, generated by passing the Class 1a bits through the shift register represented by the polynomial $D^6 + D^5 + D^3 + D^2 + D + 1$, is used for error detection. Compared with those of TCH/AFS, the channel coding rates are higher, providing lower error correction capability. The puncturing of channel-coded bits further compromises the capability. Likewise, the two-bit CMI or CMC/CMR is protected with a rate $\frac{1}{2}$ block code. Considering the reduced level of protection against transmission errors, it can be expected that the quality of AMR in TCH/AHS will degrade abruptly once the SNR falls below acceptable levels. Figure 3.10 shows the speech and radio signal processing operations of AMR for TCH/AHS.

Although AMR uses the same encoder structure as EFR, its DTX mechanism is more refined than those of EFR or other previous speech codecs. Figure 3.11 illustrates the DTX operation of AMR including the two SID types [3GPP (2000e)].

Table 3.8 Bit allocation of AMR for TCH/AHS.

Codec mode	4.75	5.15	5.90	6.70	7.40	7.95
Class 1a bits	39	49	55	55	61	67
CRC	6	6	6	6	6	6
Class 1b bits	44	42	47	55	59	56
Convolutional encoder input	89	97	108	116	126	129
Convolutional coding rate	$\frac{1}{3}$	$\frac{1}{3}$	$\frac{1}{2}$	$\frac{1}{2}$	$\frac{1}{2}$	$\frac{1}{2}$
Padding bits	2	2	1	1	1	1
Convolutional encoder output	285	303	224	240	260	266
Punctured bits	73	91	16	40	64	78
Class 2 bits	12	12	16	24	28	36
Block-coded CMI/CMC/CMR	4	4	4	4	4	4
Total	228	228	228	228	228	228

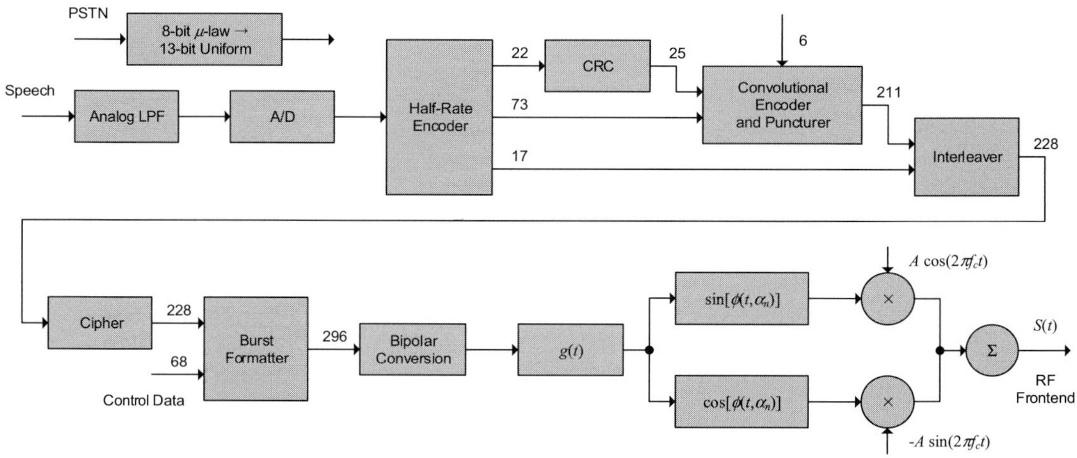

Fig. 3.10 Speech and radio signal processing operations of GSM for AMR (TCH/AHS).

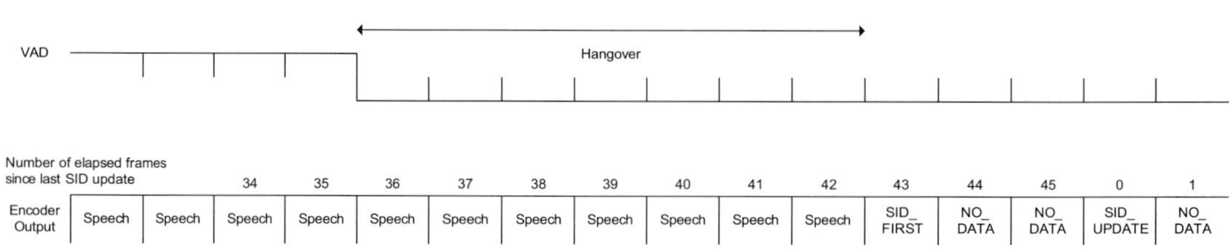

Fig. 3.11 DTX in AMR.

After the voice activity detector identifies the end of voice activity, seven consecutive frames are still encoded as speech. After this hangover period, the next frame is encoded as an SID_FIRST, which marks the end of a talkspurt, i.e., the beginning of a speech pause. An SID_UPDATE is transmitted as the third frame after an SID_FIRST, if more than 24 frames have passed since the last SID_UPDATE was transmitted.

Table 3.9 Transmission type identifiers for downlink in TFO.

	Information bits	Mode indication
SPEECH_DEGRADED	Possible errors in Class 2 bits	Current codec mode
SPEECH_BAD	Possible errors in Class 1 bits	Current codec mode
SID_BAD	Errors detected in SID	
ONSET	Beginning of a speech burst (no information included)	Codec mode of the following speech frame

DTX may generate special messages used in the downlink in Tandem-Free Operation (TFO) [3GPP (2017e)], i.e., when the encoded speech is transmitted all the way to the far-end in the compressed format without being transcoded to and from A-law or μ-law. Such messages, summarized in Table 3.9, are required to report the status of a speech frame that might have been corrupted en route. From the difference of SPEECH_DEGRADED and SPEECH_BAD, it can be seen that the loss of Class 1 bits is treated more seriously than the loss of Class 2 bits. SID_UPDATE and SID_BAD include 35 bits of comfort noise information while SID_FIRST does not contain any information bits.

3.2.3 Link Adaptation

To help the receiver identify whether a received in-band signaling message is a CMI or a CMC/CMR, a CMI is transmitted aligned to the 26-multiframe structure to ensure synchronization. For both TCH/AFS and TCH/AHS, a CMI is initially sent with speech frames having their first burst sent on TDMA frames 0, 8, 17 (modulo 26) in the uplink and TDMA frames 4, 13, 21 (modulo 26) in the downlink. For TCH/AHS, TDMA frames 1, 9, 18 (modulo 26) and TDMA frames 5, 14, 22 (modulo 26) can also be used for the uplink and downlink, respectively. In the downlink, the relative order of CMI and CMC/CMR can be switched with the Robust AMR Traffic Synchronized Control Channel (RATSCCH) message [3GPP (2017h)], which includes both CMI and CMC/CMR. Since an alternating transmission of CMI and CMC/CMR, which saves just eight bits per speech frame, would complicate the operation of the sender and the receiver significantly, such a signaling practice is not enforced when AMR is used in mobile communications systems with larger capacity for control channels.

To adapt to the variations of link quality, the signal processing operations are changed upon the request of the far-end, which bases its decision on the quality indicator, a normalized CIR. For the reference conditions represented by a typical urban 3 km/h scenario with ideal frequency hopping at 900 MHz, no further adjustment or normalization is required. The CIR is estimated by taking the signal and noise/interference measurements for each slot, typically using the equalizer algorithms. Then the ratio of the energies of the two values gives an estimate of the CIR for the slot. This value is converted to dB and filtered for smoothing and prediction. The output is an estimate of the expected CIR at the time instant for which the link (codec mode) adaptation operation becomes effective. FIR smoothing filters of lengths 100 and 50 are used for TCH/AFS and TCH/AHS, respectively.

When operating in channel conditions that are different than the reference, a normalization factor is applied, i.e., added to the filtered values, to take the differences into account. Since the receiver-driven adaptation policy of AMR does not directly depend on the quality indicator, new channel estimation techniques can be easily applied. For example, the quality indicator may be derived in a different fashion, such as by manipulating the estimated bit error rate. The MS and the BSS should continuously update the quality indicator estimates to track the variations of CIR.

Figure 3.12 illustrates the partitioning of channel conditions to map the quality indicator to the codec mode when four codec modes are included in the ACS. While partitioning the CIR range into four disjoint areas, each of which corresponds to a codec mode, would be a straightforward approach, it is also necessary to avoid unnecessary switching of the bit-rate allocation, as the filtered quality indicator may still include noise. Switching the bit-rate allocation according to the quality indicator may at least minimize the loss of speech frames or numerically minimize the combined distortion from compression and transmission. However, frequent switchings or direct changes between non-adjoining codec modes are often perceived as a new type of quality degradation not anticipated by the classical joint source-channel coding theory.

The term *hysteresis* refers to the behavior of a system where the output depends not only on the input but also on the current state of the system or the history of the input [Mayergoyz (2003), Krasnosel'skii and Pokrovskii (1989)]. For example, it is known that certain magnetic materials exhibit a delay between the application and removal of electro-magnetic field, and the manifestation of their effects. In the link adaptation of AMR, hysteresis refers to those mechanisms that resist the switching of codec mode temporarily when the normalized CIR enters the domain of other codec modes.

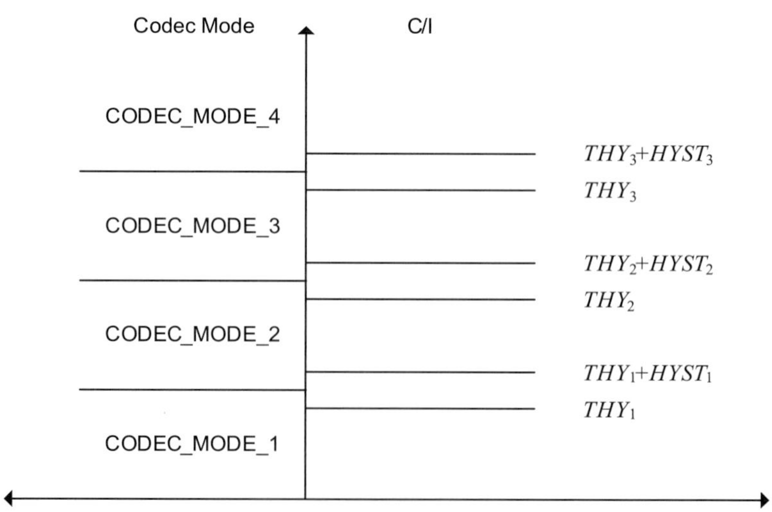

Fig. 3.12 Threshold and hysteresis for codec mode adaptation.

Table 3.10 Initial codec mode at call setup and after handover.

Number of codec modes	ICM
1	Current codec mode
2, 3	Codec mode with the lowest bit-rate
4	Codec mode with the second lowest bit-rate

In Fig. 3.12, THY_j is the threshold for switching from mode $j + 1$ to j. Suppose that the current codec mode is $j + 1$. Then the receiver asks the sender to switch down to codec mode j if the normalized CIR falls below THY_j, by means of a CMC or a CMR. $HYST_j$ is the incremental value added to THY_j, i.e., the hysteresis, to obtain the threshold for switching from codec mode j to $j + 1$. The receiver asks the sender to switch up the codec mode to $j + 1$ if the CIR exceeds $THY_j + HYST_j$. The current mode is maintained if the CIR lies between THY_j and $THY_j + HYST_j$. Therefore the level of hysteresis, the resistance to codec mode switching, can be controlled with $HYST_j$.

With four codec modes in the ACS, THY_1, THY_2, and THY_3 divide the CIR range into four domains corresponding to the four codec modes. $HYST_1$, $HYST_2$, and $HYST_3$ specify the hysteresis band between two neighboring modes. To order the domains, the following constraints are enforced: $THY_1 \leq THY_2 \leq THY_3$, and $THY_1 + HYST_1 \leq THY_2 + HYST_2 \leq THY_3 + HYST_3$. The Initial Codec Mode (ICM), to which the speech encoder is initialized when speech encoding starts or resumes after handover, can be configured using control signaling. The implicit or recommended rules when the ICM is not explicitly specified depend on the number of codec modes in the ACS, as summarized in Table 3.10.

Another constraint applied to the codec mode adaptation is the *neighborhood-only transition policy*. Suppose that CODEC_MODE_2 is currently set. Then the receiver cannot ask the sender to switch up to CODEC_MODE_4 directly even when the normalized CIR exceeds $THY_3 + HYST_3$. This constraint is enforced to avoid reducing the speech quality when the CMC/CMR bits are corrupted during transmission and an abrupt increase or decrease of codec mode is mistakenly requested or commanded. In the case of CMI, the errors are less critical as the receiver will easily detect the corruption of in-band signaling bits from a CRC check. Notice that filtering of the quality indicator, e.g., applying a moving average of the CIR values, also has the effect of decelerating codec mode switching. Under this conservative transition policy, the codec mode can be switched only by a single step at a time. From CODEC_MODE_2, the codec mode can be first switched up to CODEC_MODE_3. If the normalized CIR is still greater than $THY_3 + HYST_3$ at the next measurement opportunity, the codec mode is finally switched up to CODEC_MODE_4. Direct switching of codec mode by more than one level is, however, allowed after handover or when the AMR configuration is changed with control messages. Tables 3.11 and 3.12 show the formats of the messages to initialize and reconfigure the AMR link adaptation when there are up to three and four codec modes, respectively.

When there are up to three codec modes in the ACS, THY_j can be set to a value between 0 and 31.5 dB, in 0.5 dB increments, and $HYST_j$ can be set to a value between

Table 3.11 AMR_CONFIG_REQ message for up to 3 codec modes.

Bit	34 . . . 30	29 28	27 . . . 20	19 . . . 16	15 . . . 10	9 . . . 6	5 . . . 0
Value	00111	ICM	ACS	HYST2	THRESH2	HYST1	THRESH1

Table 3.12 AMR_CONFIG_REQ message for 4 codec modes.

Bit	34 . . . 30	29 28	27 . . . 20	19 18	17 . . . 12	11 . . . 6	5 . . . 0
Value	00111	ICM	ACS	HYSTc	THRESH3	THRESH2	THRESH1

0 and 7.5 dB, also in 0.5 dB increments. This requires six and four bits to represent THY_j and $HYST_j$, respectively. With four codec modes, a single $HYST_j$ can be used to represent all three hysteresis parameters, rather than configuring them separately with the THRESH_REQ messages. In this case, only two bits are assigned to represent $HYST_j$ values between 1.0 and 4.0 dB, in 1 dB steps. A typical configuration, when three codec modes, 5.9, 7.95, and 12.2 kbps, are included in the ACS, is $THY_1 = 6.5$ dB, $THY_2 = 11.5$ dB, and $HYST_1 = HYST_2 = 2.0$ dB.

AMR_CONFIG_REQ can be transmitted by the BTS using RATSCCH during a call to change the configuration of AMR without interrupting speech transmission. The message contains the key parameters for link adaptation: ACS, ICM, and a pair of threshold and hysteresis parameters. The thresholds and hysteresis parameters in the AMR_CONFIG_REQ message, except for the ACS and ICM, are valid only for the downlink. BTS may change the thresholds and hysteresis parameters for the uplink direction at will without informing the MS. If the downlink speech frame N is replaced with an AMR_CONFIG_REQ message, then the MS and the BTS should use the new set of link adaptation parameters beginning at frame $N + 12$. During the transition period lasting 12 frames, the configuration of AMR can be different in the uplink and downlink.

Figure 3.13 shows a block diagram of the AMR codec mode adaptation system distributed over the MS, BTS, and TRAU. At the MS, the channel decoder extracts the CMI of the downlink and informs the speech decoder of that information. If the CMI does not match any mode in the ACS, the frame is assumed to be corrupted. It also extracts the CMC or CMR, and informs the speech encoder so that the requested codec mode is used in the next encoding opportunity. The downlink channel quality is estimated from correlating the TSC bits with known sequences, and after the normalization and comparison with the thresholds, is mapped to the CMR, which is block-coded and sent to the interleaver. The block-coded CMR replaces the CMI on every second frame. Likewise, at the BTS, the channel decoder extracts the CMI and CMR from the received TDMA frames, and informs the speech decoder and the encoder of that information.

If the requested codec mode is not adjacent to the current mode, the request is not accepted. The uplink channel quality is estimated from the TSC bits, and after

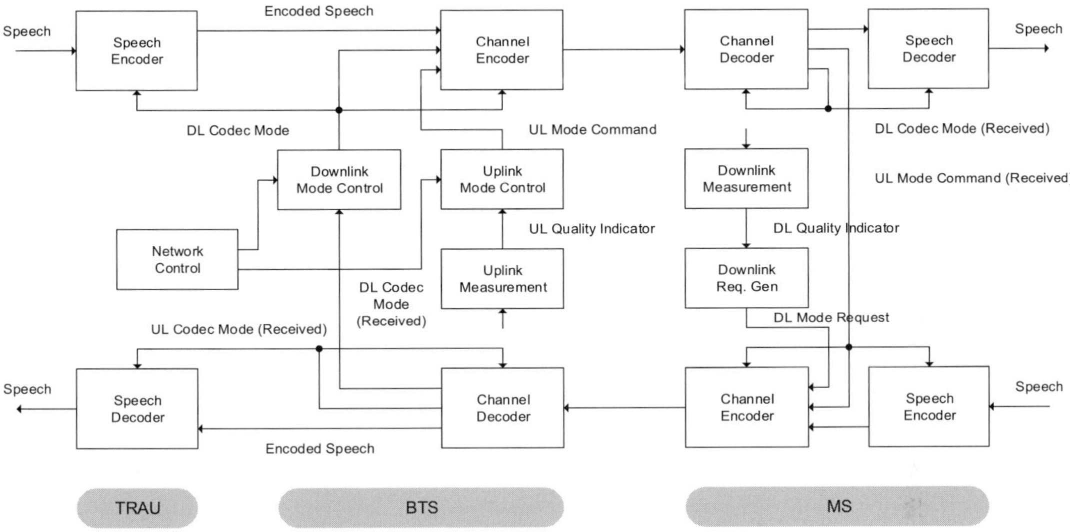

Fig. 3.13 Block diagram of AMR codec mode adaptation.

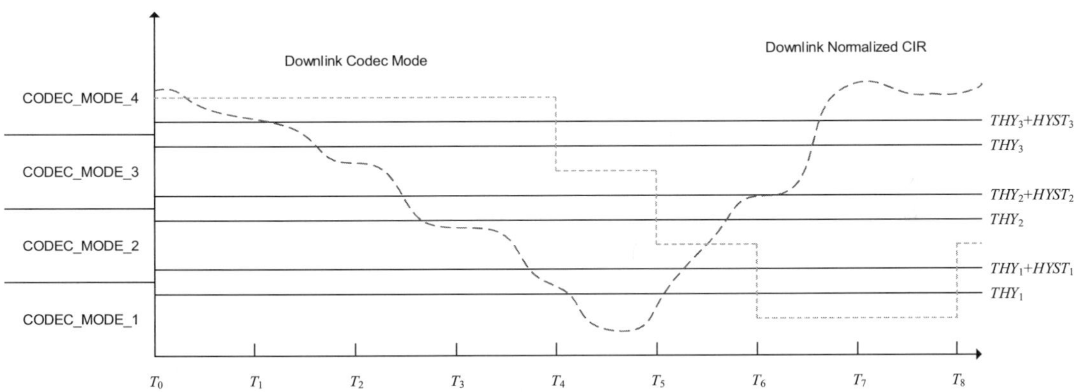

Fig. 3.14 Channel tracking in AMR codec mode adaptation.

normalization and comparison with the thresholds, is mapped to the CMC, which is block-coded and sent to the interleaver. The block-coded CMC replaces the CMI on every second frame. If the performance of link adaptation does not meet its expected level, the BTS may re-configure the codec mode adaptation parameters with the AMR_CONFIG_REQ message.

Figure 3.14 illustrates a hypothetical operation of AMR codec mode adaptation, in which the impact on speech quality from deep fading is reduced by a series of rapid codec mode switchings. At time instant T_0, the downlink codec mode is set to CODEC_MODE_4, to operate in benign channel conditions. Then the channel condition begins to degenerate, and at T_2, the normalized CIR falls below THY_3, triggering a down-switching to CODEC_MODE_3. The CMR for the new codec mode is transmitted in the uplink, which the BTS is likely to apply at its earliest opportunity. After the signal propagation and the processing delay, a speech frame in the new configuration is transmitted at T_4. The codec mode adaptation of AMR can track the variations in channel quality with a lag that is less than the two-way delay between the MS and the BTS. MS has to meet stricter timing requirements than the BTS, since the MS should apply the new CMC received for the uplink no later than three speech frames. The normalized CIR is calculated by filtering the measured CIR samples with such a delay in mind. As the channel condition deteriorates further, the codec mode is switched down to CODEC_MODE_2, and finally to CODEC_MODE_1, at T_5 and T_6, respectively. Fortunately, the channel condition begins to rebound at T_5, and the codec mode is switched up to CODEC_MODE_2 at T_8. In commercially operational GSM networks, it is often observed that the codec mode is switched once or twice per minute when the link quality fluctuates. Note that the responsibility for channel tracking is shared with other activities, including the power control, slow frequency hopping, and timing advance. The link adaptation of GSM is a rare but successful implementation of joint source-channel coding at a fixed total bit-rate. AMR and its closed-loop signaling for codec mode adaptation are classified as key features of GSM Phase 2+, following advancements such as EFR and HR in Phase 2.

3.3 Enhancements in Wireless Transmission

The previous chapters explained how AMPS could be systematically improved for higher speech quality and increased network capacity, either by applying more refined analog signal processing operations, as in N-AMPS, or by applying new digital signal processing operations, as in D-AMPS. With more room for innovation than other contemporary TDMA mobile communications systems whose origins were the first generation analog systems, GSM improved gradually with increased computational capacity and more efficient RF circuitry. One of the areas benefiting from improved computational power was speech compression, where more complex algorithms based on the ACELP principles, such as EFR and HR, exceeded FR in terms of speech quality and network capacity. Another approach to boost the performance of mobile communications systems is to reduce the minimum carrier-to-interference ratio (CIR) required for normal operations by applying more complex receiver or equalizer algorithms.

3.3.1 Downlink Advanced Receiver Performance

Interference Cancellation (IC) is an area where increased complexity can be applied directly to estimate the interference and subtract it from the received signal. This increases the likelihood of successful channel decoding at low CIR. However, unlike speech compression algorithms or radio signal processing procedures in the transmitter, algorithms in the receiver or the channel decoder are not standardized, since this

Table 3.13 Reference test scenarios for DARP.

	Interferer	Interferer relative power level (dB)	TSC	Interferer delay range (symbols)
DTS-1	Co-channel 1	0	None	None
DTS-2	Co-channel 1	0	None	None
	Co-channel 2	−10	None	None
	Adjacent 1	3	None	None
	AWGN	−17		
DTS-3	Co-channel 1	0	Random	−1 – 4
	Co-channel 2	−10	None	None
	Adjacent 1	3	None	None
	AWGN	−17		
DTS-5	Co-channel 1	0	None	74
	Co-channel 2	−10	None	None
	Adjacent 1	3	None	None
	AWGN	−17		

would limit the opportunity for differentiation and innovation among the implementors. Therefore GSM specifies only three groups of performance requirements for IC. These define the required performance in diverse situations with a variety of interferer types. The situations encompass varied correlation and delay between the desired signal and the interferer [3GPP (2017g)]. These requirements are much more stringent than those for conventional GMSK receivers. As it is an important design objective to limit the modification of existing channel structures and signaling procedures, performance of conventional MSs should not be compromised in those cells where the IC is used. Table 3.13 shows such test scenarios in which the power levels are measured relative to the strongest interferer, co-channel 1. The desired signal uses TSC 0. When TSC is indicated as none, the field is filled with random data bits. Likewise, random TSC means that the TSC is randomly selected on a slot-by-slot basis. The power level of the strongest co-channel interferer, co-channel 1, is -80 dBm. In DARP Test Scenarios (DTS) 5, the interfering signal of co-channel 1 is asynchronous with the desired signal, with a delay difference much larger than that of DTS-3.

If a minimum set of requirements is met, then the interference cancellation capability of the MS is classified as meeting the Downlink Advanced Receiver Performance (DARP) Phase 1. DARP Phase 2 requires an even higher capability to suppress the interference. The receivers meeting the DARP requirements reduce the operational CIR significantly, thereby moving the speech quality vs. channel condition curves, illustrated in Fig. 3.8, to the left. This means that a comparable speech quality can be obtained at a lower CIR, which would permit a lower frequency re-use factor. The difference between DTS-2 and 3 is that in DTS-3, TSC of the co-channel is known to the receiver. To meet the requirements, the measured frame error rate should be equal to or lower than 1% for the speech, and 5% for the control signaling in the downlink.

Table 3.14 Performance requirements of DARP test cases [CIR (dB)].

	DTS-1	DTS-2	DTS-3	DTS-5
TCH/FS	4.5	9.5	10.0	9.5
TCH/AFS 12.2	5.0	10.0	10.0	10.0
TCH/AFS 7.95	1.5	6.5	7.5	
TCH/AFS 4.75	−1.5	4.5	5.0	

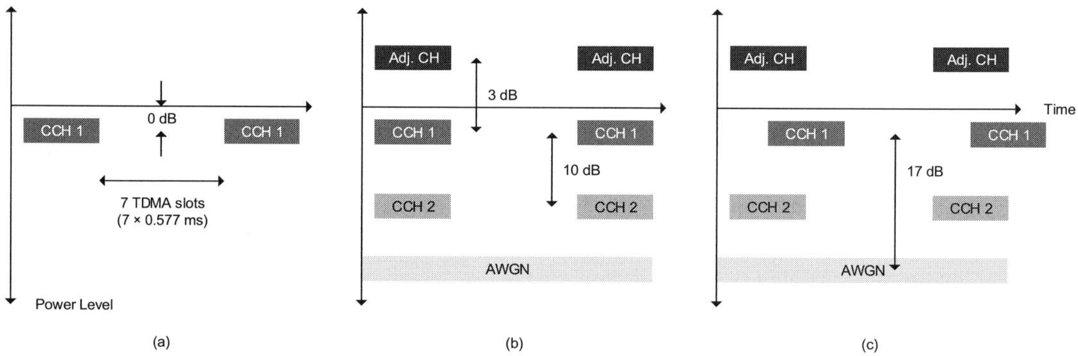

Fig. 3.15 (a) DTS-1 co-channel. (b) DTS-2, 3 synchronous tests. (c) DTS-5 asynchronous test.

Figure 3.15 illustrates two types of test cases as well as a reference, DTS-1, in which the interferer is only a co-channel. The parameters for this scenario are outlined in Table 3.13. Figure 3.15(b) describes the scenario where the channels inside and outside the cell are synchronized. Figure 3.15(c) is a more challenging but realistic case, where not all interferers are synchronized to the signal. The time slot of co-channel 1 is delayed by 50% or 74 symbols. Note that the power level of the co-channels or adjacent channels, in comparison with that of the Additive White Gaussian Noise (AWGN), is visualized as the gradation toward black. The interferers shown in darker color generate stronger interference. Table 3.14 shows the performance requirements for DARP at a fixed error rate of 10%, for FR and AMR with three codec modes. The receivers satisfying DARP should operate at much lower CIR values than conventional receivers.

Single Antenna Interference Cancellation (SAIC) is often used synonymously with the term DARP in the sense that the influence of co-channel interference is removed through complex signal processing that requires only an antenna. When applied to GMSK and fully deployed situations, SAIC is known to provide significant gains in voice capacity in the interference-limited networks, both synchronous and asynchronous [3GPP (2017b)]. For a fractionally loaded, synchronous network with $K = 1$ re-use and frequency hopping, SAIC can support voice capacity gains ranging from 37.8% to 53.1%. The gain for an asynchronous network employing $K = 3$ re-use with frequency hopping is 34.3%.

The baseband signal processing, i.e., the signal processing after RF carrier components are removed, for interference cancellation is taken at the symbol level, taking the level of synchronization between adjacent cells into account. Typical IC techniques include Joint Demodulation (JD) or Blind Interference Cancellation (BIC) [Halonen *et al.* (2007)]. JD is appropriate for synchronized networks where the training sequences of the desired and interfering BTSs are aligned in time while in BIC the receiver does not assume any structures for the interfering signals, which naturally requires more complex signal processing operations.

At a receiver supporting the IC, the interference is estimated and subtracted from the received signal for one interferer at a time until the delay budget expires or there remain no more interferers. DARP Phase 1 assumes the basic radio transmission techniques of GSM using GMSK, not requiring any support from the network. In what follows, approaches are introduced to achieve higher capacity by enhancing the radio signal processing, which has been kept largely intact while the speech codec has evolved from FR to EFR, HR, and AMR. Like HR, these focus on higher network capacity while leaving speech quality at the same or slightly lower levels. However, compared to the enhancements achievable by enhancing the speech codecs or coordinating between the MS and the network, which include changes only in the digital signal processing domain, these approaches require more careful re-design of the signal processing procedures as the RF circuitry and the timing relationships also have to be modified.

3.3.2 8-PSK Half-Rate Channel

One straightforward approach to increase the bit-rate without using additional frequency spectrum is to use higher-order modulation techniques. Increasing the symbol rate was rarely considered as a practical solution for TDMA systems, since it required re-designing and re-testing the entire frame structure. Compared with GMSK which represents a single bit per symbol, higher-order modulation techniques, such as QPSK, 8-PSK, 16QAM, and 32QAM, convey 2, 3, 4, and 5 bits per symbol, respectively. However, these higher-order modulation techniques require higher SNR, which has to be provided by either increasing the transmit power or applying advanced receiver algorithms requiring higher complexity. The former approach increases the interference and reduces network capacity while the latter requires higher computational complexity and more efficient RF circuitry, which was becoming increasingly affordable for the receivers in the MS.

A new half-rate speech channel, the Adaptive Multi-Rate Speech Channel at 8-PSK Half-Rate (O-TCH/AHS), uses 8-PSK, and transports all eight codec modes of AMR. O-TCH/AHS has a channel structure similar to that of TCH/AHS but uses 8-PSK instead of GMSK. Compared with the basic frame structure of TCH/FS, each field in the frame structure of O-TCH/AHS contains three times as many bits, as shown in Fig. 3.16, while maintaining the same timing constraints. In the O-TCH/AHS, three bits are mapped to an 8-PSK symbol, denoted as l in Table 3.15, but the symbol rate of 8-PSK is the same as that of GMSK, 270.83 ksps. Note that l is gray-coded such that two consecutive 3-tuples differ in only one bit location. Figure 3.20(a) illustrates the gray-coding of 8-PSK symbol mapping.

Table 3.16 summarizes the bit allocation of AMR for O-TCH/AHS. As in the TCH/AFS, each bit is protected by at least a channel code, and corruption of the more

Table 3.15 8-PSK symbol mapping.

Modulating bits $(d_{3i}, d_{3i+1}, d_{3i+2})$	Symbol parameter l
(1,1,1)	0
(0,1,1)	1
(0,1,0)	2
(0,0,0)	3
(0,0,1)	4
(1,0,1)	5
(1,0,0)	6
(1,1,0)	7

Table 3.16 Bit allocation of AMR for O-TCH/AHS.

Codec mode	4.75	5.15	5.90	6.70	7.40	7.95	10.2	12.2
Class 1a bits	39	49	55	55	61	75	65	81
CRC	6	6	6	6	6	6	6	6
Class 1b bits	56	54	63	79	87	84	139	163
Convolutional encoder input	101	109	124	140	154	165	210	250
Convolutional coding rate	$\frac{1}{7}$	$\frac{1}{6}$	$\frac{1}{6}$	$\frac{1}{5}$	$\frac{1}{5}$	$\frac{1}{4}$	$\frac{1}{4}$	$\frac{1}{3}$
Padding bits	4	4	3	3	2	2	2	1
Convolutional encoder output	749	690	780	730	800	684	864	768
Punctured bits	77	18	108	58	128	12	192	96
Block-coded CMI/CMC/CMR	12	12	12	12	12	12	12	12
Total	684	684	684	684	684	684	684	684

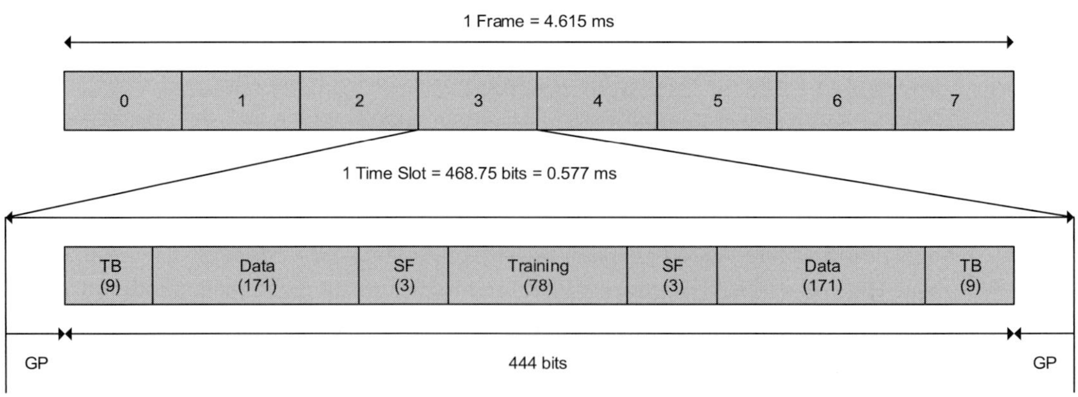

Fig. 3.16 Hierarchical frame structure of O-TCH/AHS.

important bits is detected with a 6-bit CRC, generated by passing the Class 1a bits through a shift register represented by the polynomial $D^5 + D^4 + D^3 + D^2 + D + 1$. As 8-PSK carries three times as many bits as GMSK, the channel coding rate of O-TCH/AHS can be reduced from that of TCH/AHS, which applies also to those of

Table 3.17 Block coding of in-band signal for O-TCH/AHS.

	Message	Coded message
CODEC_MODE_1	00	000000000000
CODEC_MODE_2	01	110110101110
CODEC_MODE_3	10	101101110101
CODEC_MODE_4	11	011011011011

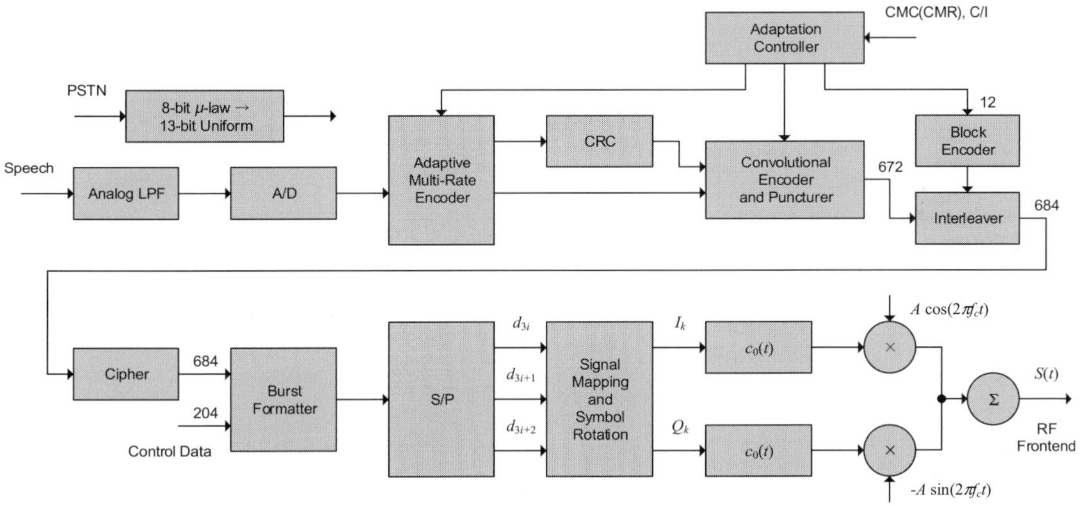

Fig. 3.17 Speech and radio signal processing operations of GSM for AMR (O-TCH/AHS).

CMI and CMC/CMR. As shown in Table 3.17, the two-bit CMI or CMC/CMR is block-coded to 12-bit messages, in contrast to the four-bit messages of TCH/AHS. The channel-coded bits are punctured in all codec modes to meet the budget of 672 bits, to which the 12-bit codec mode information is added. Therefore for each codec mode, a total of 684 bits, corresponding to 228 8-PSK symbols, is interleaved and ciphered.

Figure 3.17 shows the signal processing operations associated with AMR for O-TCH/AHS. The burst-formatted bits, $l = (d_{3i}, d_{3i+1}, d_{3i+2})$, are mapped in groups of three to 8-PSK symbols such that $S_i = e^{j2\pi l/8}$. Then the 8-PSK symbols are rotated by $3\pi/8$ radians per symbol, to $\widehat{S_i} = S_i e^{ji3\pi/8}$, and transformed into the in-phase component, I_k, and the quadrature component, Q_k, respectively. The symbol rotation steers the instantaneous vector connecting two consecutive symbols away from the region surrounding the origin, effectively limiting the dynamic range of the symbol. Otherwise, the instantaneous vector crosses the origin, which would require infinite dynamic range on the power amplifier.

The components are then input to a linear pulse-shaping filter with an impulse response $c_0(t)$ respectively, generating two waveforms. $c_0(t)$ is a Linearized GMSK

Pulse (LGP) that is the main component in a Laurent decomposition of GMSK [Laurent (1986)], where $c_0(t)$ is defined as

$$c_0(t) = \prod_{i=0}^{3} y(t + iT) \tag{3.3}$$

for $0 \leq t \leq 5T$, and 0 elsewhere. Here T is the normal symbol period of GSM, equal to $\frac{48}{13}$ μs [3GPP (2017i)], and $y(t)$ is defined as

$$y(t) = \begin{cases} \sin(\pi \int_0^t g(\tau)d\tau) & 0 \leq t \leq 4T, \\ \sin(\frac{\pi}{2} - \pi \int_0^{t-4T} g(\tau)d\tau) & 4T < t \leq 8T, \\ 0 & \text{elsewhere,} \end{cases} \tag{3.4}$$

where

$$g(t) = \frac{1}{2T}\left[Q(2\pi \cdot 0.3 \cdot \frac{t - 5T/2}{T\sqrt{\ln 2}}) - Q(2\pi \cdot 0.3 \cdot \frac{t - 3T/2}{T\sqrt{\ln 2}})\right] \tag{3.5}$$

and

$$Q(t) = \frac{1}{\sqrt{2\pi}} \int_t^{\infty} e^{-\frac{\tau^2}{2}} d\tau. \tag{3.6}$$

The waveforms are modulated to RF carriers and summed to generate $S(t)$, which is finally amplified and transmitted over the wireless channel. Detailed information on the 8-PSK modulation procedures can be found in [3GPP (2017f)].

As in TCH/AHS, O-TCH/AHS uses a time slot in every other TDMA frame so that a speech frame is interleaved over four, but eight in length, TDMA frames. 8-PSK was originally introduced to GSM not to increase the speech quality or network capacity, but to enable higher bit-rate data transmission for the Enhanced Data Rates for GSM Evolution (EDGE), whose coverage is smaller than that of the ubiquitous voice services. GMSK may be used instead if the link quality at the location of MS does not allow the use of 8-PSK. The key benefit of 8-PSK is to enable the two highest codec modes of AMR, 12.2 and 10.2 kbps, to be used in O-TCH/AHS, a half-rate channel. They are not supported with GMSK in TCH/AHS. DARP Phase 2 includes the receiver diversity for GMSK and 8-PSK, i.e., the capability to combine signals received in multiple antennas, which may have experienced different delay, gain, or phase during transmission.

3.3.3 Voice Services over Adaptive Multi-User Channels on One Slot

The fundamental design principle of TDMA is that the radio resource assigned to each MS, time slots, is both temporally and spectrally separated from those assigned to other MSs or control signaling in the uplink and downlink. Thus in TDMA, two sets of time slots belonging to two MSs are considered to be *orthogonal* such that a product of signals in two slots belonging to two MSs is zero. After the deployment of GSM, there have been many approaches to increase the network capacity but this design principle remains valid in each approach.

The EFR and AMR speech codecs increase the capacity using indirect measures, e.g., by increasing compression efficiency and error resilience, or by dynamically changing speech compression algorithms and re-assigning the total bit-rate over the elements constituting the bit-stream. These approaches result in lower CIR values required for network operation and speech quality that is comparable to that of FR in such conditions. DARP also reduces the CIR but does not modify the signal processing operations at the sender. In contrast, deploying HR is a more direct approach that actually halves the total bit-rate, albeit with limited network coverage and compromised speech quality.

Voice Services over Adaptive Multi-User Channels on One Slot (VAMOS) is another approach for higher capacity [3GPP (2017c)]. It bypasses the fundamental TDMA design principles by enabling a time slot to be used by two MSs in the same cell. Figure 3.18 illustrates the concept, where MS 1 and MS 2 are communicating with a BTS, using two *subchannels* sharing an identical ARFCN and a training sequence (TS) set. The two training sequences in the same TS set need to have a very low cross-correlation with each other if the intended time slot is to be recovered. Therefore, the operation of VAMOS depends to a large extent on the capability of the receiver to extract target signals corrupted not just by the noise or interference from other cells but by the overlapping signals from another MS in the same cell. In this regard, the orthogonality to achieve in VAMOS is extended such that an integral of the inner product of signals in two slots belonging to two MSs is zero.

Under these challenging operating conditions, the interference is classified as either Adjacent Channel Interference (ACI), which originates from the transmission of neighboring channels, or Co-Channel Interference (CCI), which originates from the transmission of other cells using the same channel. In favorable channel conditions,

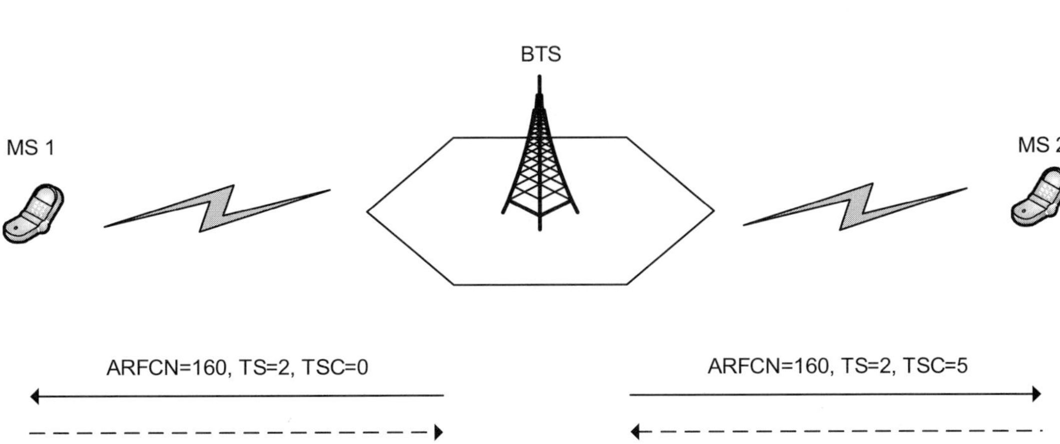

Fig. 3.18 Sharing of time slot by two MSs.

VAMOS can theoretically quadruple the network capacity, compared with the initial capacity of GSM using the full-rate channel, if all MSs in the cell use VAMOS on the half-rate channel. To implement the concept, the following performance objectives were set in the design of VAMOS.

Objective 1

The voice capacity should be increased by a factor of two per BTS, for both TCH/FS and TCH/HS with related associated signaling channels.

Objective 2

The voice capacity should be further increased by means of multiplexing at least two MSs simultaneously on the same radio resource both in the uplink and downlink, for both TCH/FS and TCH/HS with related associated signaling channels. CCI and ACI increase with the number of MSs, leading to the decrease of CIR and frequency re-use factor. The balance between low frequency re-use and high time slot re-use should be considered carefully.

Objective 3

The voice quality as perceived by the user should not be decreased. In particular, a voice quality better than that of HR should be ensured. This is due to the fact that in the case of subchannels, where speech frames of two MSs are allocated in the same time slot within the same radio frequency, the influence of the inevitable Inter-Channel Interference (ICI) on speech quality and the actual proportion of the subscribers sharing the same time slot cannot be ignored.

Other factors taken into account in the design of VAMOS include the support of *legacy*, i.e., conventional GMSK-based MSs or MSs supporting only DARP Phase 1 requirements, minimization of additional complexity for both the MS and the BSS, and minimization of the impact on network planning.

Table 3.18 and Fig. 3.19 together illustrate the slot allocations of TCH/FS, TCH/HS, and VAMOS. A key requirement for the smooth introduction of VAMOS to GSM is that for conventional MSs based on GMSK, the VAMOS channels should be seen as TCH/FS or TCH/HS but the MSs capable of VAMOS operation should be able to detect and receive the assigned subchannel correctly. This requirement necessitates that the waveform of VAMOS includes the GMSK waveform in some way. Therefore a new modulation technique that can meet the challenging requirements, Adaptive Quadrature Phase Shift Keying (AQPSK), is used to modulate two bit-streams, a_i and b_i, for two MSs. Table 3.19, and Figures 3.20(b) and 3.20(c) show the mapping procedures between the modulating bits and the quaternary symbols.

It is possible to control the ratio of transmit power between the I and Q channels with the rotation angle α. Sub-Channel Power Imbalance Ratio (SCPIR) is defined as the ratio, $SCPIR = 20\log_{10}(\tan(\alpha))$ dB. In the downlink of VAMOS, the subchannel for the other MS becomes the source of strong interference. SCPIR is controlled so that appropriate CIR is maintained in the downlink of each MS. In the uplink, such a fine adjustment is neither necessary nor possible as the two subchannels originate from different transmitters.

Table 3.18 VAMOS slot allocation.

(a)	1 full-rate
(b)	2 half-rate
(c)	1 full-rate, 1 VAMOS full-rate
(d)	1 full-rate, 2 VAMOS half-rate
(e)	2 half-rate, 1 VAMOS full-rate
(f)	2 half-rate, 2 VAMOS half-rate

Table 3.19 AQPSK symbol mapping.

Modulating bits (a_i, b_i)	AQPSK symbol
(0,0)	$e^{j\alpha}$
(0,1)	$e^{-j\alpha}$
(1,0)	$-e^{-j\alpha}$
(1,1)	$-e^{j\alpha}$

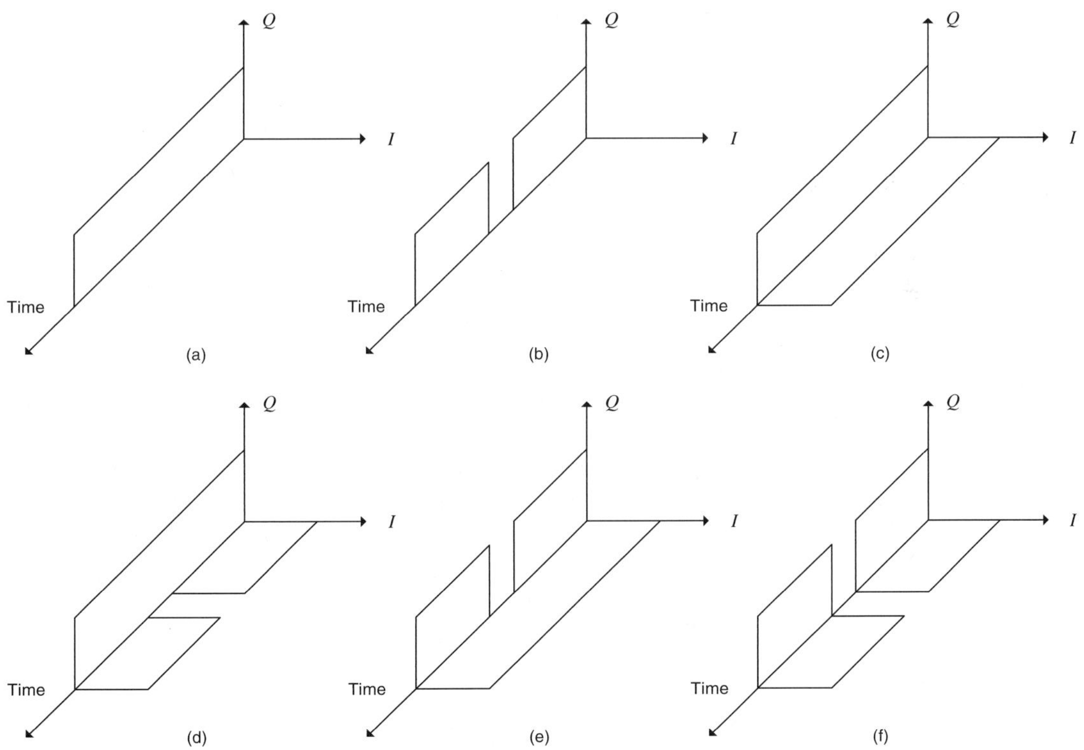

Fig. 3.19 VAMOS slot allocation.

Even with similar propagation characteristics and distances from the BTS, conventional MSs using GMSK may require a higher CIR level since the VAMOS-enabled MSs, which must support at least DARP Phase 1 requirements, can extract the signal at lower CIR level. Figure 3.20(b) illustrates the case when higher power is assigned to

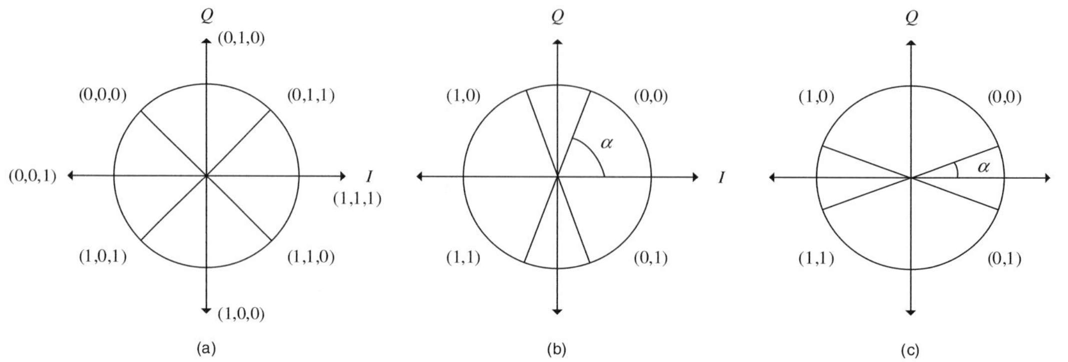

Fig. 3.20 (a) 8-PSK symbol mapping. (b) AQPSK case 1. (c) AQPSK case 2.

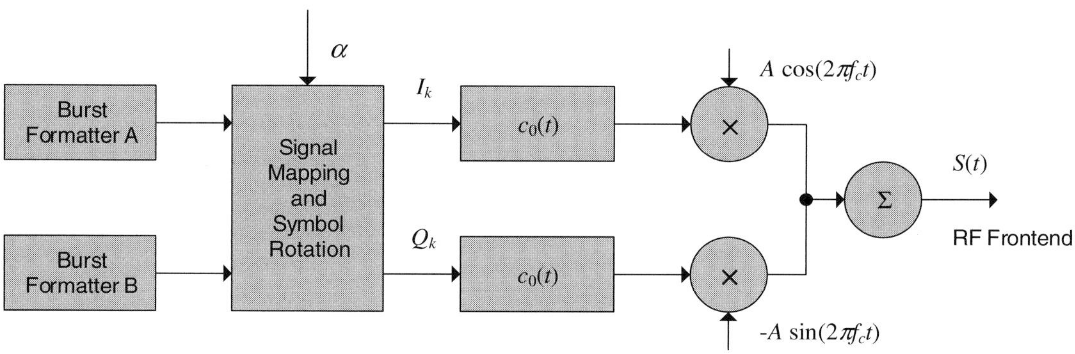

Fig. 3.21 VAMOS downlink modulation.

bit-stream b_i while 3.20(c) illustrates the opposite case. SCPIR is required to meet the constraint, $20\log_{10}|\tan\alpha| \leq 10$, i.e., $|\alpha| \leq \tan^{-1}\sqrt{10} = 1.26$ radians (72.45 degrees), to limit the level of imbalance. The interference needs to be distributed between the two subchannels as evenly as possible as long as they can be decoded by the two MSs.

The symbols, whose rates are the same as that of GMSK, 270.83 ksps, are continuously rotated with $\varphi = \frac{\pi}{2}$ radians per symbol, to $\widehat{S_i} = S_i e^{ji\pi/2}$, and transformed into the in-phase component, I_k, and the quadrature component, Q_k, which are input to a linear pulse-shaping filter to drive $c_0(t)$, the same LGP used in 8-PSK. The waveforms are modulated to RF carriers and summed to generate $S(t)$, which is amplified and transmitted over the wireless channel. Figure 3.21 shows the downlink modulation procedures of VAMOS. For the uplink, the two MSs use the original GMSK modulation with different training sequences. Both subchannels are simultaneously received by the BTS that uses the receiver diversity and interference cancellation to separate the orthogonal subchannels. Further information on the AQPSK modulation can be found in [3GPP (2017f)].

Each subchannel of the AQPSK-modulated waveforms can be decoded without consideration of the other subchannel. Let y_i be the ith received sample, corresponding to the modulated symbols a_i and b_i. Assuming a complex channel gain of $\alpha_i e^{j\beta_i}$ and additive noise h_i, y_i can be represented as $y_i = \alpha_i e^{j\beta_i}(a_i + jb_i) + h_i$. Then the maximum likelihood decision metric for the ith symbol is

$$\arg\max_{(a_i,b_i)} \ln[p(y_i|a_i, b_i, \alpha_i, \beta_i, h_i)] = \arg\max_{(a_i,b_i)} -|y_i - \alpha_i e^{j\beta_i}(a_i + jb_i) - h_i|^2. \quad (3.7)$$

This metric can be expanded as

$$\begin{aligned}\arg\max_{(a_i,b_i)} \ln[p(y_i|a_i, b_i, \alpha_i, \beta_i, h_i)] &= \arg\max_{(a_i,b_i)} [-\alpha_i^2 a_i^2 - \alpha_i^2 b_i^2 + \alpha_i e^{j\beta_i}(y_i^* - h_i^*)a_i + \alpha_i e^{-j\beta_i}(y_i - h_i)a_i \\ &\quad + j\alpha_i e^{j\beta_i}(y_i^* - h_i^*)b_i - j\alpha_i e^{-j\beta_i}(y_i - h_i)b_i] \\ &= \arg\max_{(a_i,b_i)} [-\alpha_i^2 a_i^2 + \alpha_i[e^{j\beta_i}(y_i^* - h_i^*) + e^{-j\beta_i}(y_i - h_i)]a_i] \\ &\quad + \arg\max_{(a_i,b_i)} [-\alpha_i^2 b_i^2 + j\alpha_i[e^{j\beta_i}(y_i^* - h_i^*) - e^{-j\beta_i}(y_i - h_i)]b_i].\end{aligned} \quad (3.8)$$

Since a_i and b_i are statistically independent,

$$\begin{aligned}\arg\max_{(a_i,b_i)} \ln[p(y_i|a_i, b_i, \alpha_i, \beta_i, h_i)] &= \arg\max_{a_i}[-\alpha_i^2 a_i^2 + \alpha_i[e^{j\beta_i}(y_i^* - h_i^*) + e^{-j\beta_i}(y_i - h_i)]a_i] \\ &\quad + \arg\max_{b_i}[-\alpha_i^2 b_i^2 + \alpha_i[je^{j\beta_i}(y_i^* - h_i^*) - je^{-j\beta_i}(y_i - h_i)]b_i].\end{aligned} \quad (3.9)$$

Since a_i, b_i, α_i, and β_i are real variables, the decision metric can be decomposed into two real metrics which can be independently computed. Therefore, with appropriate equalization and interference cancelation enabled by the training sequence, each MS in VAMOS mode can extract its bits while regarding the signal of another subchannel as CCI. Likewise, the BTS can extract both bit-streams in the uplink subchannels. This concept is called the Orthogonal Sub-Channel (OSC). A new set of training sequences, TS 2, was introduced to reduce the correlation between the I and Q channels.

The training sequences in TS 2 were determined through exhaustive search to have the lowest correlation when paired with its counterpart sequence in TS 1 with the same sequence number. For the BTS to initiate the VAMOS mode of operation, at least one MS should be VAMOS-capable, supporting both TSs 1 and 2, as well as DARP Phase 1 requirements. As in DARP, the capability of VAMOS comes in two classes, VAMOS I and II, depending on the capability of the receiver. The MS can indicate to the BTS its DARP and VAMOS capability with control signaling. The BTS may operate VAMOS-capable MSs in conventional fashion when their link quality is not good and orthogonality between subchannels is not maintained enough for sharing slots.

3.3.4 Adaptive Pulse Shaping

Since the performance of GSM is optimized using adaptation of its key parameters for the time-varying environments, the VAMOS operation can exploit the adaptation in several domains, such as the codec mode, channel type, or frequency hopping. During the design of VAMOS, a new type of adaptation proposed for the downlink was the adaptation of pulse shaping, where the $c_0(t)$ of one or two subchannels may be replaced with other pulse shapes during a call if higher performance can be obtained. This class of techniques, Adaptive Pulse Shaping (APS), was not adopted for the finalized VAMOS standard but showed potential as a new opportunity for improvement [3GPP (2017c)].

APS is based on the benefit that both MSs can improve their performance, i.e., reduce the required CIR, by a couple of dB when there is a 5 dB power imbalance on TCH/FS, if the MS with the weak RF signal uses a pulse of broader bandwidth while the other MS sharing the same slot uses a pulse of narrower bandwidth. Therefore a power imbalance between two VAMOS subchannels is an opportunity for adaptive pulse shaping. A key requirement for this adaptation technique is that the sum of both pulses must be within the GMSK mask so that the cell planning of the GSM networks is not influenced. Otherwise, the GSM networks may not operate properly in transmission modes other than the VAMOS.

APS can help VAMOS make a much better use of the available radio resources with little additional cost and complexity. GMSK and LGP can be considered as the basic pulse shaping methods the BTS can use. To adapt to a wide range of power imbalance, a family of Root-Raised Cosine (RRC) pulses with different bandwidths may be used for VAMOS. For example, suppose that MS A is at a greater distance from the BTS than MS B, whose signal strength is higher than that of MS A by 7 dB. Then the BTS may use an RRC pulse with a wider bandwidth in the subchannel to MS A while the subchannel to MS B is modulated with a narrower RRC pulse or LGP.

3.4 Performance Evaluation

In this section, the compression efficiency and error resilience of the speech codecs discussed so far are compared for several channel conditions and channel types. To design the link adaptation algorithms of AMR, the quality of each codec mode is evaluated over a wide range of static channel conditions and also compared with the quality of the fixed bit-rate speech codecs. Based on the testing results, three configurations of AMR, consisting of a set of codec modes, thresholds, and hysteresis parameters, are defined for TCH/AFS and TCH/AHS, respectively. Then the speech quality of each AMR configuration is evaluated under dynamic channel conditions and compared with the quality of the fixed bit-rate codecs or other AMR configurations. Finally, the improvement of network capacity from applying VAMOS is analyzed.

Table 3.20 summarizes the complexity of the FR, HR, and EFR speech codecs, and their quality measured by their MOS values for clean channels [Redl *et al.* (1998)]. The complexity is measured by the maximum number of instructions executed when the codec is running, measured in the Million Instructions Per Second (MIPS), storage for program and data, and memory space required to run the codec. It should be noted that EFR and HR employ much more complex speech compression algorithms than FR, to achieve higher speech quality or reduce the bit-rate.

3.4.1 Speech Compression and Transmission Performance

To evaluate the performance in real environments, each combination of speech codec and traffic channel type is subjectively tested for a range of CIR values.

Table 3.20 Complexity and quality of speech codecs.

	Maximum MIPS	Program and data ROM (kWords)	RAM (kWords)	MOS (clean channel)
FR	2.5–4.5	4–6	1–2	3.50
HR	17.5–22	16–20	5	3.35
EFR	17–22	15–20	5	4.21

Table 3.21 Test results (MOS) of EFR and AMR in full-rate stationary channel.

C/I (dB)	1	4	7	10	13	16	Clean channel
EFR		1.53	3.05	3.65	4.01		4.01
AMR 12.2		1.46	3.44	3.93	4.13	4.06	4.01
AMR 10.2		2.04	3.80	4.05	3.96		4.06
AMR 7.95	1.43	3.26	3.96	4.08	4.01		3.91
AMR 7.40	1.39	3.11	3.84	3.98	3.94		3.83
AMR 6.70	1.87	3.29	3.86	3.80			3.77
AMR 5.90	2.20	3.59	3.69				3.72
AMR 5.15	2.43	3.44	3.58				3.50
AMR 4.75	2.66	3.43	3.52				3.50

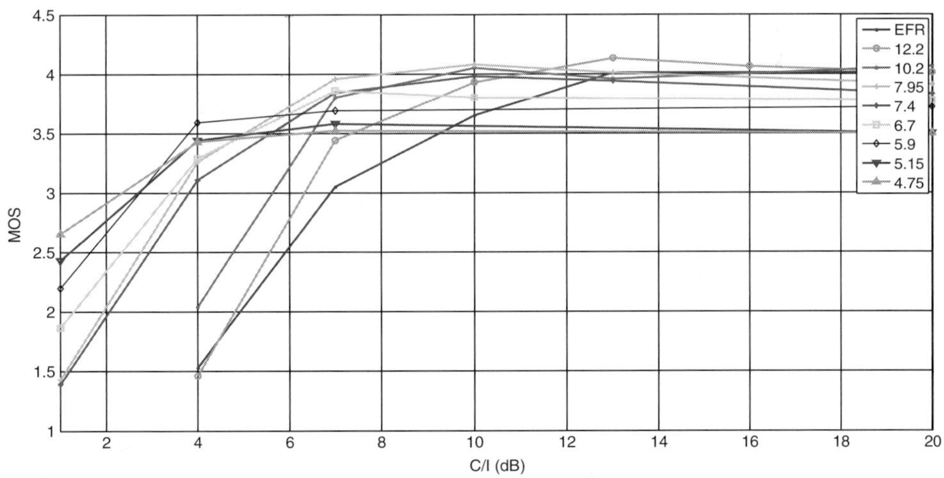

Fig. 3.22 Test results of EFR and AMR in full-rate stationary channel.

Full-Rate Channel under Stationary Channel Conditions

Table 3.21 and Fig. 3.22 show the MOS values of each codec mode of AMR and EFR [3GPP (2017a)] for TCH/AFS and TCH/EFS, respectively. MOS is measured under various listening conditions including languages, cultural backgrounds, and listening subjects. Listening tests with different conditions would be expected to result in different performance.

In a clean (error-free) channel, the CRC and channel code are redundant as the bits are not lost during transmission, and any degradation of quality comes from the speech compression. In these ideal conditions, the higher codec modes of AMR naturally achieve higher MOS values, as the total bit-rate is fixed and more bits are assigned for the speech data in these modes. EFR and AMR 12.2 kbps achieve the same scores of 4.01 while the 4.75 kbps results in a score of 3.50. Notice that the 10.2 kbps achieves a higher MOS value than the 12.2 kbps in this test, although in many cases the 12.2 kbps would outperform the 10.2 kbps at this ideal channel condition. Therefore a MOS difference of 0.05 in the benign channel conditions should not be considered statistically significant. It is shown that at a CIR of 13 dB, EFR and AMR reach the error-free status and the speech quality saturates, i.e., does not improve further, for the CIR values beyond this level.

On the other hand, in the worst channel conditions corresponding to CIR=1 dB, degradation of quality comes from the speech compression to some extent but to a greater extent from the transmission. Under these hostile channel conditions, the 4.75, 5.15, and 5.9 kbps still maintain MOS values higher than 2, exceeding the other higher codec modes of AMR or EFR. This can be explained because the degradation of speech quality from the transmission is much larger than that from the compression. As a result, the lower codec modes, i.e., codec modes with higher bit-rates for the CRC and channel code, result in higher speech quality, even if lower bit-rates are assigned for the speech.

In these situations, the perceived quality of EFR, and the 12.2 and 10.2 kbps of AMR fall to unrecognizable levels, rendering the measurement of MOS meaningless. Although EFR and the 12.2 kbps use identical speech compression algorithms, they exhibit slightly different performance when the channel is not clean since different channel coding procedures are applied. It has been shown that the 12.2 kbps of AMR outperforms EFR in most channel conditions except at CIR= 4 dB.

In the intermediate channel conditions, for example, at CIR=7 and 10 dB, the codec modes which are more balanced in spending their bit-rate on both the speech compression and the channel coding outperform the codec modes which excel in either the error-free or the lowest CIR conditions. This can be understood since in these more practical situations, the intermediate codec modes maintain a better balance in the bit-rate allocation over the speech, CRC, and channel code so that the overall distortion occurring in the entire end-to-end paths is minimized. For example, at CIR=10 dB, the 10.2 and 7.95 kbps outperform EFR, and the 12.2, 7.5, and 6.7 kbps. At CIR=7 dB, the 10.2, 7.95, 7.4, and 6.7 kbps achieve higher MOS values than the modes with higher or lower speech bit-rates.

It is shown that the four highest codec modes, the 12.2, 10.2, 7.95, and 7.4 kbps, provide similar speech quality as EFR in the error-free conditions, but outperform EFR when the channel condition deteriorates until CIR=7 dB. The speech quality of the codec modes drops rapidly once the CIR falls below 7 dB. The three lowest codec modes, the 4.75, 5.15, and 5.9 kbps, remain more resilient against transmission errors and correspondingly stay more competitive until CIR= 4 dB, beyond which they cannot match the codec modes with higher bit-rates for the speech.

Table 3.22 Test results (MOS) of EFR and AMR envelope in full-rate stationary channel.

CIR (dB)	1	4	7	10	13	16	Clean channel
EFR		1.53	3.05	3.65	4.01		4.01
AMR envelope	2.66	3.59	3.96	4.08	4.13	4.06	4.06

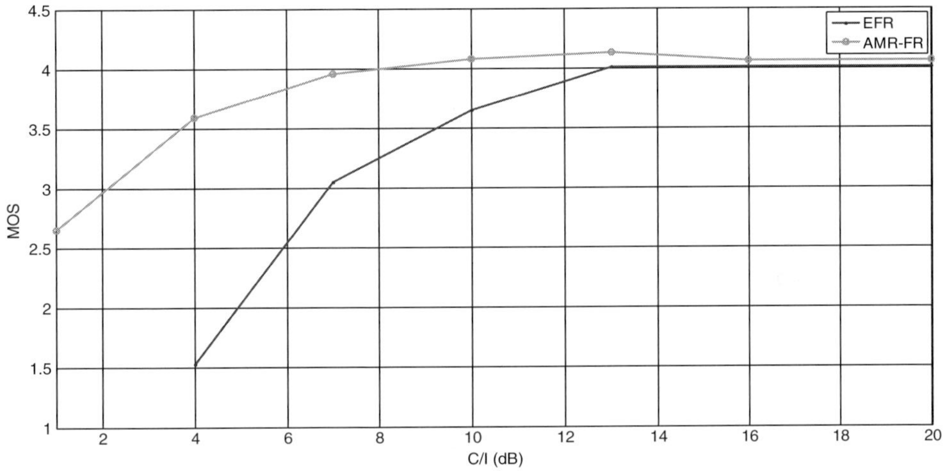

Fig. 3.23 Test results of EFR and AMR envelope in full-rate stationary channel.

Table 3.22 and Fig. 3.23 compare the envelope generated by connecting the MOS values of the best performing AMR codec mode at each CIR value, with the performance curve of EFR. The comparison partially proves the feasibility of the AMR concept visualized in Fig. 3.8, since the synthetic curve of the AMR envelope is located above that of EFR in most channel conditions. However, the AMR concept would be fully justified only when the bit allocation can be switched as optimally under time-varying channel conditions as envisioned with the results under stationary channel conditions.

Half-Rate Channel under Stationary Channel Conditions

A similar evaluation of the fixed bit-rate speech codecs or AMR codec modes can be made for the half-rate channels. Table 3.23 and Fig. 3.24 show the testing results of HR for TCH/HS, and the six lower codec modes of AMR for TCH/AHS. For a performance comparison, the results of FR and EFR are included in the test using their respective full-rate channels.

Under error-free conditions, speech quality of the 7.95 kbps exceeds that of FR by a large margin, even approaching the performance of EFR. In the worst channel condition with CIR= 4 dB, the 4.75 kbps outperforms not only other higher modes or HR but also FR and EFR. It is notable that the 4.75 kbps consistently outperforms FR and HR when the CIR is less than 13 dB. On the other hand, the 7.95 kbps, which receives the lowest

Table 3.23 Test results (MOS) of FR, HR, EFR, and AMR in half-rate stationary channel.

CIR (dB)	4	7	10	13	16	19	Clean channel
EFR	1.58	3.34	3.74	4.21			4.21
AMR 7.95		1.6	2.53	3.37	3.96	4.04	4.11
AMR 7.40		1.78	2.74	3.52	3.95	3.93	3.93
AMR 6.70	1.21	2.22	3.10	3.53	3.90		3.94
AMR 5.90	1.33	2.57	3.19	3.72	3.82		3.68
AMR 5.15	1.84	2.85	3.38	3.60	3.60		3.70
AMR 4.75	2.00	3.10	3.30	3.42	3.46		3.59
FR	1.50	2.74	3.14	3.50			3.50
HR	1.92	2.80	3.24				3.35

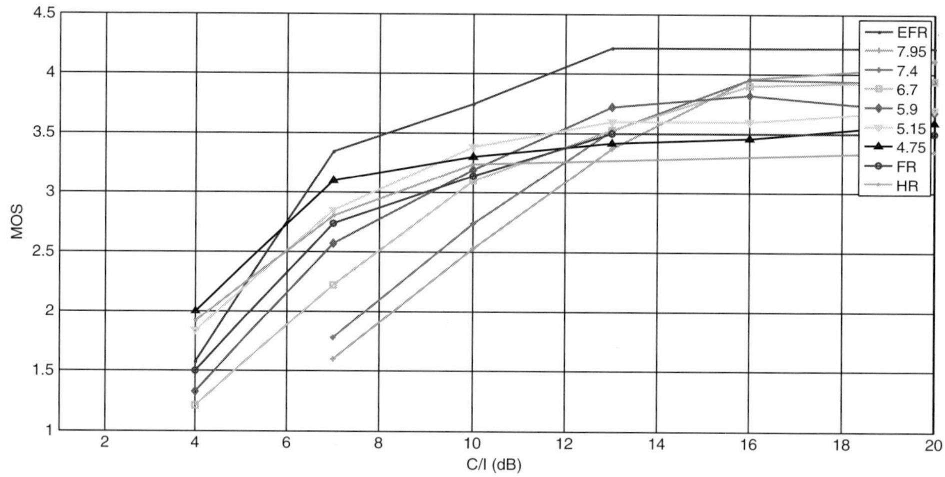

Fig. 3.24 Test results of FR, HR, EFR, and AMR in half-rate stationary channel.

level of error protection of the six AMR codec modes, fails to provide a perceivable speech quality at CIR = 4 dB.

As in the case of the full-rate channels, when the channel conditions are neither too benign nor too hostile, e.g., at CIR=13 and 16 dB, the intermediate codec modes outperform the extreme codec modes. At CIR=13 dB, the 5.9 kbps outperforms all the other AMR codec modes, in addition to FR and HR. Likewise, at CIR=10 dB, the 5.15 kbps achieves an MOS value that is higher than those of other codec modes with higher or lower speech bit-rates as well as FR and HR. The results clearly show that the bits are partitioned with enough differentiation, and each codec mode of AMR is better than the other codec modes at certain channel conditions, for both the full-rate and half-rate channels.

Table 3.24 and Fig. 3.25 compare the envelope generated by connecting the MOS values of the best performing AMR codec mode at each CIR value with the performance curves of EFR, FR, and HR. The synthetic curve of the AMR envelope is located above those of FR and HR under most channel conditions. Whether the channel frequency is

Table 3.24 Test results (MOS) of FR, HR, EFR, and AMR envelope in half-rate stationary channel.

CIR (dB)	4	7	10	13	16	19	Clean channel
EFR	1.58	3.34	3.74	4.21			4.21
FR	1.50	2.74	3.14	3.50			3.50
HR	1.92	2.80	3.24				3.35
AMR envelope	2.00	3.10	3.38	3.72	3.96	4.04	4.11

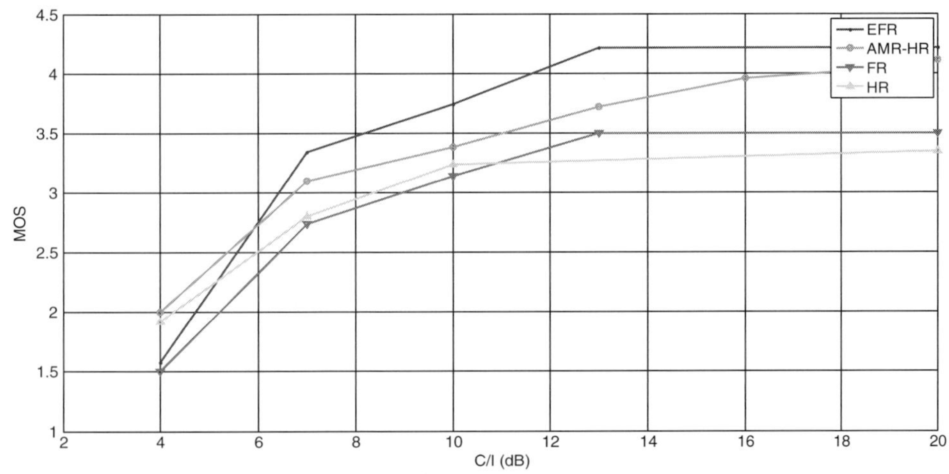

Fig. 3.25 Test results of FR, HR, EFR, and AMR envelope in half-rate stationary channel.

hopped or not, the channel condition in GSM is highly dynamic and the gains achieved by AMR in TCH/AHS at fixed CIR should be confirmed through the switching of the codec modes under time-varying channel conditions. The experimental results under the stationary channel conditions for the full-rate and half-rate channels can be taken into account in the determination of the thresholds and hysteresis parameters for the AMR operation.

It is necessary to estimate the expected condition of the next channel, in the same or different frequency, and execute the signaling and switching within a time limit. In [3GPP (2017h)], FIR filters of orders 100 and 50 are introduced for the full-rate and half-rate channels respectively. These filters smooth the input CIR values and output estimates of the expected CIR values at the time instant for which the link adaptation operation, i.e., the codec mode selection, becomes effective. The estimates are compared with the thresholds and mapped to the CMR bits.

Full-Rate Channel under Dynamic Channel Conditions

To evaluate the codec mode adaptation performance of AMR in TCH/AFS, three sets of codec modes are defined, which differ in their number of codec modes and the bit-rate range for adaptation. Mode set (a) consists of the 12.2, 7.95, and 5.9 kbps while mode set (b) consists of the 12.2 and 7.95 kbps. Mode set (c) includes the maximum number

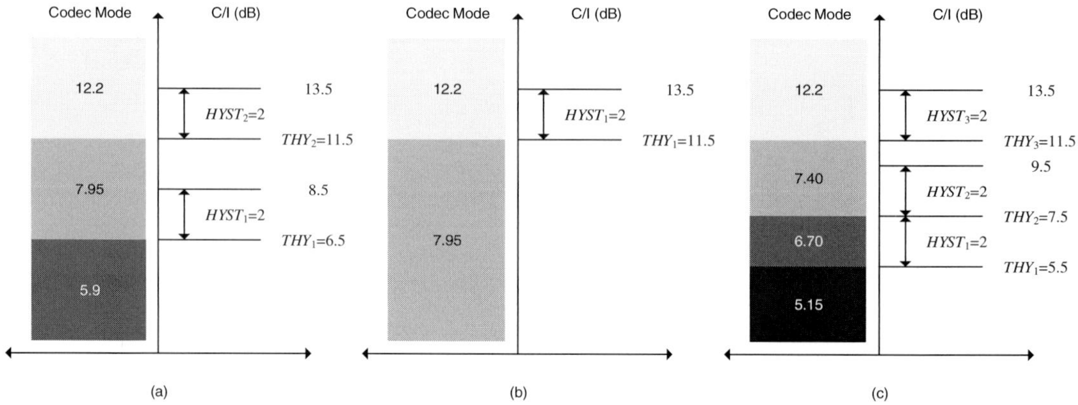

Fig. 3.26 AMR configurations in full-rate dynamic channel.

of codec modes: 12.2, 7.40, 6.70, and 5.15 kbps. For each set containing n codec modes, $n-1$ thresholds and $n-1$ hysteresis parameters are defined, respectively. Note that the thresholds need to be distanced enough from each other, and an identical hysteresis parameter can be used for multiple thresholds. In mode set (a), $THY_1 = 6.5$ dB, $THY_2 = 11.5$ dB, and $HYST_1 = HYST_2 = 2$ dB. In mode set (b), $THY_1 = 11.5$ dB and $HYST_1 = 2$ dB. In mode set (c), $THY_1 = 5.5$ dB, $THY_2 = 7.5$ dB, and $THY_3 = 11.5$ dB. An identical hysteresis parameter, $HYST_1 = HYST_2 = HYST_3 = 2$ dB, is applied to all three thresholds of mode set (c). Figure 3.26 illustrates the thresholds and hysteresis parameters defined for each codec mode set.

To update the parameters of uplink codec mode adaptation, mode sets (a) and (b) require the AMR_CONFIG_REQ message type, shown in Table 3.11, while mode set (c) requires the message type in Table 3.12. By comparing the adaptation performance of the codec mode sets under dynamic channel conditions where a fixed CIR is no longer maintained, it can be confirmed whether more codec modes contribute to higher speech quality, or whether quality improvement from codec mode adaptation saturates once the number of modes exceeds a certain value. Such dynamic channel conditions can be generated by simulating different behavior of radio channels and slow fading effects. An appropriate signaling delay between channel estimation and codec mode switching is assumed.

Figure 3.27 compares the testing results of the three codec mode sets of AMR in TCH/AFS and EFR in TCH/EFS. Unlike the stationary channel conditions shown in Figures 3.22 and 3.24, each channel type of Fig. 3.27 exhibits time-varying CIR conditions that can statistically trigger the switching of codec modes. An identical channel type is used in CH5, CH6, CH6+DTX, and CH7 to generate multiple error patterns representative of diverse frequency hopping operation modes: ideal frequency hopping, non-ideal frequency hopping, non-ideal frequency hopping with DTX, and no frequency hopping. Non-ideal frequency hopping, in which the channel frequency is changed among four statistically-related channels, refers to the situations where statistical properties of channel frequencies are not completely independent. Such a non-ideal frequency hopping

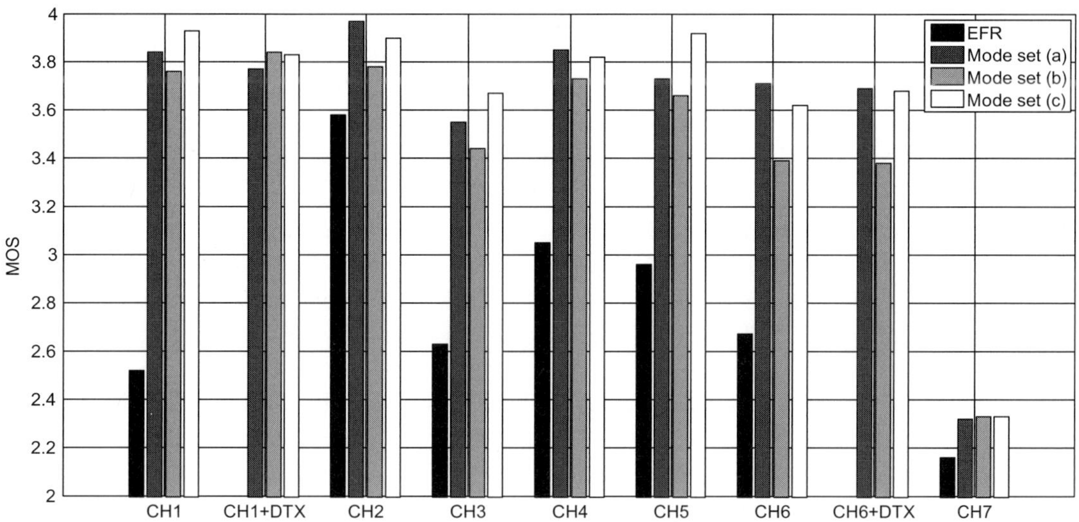

Fig. 3.27 Test results of EFR and AMR in full-rate dynamic channel.

can occur when the new channel frequency is still within the influence of fading that covered the previous frequency. The effects of DTX on the speech quality while codec mode adaptation is activated are evaluated in CH1 and CH6.

The test results of CH1, CH2, CH3, and CH4 show that the codec mode adaptation of AMR consistently outperforms EFR. In some cases, the quality difference in MOS is even larger than 1, which is significant. It can be seen that the speech quality of codec mode sets with three and four modes is consistently higher than that with only two modes, but the difference is relatively small. In addition, little gain is observed when the number of codec modes is increased from three to four. Therefore it is reasonable to hypothesize that as long as the codec mode sets include at least two modes and share the same highest bit-rate, the 12.2 kbps, the difference in speech quality from codec mode adaptation is not clearly noticeable.

Notice that both DTX and codec mode adaptation interrupt the operation of the speech encoder, albeit in different ways, and may send conflicting requests to the speech encoder simultaneously. For example, the CMC from a BTS may command the speech encoder of an MS whose link quality is favorable to increase the speech bit-rate, but the absence of voice activity may drive the bit-rate down to a negligible level, allowing only periodic transmission of SIDs. In such cases, the local decisions on DTX will always override the feedback-based action of codec mode adaptation on the speech encoder, whether the incoming signal is CMC or CMR. Test results of CH1 and CH6, with and without DTX, show that the impact of using DTX is negligible even if the bit-rate of the speech encoder fluctuates more abruptly when DTX is activated.

It is seen that the use of non-ideal frequency hopping in CH6 reduces the quality of EFR and the codec mode sets with two and four modes more than that of the codec mode set with three modes. Finally, the contribution of slow frequency hopping to the performance of codec mode adaptation is clarified since without frequency hopping, the quality of EFR and the three codec mode sets is reduced to unacceptable levels. Therefore it can be concluded that in TCH/AFS, codec mode adaptation and frequency hopping contribute significantly to the perceived speech quality while DTX or the correlation of channels has a smaller impact. Considering the reduction of the average bit-rate and the corresponding interference, it is advantageous to employ DTX in any case.

Half-Rate Channel under Dynamic Channel Conditions

As in the case of TCH/AFS, three sets of codec modes are defined to evaluate the codec mode adaptation performance of AMR in TCH/AHS. Mode set (a) consists of the 7.95, 6.7, 5.9, and 5.15 kbps while mode set (b) consists of the 6.7, 5.9, and 4.75 kbps. Mode set (c) includes the 7.40 and 5.15 kbps. For each set with n codec modes, $n-1$ thresholds and $n-1$ hysteresis parameters are defined. In mode set (a), $THY_1 = 11.0$ dB, $THY_2 = 12.5$ dB, and $THY_3 = 15.0$ dB, $HYST_1 = 2$ dB, $HYST_2 = 2.5$ dB, and $HYST_3 = 2$ dB. In mode set (b), $THY_1 = 11$ dB, $THY_2 = 12.5$ dB, and $HYST_1 = HYST_2 = 2$ dB. In mode set (c), $THY_1 = 13.5$ dB and $HYST_1 = 2$ dB. Figure 3.28 illustrates the thresholds and hysteresis parameters defined for each codec mode set. Note that shorter bit-rate ranges are used than in TCH/AFS.

A set of dynamic channel conditions is generated by simulating different behaviors of the radio channels and slow fading effects. An identical channel type is used in CH5, CH6, CH6+DTX, and CH7, to generate multiple error patterns representative of diverse frequency hopping operation modes: ideal frequency hopping, non-ideal frequency hopping, non-ideal frequency hopping with DTX, and no frequency hopping. Non-ideal frequency hopping is limited to four channel frequencies. The effects of DTX are evaluated in CH1 and CH6. Compared with the evaluations in TCH/AFS, 3 dB higher CIR is applied to CH1, CH3, CH5, CH6, and CH7, and 6 dB higher CIR is applied to CH2 and CH4, respectively, to compensate for the lower error resilience of the half-rate channels. As TCH/AHS is deployed in an area inside each cell much smaller than the cell boundary covered by TCH/AFS, an ongoing call on TCH/AHS is often handed over to another channel of the BTS currently connected, i.e., handled as an intra-cell handover.

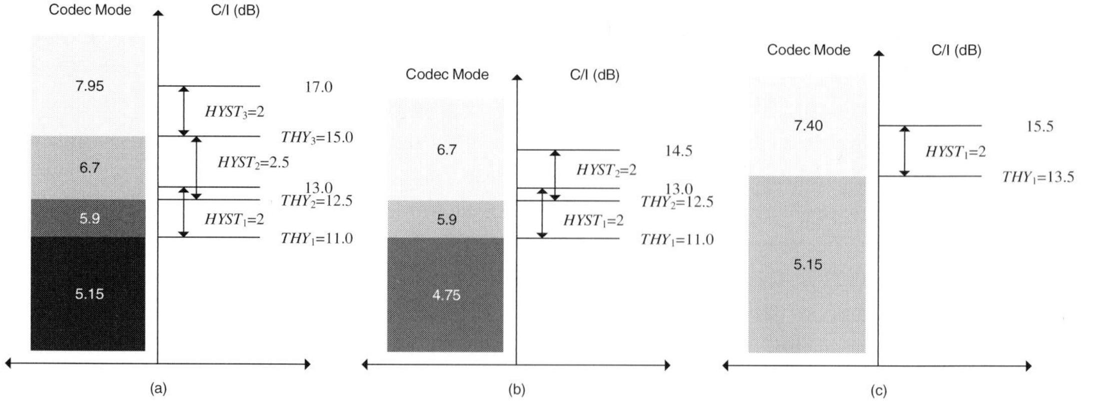

Fig. 3.28 AMR configurations in half-rate dynamic channel.

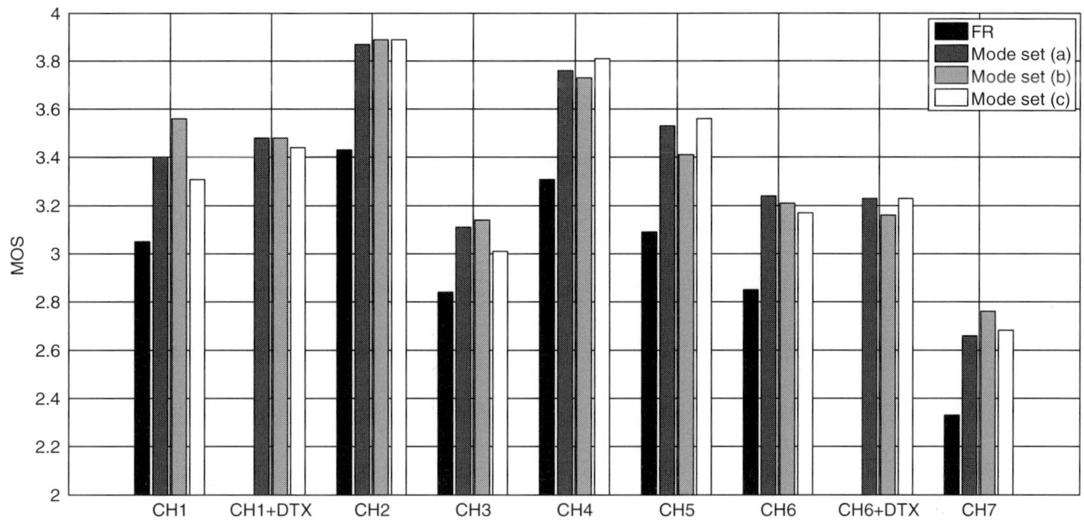

Fig. 3.29 Test results of FR and AMR in half-rate dynamic channel.

Figure 3.29 shows the testing results of the three codec mode sets of AMR in TCH/AHS and FR in TCH/FS. From the testing results in stationary channel conditions outlined in Table 3.24, it is anticipated that AMR in TCH/AHS, with codec mode adaptation, would outperform FR and HR but not exceed EFR. The quality of FR is considered relatively close to that of AMR in TCH/AHS, which is analyzed in further detail in the testings under dynamic channel conditions and frequency hopping. The testing results of CH1, CH2, CH3, and CH4 show that the codec mode adaptation of AMR in TCH/AHS consistently outperforms FR, albeit using only a half of the TDMA frames. As in the testings for AMR in TCH/AFS, the number of codec modes does not have a noticeable effect on the performance. Likewise, DTX has little impact on the performance of AMR in TCH/AHS. Note that in contrast to the case of TCH/AFS, in the tests of TCH/AHS, each codec mode set not only has a different number of codec modes but also has a different maximum bit-rate, which can influence the performance especially in favorable channel conditions where the speech bit-rate determines the quality.

It can be seen that non-ideal frequency hopping reduces the quality of AMR in TCH/AHS and FR in TCH/FS, which is further degraded if the channel frequency is not hopped, as in CH7. From the testing results, as in the case of AMR in TCH/AFS, it can be concluded that codec mode adaptation and frequency hopping contribute significantly to the perceived speech quality while DTX or correlated channels has a smaller impact. Note that DTX contributes to both quality and capacity indirectly, by reducing the interference and saving time slots. It is also shown that AMR in TCH/AHS requires a higher CIR than in TCH/AFS for a similar quality, matching the situations where higher network capacity is required but higher CIR is fortunately available. As an example of such situations, MSs located near a BTS at a busy hour may be served with AMR in TCH/AHS, saving network capacity without compromising speech quality. Detailed information on the testing procedures and performances of AMR in TCH/AFS and TCH/AHS can be found in [3GPP (2017a)].

3.4.2 Live Call Analysis

Figure 3.30, consisting of six plots, shows the time-varying characteristics of the key parameters measured during a live GSM call using AMR in its full-rate mode. The first and second plots show the codec modes of the downlink and uplink, consisting of 4.75, 5.90, 7.40, and 12.2 kbps in the ACS. DTX and those periods where information is not available are represented by not plotting the codec mode. It can be seen that in either link, the initial codec mode (ICM) is set to the highest mode of the ACS, 12.2 kbps. As the link quality deteriorates, the codec mode gradually switches down to save bits in order to protect the speech more. Compared with the adaptation trajectory of the uplink, where the neighborhood-only transition policy is faithfully followed and the codec mode does not increase or decrease by more than one level in the ACS per transition, in the downlink the BTS sometimes adjusts the codec mode to non-neighboring modes in a single step.

With the strict master–slave relationship between the MS and the BTS, it is clear that the BTS can override the codec mode request from the MS when its estimate of the expected channel condition or speech quality differs from that of the MS. In the plots, the estimated CIR values indicate a disturbance of link quality. This can be seen in the continuous reduction of the received signal power and its level, RXLEV, shown in the third and fifth plots, respectively. However, the received signal quality, RXQUAL, is maintained at relatively high values, which might be because of the capability of the receiver to maintain the CIR when the received signal level decreases. It can also be seen that the BTS asked the MS to advance the transmission time by three bit periods. Notice that the frequency of codec mode switching is low when the MS is stationary, and also that the frequency of issuing timing advance commands is lower than that of other control signaling.

3.4.3 VAMOS Operation

In the previous sections, performance of the speech and radio signal processing was expressed using the MOS values, as the analysis was focused on the speech quality over the wireless channel. It was shown that speech compression and transmission using EFR and AMR outperform that using FR, with higher MOS scores at similar CIRs or lower CIRs at similar MOSs. The approaches targeting higher network capacity rather than higher speech quality, such as HR or TCH/AHS, were also studied based on similar criteria. In the following analysis of VAMOS, the performance is represented with an emphasis on the network capacity, by comparing the number of MSs with satisfactory quality per unit frequency spectrum per transmitter and receiver (transceiver). Notice that each BTS typically has multiple, e.g., three or six, transceivers, each covering 120 or 60 degrees.

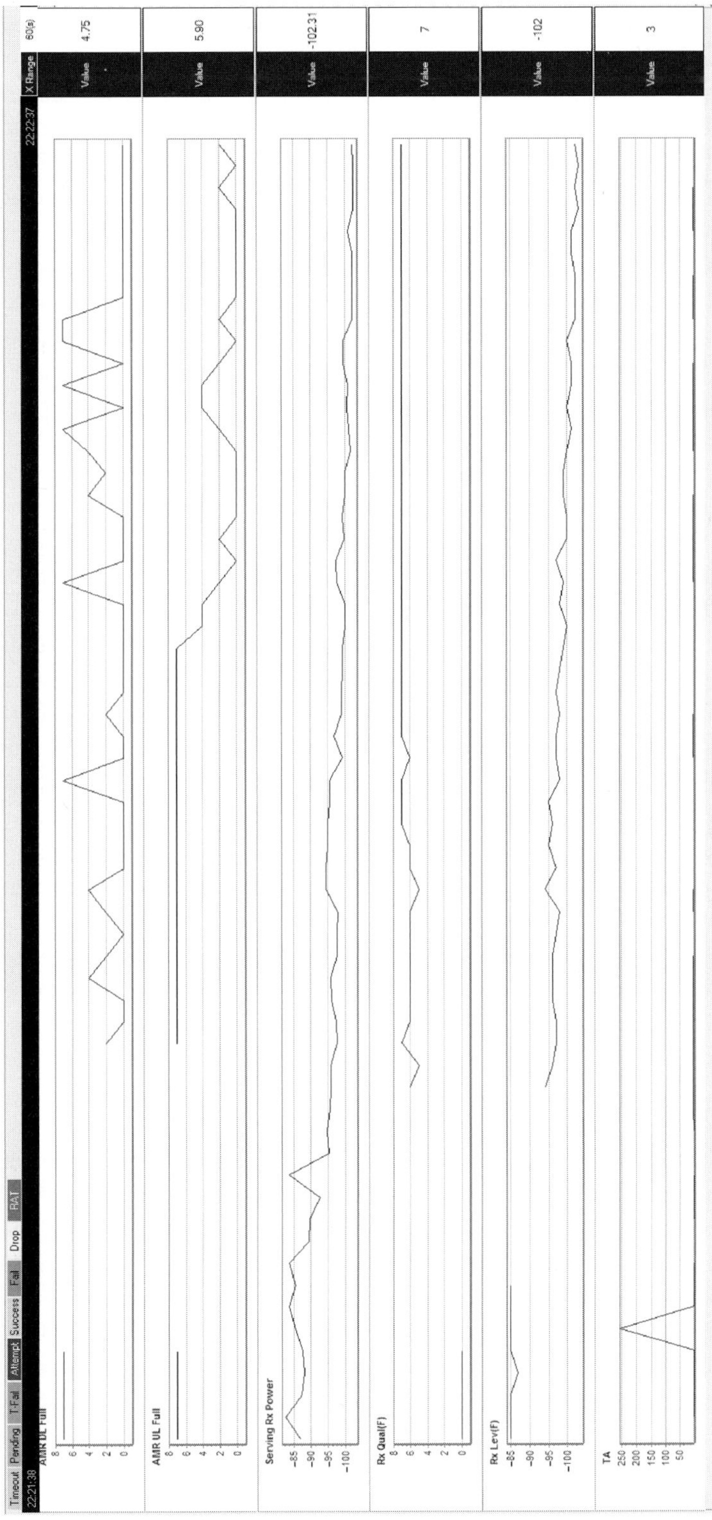

Fig. 3.30 GSM-AMR call analysis (courtesy of Innowireless Co., Ltd.).

In circuit-switched networks, the voice capacity is typically measured in *Erlangs*, a dimensionless quantity defined as the call arrival rate multiplied by the call duration for a unit of communications links or a trunk, based on the classic teletraffic theory [Lee and Miller (1998)]. The links are assigned to calls whose duration and occurrence are set using some statistical models. The number of call arrivals is typically assumed to be a Poisson process while the call duration is assumed to be a random variable whose statistical properties follow an exponential distribution. For economic reasons, fewer links are typically available than the maximum number of call arrivals expected during the busiest time. Figure 3.3 illustrates such times during which using HR in TCH/HS can temporarily boost the network capacity. Then the Erlang capacity is estimated as the maximum usage of the communications links while the probability of blocking incoming call requests is maintained within an acceptable level.

It is often observed that increasing the number of links by a certain amount results in a larger gain in the Erlang capacity, a phenomenon called *trunking efficiency*. There are several formulas defined to compute the Erlang capacity, based on the strategy to deal with calls that cannot be served for the lack of capacity, e.g., dropping the excessive calls and assuming the calls are not re-established, or putting the calls into a waiting list assuming optimistically that the caller is patient enough to wait indefinitely. As a capacity metric for mobile communications systems, a key limitation of the Erlang capacity is that it does not take into account the possibility of call drop due to link failure, whose occurrence ratio is no less than that of capacity shortage in mobile communications. Historically, the Erlang capacity was formulated for the PSTN where link failure rarely happens. In mobile communications systems, as the number of calls in progress or the cell loading level elevates, so does the interference level, making the wireless links unstable. In the TDMA or other multiple access systems with a fixed channel width, the trunking efficiency is suppressed by the interference in the wireless links and the limited number of control channels such that the capacity is proportional to the number of channels. Therefore, in this text the voice capacity of mobile communications systems is defined as the maximum number of calls per unit bandwidth while the call block ratio, call drop ratio, and (speech) frame error ratio are kept within acceptable ranges.

DARP Phase 1 Performance

Table 3.25 compares the required CIR values for several combinations of pulse types and training sequences, with a target speech quality of FER = 1%. The reception performance of an MS is evaluated as the additional CIR in comparison with the case when a time slot is filled by an MS using GMSK. It is assumed that the two MSs in the VAMOS operation are of the same type. It can be seen that a higher CIR is inevitably necessary in VAMOS, although the receivers meet the DARP Phase 1 requirements. However, the loss in link quality can be reduced with less-correlated training sequences. A larger gain can be achieved by switching the pulse, from LGP to RRC with a width of 270 kHz.

Test Conditions

Table 3.26 summarizes the channel modes used in the estimation of the downlink voice capacity when VAMOS is employed. The estimated numbers of MSs or channels per

Table 3.25 Link performance and loss to reference at AMR 5.9 kbps for TCH/AFS (DARP Phase 1).

Configuration	CIR for FER= 1%(dB)	Loss to reference (dB)
GMSK	1.9	
RRC 270 kHz, TSC 2 pair	5.1	3.2
RRC 270 kHz, TSC 1 pair (lowest correlation)	5.4	3.5
LGP, TSC 2 pair	7.0	5.1
LGP, TSC 1 pair (highest correlation)	7.9	6.0

Table 3.26 Channel mode definitions.

A0	GSM HR
A1	GSM HR $\Longleftrightarrow$ MUROS HR
B0	AFS 12.2
B1	AFS 12.2 $\Longleftrightarrow$ MUROS AFS 12.2
C0	AFS 5.9
C1	AFS 5.9 $\Longleftrightarrow$ MUROS AFS 5.9
D0	AHS 5.9
D1	AHS 5.9 $\Longleftrightarrow$ MUROS AHS 5.9

unit bandwidth per transceiver, in MSs/MHz/TRX, are compared for cases with and without VAMOS. Three types of MSs are considered in the evaluations: conventional MS, DARP Phase 1 MS, and VAMOS-enabled MS. The ITU-T Terrestrial Urban (TU) environments, an MS speed of 3 km/h, and single antenna of MS are assumed in the system simulations. AMR codec mode adaptation is activated whenever available. The voice calls are generated based on Poisson distributions where the call arrival rate is set based on the cell loading level and the mean call duration is set to 90 seconds with a minimum call duration of 5 seconds.

Detailed information on the simulation conditions can be found in [3GPP (2017c)]. Note that in Table 3.27, an interim term Multi-User Re-using One Slot (MUROS), a term used during the development of VAMOS, is used. Unlike in the MOS measurements of codec performance, the speech is not actually encoded and decoded in this type of system-level capacity estimation simulating the dynamic operation of multiple cells. Instead, the number of MSs which meet certain objective metrics is counted. An MS is considered to be in satisfactory service status if the following criteria are met simultaneously:

Criterion 1
Fewer than 2% of calls should be blocked.
Criterion 2
In the case of TCH/AFS, the FER should be less than 2% for at least 95% of the MSs. However, in the case of TCH/AHS, the FER should be less than 3% for at least 95% of MSs, as TCH/AHS is typically employed when the cell is highly loaded and the level of interference is high.

Table 3.27 Network conditions for performance evaluation.

	MUROS-1	MUROS-2	MUROS-3 (a)	MUROS-3 (b)
Frequency band (MHz)	900	900	1800	1800
Cell radius (m)	500	500	500	500
Bandwidth (MHz)	4.4	11.6	2.6	2.6
Guard band (MHz)	0.2	0.2	0.2	0.2
Number of channels (without guard band)	21	57	12	12
Number of TRX	4	6	4	4
BCCH frequency re-use factor	4/12	4/12		
TCH frequency re-use factor	1/1	3/9	1/3	1/1
Frequency hopping	Synthesized	Baseband	Synthesized	Synthesized
Length of MA (number of FH frequencies)	9	5	4	4
Fast fading type	TU	TU	TU	TU
BCCH or TCH under interest	Both	Both TCH	TCH	
Network sync mode	Synchronized	Synchronized	Synchronized	Synchronized

Criterion 3
When the VAMOS is used, the relative performance of the associated signalling channels compared against the traffic channel, as derived in the link performance evaluation for a reference scenario, shall be maintained.

Although essential in the performance evaluation of mobile communications systems, the criterion of call drop ratio is not considered, as the third criterion assumes sufficient performance of the associated signaling channels. However, failure of the signaling channels may occur from insufficient transmit power or the absence of radio resources for calls handed over, which may result in dropped calls. The conditions of the GSM networks used in the capacity estimation of VAMOS are outlined in Table 3.27. Although the codec mode adaptation of AMR is not used, the channel mode can be adapted between VAMOS and non-VAMOS, based on the cell loading level and the measured link quality. In the VAMOS mode, the quality degradation from increased correlation of training sequences is not considered as it is assumed that a pair of sequences sufficiently uncorrelated is assigned to two MSs sharing time slots.

Voice Capacity

Table 3.28 compares the estimated spectral efficiency, defined as the maximum number of MSs per MHz per TRX of each channel mode in Table 3.26. It is assumed that all MSs in the cells are of the same type. However, not all MSs are in the same mode during the evaluation, as a VAMOS-enabled MS supports the conventional GMSK mode, DARP

Table 3.28 Capacity enhancements (MSs/MHz/TRX).

Channel mode	MUROS-1	MUROS-2	MUROS-3 (a)	MUROS-3 (b)
A0	36.11	21.16	73.54	73.13
A1	44.79	37.71	59.28	91.93
B0	15.46	9.64	32.86	32.76
B1	17.31	12.67	34.80	37.92
C0	15.4	9.66	32.85	32.88
C1	26.18	20.74	41.72	51.10
D0	36.13	21.14	73.01	72.97
D1	36.21	24.50	72.19	77.67

Phase 1, and VAMOS. The network switches the channel type between GMSK and VAMOS in channel modes A1, B1, C1, and D1. In the MUROS-1 condition, spectral efficiency is increased when VAMOS is enabled. It can be seen that noticeable gains are achieved when VAMOS is applied to HR and AMR 5.9 kbps in TCH/AFS. As the 5.9 kbps in TCH/AFS spends only 118 bits out of 456 bits for the speech, more than enough bits are used for the CRC and channel code. Sharing a time slot can be facilitated when the error resilience of data or the CIR is high.

The spectral efficiency of HR in TCH/HS is already quite high, as 112 bits out of 228 bits are used for the speech. However, HR requires a high CIR to meet the quality criteria, due to its insufficient error resilience, as shown in Table 3.23. Two VAMOS half-rate slots can be packed into one half-rate slot under such conditions. Therefore, VAMOS can be applied to channel modes with inefficient bit-rate allocations for the speech, CRC, and the channel code, which are unfit for the situations. Notice that the spectral efficiency is not doubled even in the case of the 5.9 kbps in TCH/AFS, which may be due to the shortage of control channels or the fallback of some VAMOS slots to TCH/AFS as the time slots cannot hold two simultaneous speech sub-channels steadily.

As the 12.2 kbps in TCH/AFS and the 5.9 kbps in TCH/AHS are quite efficient in terms of their minimum required CIR, the gain from applying VAMOS is not significant. A key difference between VAMOS and non-VAMOS modes is the factor that limits the capacity. In the non-VAMOS mode, the spectral efficiency cannot be increased further as the first criterion is not met where the network starts blocking more calls than allowed while the quality of ongoing calls is satisfactory. In the VAMOS mode, the capacity is limited by link failure, as allowing further calls renders the quality of new or ongoing calls unsatisfactory. Similar observations can be made in the other three network configurations. Although the performance of VAMOS is estimated assuming that all MSs in the cell are of an identical type, this will likely not be true in practical network deployments. As VAMOS-enabled MSs are gradually introduced into the network, the capacity will be gradually increased. Even before the proportion of VAMOS-enabled MSs becomes significant, its effects will be exhibited in the busy hour situations. A set of capacity estimations for mixed MS type situations can be found in [3GPP (2017c)].

3.5 References

3GPP. 2000a. TS 06.20 V8.0.1 Half Rate Speech Transcoding. November.

3GPP. 2000b. TS 06.22 V8.0.1 Comfort Noise Aspects for Half Rate Speech Traffic Channels. November.

3GPP. 2000c. TS 06.60 V8.0.1 Enhanced Full Rate Speech Transcoding. November.

3GPP. 2000d. TS 06.61 V8.0.1 Substitution and Muting of Lost Frames for Enhanced Full Rate Speech Traffic Channels. November.

3GPP. 2000e. TS 06.93 V7.5.0 Discontinuous Transmission (DTX) for Adaptive Multi-Rate Speech Traffic Channels. December.

3GPP. 2017a. TR 26.975 V14.0.0 Performance Characterization of the Adaptive Multi-Rate (AMR) Speech Codec. March.

3GPP. 2017b. TR 45.903 V14.0.0 Feasibility Study on Single Antenna Interference Cancellation (SAIC) for GSM Networks. March.

3GPP. 2017c. TR 45.914 V14.0.0 Circuit Switched Voice Capacity Evolution for GSM/EDGE Radio Access Network (GERAN). March.

3GPP. 2017d. TS 26.090 V14.0.0 Mandatory Speech Codec Speech Processing Functions; Adaptive Multi-Rate (AMR) Speech Codec; Transcoding Functions. March.

3GPP. 2017e. TS 28.062 V14.0.0 Inband Tandem Free Operation (TFO) of Speech Codecs; Service Description; Stage 3. March.

3GPP. 2017f. TS 45.004 V14.0.0 GSM/EDGE Modulation. March.

3GPP. 2017g. TS 45.005 V14.0.0 GSM/EDGE Radio Transmission and Reception. March.

3GPP. 2017h. TS 45.009 V14.0.0 GSM/EDGE Link Adaptation. March.

3GPP. 2017i. TS 45.010 V14.0.0 GSM/EDGE Radio Subsystem Synchronization. March.

Bruhn, S., Blöcher, P., Hellwig, K., and Sjöberg, J. 1999. Concepts and Solutions for Link Adaptation and Inband Signaling for the GSM AMR Speech Coding Standard. *IEEE Vehicular Technology Conference*, **3**(May).

Corbun, O., Almgren, M., and Svanbro, K. 1998. Capacity and Speech Quality Aspects using Adaptive Multi-Rate (AMR). *The 9th IEEE International Symposium on Personal, Indoor and Mobile Radio Communications*, **3**(September).

Hagenauer, J. 1988. Rate-Compatible Punctured Convolutional Codes (RCPC Codes) and their Applications. *IEEE Transactions on Communications*, **36**(4).

Halonen, T., Romero, J., and Melero, J. 2007. *GSM, GPRS, and EDGE Performance: Evolution Towards 3G/UMTS*. 2nd edn. Wiley.

Krasnosel'skii, M. A., and Pokrovskii, Aleksei V. 1989. *Systems with Hysteresis*. Springer-Verlag.

Laurent, P. A. 1986. Exact and Approximate Construction of Digital Phase Modulation by Superposition of Amplitude Modulated Pulses (AMP). *IEEE Transactions on Communications*, **34**(2).

Lee, J. S., and Miller, L. E. 1998. *CDMA Systems Engineering Handbook*. Artech House Publishers.

Massey, J. L. 1978. Joint Source and Channel Coding. In: Skwirzynski, J. K. (ed), *Communication Systems and Random Process Theory*. Sijithoff and Nordhoff.

Mayergoyz, I. D. 2003. *Mathematical Models of Hysteresis and their Applications*. 2nd edn. Academic Press.

Redl, S., Weber, M., and Oliphant, M. W. 1998. *GSM and Personal Communications Handbook*. Artech House Publishers.

Shannon, C. E. 1948. A Mathematical Theory of Communication. *The Bell System Technical Journal*, **27**(July).

Uvliden, A., Bruhn, S., and Hagen, R. 1998. Adaptive Multi-Rate: A Speech Service Adapted to Cellular Radio Network Quality. *The 32nd Asilomar Conference on Signals, Systems and Computers*, November.

Vembu, S., Verdú, S., and Steinberg, Y. 1995. The Source-Channel Separation Theorem Revisited. *IEEE Transactions on Information Theory*, **41**(1).

4 Signal Processing in CDMA Systems

As GSM and its signal processing techniques based on the TDMA principles became dominant in mobile communications, new approaches using radically different types of signal processing started to emerge for higher network capacity. The key reasons for the success of TDMA were its heritage from analog systems, whose basic operation of planning cells and assigning channels could be re-used after the migration of networks, and relatively low requirements for complexity that were essential for the first generation of digital systems with limited capability for signal processing. We begin this chapter with a short discussion on the weakness of TDMA, which is followed by an outline of spread spectrum theory that has been developed to avoid these shortcomings. A brief tutorial on pseudo-noise theory is presented as the signal processing techniques to artificially create randomness are used in many areas of mobile communications systems. Then we discuss the speech and radio signal processing of a new Code Division Multiple Access (CDMA) mobile communications system, the Interim Standard 95 (IS-95).

4.1 TDMA Limitations

TDMA provided a straightforward evolutionary path from analog to digital mobile communications systems that worked well within the limited signal processing capability available at the time. However, the commercial success of TDMA systems, such as GSM, led to serious concerns about meeting the network demands that were greater than expected during their development. Reducing the number of time slots for each MS to a half of their original value was a common technique to double the capacity, as used in D-AMPS and GSM, but its impact on speech quality and service coverage limited the use of half-rate channels to urgent situations when the capacity was considered to be more important than the quality. In [Viterbi (1995), Ross and Gilhousen (1996)], the technical shortcomings of TDMA and FDMA, expected when further increase in network capacity was required, are outlined.

4.1.1 Guard Time and Guard Band

In the frame structure of D-AMPS, a time slot carries 324 bits of information, of which 260 bits are speech data and the other 64 bits are used for control and synchronization. Likewise, in GSM, only 114 bits out of each 156.25 bits sent in a slot are speech data. Reducing the proportion of control and synchronization bits increases the network capacity as long as the link quality does not suffer from the reduction. Another opportunity to exploit reduction of the guard time between time slots may also contribute to higher capacity. However, in TDMA, the guard time is related to the maximum cell radius, which cannot be reduced arbitrarily.

In the frequency domain, an opportunity for higher capacity lies in reducing the guard band inserted between two channels. Each channel frequency is separated from its neighboring frequencies by 30 kHz in D-AMPS and by 200 kHz in GSM. Although it can be very tricky to quantify the proportion of frequency spectrum for speech data to the overall channel bandwidth, increasing that proportion is likely to increase the network capacity. Therefore, it can be said that radio transmission techniques and multiple access schemes that differentiate the information of each MS with measures other than guard time or guard band have the potential for higher network capacity.

4.1.2 Fixed Bit-Rate Speech Coding

In TDMA, each MS is assigned a periodic series of time slots by the network. During the periods when voice activity is absent, the transmitter is turned off, and a small amount of information concerning the background noise level is transmitted to update the noise parameters used at the far-end. Because of the signaling delay taken for the interaction between the MS and the BSS, the time slots not used during the silence cannot be immediately re-assigned to other MSs.

Additional opportunities may be found in the speech analysis and synthesis procedures. More bit-rate levels than the two-level DTX operations of TDMA may be applied to speech compression based on a more refined acoustic analysis of the speech, e.g., as transient, unvoiced, or voiced. If the speech quality of this variable bit-rate (VBR) coding strategy is comparable to that of the fixed bit-rate (FBR) coding, such as AMR, but can be achieved at lower average bit-rates, this reduction may be exploited for higher network capacity.

A key requirement for VBR operation in mobile communications systems is that the radio signal processing is able to adapt to the fast variation of speech bit-rate without communicating with the network. If the operation needs an approval or a guide from the BTS or the BSC in the control of channel bit-rate or transmit power, the radio signal processing cannot track the variation fast enough, because of the round-trip and processing delay. As can be seen in the delay partitioning of GSM, shown in Tables 2.10 and 2.11, and the variation of speech signal, as shown in Fig. 2.3, the acoustic nature of a speech signal changes much faster than the interaction between the MS and the BSS. Note that even if they can interact with each other fast enough, an error in the control signaling can induce the collisions of slots from multiple MSs as the coordination of slow frequency hopping fails.

4.1.3 Frequency Re-Use Factor

In FDMA or TDMA, it is not possible for an MS to use all of the channels in a cell simultaneously, since a fixed number of channels have to be geographically allocated for multiple cells. The time slots assigned to a cell cannot be used by contiguous cells, as illustrated in Fig. 1.9 as hexagons, at the same time. In these situations, the frequency re-use factor, the number of cells around a cell which do not use the channels assigned to the center cell, should be greater than one. This condition requires careful planning in the determination of the cell-site locations and the number of channels that meet the expected traffic level of each cell. Therefore reducing the frequency re-use factor is likely to increase the network capacity, but this necessitates radio transmission techniques and multiple access schemes that are more resilient to the interference. In other words, to increase the network capacity, it is necessary to assign as many channels as possible to an MS or to a BTS when the situation requires. However, such assignments should not be made in a centralized fashion that requires prior arrangements that cannot handle time-varying situations.

4.1.4 Wideband Multipath Fading

Since multipath fading can induce rapid fluctuations in the received signal envelopes, transmission of information over narrowly-confined slices of the frequency spectrum can be susceptible to the influences of fading over a band that is wider than the information bandwidth. In GSM, measures against wideband multipath fading are more systematically provided than in D-AMPS. For example, a wider channel bandwidth of 200 kHz provides a higher resistance to fading than the 30 kHz bandwidth of D-AMPS. In addition, slow frequency hopping (SFH) enables the impact of fading to be spread over multiple MSs. However, as can be seen in the influence of non-ideal SFH on the performance of the AMR codec mode adaptation, switching channels in a cyclic or pseudo-random fashion has limitations. Considering the limited opportunity for channel estimation, as illustrated in Fig. 2.37, switching channels in an adaptive fashion will not be competitive either, due to an insufficient time window and the required control signals. A more fundamental approach may be to extend the information bandwidth to a very large extent so that typical wideband multipath fading impacts only a portion of the information bandwidth, which can be recovered by the receiver using redundancy sent with the information. To spread the information bandwidth, the symbol rate should be increased, for example, by multiplying a high-frequency signal by the information waveform. Such a wide information bandwidth should be shared by as many MSs as possible, to achieve higher capacity than TDMA.

4.2 CDMA Principles

Code Division Multiple Access (CDMA) is a class of radio access, i.e., allocation of frequency spectrum, techniques designed to provide more channel capacity than TDMA. Figure 4.1 illustrates the concept of CDMA and compares it with FDMA and TDMA, using a frequency-, time-, and power-domain representation. In the figure, a CDMA channel is seen to be as wide as several FDMA or TDMA channels. It uses neither a guard band nor a guard time to separate the information of each MS. Recall that in the

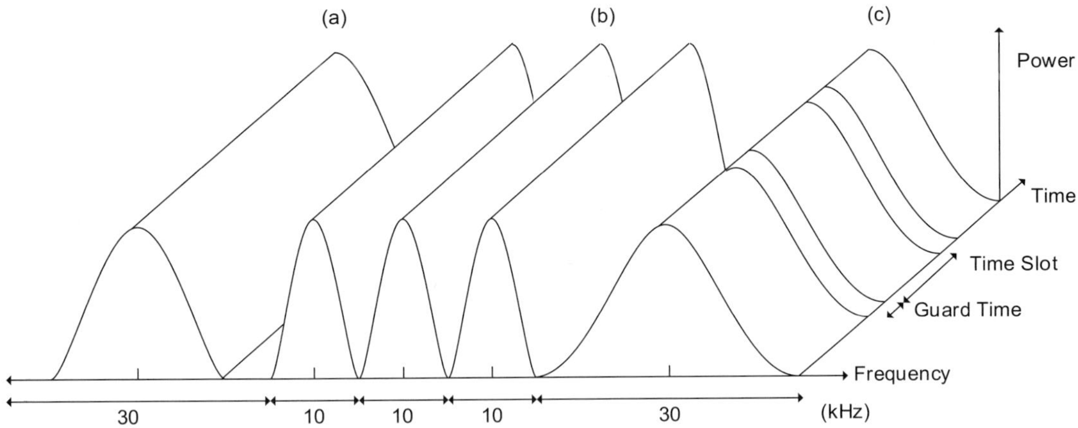

Fig. 4.1 Frequency-, time-, and power-domain representation. (a) FDMA. (b) TDMA. (c) CDMA.

VAMOS mode of GSM, the AQPSK-based orthogonality provides a mechanism for the advanced receiver to separate two sub-channels that overlap in the same time slots.

To provide orthogonality for a large number of MSs, CDMA uses a more systematic approach employing the Pseudo Noise (PN) sequences. These may be considered to be a more generalized form of the training sequences used in TDMA for synchronization. Although drawn in a comparable scale in Fig. 4.1, the transmit power of TDMA is generally much lower than that of FDMA. CDMA increases the network capacity by reducing transmit power even further but uses greater computational complexity for a wider spectrum.

4.2.1 Spread Spectrum Theory

CDMA is a member of a class of spread spectrum techniques which traditionally have been applied to military applications to counter eavesdropping or jamming [Simon *et al.* (1985)]. The slow frequency hopping of GSM is also a spread spectrum technique. In GSM, the information bandwidth is tied to the channel bandwidth, as each 200 kHz channel carries up to 270.83 kbps, corresponding to 1.35 bps/Hz. In this book, CDMA refers to the Direct Spread (DS) CDMA, which directly spreads the information bandwidth over a much wider channel bandwidth, e.g., larger than 1 MHz. Figures 4.2 and 4.3 illustrate the signal processing operations by which the information bandwidth is spread, transmitted, and recovered in the presence of additive white noise.

Figure 4.2(a) shows the Power Spectral Density (PSD) of a baseband signal, e.g., the bit-stream of compressed and channel-coded speech whose information bandwidth is B Hz and PSD is A_0 Watts per Hz. By multiplying this by a high-frequency signal whose frequency is N times that of the baseband signal, the information bandwidth is spread to NB but the PSD amplitude is reduced by a factor of $\frac{1}{N}$ compared to the baseband signal. Then the signal is modulated to an RF carrier, amplified, and transmitted over the channel. Figure 4.3(a) shows the PSD measured at the far-end receiver, where the transmitted signal is received with additive white noise. By demodulating and low-pass

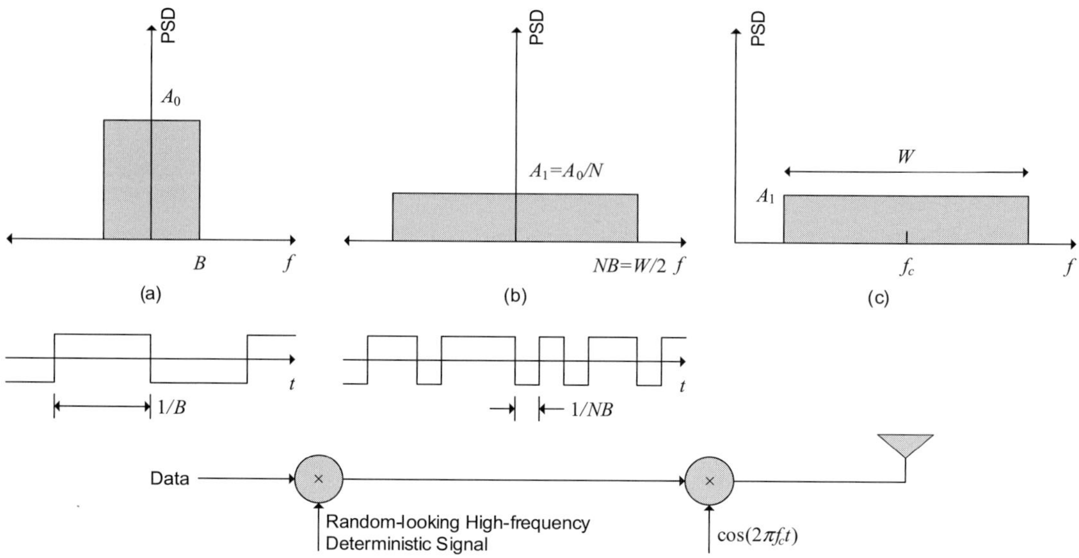

Fig. 4.2 Spectrum spreading and modulation.

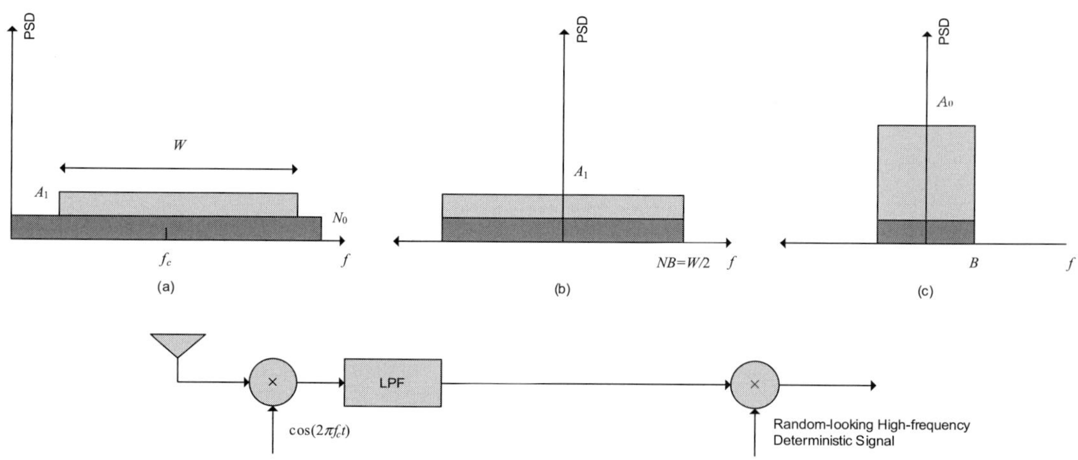

Fig. 4.3 Demodulation and spectrum despreading.

filtering, the baseband signal is reconstructed, with white noise still present. When this signal is multiplied by the same high-frequency modulating signal, which has been synchronized to the transmitted signal, the original baseband signal can be reconstructed. The PSD of white noise does not change during despreading or low-pass filtering but the reconstruction process can be facilitated with increased SNR.

If the wide bandwidth, W, is to be shared by as many MSs as possible, MSs without accurate information on the high-frequency modulating signal would not be able to discriminate the received wideband signal from noise. For this to happen, the high-frequency signal used to spread the information bandwidth should have an autocorrelation function which decreases rapidly with delay so that the receivers not synchronized with the transmitter cannot reconstruct the baseband signal correctly. Ideally the autocorrelation function of such signals should have an impulse-like shape. The high-frequency modulating signal also needs to be accurately representable and identically generated at will, so that it can be shared by the transmitter and the receiver.

The received waveform $y(t)$ of a CDMA system with K MSs can be mathematically modeled as [Verdú (1998)]

$$y(t) = \sum_{k}^{K} M_k b_k s_k(t) + \sigma n(t), \tag{4.1}$$

where $s_k(t)$ is a deterministic signature waveform, i.e., a high-frequency signal, assigned to the kth MS. $s_k(t)$ is normalized to have unit energy over a period T, and orthogonal to the other signature waveforms such that

$$\begin{aligned} \rho_{jk} &= \int_0^T s_j(t)s_k(t)\,dt, \\ \rho_{jj} &= 1, \\ \rho_{jk} &= 0, j \neq k. \end{aligned} \tag{4.2}$$

Here M_k is the received signal amplitude of the kth MS, and b_k is the bit transmitted by the kth MS. σ^2 is the noise variance, and $n(t)$ is a zero-mean Gaussian random process with unit variance. In CDMA, separation of signals from different MSs is achieved not by using a guard band or a guard time but by using the orthogonality of signature waveforms.

Although conceptually straightforward, implementation of CDMA requires digital circuitry in the transmitter and the receiver whose complexity is much higher than that used in TDMA. In particular, rapid synchronization to such a high-frequency signal and high-speed control of the transmit power were considered to be difficult to realize at the time that other second generation systems were being developed. For this reason CDMA was considered to be less cost-effective than the alternatives when D-AMPS or GSM was initially designed. The first mobile communications system based on CDMA, IS-95, was commercially deployed several years after GSM. The potential of CDMA for higher network capacity, however, became a powerful stimulus for later developments of GSM such as Phase 2.

4.2.2 Pseudo Noise Sequence

To restore the transmitted information, the high-frequency signal in Figures 4.2 and 4.3 should be a random-looking but ultimately deterministic signal that is long enough to frustrate trial-and-error deciphering attempts using high-speed computers. In addition, it should be possible to generate such signals using a small set of parameters, as the alternative of storing a large number of long sequences at the transmitter and the receiver would make their management difficult. While many types of such Pseudo Noise (PN) sequences have been mathematically defined, we focus on the major attributes of *pseudo-randomness* as classified in [Golomb (1981)]. To be qualified as a PN sequence, the following three properties should be met simultaneously:

Balanced Property

Each PN sequence should have an equal number of zeros and ones. If a PN sequence is biased to zero, i.e., contains more zeros than ones, then such an imbalance can be exploited by an eavesdropper to estimate the sequence.

Run-Length Property

Intuitively, in a truly random sequence, a long series of zeros or ones will have a lower probability of occurrence than a short series. Therefore, in each PN sequence, probabilities of consecutive strings of zeros or ones should follow the typical expectations in the experiments of flipping fair coins, where a half of the run-lengths are length one, a quarter of the run-lengths are length two, and so on. In other words, the fraction $\frac{1}{2^n}$ of the run-lengths are of length n.

Delay and Add Property

If a PN sequence is cyclically shifted in either direction, the resulting sequence should have an equal number of agreements and disagreements with the original sequence when the two sequences are compared bit by bit. This property is required in order to have impulse-like autocorrelation functions. If this property is met by a sequence, the autocorrelation function of the sequence will exhibit a peak value when the sequence is correlated with itself and be zero otherwise, regardless of the amount or direction of a circular shift.

4.2.3 Generation of PN Sequence

PN sequences meeting the three requirements very closely can be systematically generated using the following procedure. Begin with an irreducible binary polynomial,

$$f(x) = 1 + c_1 x + c_2 x^2 + \cdots + c_{n-1} x^{n-1} + x^n, \tag{4.3}$$

where the coefficients $c_i \in \text{GF}(2)$. GF(2) is the Galois field of binary elements $\{0, 1\}$, and $f(x)$ cannot be factored into lower-degree polynomials. If α is a root of $f(x) = 0$ and $P = 2^n - 1$ is the minimum integer that satisfies $\alpha^P = 1$, then $0, \alpha^0, \alpha^1, \ldots, \alpha^{P-1}$ are distinct, constituting an extended Galois field GF(2^n). Therefore α is a primitive element of GF(2^n) and $f(x)$ is a primitive polynomial. PN sequences of length $P = 2^n - 1$ can be generated with the following shift register circuitry, shown in Fig. 4.4.

If $c_i = 0$, then the ith branch is disconnected so that the value stored in register R_n is not fed back to the ith adder. To generate a unique PN sequence, registers $R_1, R_2, \ldots, R_n$ are loaded with a unique non-zero n-tuple, the initial state. At each clock increment, one bit is output from R_n. After $2^n - 1$ clock increments, a full period of the PN sequence has been generated, with $R_1, R_2, \ldots, R_n$ returning to the initial state. Loading different initial states only changes the location of the first output bit in the PN sequence. Therefore two PN sequences generated with the same primitive polynomial but using different initial states, i.e., *phases*, are two circularly-shifted versions of the same PN sequence. This procedure for generating PN sequences is called Modular Shift Register Generation (MSRG).

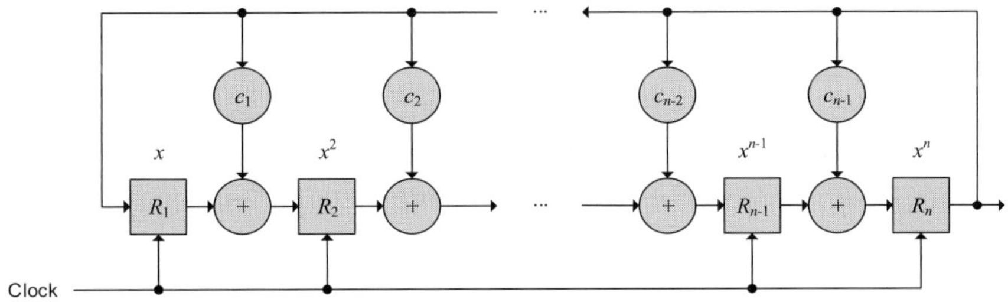

Fig. 4.4 Modular shift register generation of PN sequence.

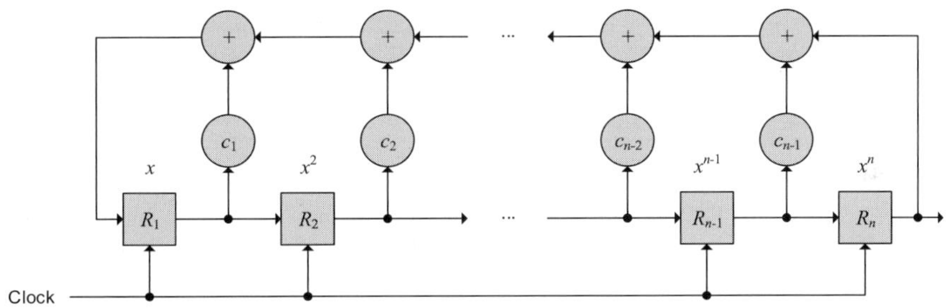

Fig. 4.5 Simple shift register generation of PN sequence.

Another configuration of the shift registers, Simple Shift Register Generation (SSRG), generates another family of PN sequences from the same primitive polynomial. Figure 4.5 shows the SSRG implementation of $f(x)$, which includes the same numbers of registers and adders as those of MSRG.

MSRG and SSRG are related in the sense that the MSRG implementation of the reciprocal polynomial of $f(x)$, defined as

$$f^*(x) = x^n f(x^{-1}) \tag{4.4}$$
$$= x^n(1 + c_1x^{-1} + c_2x^{-2} + \cdots + c_{n-1}x^{-n+1} + x^{-n}) \tag{4.5}$$
$$= 1 + c_{n-1}x + c_{n-2}x^2 + \cdots + c_1x^{n-1} + x^n, \tag{4.6}$$

generates the same PN sequence, possibly in a different phase, as the SSRG implementation of $f(x)$. The reciprocal polynomial of $f^*(x)$ is $f(x)$. We leave the detailed mathematical analysis on the relationship of MSRG and SSRG to [Lee and Miller (1998)] but show that the PN sequences generated with these methods meet the three requirements of PN sequences very closely. Figure 4.6 shows the MSRG implementation of a primitive polynomial, $f(x) = 1 + x^3 + x^4$.

The initial state of $R_1R_2R_3R_4$ is 1000 and the value of R_4 is the output at each clock pulse. It can be seen that after 15 clock pulses, the PN sequence generated is 000111101011001. The period of this PN sequence is $P = 2^4 - 1 = 15$. Whether the MSRG-based approach can generate PN sequences satisfying the three properties can be checked with the sequence.

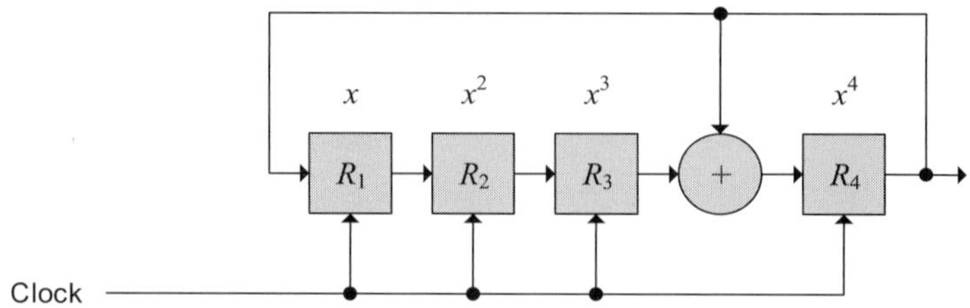

Fig. 4.6 MSRG implementation of $f(x) = 1 + x^3 + x^4$.

Satisfaction of Balanced Property

In this example, 000111101011001 contains eight ones and seven zeros. For a PN sequence of length $2^n - 1$ generated with the shift register methods, the probability of output one at each clock interval is $\frac{2^{n-1}}{2^n-1}$ while that of output zero is $\frac{2^{n-1}-1}{2^n-1}$. The output one has just one more occurrence since the all-zero state is excluded as an initial state. As the PN sequences become longer, the balanced property will be met with higher accuracy. With PN sequences whose n is larger than 10, the probabilities of outputs one and zero are close to a half.

Satisfaction of Run-Length Property

Table 4.1 presents an analysis of a PN sequence 000111101011001 in terms of the run-length property. It can be shown that the relative frequency of run-length m (zeros or ones) is $\frac{1}{2^m}$ for $m \leq n - 1$ and $\frac{1}{2^{n-1}}$ for $m = n$. Therefore, the only case for which this property is not satisfied is when $m = n$. However, as in the case of the balanced property, as n becomes larger, the run-length property is satisfied more closely. For each PN sequence generated with MSRG or SSRG, there will be only a run-length $n - 1$ of zero and a run-length n of one.

Satisfaction of Delay-and-Add Property

Let 001111010110010 be a circular shift of a PN sequence 000111101011001. These two sequences are identical except for a phase difference of one bit. Comparison of the two sequences reveals that there are seven agreements and eight disagreements over the 15 bit locations. Comparisons with other circular shifts, e.g., 011110101100100 or 111101011001000, result in the same numbers of agreements and disagreements.

Since the shift register operation is linear, bit-by-bit modulo-2 summation of two PN sequences can be implemented with the summation of two initial states as a new initial state. From the balanced property, there will always be one more output one (disagreement) than output zero (agreement). Therefore the normalized autocorrelation function of PN sequences generated with MSRG or SSRG, with zero mapped to +1 and one mapped to -1, will be 1 when the delay is zero and $-\frac{1}{2^n-1}$ otherwise, approximating the shape of an impulse.

4.2.4 Phase Shift of PN Sequence

MSRG generates PN sequences with good pseudo-random properties, using simple logic circuitry and a small number of parameters. To spread the information of each

Table 4.1 Statistics of PN sequence 000111101011001.

Run-length	Number of 0	Number of 1	Frequency
1	2	2	$\frac{1}{2}$
2	1	1	$\frac{1}{4}$
3	1	0	$\frac{1}{8}$
4	0	1	$\frac{1}{8}$

Table 4.2 Number of degree *n* primitive polynomials.

n	2	3	4	5	6	7	8	9	10	11	12	13	14	15	16	17	18
$\frac{\phi(2^n-1)}{n}$	1	2	2	6	6	18	16	48	60	176	144	630	756	1800	2048	7710	7716

MS, measures to systematically generate a large number of PN sequences are required. One straightforward strategy is to assign a unique primitive polynomial to each MS; this requires that there be enough primitive polynomials. The number of primitive polynomials of degree n can be computed as $\frac{\phi(2^n-1)}{n}$, where $\phi(2^n - 1)$ is the number of positive integers, including 1, which are relatively prime to the numbers less than $2^n - 1$. Table 4.2 shows the number of primitive polynomials for degree n from 1 to 15. It can be seen that for a large value of n, there will be enough primitive polynomials for MSs.

However, practical implementations limit the use of multiple primitive polynomials. For example, it is necessary to be able to synchronize to a PN sequence within a limited time interval; this necessitates multiple circuits to simultaneously track the received sequences, each of which is the PN sequence in a different phase. Thus, the use of multiple primitive polynomials may drive the computational complexity of the receiver to impractical levels. From the delay-and-add property, a circular shift of a PN sequence can be considered as another PN sequence. To synchronize to a new PN sequence, if the receiver has already established synchronization with a PN sequence, only the information about phase difference and appropriate mechanisms to shift the phase of a PN sequence on the fly are required.

In MSRG, output bits from the shift-register circuitry can be represented by $\frac{g(x)}{f(x)}$, where $f(x)$ is a primitive polynomial and $g(x)$ is a function related to the initial state of the shift registers. It can be shown that, in general, $g(x) = x^{n-1}s(x^{-1})$ where $s(x)$, whose degree is lower than n, is the polynomial representing the initial loading of the registers. For example, with the initial state $R_1R_2R_3R_4 = 1000$ of the MSRG shown in Fig. 4.6, $s(x) = 1$. Notice that the MSRG is constructed based on $f^*(x) = 1 + x^3 + x^4$ and $f(x) = x^4(1 + x^{-3} + x^{-4}) = 1 + x + x^4$. The PN sequence can then be computed by

$$g(x) = x^{n-1} = x^3, \tag{4.7}$$

$$\frac{g(x)}{f(x)} = \frac{x^3}{1 + x + x^4} = x^3 + x^4 + x^5 + x^6 + x^8 + x^{10} + x^{11} + x^{14} + \cdots, \tag{4.8}$$

whose coefficients match the original sequence, 000111101011001, computed with the shift register operations.

The idea behind *masking* is the fact that a PN sequence with a fixed relative phase shift can be generated from a linear combination of the values of shift registers that are running to generate the original PN sequence. With n shift registers, the mask can be represented as a polynomial with a degree lower than n. To compute the phase shift of a PN sequence, it is first necessary to determine the initial phase shift of the original PN sequence, q, which is defined as the relative location of the first one after $n - 1$ consecutive zeros. Therefore q is the absolute phase of the original sequence. For PN sequences generated with MSRG or SSRG, there is only a single run of zeros of length $n - 1$ and the first one after this is defined as the start of the PN sequence. Figure 4.7(a) illustrates the case when $q = n - 1$, which can occur with an MSRG initialized as $R_1 R_2 \cdots R_n = 10 \cdots 0$. In the MSRG implementation of Fig. 4.6, $q = 3$ as a one is generated after three zeros.

Let k be the phase shift of the sequence. Then the mask can be computed as the terms of $m(x)$ whose degree is lower than n,

$$m(x) = \frac{x^{k-q+n-1} \bmod f(x)}{f(x)}. \tag{4.9}$$

Figure 4.8 shows an MSRG with a mask. The clock that drives the periodic shift is not shown in the figure.

Suppose it is necessary to generate a PN sequence with a phase shift of 12 from the MSRG shown in Fig. 4.6. Two PN sequences are computed with two initial states

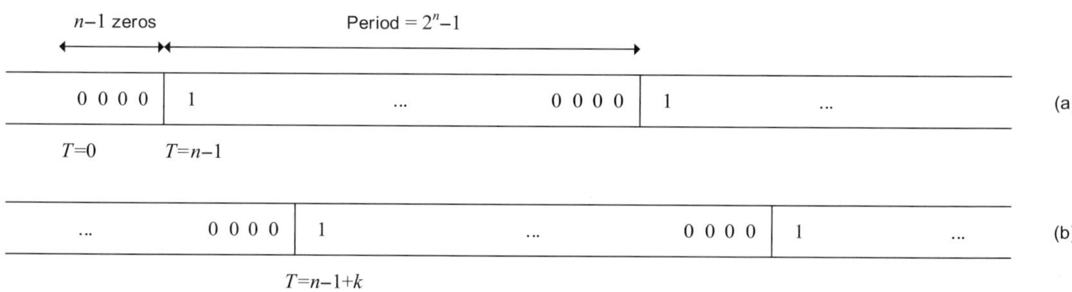

Fig. 4.7 Equivalence of masking shift registers.

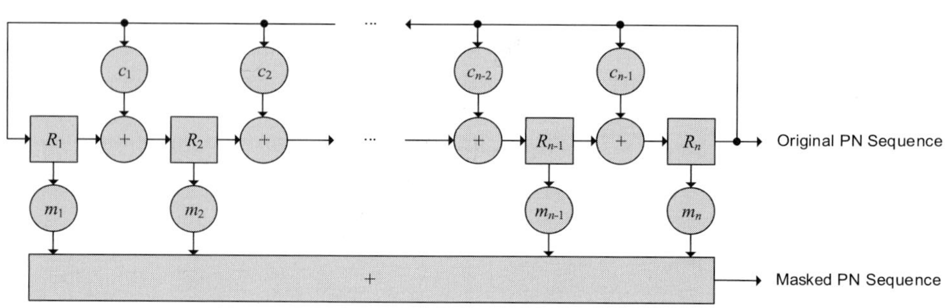

Fig. 4.8 Masking of MSRG.

and two masks, to show that an identical PN sequence can be generated from both conditions. With an initial state of $R_1R_2R_3R_4 = 1000$ and $q = 3$, the mask to generate a PN sequence with $k = 12$ is

$$m(x) = \frac{x^{k-q+n-1} \bmod f(x)}{f(x)} = \frac{x^{12-3+3} \bmod f(x)}{f(x)} \tag{4.10}$$

$$= \frac{x^{12} \bmod f(x)}{f(x)}, \tag{4.11}$$

$$(x^4)^3 \bmod (1 + x + x^4) = (1 + x)^3 = 1 + 3x + 3x^2 + x^3 = 1 + x + x^2 + x^3, \tag{4.12}$$

$$m(x) = \frac{1 + x + x^2 + x^3}{1 + x + x^4} = 1 + x^2 + x^4 + \cdots \tag{4.13}$$

$m(x) = 1 + x^2$ since its degree should be lower than $n = 4$. Table 4.3 shows the state of the shift registers at each clock and the PN sequence generated from the mask, 101011001000111. Note that the one after $n - 1 = 3$ zeros appears at clock= 12.

With an initial state of $R_1R_2R_3R_4 = 1001$, $s(x) = 1 + x^3$ and the output PN sequence can be computed as

$$g(x) = x^3(1 + x^{-3}) = 1 + x^3, \tag{4.14}$$

$$\frac{g(x)}{f(x)} = \frac{1 + x^3}{1 + x + x^4} = 1 + x + x^2 + x^4 + x^6 + x^7 + x^{10} + x^{14} + \cdots, \tag{4.15}$$

which corresponds to 111010110010001. $q = 14$ since at clock = 14 a one appears after $n - 1 = 3$ zeros. Therefore the mask is computed as

$$m(x) = \frac{x^{k-q+n-1} \bmod f(x)}{f(x)} = \frac{x^{12-14+3} \bmod f(x)}{f(x)} \tag{4.16}$$

$$= \frac{x \bmod f(x)}{f(x)} = \frac{x}{1 + x + x^4} = x + x^2 + x^3 + \cdots \tag{4.17}$$

Thus the mask is 0111. Table 4.3 shows the state of the shift registers at each clock and the PN sequence generated from the mask, which is identical to the PN sequence with an initial state of $R_1R_2R_3R_4 = 0001$ and mask = 1010. This example shows that PN sequences with different phase shifts can be simultaneously generated from an original sequence with different masks, whose implementation requires no more than simple adder circuitry.

4.2.5 Decimation of PN Sequence

Suppose that a PN sequence, a_i, of period $P = 2^n - 1$, is sampled every k digits. Then the sequence consisting of the kth samples, b_j, is another PN sequence with

$$b_0, b_1, b_2, \ldots = a_0, a_k, a_{2k}, \ldots \tag{4.18}$$

The relationship between a_i and b_j depends on the values of k and P. If k and P are relatively prime, then a_i and b_j are two distinct PN sequences generated from two different primitive polynomials. However, if $k = 2^m$ for $m \leq n - 1$, then a_i and b_j are based on

Table 4.3 Phase shift of PN sequence.

Clock	Mask = 1010			Mask = 0111		
	$R_1R_2R_3R_4$	Original	Masked	$R_1R_2R_3R_4$	Original	Masked
0	1000	0	1	1001	1	1
1	0100	0	0	1101	1	0
2	0010	0	1	1111	1	1
3	0001	1	0	1110	0	0
4	1001	1	1	0111	1	1
5	1101	1	1	1010	0	1
6	1111	1	0	0101	1	0
7	1110	0	0	1011	1	0
8	0111	1	1	1100	0	1
9	1010	0	0	0110	0	0
10	0101	1	0	0011	1	0
11	1011	1	0	1000	0	0
12	1100	0	1	0100	0	1
13	0110	0	1	0010	0	1
14	0011	1	1	0001	1	1

the same primitive polynomial, possibly differing only in the phase. It is not possible to control the phase as precisely as the masking, but decimation of a PN sequence can be used to generate a phase-shifted version of the same sequence with a lower symbol rate.

4.2.6 Rake Receiver Theory

For TDMA, its relatively low symbol rate and the isolation of signals with guard time and guard band mean that the received signal is a sum of delayed and phase-shifted versions of the original signal with added noise. The magnitudes of received signals are attenuated, depending on the paths of wave propagation. Since the delay spread is typically smaller than the symbol duration, because of the low symbol rate, e.g., 270.83 kbps of GSM, its impact can be easily normalized using equalization, which simplifies the receiver. In CDMA where the frequency of signals used to spread the information bandwidth, the chip frequency, is typically higher than 1 MHz, the delay spread is much larger than the chip duration. The receiver has to reconstruct the information from multiple copies of signals whose delay is larger than an integer number of chips.

Figure 4.9 shows the typical structure of a *rake receiver* used in a CDMA system. It consists of L processing units, called *fingers*, that track the signal of each transmission path. A rake receiver compensates for the channel effects of each path independently and combines the signal from each finger in such a way that maximizes the SNR. In the operation of a rake receiver, one finger tracks the dominant signal and establishes synchronization while the other fingers identify the delayed or weaker signals and estimate

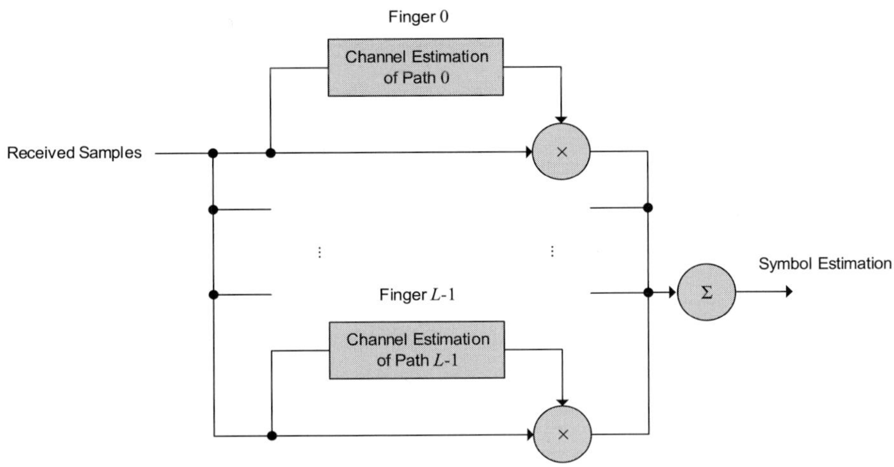

Fig. 4.9 MRC diversity reception with rake receiver.

their relative differences in amplitude and phase, i.e., the channel gain, with the dominant signal. After reversing the effects of transmission, signals from L paths can be combined coherently.

The principles of a rake receiver can be described mathematically as follows. Let the signal received from path l be

$$y_l(t) = s_l(t) + n_l(t) \tag{4.19}$$

$$= \alpha_l A_l \cos[2\pi f_c t + \phi_l(t)] + n_l(t), \tag{4.20}$$

where α_l is the attenuation factor, A_l is the signal amplitude at the transmitter, $n_l(t)$ is the noise component, and $\phi_l(t)$ is the signal phase, for path l, $l = 0, 1, \ldots, L-1$. Then the maximum SNR can be achieved by the Maximal Ratio Combining (MRC), in which the phases of signals from L paths are aligned and their envelopes are weighted in proportion to their respective SNR values. Then the optimum SNR at the combiner is equal to the sum of the SNR values in the L paths,

$$SNR_{\text{optimum}} = \sum_{l=0}^{L-1} SNR_l. \tag{4.21}$$

An optimality proof for the diversity reception of MRC can be found in [Viterbi (1995)].

4.3 Interim Standard 95

The Interim Standard 95 (IS-95) was the first mobile communications system based on the CDMA principles [EIA/TIA (1995)]. It provided a smooth migration path from AMPS or D-AMPS, as probably the only second-generation digital system that could rival GSM in some areas. Unlike IS-54 which is often called D-AMPS or TDMA, the

Table 4.4 System parameters of IS-95.

Multiple access	FDMA/CDMA
Modulation	OQPSK (reverse link), QPSK (forward link)
Channel bandwidth (MHz)	1.25
Chip rate (Mcps)	1.2288

official name of IS-95 is commonly used when its radically different signal processing operations are compared with the more conventional ones of TDMA networks. As GSM became enhanced for higher speech quality and larger network capacity, and for new services such as data access through innovations ranging from speech compression to wireless transmission, so did IS-95, which evolved to the next generation as cdma2000.

Table 4.4 summarizes the key features of IS-95, whose channel bandwidth is 1.25 MHz, much larger than the 30 kHz of D-AMPS or the 200 kHz of GSM. The high-frequency signal, multiplied by the baseband signal to spread the information spectrum, is called the *chip*, with a rate of 1.2288 Million Chips Per Second (Mcps). Although IS-95 was designed primarily to replace AMPS or D-AMPS, its chip rate did not depend on the signal processing procedures of the previous systems. Instead, the 1.2288 Mcps was determined as a compromise between easy implementation of the receiver and maximum re-use of the fragmented AMPS spectrum in the US Note that unlike the system parameters of GSM, shown in Table 2.4, the number of voice channels per carrier is not specified, as the separation of voice channels from a PN sequence is not as clear as that from guard time. In this regard, network capacity of CDMA systems is often considered as *soft*, as increasing the number of voice channels simultaneously influences the quality of ongoing and incoming calls more noticeably than in TDMA systems.

In Fig. 1.4, a 25 MHz spectrum is allocated to a non-wire-line service provider (A) and a wire-line service provider (B) for AMPS operation in the US. After an initial assignment of 10 MHz to each service provider, an additional 2.5 MHz was given to each to meet the surging demands for mobile communications. The wire-line service provider received a contiguous band of 2.5 MHz, while the non-wireline service provider was given the same amount of bandwidth less favorably, in two separate bands of 1 and 1.5 MHz. The 1.5 MHz band was located between the two bands of the wire-line service provider, thus becoming a key limiting factor in the design of new mobile communications systems to be deployed over the spectrum in the US. Therefore 1.25 MHz was determined for IS-95 to be the maximum channel bandwidth that could be located inside the 1.5 MHz band, without influencing the neighboring bands.

Table 4.5 shows the frequency ranges of AMPS illustrated in Fig. 1.4 where the center frequency of a 1.25 MHz channel of IS-95 can be located. This assignment strategy can put up to five channels in the spectrum of the non-wire-line (A) operator and six channels in the spectrum of the privileged wire-line (B) operator. Notice that in the leftmost column for the A operator, the channel number reaches 1023 and re-starts from 1. While GSM achieves higher capacity than AMPS or D-AMPS with a lower operational SNR,

Table 4.5 IS-95 center frequency location in AMPS spectrum.

Channel number	1013–311 (A)	356–644 (B)	689–694 (A)	739–777 (B)
Reverse (MHz)	824.7–834.33	835.68–844.32	845.67–845.82	847.17–848.31
Forward (MHz)	869.7–879.33	880.68–889.32	890.67–890.82	892.17–893.31

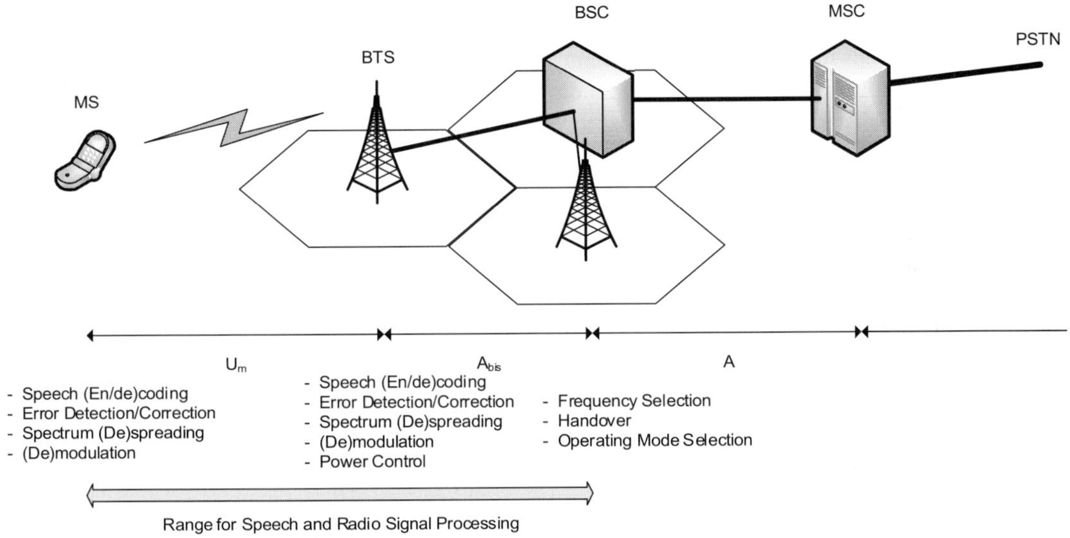

Fig. 4.10 Network architecture of IS-95.

IS-95 reduces the value further, achieving the unprecedented frequency re-use factor of 1. Thus, in IS-95 all channels can be shared by the neighboring cells. Such a feature of CDMA is achieved because a measure for orthogonality rather than a guard band or a guard time is used to differentiate the information of each MS on the channel. Like GSM, IS-95 was designed without much consideration for the previous systems, thereby providing enough opportunity for further advances.

4.3.1 Network Architecture

Figure 4.10 shows that IS-95 shares the same architecture as circuit-switched networks such as D-AMPS and GSM, where the speech is compressed, transmitted, and reconstructed between the MS and the BTS. The BTS and the BSC are equipped with measures to control the tradeoff between speech quality and network capacity, using means such as power control or handover. From an architectural point of view, a key characteristic of IS-95 is the *soft handover*.

In the forward link, a term used in IS-95 and its descendant systems to represent the downlink, when an MS enters the soft handover mode, multiple BTSs transmit identical signals for the MS, which combines the received signals at the chip level before decoding the speech frames. In the reverse link, i.e., the uplink, the signal transmitted from an MS is demodulated and decoded by multiple BTSs independently. Then the BSC selects the one signal with the highest quality. This strategy is in contrast to the *hard handover* of TDMA, where each MS can communicate with only one BTS at a time and the previous link with a BTS must be released before the MS establishes a new link with another BTS. Soft handover is enabled by the low frequency re-use factor of CDMA. Soft handover reduces the quality loss around handover by decreasing the probability of a call drop. In return, soft handover reduces the network capacity since during soft handover, a call consumes the radio resources of multiple cells simultaneously.

4.3.2 QCELP Speech Codec

Figure 4.11 shows a simplified block diagram of the Qualcomm Code-Excited Linear Prediction (QCELP) speech encoder [EIA/TIA (1990)], named after the company that developed the first speech codec for IS-95. Code-Excited Linear Prediction (CELP) is a generic speech compression principle in which the linear prediction coefficients are computed and quantized for a short period of speech, which is followed by the search of an adaptive (pitch) codebook and a fixed (innovation) codebook, and by the removal of their contributions [Schroeder and Atal (1985)]. The Algebraic Code-Excited Linear Prediction (ACELP) used in EFR and AMR is also a flavor of CELP. The output possibilities of QCELP, 171, 80, 40, and 16 bits, are called Rates 1, $\frac{1}{2}$, $\frac{1}{4}$, and $\frac{1}{8}$. The terms are also used in other speech codecs of IS-95 with the same set of output possibilities. For Rates 1, $\frac{1}{2}$, and $\frac{1}{4}$, the input signal is considered to be speech, and a fixed codebook is searched to find an excitation signal for the synthesis filter. For Rate $\frac{1}{8}$, the input signal is assumed to be noise, and a gain-scaled pseudo-random number is used instead.

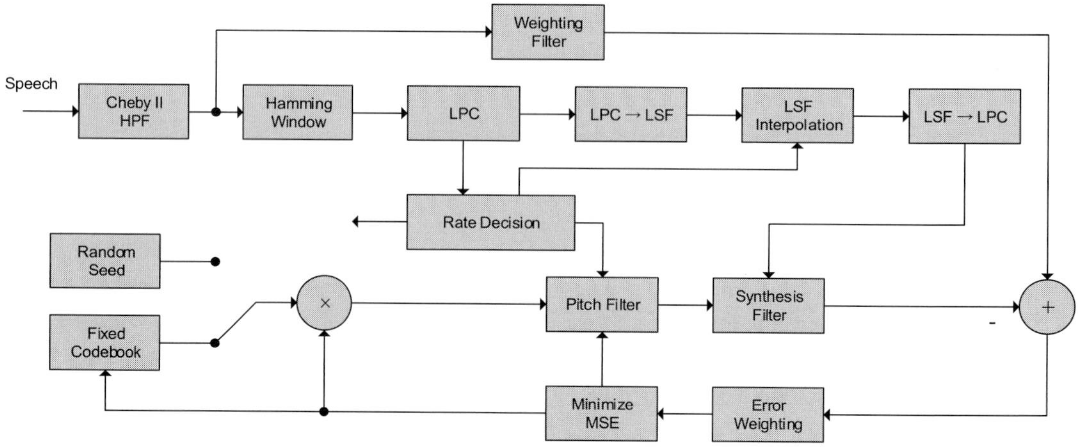

Fig. 4.11 Block diagram of QCELP speech encoder.

Table 4.6 Bit allocation of QCELP speech codec.

	Rate 1	Rate $\frac{1}{2}$	Rate $\frac{1}{4}$	Rate $\frac{1}{8}$
LPC updates / frame	1	1	1	1
Samples / LPC update L_A	160 (20 ms)	160 (20 ms)	160 (20 ms)	160 (20 ms)
Bits / LPC update	40	20	10	10
Pitch updates / frame	4	2	1	0
Samples / pitch subframe L_P	40 (5 ms)	80 (10 ms)	160 (20 ms)	
Bits / pitch update	10	10	10	
Codebook updates / frame	8	4	2	1
Samples / codebook subframe L_C	20 (2.5 ms)	40 (5 ms)	80 (10 ms)	160 (20 ms)
Bits / codebook update	10	10	10	6
CRC	11	0	0	0
Total	171	80	40	16

In TDMA mobile communications systems such as D-AMPS or GSM, the speech encoder determines whether the input signal is speech or background noise, by comparing the energy of input signal with an estimate of the background noise energy, which is also updated for each frame. In IS-95, the speech encoder has to decide not only whether the input signal is speech or noise, but also at which bit-rate to encode the frame when the input signal is diagnosed as speech. As a straightforward approach, a set of three thresholds can be used to determine the rate from four candidates. To adjust the relative frequency of encoding, each rate can be a more flexible and faster mechanism to control the tradeoff between speech quality and network capacity than to switch between the full-rate and half-rate channels of TDMA. As the decision on speech bit-rate is made at the MS, it is not necessary for the MS to communicate with the network.

Table 4.6 summarizes the bit allocation of the QCELP speech codec. In Rate 1, the speech is encoded with 160 bits and an 11-bit CRC is computed to detect errors in the most perceptually significant 18 bits. Ten parity check bits are generated, with a (28,18) BCH code, from the 18 bits and a single parity bit is computed and added. For the other three rates, the CRC is not computed. Notice that although the rates are specified as Rates 1, $\frac{1}{2}$, $\frac{1}{4}$, and $\frac{1}{8}$, the total numbers of bits of the QCELP speech codec do not match these ratios at this point.

In the rate-selection mechanisms of QCELP, the energy of each frame is estimated using $R(0)$, the zero lag value of the autocorrelation function

$$R(k) = \sum_{m=0}^{L_A-1-k} s_W(m)s_W(m+k). \tag{4.22}$$

$s_W(m)$ is the input signal of 160 samples filtered by a Hamming window, and $L_A = 160$. Then the computed $R(0)$ is compared with three thresholds, $T_1(B_i)$, $T_2(B_i)$, and $T_3(B_i)$, which are calculated based on B_i, an estimate of the background noise level for the ith frame defined as

$$B_i = \min(R(0)_{\text{prev}}, 5059644, \max(1.00547B_{i-1}, B_{i-1}+1)). \tag{4.23}$$

Where $R(0)_{\text{prev}}$ is the energy of the previous frame and B_i is set to 5059644 for $i = 1$. In GSM, the thresholds of the voice activity detector (VAD) are continuously adapted to reflect the local acoustic environments [3GPP (2000)]. A similar strategy is employed in the QCELP speech codec.

For $B_i \leq 160000$, the thresholds are updated as

$$T_1(B_i) = -(5.544613 \times 10^{-6})B_i^2 + 4.047152B_i + 362, \tag{4.24}$$

$$T_2(B_i) = -(1.529733 \times 10^{-5})B_i^2 + 8.750045B_i + 1136, \tag{4.25}$$

$$T_3(B_i) = -(3.957050 \times 10^{-5})B_i^2 + 18.89962B_i + 3347. \tag{4.26}$$

For $B_i > 160000$, the thresholds are updated as

$$T_1(B_i) = (9.043945 \times 10^{-8})B_i^2 + 3.535748B_i - 62071, \tag{4.27}$$

$$T_2(B_i) = -(1.986007 \times 10^{-7})B_i^2 + 4.941658B_i + 223951, \tag{4.28}$$

$$T_3(B_i) = -(4.838477 \times 10^{-7})B_i^2 + 8.63002B_i + 645864. \tag{4.29}$$

Figure 4.12 shows $T_1(B_i)$, $T_2(B_i)$, and $T_3(B_i)$ over a wide range of values for B_i. For any value of B_i, $T_3(B_i) > T_2(B_i) > T_1(B_i)$. The rate-decision logic operates as follows: if $R(0)$ is greater than $T_3(B_i)$, Rate 1 is selected. If $R(0)$ is less than $T_1(B_i)$, Rate $\frac{1}{8}$ is selected. Rates $\frac{1}{2}$ and $\frac{1}{4}$ are selected when $R(0)$ is between $T_3(B_i)$ and $T_2(B_i)$, and between $T_2(B_i)$ and $T_1(B_i)$, respectively. The decision logic is similar to those in the codec mode adaptation of GSM with four modes but the hysteresis is not used. Instead the history of previous frames has an indirect influence on the determination of the current frame via $R(0)_{\text{prev}}$.

Just as the neighborhood-only transition policy or hysteresis is enforced in the AMR codec mode adaptation of GSM, similar constraints are applied to the rate determination

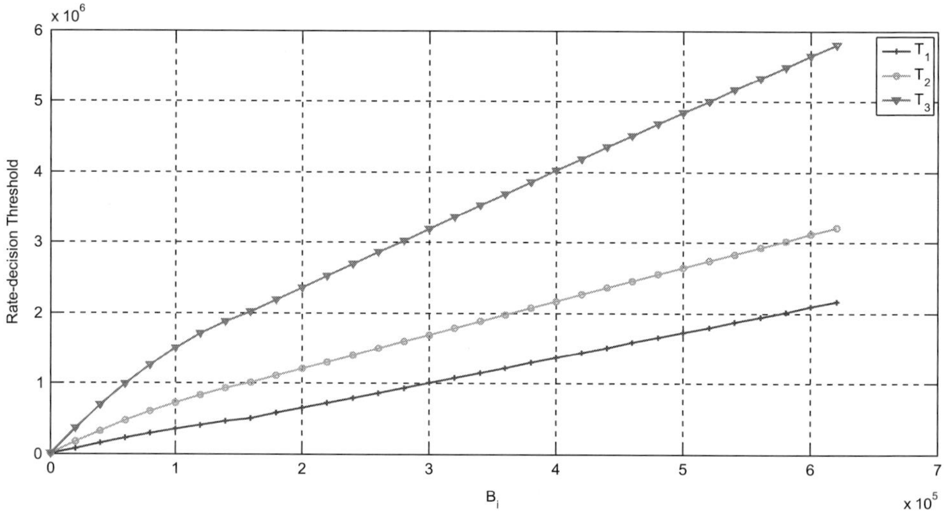

Fig. 4.12 Noise-adaptive thresholds for rate decision.

of QCELP in IS-95. A key difference is that in QCELP, the neighborhood-only transition policy is applied only for rate reduction. For example, if the algorithm decides to encode an input frame as Rate $\frac{1}{4}$, after encoding a Rate 1 frame, then the frame is encoded at Rate $\frac{1}{2}$, instead of Rate $\frac{1}{4}$. If the next input frame is decided as Rate $\frac{1}{4}$, then it is possible to encode that frame as decided. Likewise, if $R(0)$ is less than $T_1(B_i)$ but the QCELP speech encoder just encoded a Rate $\frac{1}{2}$ frame, a Rate $\frac{1}{4}$ frame is encoded instead of a Rate $\frac{1}{8}$ frame. This constraint is enforced as a kind of hysteresis to limit the reduction of speech quality from an abrupt reduction of the bit-rate. In contrast, no constraints are applied to an increase or maintenance of the bit-rate.

Although the VBR speech compression algorithms of QCELP apply different encoding methods based on the acoustic nature of the input signal, the rate-decision mechanisms resort to a relatively simple logic, by comparing the estimated signal energy against time-varying thresholds. Rate 1 is most likely to be assigned to voiced speech while unvoiced speech or sounds of other acoustic origins are encoded at Rates $\frac{1}{2}$ and $\frac{1}{4}$. Rate $\frac{1}{8}$ is in most cases used to represent background noise. If necessary, the network or the MS can override the decision of rate-decision algorithms to make room for urgent control signaling. For example, a limited capacity for important data transmission can be secured by temporarily prohibiting the speech encoder from encoding Rate 1 frames.

At any time the network can apply new rules for the update of thresholds or new constraints on the rate decisions for the QCELP speech encoder at the transcoder and rate adaptation unit (TRAU), which might be located at the BTS or the BSC. These rules can control the tradeoff between speech quality and network capacity, but such a control does not require any new signaling. How to realize this control over the forward link is left to the discretion of the implementation. On the other hand, more limited measures are available to modify the speech encoding strategy of the MS. Instead of using complex signaling to control the relative frequencies of Rates 1 and $\frac{1}{2}$ frames, the network might transmit only a few signaling bits to the MS to control its encoding behavior for the reverse link, in the form of a few pre-defined operating modes.

Table 4.7 summarizes the five-level control of the average reverse link bit-rate using the 3-bit RATE_REDUC message. When RATE_REDUC is set to 000, the network does not influence the speech encoding of the MS. If RATE_REDUC is set to 100, those input frames determined as Rate 1 are encoded at Rate $\frac{1}{2}$ instead. With RATE_REDUC of 001, 010, and 011, a fraction of Rate 1 frames is converted to Rate $\frac{1}{2}$, on average, by 25%, 50%, and 75% respectively.

In practice, the relative frequencies of Rates 1 and $\frac{1}{2}$ frames cannot be controlled accurately. Table 4.8 introduces simple encoding strategies to meet the target distributions as closely as possible. Each input frame whose $R(0)$ is greater than $T_3(B_i)$ is encoded at Rate 1 if RATE_REDUC = 000, and at Rate $\frac{1}{2}$ if RATE_REDUC = 100. With RATE_REDUC = 001, 010, and 011, the first L frames in a sequence of length N are allowed to be encoded at Rate 1 if determined as such, but the next $N-L$ frames are forced to be encoded at Rate $\frac{1}{2}$. Whenever the rate-decision algorithms determine a rate other than Rate 1, the sequence is reset, to ensure that the first frame in the onset of a speech waveform will be encoded at Rate 1, unless RATE_REDUC is set to 100 or the network commands the MS otherwise.

Table 4.7 Reducing average bit-rate by shifting Rate 1 down to Rate $\frac{1}{2}$.

RATE_REDUC	Fraction of Rate 1 frames encoded at Rate 1	Fraction of Rate 1 frames encoded at Rate $\frac{1}{2}$
000	1	0
001	$\frac{3}{4}$	$\frac{1}{4}$
010	$\frac{1}{2}$	$\frac{1}{2}$
011	$\frac{1}{4}$	$\frac{3}{4}$
100	0	1

Table 4.8 Rate control messages for reverse link.

RATE_REDUC	Sequence length (N)	Maximum number of contiguous Rate 1 frames (L)	Number of contiguous Rate $\frac{1}{2}$ frames ($N-L$)
000	1	1	0
001	4	3	1
010	2	1	1
011	4	1	3
100	1	0	1

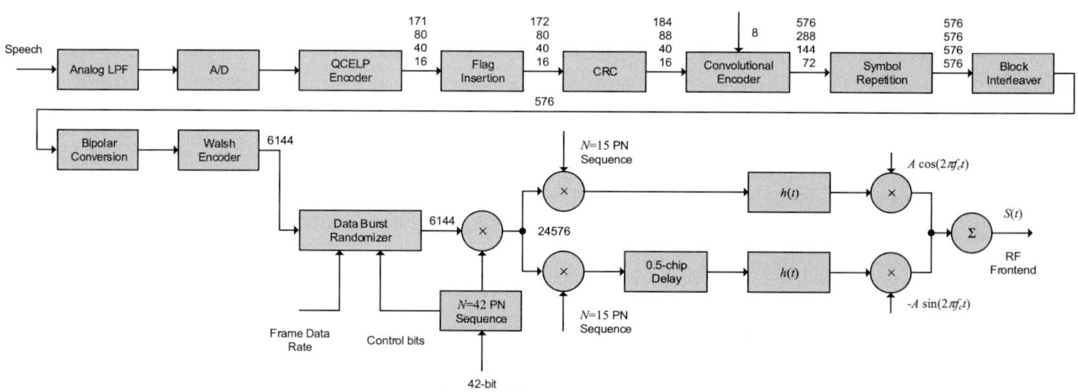

Fig. 4.13 Speech and radio signal processing operations of IS-95 reverse link for QCELP (RC 1).

4.3.3 Reverse Link Signal Processing

Figure 4.13 shows the speech and radio signal processing operations for the IS-95 reverse link. In GSM, the same signal processing operations are used in the uplink and downlink, whereas in D-AMPS the frame structures of the uplink and downlink are slightly different but offer the same transmission capability. In contrast, the signal processing operations in the reverse and forward links of IS-95 differ significantly. Such an asymmetrical channel structure can also be seen in other mobile communications systems based on the CDMA principles, such as W-CDMA.

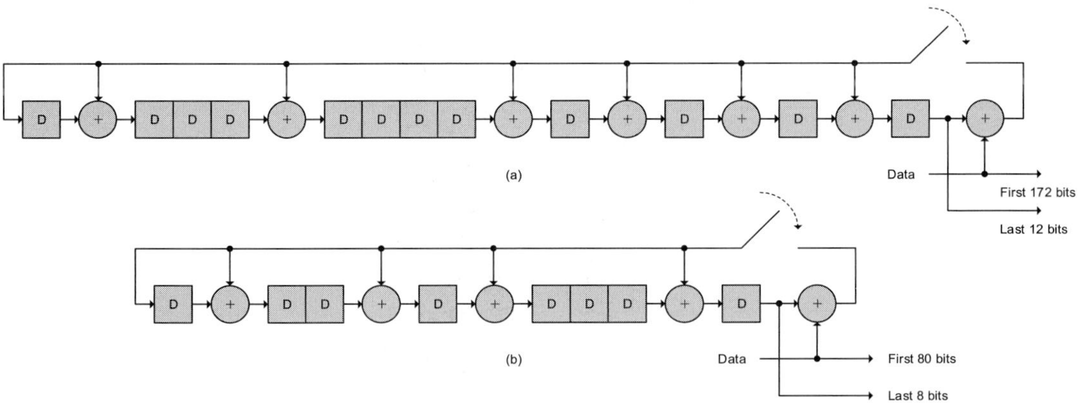

Fig. 4.14 CRC generation. (a) Rate 1. (b) Rate $\frac{1}{2}$.

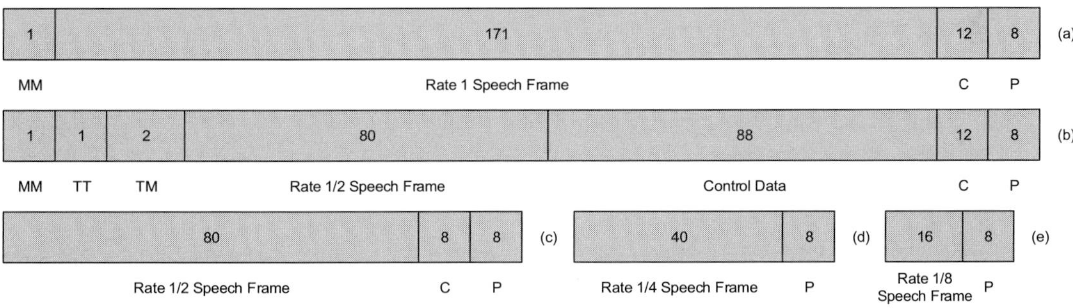

Fig. 4.15 Speech frames complemented with CRC (C) and padding (P) bits.

For each 20 ms period in the reverse link of IS-95, the speech is digitized 160 times in the 13-bit Uniform format and compressed over the period. Unlike in D-AMPS or GSM, the encoded speech bits are not classified according to their relative perceptual importance. Unequal error detection (UED) is applied inside the QCELP speech encoder only for Rate 1, where an 11-bit CRC is generated from the most significant 18 bits. In the other signal processing operations, each bit is treated equally. In addition, a 1-bit flag, Mixed Mode (MM), is added to the encoded frame to indicate whether the 171 bits are encoded speech or 80 bits of encoded speech plus 91 bits of data. Then a 12-bit CRC is computed from the 172 bits, which can be generated from the shift-register circuitry shown in Fig. 4.14(a).

For Rate $\frac{1}{2}$, an 8-bit CRC is computed from the 80 bits with the shift-register circuitry shown in Fig. 4.14(b). For Rates $\frac{1}{4}$ and $\frac{1}{8}$, a CRC is not computed. Thus, it can be said that in IS-95, UED is applied not at the bit level but at the frame level. Then padding bits consisting of eight zeros are added before the bits of each rate are applied to a rate $\frac{1}{3}$ convolutional encoder. The total number of bits complemented with a CRC and padding bits are 192, 96, 48, and 24, corresponding to Rates 1, $\frac{1}{2}$, $\frac{1}{4}$, and $\frac{1}{8}$, respectively. It is at this step that the number of bits at each rate finally reaches the ratio of eight, four, two, and one. Figure 4.15 illustrates the bit-stream structure of each rate. Figure 4.15(b)

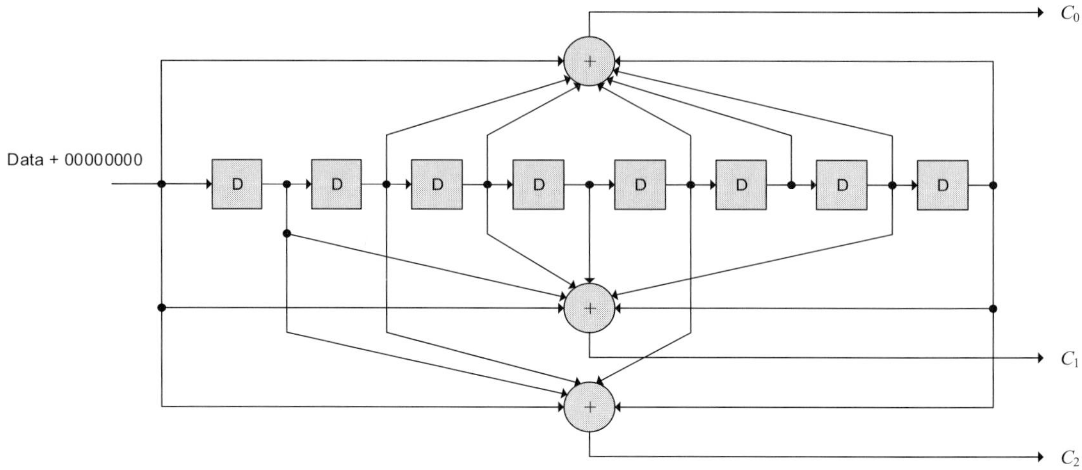

Fig. 4.16 Rate $\frac{1}{3}$ convolutional encoding.

is the case when the capacity of a Rate 1 frame is divided between a Rate $\frac{1}{2}$ frame and control information, in addition to the Traffic Type (TT) and Traffic Mode (TM) bits. The number of control information bits can be increased by reducing the rate further, to Rate $\frac{1}{4}$ or $\frac{1}{8}$.

In the AMR speech codec operating in GSM, each codec mode results in a different number of bits after adding a CRC and padding bits, and then applying a mode-dependent channel coding rate to the added bits. The redundant bits are punctured, however, to leave the same number of symbols for all rates. In contrast, in the reverse link of IS-95, an identical channel coding of rate $\frac{1}{3}$ is applied to the bits of each rate, generating 576, 288, 144, and 72 symbols for Rates 1, $\frac{1}{2}$, $\frac{1}{4}$, and $\frac{1}{8}$, respectively. The coding rate of the convolutional encoder shown in Fig. 4.16 is comparable to those of GSM but its constraint length $K = 9$ is longer than that of GSM, i.e., $K = 5$ of TCH/FS. A longer constraint length requires a more complex channel decoder but in general provides a higher level of error correction capability. Moreover, since the convolutional decoder for the reverse link resides in the BTS, increasing K is an efficient approach to provide higher error resilience with a controllable impact on the computational complexity and the power consumption.

After convolutional encoding, symbols of rates lower than Rate 1 are repeated to make the same number of symbols as that of Rate 1. To have 576 symbols, the 288 symbols of Rate $\frac{1}{2}$ are repeated once and the 144 symbols of Rate $\frac{1}{4}$ are repeated three times. The 72 symbols of Rate $\frac{1}{8}$ are repeated seven times. After the repetition, 576 symbols are generated for all four rates.

The repeated symbols are then entered into the interleaver, an 18 (column) by 32 (row) array, where the data is written in column-wise and read out row-wise, to spread the effects of bit errors over the bit-stream as widely as possible. Figures 4.17 and 4.18 show the four interleavers used in the reverse link. For each rate, the rows are read out in different orders, as outlined in Table 4.9, to maximize the effects of interleaving and

Table 4.9 Read-out order of reverse link interleaver.

Rate	Read-out order
1	0, 1, 2, 3, 4, 5, 6, 7, 8, 9, 10, 11, 12, 13, 14, 15, 16 17, 18, 19, 20, 21, 22, 23, 24, 25, 26, 27, 28, 29, 30, 31
$\frac{1}{2}$	0, 2, 1, 3, 4, 6, 5, 7, 8, 10, 9, 11, 12, 14, 13, 15, 16, 18, 17, 19, 20, 22, 21, 23, 24, 26, 25, 27, 28, 30, 29, 31
$\frac{1}{4}$	0, 4, 1, 5, 2, 6, 3, 7, 8, 12, 9, 13, 10, 14, 11, 15, 16, 20, 17, 21, 18, 22, 19, 23, 24, 28, 25, 29, 26, 30, 27, 31
$\frac{1}{8}$	0, 8, 1, 9, 2, 10, 3, 11, 4, 12, 5, 13, 6, 14, 7, 15, 16, 24, 17, 25, 18, 26, 19, 27, 20, 28, 21, 29, 22, 30, 23, 31

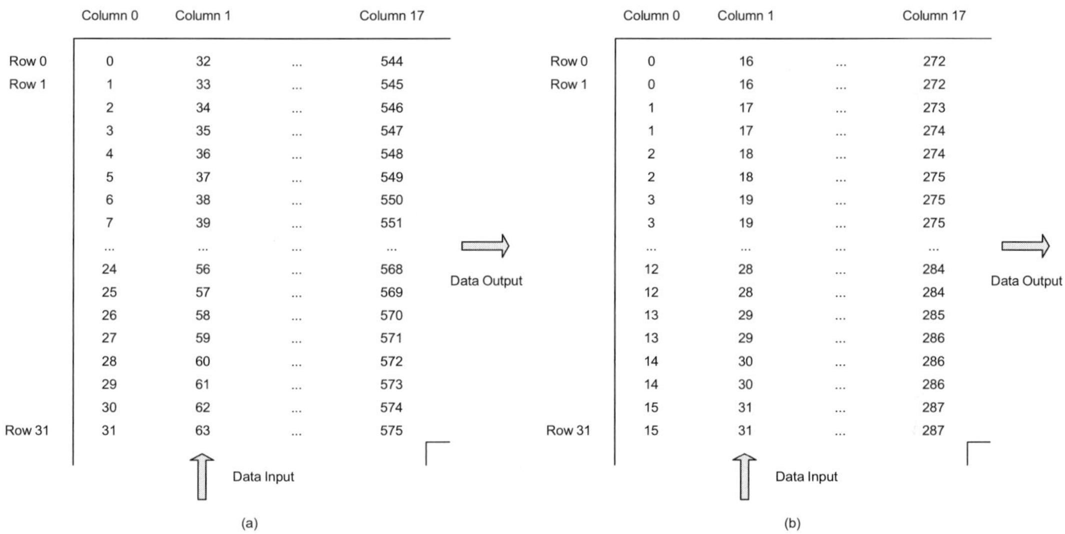

Fig. 4.17 Reverse link interleaving. (a) Rate 1. (b) Rate $\frac{1}{2}$.

facilitate the *data burst randomizing*, a series of signal processing procedures using a PN sequence for the combined objective of data encryption and power control.

After interleaving, each group of six interleaved symbols is mapped to a Walsh sequence of length 64. Walsh sequences or codes are a set of binary sequences that are, after the conversion of binary zeros and ones to $+1$ and -1, respectively, mutually orthogonal, i.e., they generate zero if two distinct Walsh sequences are correlated over a period. Such bi-level sequences are also called Walsh functions. Let the correlation, $R(i,j)$, of two Walsh functions be

$$R(i,j) = \sum_{n=0}^{2^n-1} W_i(n)W_j(n). \tag{4.30}$$

Then $R(i,j) = 2^n$ for $i = j$ and $R(i,j) = 0$ otherwise. Figure 4.19 shows eight Walsh functions of length 8.

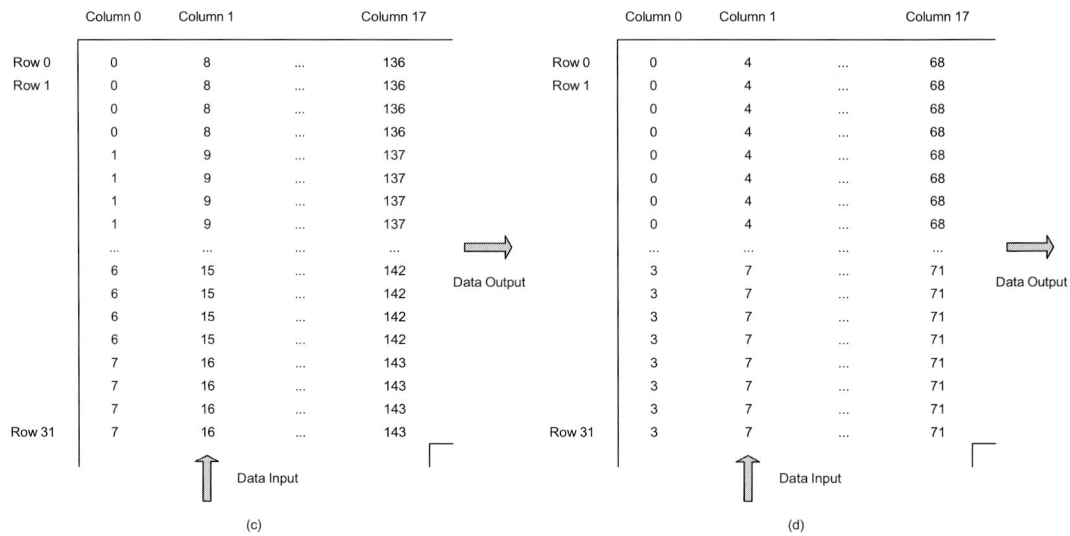

Fig. 4.18 Reverse link interleaving. (a) Rate $\frac{1}{4}$. (b) Rate $\frac{1}{8}$.

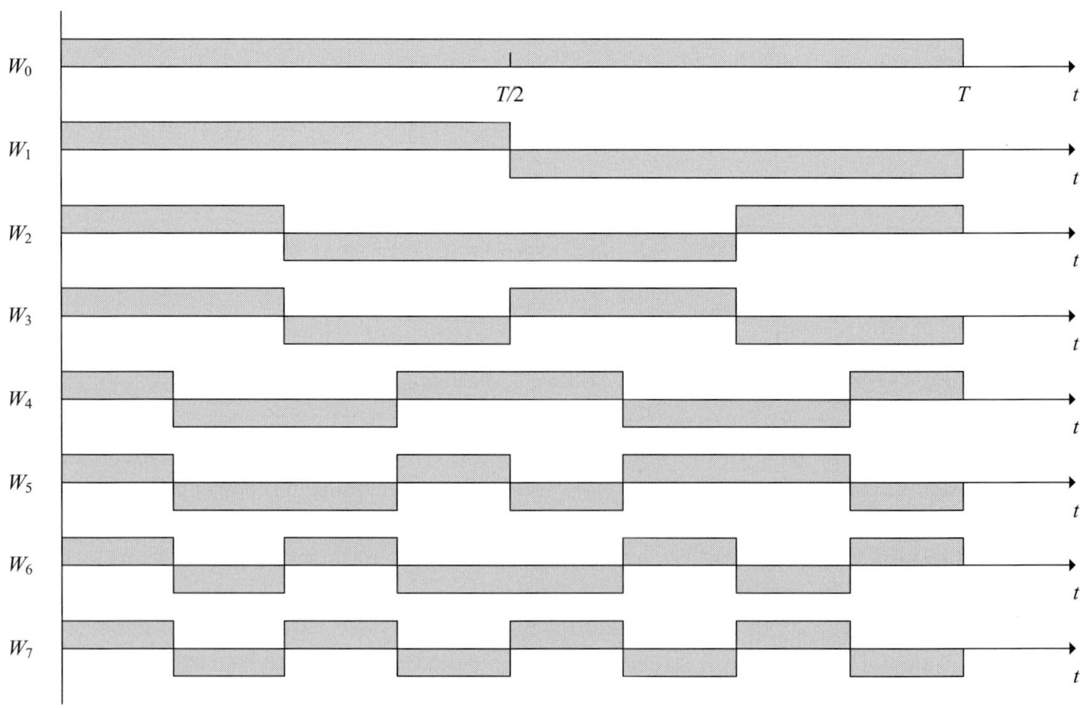

Fig. 4.19 Walsh functions of length 8.

Walsh functions can be systematically generated by repeatedly expanding Hadamard matrices. The Hadamard matrix of order 2 is defined as

$$H_2 = \begin{bmatrix} 1 & -1 \\ 1 & 1 \end{bmatrix}. \tag{4.31}$$

The first and second rows are mutually orthogonal, constituting a minimum set of Walsh functions. The complement of H_2 is defined as

$$\overline{H_2} = \begin{bmatrix} -1 & 1 \\ -1 & -1 \end{bmatrix}, \tag{4.32}$$

where the sign of each element has been reversed. It can be shown that a Hadamard matrix of order $2n$ can be constructed from four Hadamard matrices of order n as follows,

$$H_{2n} = \begin{bmatrix} H_n & H_n \\ H_n & \overline{H_n} \end{bmatrix}, \tag{4.33}$$

for $n = 1, 2, 3, \ldots$

Walsh functions of length 8 in Fig. 4.19 show that the index k of each function W_k is equal to the number of sign changes during a period. For example, in W_2 and W_5, conversion from $+1$ to -1 or vice versa happens two and five times respectively. Walsh functions generated by the expansion of a Hadamard matrix are not indexed according to the number of sign changes but the row number of the Hadamard matrix that corresponds to each Walsh function W_k can be recursively computed.

The mapping of six symbols to a Walsh sequence of length 64 is referred to as Walsh encoding or 64-ary orthogonal modulation. It can be considered to be a form of block coding. Walsh encoding provides the BTS a limited measure for synchronization over a duration of six symbols. Unlike the case in D-AMPS or GSM, where training sequences are separated from data in time slots, dedicated assistance for synchronization is not available in the reverse link of IS-95. Walsh encoding is a more cost-effective method than providing a separate reverse channel for synchronization when the transmit power or number of PN sequences is limited. At the BTS, the timing reference for demodulation can be found by correlating the received symbols with 64 Walsh sequences. The index number of the Walsh sequence is selected based on the following rule,

$$i = c_0 + 2c_1 + 4c_2 + 8c_3 + 16c_4 + 32c_5, \tag{4.34}$$

where i is the index number and $c_0c_1c_2c_3c_4c_5$ are the binary symbols from the interleaver. Figure 4.20 illustrates the Walsh encoding process and the corresponding increase of the symbol rate, from 28.8 ksps to 307.2 kcps.

Considering that the total bit-rate after channel coding is 16.2 ksps in D-AMPS and 22.8 ksps in GSM, radio signal processing of IS-95 increases the total bit-rate more than in the TDMA systems. A notable difference is that although the QCELP speech encoder outputs different types of speech frames, information on the frame type or rate, such as

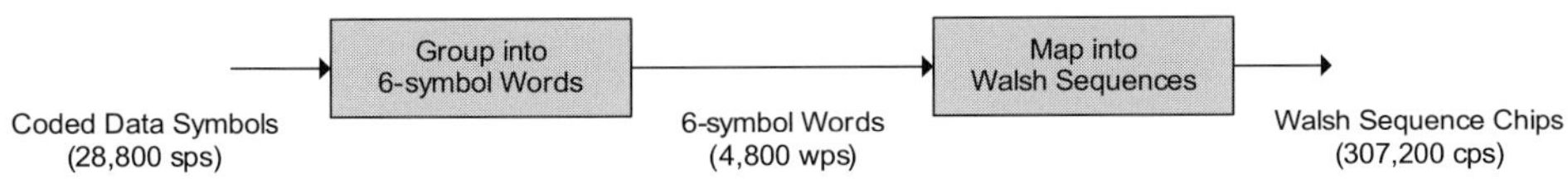

Fig. 4.20 Walsh encoding.

the CMI bits of AMR or the Mode bits of HR, is not built into the bit-stream structure. The receiver identifies the rate of each frame by an exhaustive search, i.e., by applying the received frame to the channel decoder and the error detector for each rate, selecting the rate with no errors or the one with the highest likelihood.

As long as sufficient link quality is provided by the combination of channel coding, modulation, and power control, this brute-force approach has the advantage of saving at least two bits for indicating rates, which was considered more valuable than the computational cost for simultaneously running multiple sets of channel decoders and CRC re-generators. In D-AMPS or GSM, ciphering is a separate step after interleaving but in IS-95, multiplication of long PN sequences to the symbols replaces ciphering in addition to spreading the information spectrum further.

Since the interleaver is an 18 (column) by 32 (row) array which is read out row-wise, each row generates three Walsh sequences or symbols. Therefore in a frame interval of 20 ms, 96 Walsh symbols are generated. Because of the read-out order, the Walsh symbols occur in alternating groups of six, and each group is followed by $6(n-1)$ repeated Walsh symbols, where $n = 1, 2, 4,$ and 8 for Rates 1, $\frac{1}{2}$, $\frac{1}{4}$, and $\frac{1}{8}$, respectively. The 576 symbols corresponding to an encoded speech frame are organized into 16 groups of six Walsh symbols, the Power Control Groups (PCG).

Figures 4.17 and 4.18, and Table 4.9 show that in Rate 1, six Walsh symbols from rows 0 and 1 are followed by six non-repeating symbols from rows 2 and 3. In Rate $\frac{1}{2}$, six Walsh symbols from rows 0 and 2 are followed by six repeating symbols from rows 1 and 3. In Rate $\frac{1}{4}$, six Walsh symbols from rows 0 and 4 are followed by three groups of six repeating symbols from rows 1 and 5, rows 2 and 6, and rows 3 and 7. Finally, in Rate $\frac{1}{8}$, six Walsh symbols from rows 0 and 8 are followed by seven groups of six repeating symbols from rows 1 and 9, rows 2 and 10, ..., and rows 7 and 15. To reduce the interference, the transmitter is turned off for the repeated Walsh symbols. Therefore, 16, 8, 4, and 2 PCGs are transmitted for Rates 1, $\frac{1}{2}$, $\frac{1}{4}$, and $\frac{1}{8}$. To reduce the impact of fading or bursty errors, transmissions of PCGs are pseudo-randomized with a PN sequence.

Two types of PN sequences are used in the reverse link of IS-95. A long PN sequence with $n = 42$ is used to encrypt the data of each MS at a frequency of four chips per Walsh chip, which corresponds to 1.2288 Mcps. Each MS is assigned a unique 42-bit mask, and the period of the long PN sequence is $P = 2^{42} - 1 \approx 4.4 \times 10^{12}$, which lasts over 41 days. Thus even if an illegal receiver successfully establishes synchronization with the long PN sequence and locates its beginning, i.e., the first one after 41 zeros, it will be virtually impossible to identify the phase shift of the sequence without accurate information on the mask used.

The long PN sequence is generated based on the primitive polynomial,

$$f(x) = 1 + x^7 + x^9 + x^{11} + x^{15} + x^{16} + x^{17} + x^{20} + x^{21} + x^{23} + x^{24} + x^{25} + x^{26} + x^{32} + x^{35} + x^{36} + x^{37} + x^{39} + x^{40} + x^{41} + x^{42}. \tag{4.35}$$

Its reciprocal is

$$f^*(x) = 1 + x + x^2 + x^3 + x^5 + x^6 + x^7 + x^{10} + x^{16} + x^{17} + x^{18} + x^{19} + x^{21} + x^{22} + x^{25} + x^{26} + x^{27} + x^{31} + x^{33} + x^{35} + x^{42}. \tag{4.36}$$

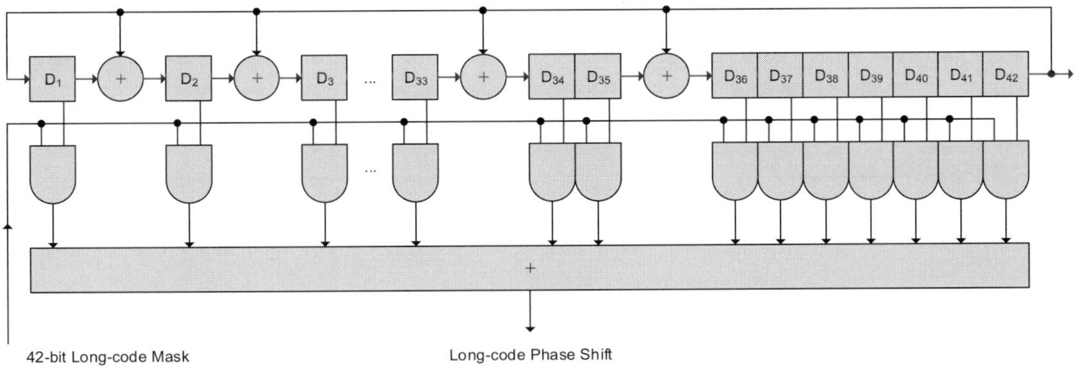

Fig. 4.21 Masking of long PN sequence.

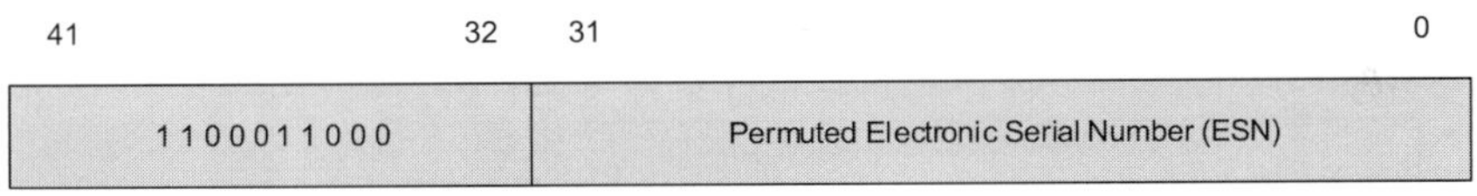

Fig. 4.22 Generation of 42-bit mask from ESN.

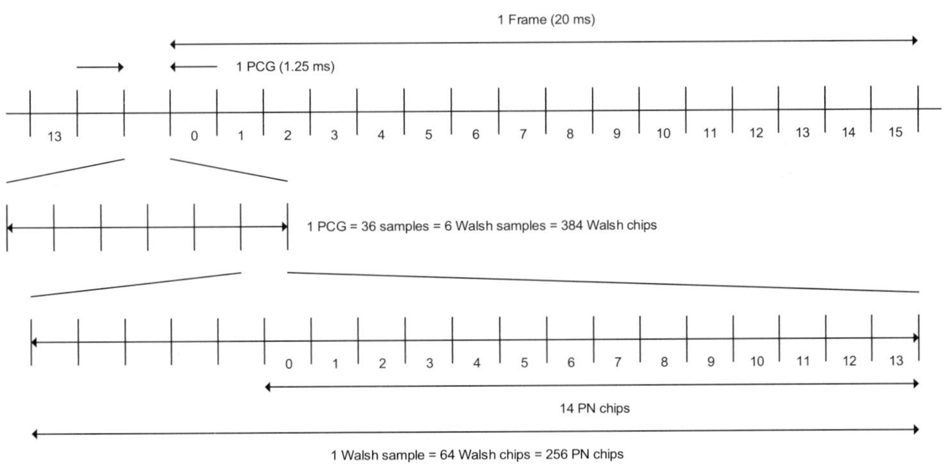

Fig. 4.23 Timing relationship of data burst randomizing.

Figure 4.21 shows the MSRG implementation of $f^*(x)$. The mask is a 42-bit sequence in which bits 41–32 are set to 1100011000 while the remaining 32 bits are set as a permutation of the Electronic Serial Number (ESN) of the MS. The ESN, as shown in Fig. 4.22, consists of the manufacturer code (8 bits), reserved bits (6 bits), and the serial number (18 bits). Identical masks and phase shifts are used in the reverse and forward links.

Figure 4.23 illustrates the timing relationship of the PCGs, Walsh samples, and PN chips. A Walsh sample consists of 64 Walsh chips, each of which is covered by four PN chips. Which PCGs are gated off, i.e., given no transmit power, depends on the chips

Table 4.10 Gating of PCGs.

Rate	Transmitted PCGs	Number of PCGs
1	$0, \ldots, 15$	16
$\frac{1}{2}$	$b_0, 2+b_1, 4+b_2, 6+b_3, 8+b_4, 10+b_5, 12+b_6, 14+b_7$	8
$\frac{1}{4}$	b_0 if $b_8 = 0$, or $2+b_1$ if $b_8 = 1$ $4+b_2$ if $b_9 = 0$, or $6+b_3$ if $b_9 = 1$ $8+b_4$ if $b_{10} = 0$, or $10+b_5$ if $b_{10} = 1$ $12+b_6$ if $b_{11} = 0$, or $14+b_7$ if $b_{11} = 1$	4
$\frac{1}{8}$	b_0 if $(b_8,b_{12}) = (0,0)$, or $2+b_1$ if $(b_8,b_{12}) = (1,0)$, or $4+b_2$ if $(b_9,b_{12}) = (0,1)$, or $6+b_3$ if $(b_9,b_{12}) = (1,1)$, or $8+b_4$ if $(b_{10},b_{13}) = (0,0)$, or $10+b_5$ if $(b_{10},b_{13}) = (1,0)$, or $12+b_6$ if $(b_{11},b_{13}) = (0,1)$, or $14+b_7$ if $(b_{11},b_{13}) = (1,1)$	2

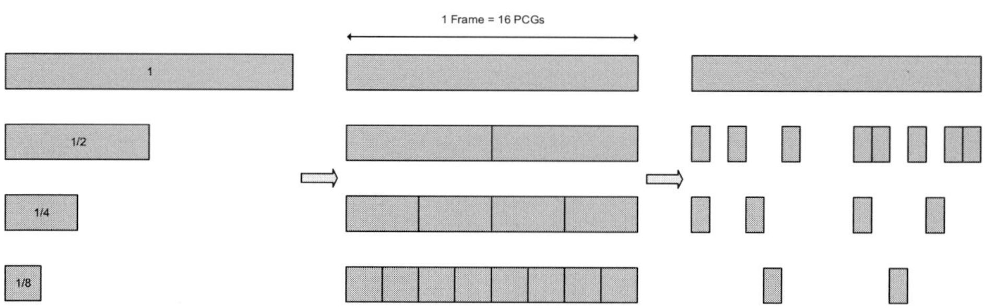

Fig. 4.24 Data burst randomizing.

of the long PN sequence that span the previous speech frame. The 14 last PN chips, $b_0, b_1, \ldots, b_{13}$, of the 15th PCG of the previous frame determine the pattern of gating off, as outlined in Table 4.10.

Note that a complete period of one PCG is necessary to read the PN chips and apply the computed values for data burst randomizing. The rate of speech frame reported by the QCELP speech encoder and the 14 chips determine the PCGs to be gated off. From the 16 PCGs of each speech frame, 16, 8, 4, and 2 PCGs are transmitted for Rates 1, $\frac{1}{2}$, $\frac{1}{4}$, and $\frac{1}{8}$, respectively. Figure 4.24 illustrates the process of data burst randomizing, where for each rate, only a copy of the processed speech data is transmitted and the temporal location where PCGs are gated off is pseudo-randomly determined. Note that with data burst randomizing, the same energy per coded symbol is maintained for all rates, as long as the transmit power is not adjusted by the network.

After data burst randomizing, the symbols are separated into two streams, the in-phase and quadrature components, and multiplied by two short PN sequences with $n = 15$. Let the randomized symbols be $s_0, s_1, s_2, s_3, s_4, s_5, \ldots$ Then the even symbols are fed to the upper branch of Fig. 4.13 as the in-phase components, and the odd symbols are fed to the lower branch, as the quadrature components. The period of the short PN sequences is $2^{15} - 1 = 32767$ chips, which corresponds to 26.67 ms. By observing the chips and locating 14 consecutive zeros, synchronization to the short PN sequences can be easily established. Therefore the short PN sequences will be of little help in the encryption of data. Instead, the phase of the short PN sequences is used to identify the BTSs.

A phase difference of 64 chips is considered as the minimum distance required for the correlator to identify each BTS. To align the period of the short PN sequences with an integer multiple of the minimum phase difference, a zero is inserted at the end of 14 consecutive zeros, which generates a set of $\frac{2^{15}}{64} = 512$ phases. Apparently it is not enough to cover the entire BTSs of typical mobile communications systems with only 512 phases. However, if the phases are re-used geographically in a manner similar to the channels of TDMA, there will not be a significant shortage of phases. The two short PN sequences of IS-95 are generated with the following primitive polynomials,

$$\begin{aligned} f_I(x) &= 1 + x^2 + x^6 + x^7 + x^8 + x^{10} + x^{15}, \\ f_Q(x) &= 1 + x^3 + x^4 + x^5 + x^9 + x^{10} + x^{11} + x^{12} + x^{15}. \end{aligned} \tag{4.37}$$

Figure 4.25 shows the MSRG implementations of their reciprocal polynomials.

In IS-95, the BTS periodically reduces the transmit power of the forward link to each MS, which reports the measured frame error rate to the BTS. Based on this information and other factors, such as the available power headroom, cell loading level, or service policy, the BTS controls the transmit power. The transmit power of the forward link is controlled more slowly over a smaller range than that of the reverse link. IS-95 does not specify the procedures of forward link power control in detail but leaves them to the discretion of the implementation.

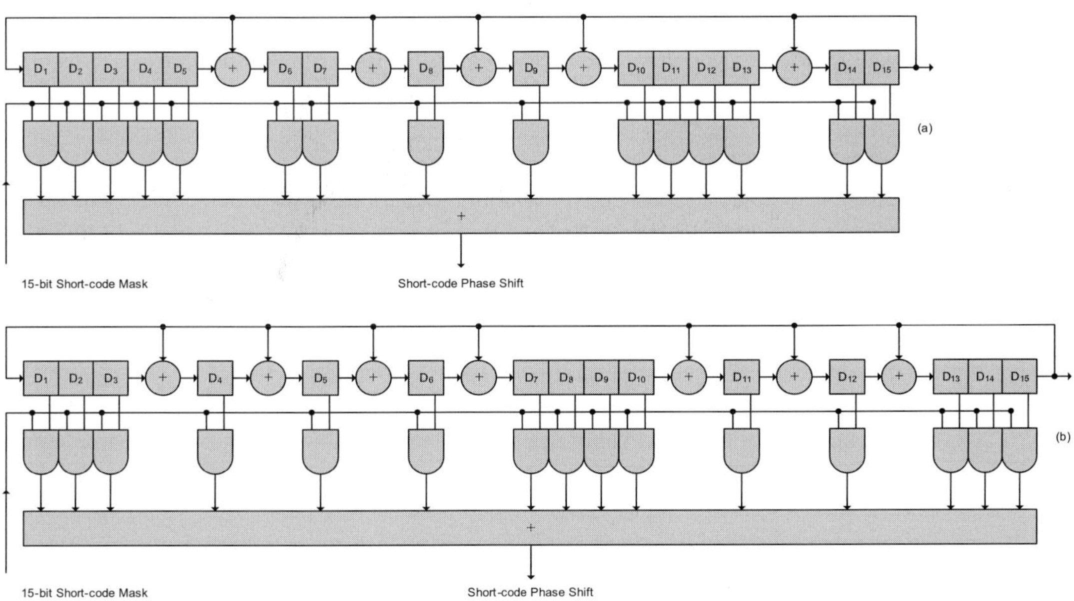

Fig. 4.25 Generation and making of short PN sequences. (a) MSRG implementation of $f_I(x)$. (b) MSRG implementation of $f_Q(x)$.

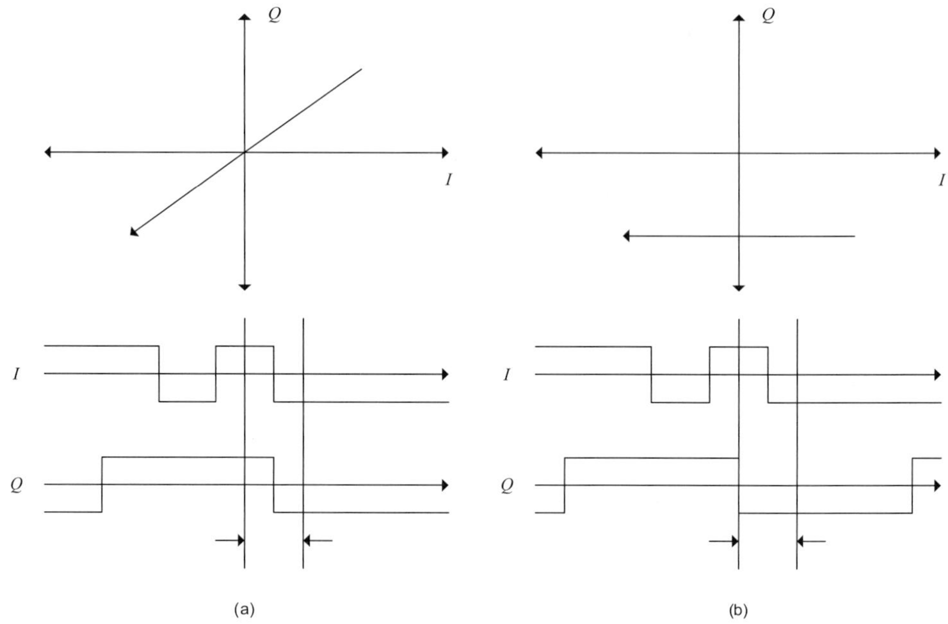

Fig. 4.26 Phase transition. (a) QPSK. (b) Offset QPSK.

Before the in-phase and quadrature components are entered into their respective low-pass filters, the quadrature components are delayed by a half chip. Then the symbols of each branch, represented in binary format, are converted to analog signals with appropriate channel gains. This modulation technique is called Offset QPSK (OQPSK). Unlike QPSK, OQPSK does not allow crossing of the origin between two consecutive symbols, which degrades the efficiency of the low-quality nonlinear power amplifiers used by typical MSs. Figure 4.26 compares the phase transitions in QPSK and OQPSK. In QPSK, it is possible for the in-phase and quadrature components to change the sign between two consecutive symbols simultaneously. This requires the power amplifier to exhibit linearity in the low power range. On the other hand, in OQPSK, either the in-phase or quadrature component, but not both, can change the sign at one time.

The low-pass filter, $h(t)$, is designed to shape the waveform spectrum so that the channel bandwidth constraints are met and the Inter-Symbol Interference (ISI) is minimized. Figure 4.27 shows the frequency response of $h(t)$, in which the filter coefficients are represented with a high precision, using 64 bits per coefficient. In practical implementations, they will be quantized into a lower resolution and the frequency response will deviate from the ideal shape. For the spectral mask of typical IS-95 implementations, ten or more bits per filter coefficient are sufficient to meet the requirements [Do and Feher (1996)].

Finally, the waveforms are modulated to RF carriers after an appropriate channel gain A is applied, and summed to generate $S(t)$, which is amplified and transmitted over the wireless channel. In IS-95, the Radio Configuration (RC) represents a set of

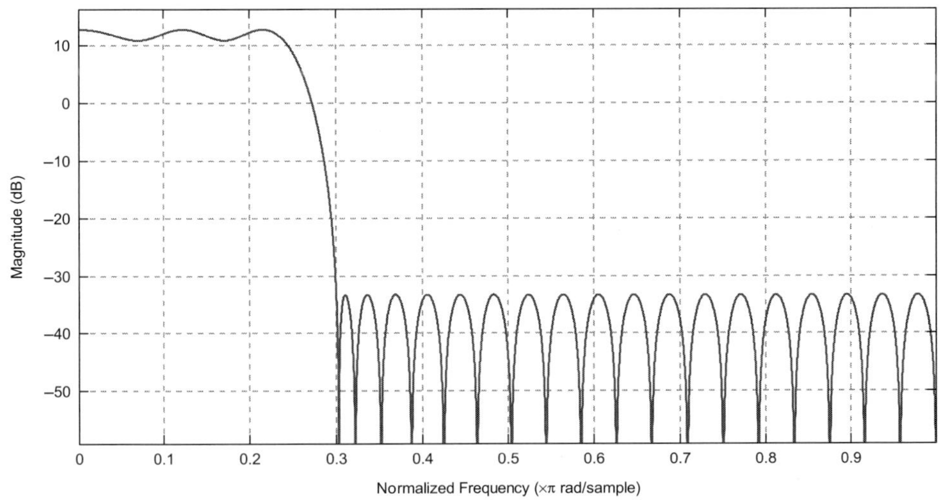

Fig. 4.27 Frequency response of 48 coefficient low-pass filter $h(t)$.

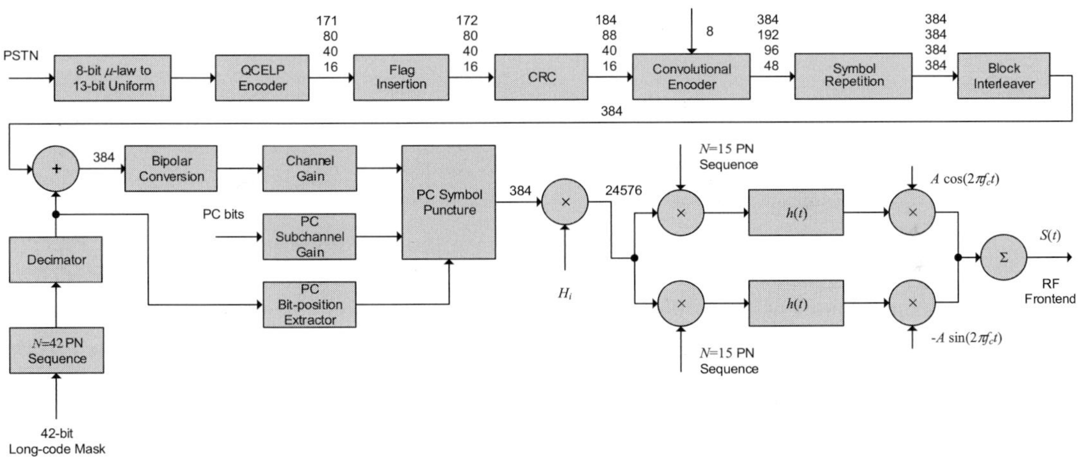

Fig. 4.28 Speech and radio signal processing operations of IS-95 forward link for QCELP (RC 1).

forward and reverse traffic channels characterized by physical layer parameters. Figure 4.13 corresponds to the Reverse Fundamental Channel (R-FCH) for RC 1.

4.3.4 Forward Link Signal Processing

Figure 4.28 shows the signal processing operations for the IS-95 forward link. The speech signal in the initial 64 kbps PCM format is expanded to the 13-bit Uniform format and input to the QCELP speech encoder. As in the case of the reverse link, it outputs 171, 80, 40, and 16 bits, depending upon the acoustic nature of the input signal, as Rates 1, $\frac{1}{2}$, $\frac{1}{4}$, and $\frac{1}{8}$. For Rate 1, a 1-bit mixed mode (MM) flag is attached to the speech frame.

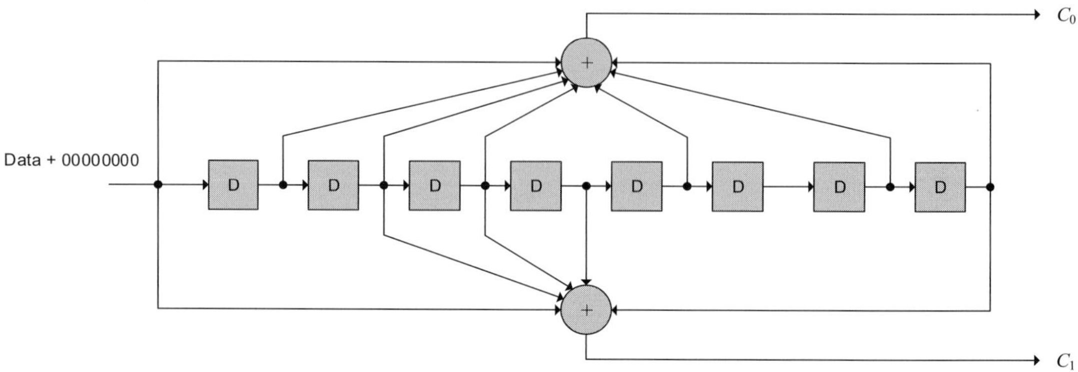

Fig. 4.29 Rate $\frac{1}{2}$ convolutional encoding.

For Rates 1 and $\frac{1}{2}$, a 12-bit CRC and an 8-bit CRC are computed from 172 and 80 bits, respectively, using the same shift register circuitry as in the reverse link. Likewise, for Rates $\frac{1}{4}$ and $\frac{1}{8}$, a CRC is not computed. Then padding bits consisting of eight zeros are attached to each frame before the bits of each rate are input to a rate $\frac{1}{2}$ convolutional encoder with a constraint length $K = 9$, as shown in Fig. 4.29. The forward link is more likely to have LOS paths, due to the elevated position of BTS antenna, than the reverse link. As a result, in the forward link, a lower level of error protection is sufficient to maintain the link quality. Power control for the forward link is also more relaxed than it is for the reverse link.

After convolutional encoding, 384, 192, 96, and 48 symbols are generated for Rates 1, $\frac{1}{2}$, $\frac{1}{4}$, and $\frac{1}{8}$, respectively. The symbols of rates lower than Rate 1 are repeated to equalize the number of symbols. Then the symbols are input to the interleaver, a 16 (column) by 24 (row) array, column-wise but read out in a more complex fashion than in the reverse link. Figure 4.30 shows the read-out order of the 16 by 24 array, based on the so-called *bit-reversal interleaving* strategy.

Let the size of the interleaver be represented as a J (column) by 2^m (row) array, where $J = m = 6$. Then the read-out order of the ith bit is computed as

$$2^m(i \bmod J) + \text{bit\_reversal}(\text{floor}((i/J), m)), \tag{4.38}$$

where floor(x) truncates x to the largest integer not exceeding x. bit_reversal(x,m) converts a non-negative integer x to an m-bit binary number and reverses the order of bits before reconverting the number into a decimal format. Therefore the read-out order of bits $0, 1, 2, 3, 4, 5, 6, \ldots,$ is $0, 64, 128, 192, 256, 320, 32, \ldots,$ respectively.

In the reverse link, repeated PCGs are gated off to maintain the same energy per encoded symbol for all rates. In the forward link, transmit powers of rates with repeated symbols are attenuated instead. For example, transmit powers of Rates $\frac{1}{2}$, $\frac{1}{4}$, and $\frac{1}{8}$ are reduced to a half, a quarter, and a one-eighth, respectively. Figure 4.31 illustrates the variation of forward link transmit power according to the variation of speech rate. Notice that as the power is controlled by the BTS, transmit power for each rate is also correspondingly adjusted to maintain the same ratios.

	Column 0	Column 1											Column 15
Row 0	0	8	4	12	2	10	...	5	13	3	11	7	15
Row 1	64	72	68	76	66	74	...	68	77	66	75	71	79
	128	135	132	140	130	138	...	133	141	131	139	135	143
	192	200	196	204	194	202	...	197	205	195	203	199	207
	256	264	260	268	258	266	...	261	269	259	267	263	271
	320	328	324	332	322	330	...	325	333	323	331	327	335
	32	40	36	44	34	42	...	37	45	35	43	39	47
	96	104	100	108	98	106	...	101	109	99	107	103	111
	...	...	...	...	...	...	...	...	...	...	...	...	...
	272	280	276	284	274	282	...	277	285	275	283	279	287
	336	344	340	348	338	346	...	341	349	339	347	343	351
	48	56	52	60	50	58	...	53	61	51	59	55	63
	112	120	116	124	114	122	...	117	125	115	123	119	127
	176	184	180	188	178	186	...	181	189	179	187	183	191
	240	248	244	252	242	250	...	245	253	243	251	247	255
	304	312	308	316	306	314	...	309	317	307	315	311	319
Row 23	368	376	372	380	370	378	...	373	381	371	379	375	383

Fig. 4.30 Read-out order of forward link interleaver.

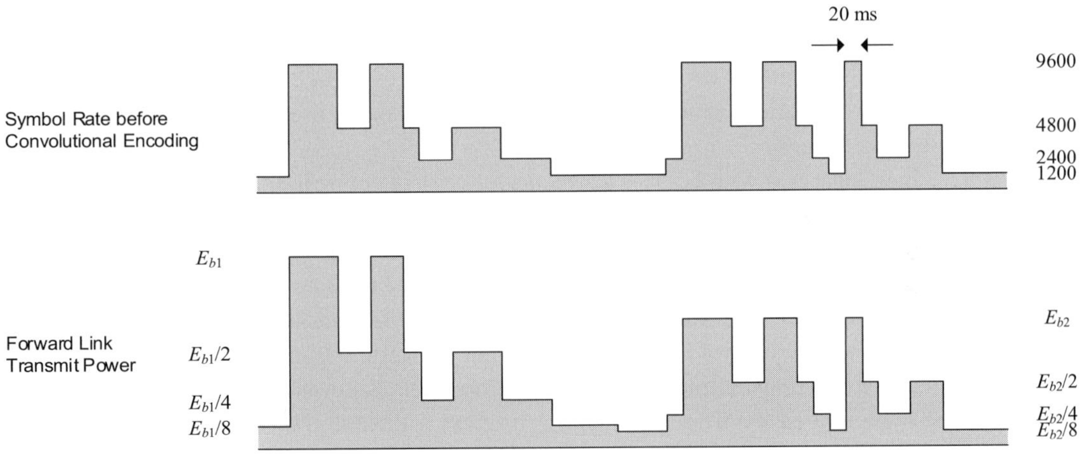

Fig. 4.31 Rate-driven power scaling.

Notice also that in IS-95, during the interleaving and frame construction process, symbols of an encoded speech frame are not mixed with those of the previous or following frames, as in D-AMPS or GSM where the two spaces in each time slot are filled with bits from two consecutive speech frames. The VBR speech coding and allocation of the same energy for each coded symbol limits the allowed range of interleaving within each speech frame.

In IS-95, transmit power of the reverse link is strictly controlled to align the levels of received power from the MSs to similar values. Therefore, power control of the reverse link is more important in CDMA than in TDMA since the BTS of CDMA has to identify the signal of each MS without the support of either guard band or guard time. Moreover, as the range of power control is much wider in CDMA than in TDMA, strong signals of some MSs can overwhelm the weak signals of other MSs, rendering it difficult for the BTS to isolate the weak signals. This phenomenon is called the *near-far effect*.

The same long PN sequence used in the reverse link is used in the power control of the forward link, albeit in a different fashion. To match the symbol rate from the interleaver, the long PN sequence is sampled at every 64th chip, generating a phase-shifted version of the original sequence at a lower rate. The decimated PN sequence is then multiplied by the interleaved symbols for encryption. The binary signals, 0 and 1, are converted at this step to $+1$ and -1 before the channel gain is adjusted. As in the reverse link, each 20 ms frame period is divided into 16 PCGs, each of which is spanned by 24 chips of the decimated samples. Two symbols of each PCG are replaced with power control bits, 00 or 11, for the reverse link. The start location of the power control bits is designated by the last four PN chips of the previous PCG, which can specify only from the first to the 16th symbol of each PCG. The update frequency of the reverse link power control is therefore $\frac{1}{1.25\times10^{-3}} = 800$ Hz, which is significantly higher than the frequency of SACCH in GSM which carries power control messages in that system.

In Fig. 4.32, the last four PN chips of the previous PCG are $c_0c_1c_2c_3 = 0111$. Then the start location of the power control bits is $c_3 * 2^3 + c_2 * 2^2 + c_1 * 2^1 + c_0 = 2^3 + 2^2 + 2 + 0 = 14$. The 17th and following bits are not used for the start location of power control. To minimize the processing delay, channel coding is not applied to the power control bits. Moreover, the bits are punctured over, i.e., replace, the existing symbols. Puncturing of power control bits obliterates a part of speech data but the loss is within the error correction capability of the convolutional coding used, if the energy of received signal is enough.

In comparison with GSM where a power control command or request is transmitted using SACCH on every 13th frame of a 26-multiframe, i.e., at a maximum frequency of 8.3 Hz, in IS-95, the information is punctured over speech data. The former approach for control signaling has the advantage of efficiently packing a set of information into a periodically reserved transmission opportunity. In this classic signaling method, introduction of new control information does not impact the frame structure. However, the frequency of transmitting urgent information, such as power control command, cannot be increased easily. The latter approach offers a flexible transmission opportunity for short, but urgent, control information without reducing the speech quality significantly.

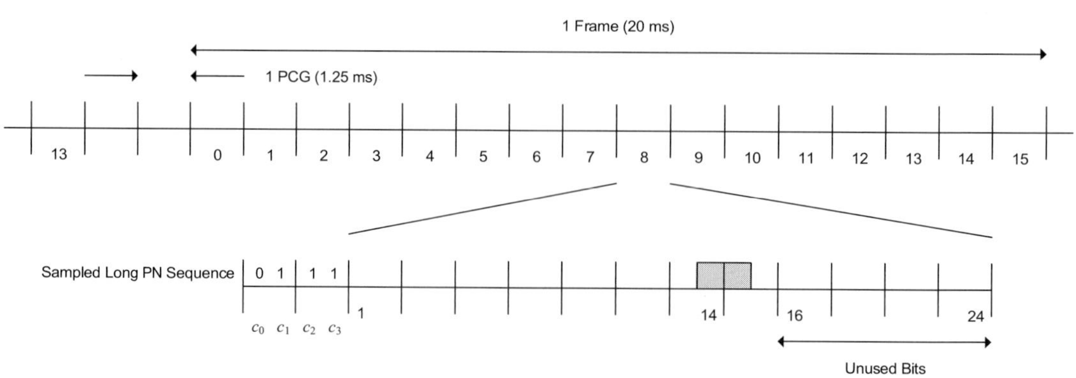

Fig. 4.32 Power control of reverse link.

Then multiplying a Walsh function of length 64 to each power-controlled symbol spreads the information spectrum to 1.2288 MHz, which also provides orthogonality for each channel. Although 64 forward link channels can be differentiated with the Walsh functions, only up to 55 channels can be assigned for voice or data services, leaving the remaining channels for the control purposes, such as the pilot or broadcast channels.

For modulation, conventional QPSK is used in the forward link. After spreading the information spectrum, the even and odd symbols are separated into two streams, the in-phase and quadrature components, and multiplied by two short PN sequences with $n = 15$ before they are applied to two low-pass filters. Unlike in the reverse link which uses binary Walsh sequences and binary short PN sequences, bi-level Walsh functions and bi-level short PN sequences, i.e., represented in -1 or 1, are used in the forward link. Then the waveforms are modulated to RF carriers, after an appropriate channel gain A is applied, and summed to generate $S(t)$, which is amplified and transmitted over the wireless channel.

Notice that in this case, the transmit power is controlled at two steps. The first channel gain, applied after the bipolar conversion, is used to control the relative power level of the channels to an MS while A is used to control the relative power level of the channels to the MSs. Figure 4.28 shows the number of bits for each rate at each step in the speech and radio signal processing procedures of the IS-95 forward link, which is the Forward Fundamental Channel (F-FCH) for RC 1. Radio configurations with different indexes can be simultaneously used in the reverse and forward links.

4.4 References

3GPP. 2000. TS 06.32 V8.0.1 Voice Activity Detection (VAD). November.

Do, G. L., and Feher, K. 1996. Efficient Filter Design for IS-95 CDMA Systems. *IEEE Transactions on Consumer Electronics*, **42**(4).

EIA/TIA. 1990. Interim Standard IS-96: Speech Service Option Standard for Wideband Spread Spectrum Digital Cellular System. April.

EIA/TIA. 1995. Interim Standard IS-95-A: Mobile Station – Base Station Compatibility Standard for Dual-Mode Wideband Spread Spectrum Cellular System. May.

Golomb, S. W. 1981. *Shift Register Sequences*. Aegean Park Press.

Lee, J. S., and Miller, L. E. 1998. *CDMA Systems Engineering Handbook*. Artech House Publishers.

Ross, A. H. M., and Gilhousen, K. S. 1996. CDMA Technology and the IS-95 North American Standard. In: Gibson, J. D. (ed), *The Mobile Communications Handbook*. CRC Press.

Schroeder, M. R., and Atal, B. S. 1985. Code-Excited Linear Prediction (CELP): High-Quality Speech at Very Low Bit Rates. *IEEE Conference on Acoustics, Speech, and Signal Processing*, **10**(April).

Simon, M. K., Omura, J. K., Scholtz, R. A., and Levitt, B. K. 1985. *Spread Spectrum Communications*. Computer Science Press.

Verdú, S. 1998. *Multiuser Detection*. Cambridge University Press.

Viterbi, A. J. 1995. *CDMA: Principles of Spread Spectrum Communication*. Addison-Wesley.

5 Evolution of CDMA Systems

Like GSM, IS-95 was gradually but continuously enhanced to improve speech quality and network capacity. The system was renamed cdma2000 after significant changes were made over the signal processing chains. As with the evolution of GSM, enhanced speech compression, robust wireless transmission, and dynamic coordination of compression and transmission were the key opportunities for renovation, but other changes exploited the characteristics of the CDMA and VBR principles. In this chapter, we discuss these approaches and analyze the effectiveness of them with the experimental and field results. In addition, upper bounds for the capacity of generic CDMA systems are derived for the reverse and forward links.

5.1 Enhancements in Speech Compression

Just as the Full-Rate speech codec was replaced by the Enhanced Full-Rate for improved speech quality and increased network capacity, the initial speech codec for IS-95, QCELP, exhibited limitations particularly under poor channel conditions, necessitating a higher-performance speech codec. In GSM, where the fixed structures of frames and slots limited only the total bit-rate, higher quality could be achieved at similar or even lower bit-rates by employing more sophisticated speech compression algorithms. New speech codecs with bit-rates that were different from that of the Full-Rate were introduced relatively easily, since the same total bit-rate could be maintained by re-designing the CRC generation, channel coding, and interleaving procedures. In IS-95, by way of contrast, the signal processing operations are tightly integrated with the processing of either Walsh sequences or PN sequences, which makes it more difficult to introduce new speech codecs with bit-rates that are different from those of QCELP.

Under these constraints, several approaches were considered to improve the speech quality and network capacity of IS-95. The first and simplest approach was to design a higher bit-rate speech codec and increase the channel coding rate, $\frac{k}{n}$, or apply the same channel coding rate while trimming some of the symbols, so that the same symbol rate would be maintained. In this case, low-complexity operations such as CRC computation or bit manipulation need to be modified but major signal processing procedures including PN sequence generation and synchronization, which require more elaborate development and testing caused by the high symbol rate and analog signal processing, could be recycled. Another approach would be to design a new speech codec with the same number of bits for each rate. The quality would be improved from more advanced speech compression algorithms but all of the existing signal processing procedures could be re-used without modifications. A third approach would be to keep the same speech codec but employ more efficient radio transmission techniques, such as stronger channel coding or enhanced modulation. As a matter of course, the third approach is considered only when the outcomes of the first two are not satisfactory, as it is more difficult to modify the radio signal processing in CDMA than in TDMA, due to the higher symbol rates.

5.1.1 QCELP-13 Speech Codec

Because the available computational complexity was limited but improving the speech quality was an urgent issue, the first approach mentioned above was followed, resulting in the QCELP-13 speech codec [3GPP2 (2004a)]. This codec generates four types of encoded speech frames, Rates 1, $\frac{1}{2}$, $\frac{1}{4}$, and $\frac{1}{8}$, at higher bit-rates than those of QCELP. QCELP-13 shares the same architecture with QCELP, as shown in Fig. 4.11, but differs in its parameter quantization and rate decisions to meet the new bit-rates.

For QCELP-13, 266 bits are generated for Rate 1, and 124, 54, and 20 bits are generated for the lower rates. Table 5.1, which has the same set of parameters as Table 4.6 for QCELP, summarizes the bit allocation for each rate. Compared with QCELP, the bits for the LPC update of Rate 1 are decreased from 40 to 32 bits but those for Rates $\frac{1}{2}$ and $\frac{1}{4}$ are increased from 20 and 10 bits, respectively, to 32 bits. The pitch is not updated in Rate $\frac{1}{4}$ but is more frequently updated in Rate $\frac{1}{2}$. The codebook is more frequently updated in Rates 1 from 8 to 16 times per frame. The 11-bit CRC internally computed for Rate 1 is removed in QCELP-13. There are 2–4 unused bits reserved in Rates 1, $\frac{1}{4}$, and $\frac{1}{8}$. Note that on average, 11.75 bits are spent for codebook updating in Rate 1 as the rate uses 12 bits per codebook update in 12 of 16 codebook subframes, and 11 bits per codebook update in the remaining four codebook subframes.

In QCELP, signaling mechanisms based on the RATE_REDUC message are available to control the relative frequencies of Rates 1 and $\frac{1}{2}$ frames, as a limited measure to control the tradeoff between speech quality and network capacity of the reverse link. In QCELP-13, this approach is more refined in that a Rate 1 frame can be replaced not only with a Rate $\frac{1}{2}$ frame but also with a Rate $\frac{1}{4}$ frame, providing a more graceful control of the tradeoff. Table 5.2 shows that the average bit-rate of QCELP-13, including a CRC and padding bits, can be controlled by 1–2 kbps with this message, which is comparable with the bit-rate resolution of AMR.

Figure 5.1 shows the number of bits for each rate at each step in the chain of signal processing operations of the IS-95 reverse link with QCELP-13, which is the Reverse Fundamental Channel (R-FCH). While many steps of the reverse link of QCELP are retained, key differences can be found in the insertion of flag bits, computation of the CRC, and convolutional coding. A one-bit Mixed Mode (MM) flag and a one-bit Erasure Indicator Bit (EIB) are attached to the output bits of all rates. Historically, QCELP-13 was deployed as the Personal Communications Service (PCS), a couple of years after the deployment of original IS-95 using QCELP. The target frequency band of PCS was 1.7–1.9 GHz, which exhibits a higher level of propagation loss and requires more frequent

Table 5.1 Bit allocation of QCELP-13 speech codec.

	Rate 1	Rate $\frac{1}{2}$	Rate $\frac{1}{4}$	Rate $\frac{1}{8}$
LPC updates / frame	1	1	1	1
Samples / LPC update L_A	160 (20 ms)	160 (20 ms)	160 (20 ms)	160 (20 ms)
Bits / LPC update	32	32	32	10
Pitch updates / frame	4	4	0	0
Samples / pitch subframe L_P	40 (5 ms)	40 (10 ms)		
Bits / pitch update	11	11		
Codebook updates / frame	16	4	5	1
Samples / codebook subframe L_C	10 (1.25 ms)	40 (5 ms)	32 (4 ms)	160 (20 ms)
Bits / codebook update	11.75	12	4	6
Reserved	2		2	4
Total	266	124	54	20

Table 5.2 Reducing average bit-rate by shifting Rate 1 down to Rates $\frac{1}{2}$ and $\frac{1}{4}$.

RATE_REDUC	Average bit rate (kbps)	Fraction of Rate 1 encoded at Rate 1	Fraction of Rate 1 encoded at Rate $\frac{1}{2}$	Fraction of Rate 1 encoded at Rate $\frac{1}{4}$
000	14.4	1	0	0
001	12.2	0.7	0.3	0
010	11.2	0.7	0	0.3
011	9.0	0.4	0.3	0.3
100	7.2	0	1	0

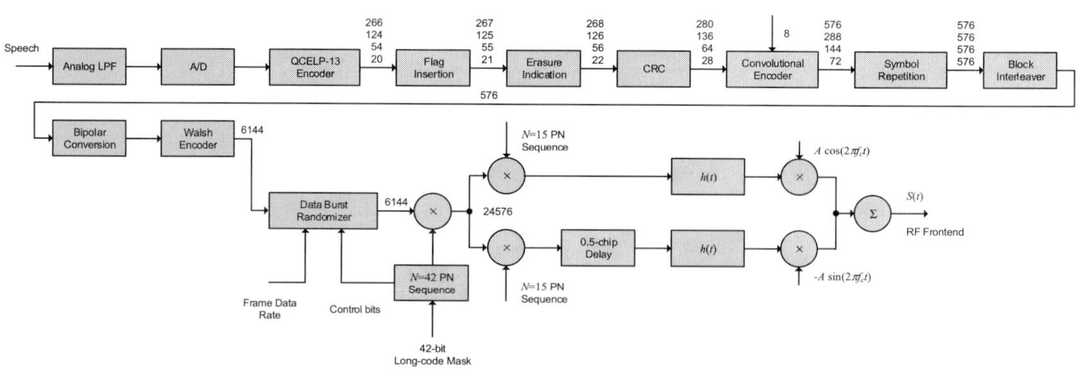

Fig. 5.1 Speech and radio signal processing operations of IS-95 reverse link for QCELP-13 (RC 2).

control of transmit power than in the 800 MHz band, although the cost for the frequency spectrum was lower.

EIB reports to the BTS whether or not the previous speech frame received on the forward link contained errors. Compared with the Power Measured Report Message (PMRM), which is typically transmitted by the MS when the measured FER exceeds

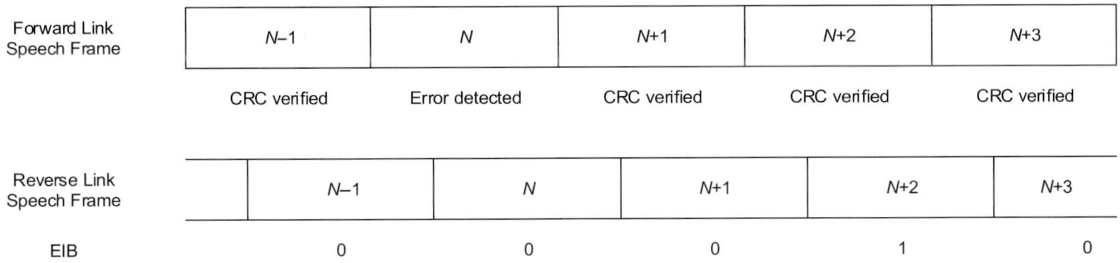

Fig. 5.2 Erasure detection and reporting.

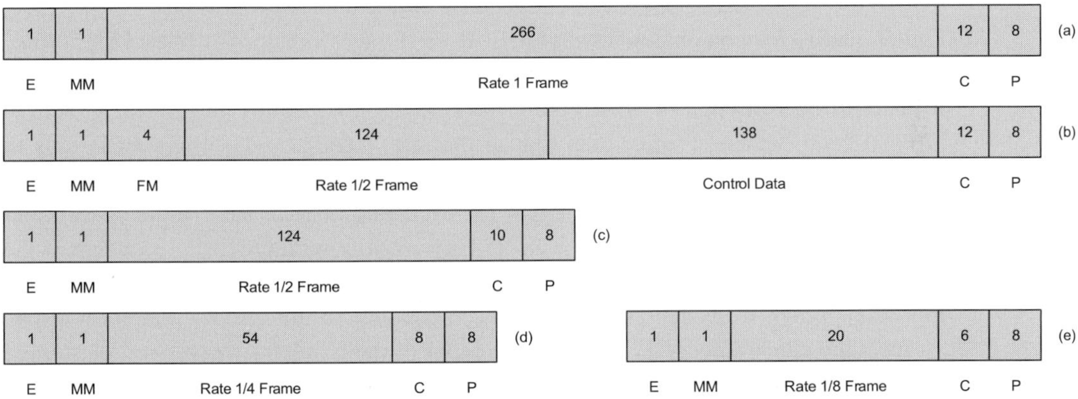

Fig. 5.3 Speech frames complemented with CRC (C) and padding (P) bits.

a threshold, the EIB continuously supplies frame reception information at a minimal transmission cost. Figure 5.2 illustrates the procedures for decoding received speech frames and reporting the results with EIB. Although the EIB is expected to signal the error detection result of a speech frame received within the past two frames, mechanisms to align the timing of reception and transmission, such as the timing advance of D-AMPS or GSM, are not necessary in IS-95, as there is neither guard time nor guard band. In the typical usages of EIB, e.g., power control, what matters to the operation of BTS is the average ratio of frame loss, rather than the detailed information on when and which frames were lost. Note that even if a frame fails in the CRC check, its speech quality may not be completely lost as the damaged frame can be replaced with an interpolated or extrapolated one, or some bits can be salvaged from the frame.

In QCELP, a CRC is applied only to Rates 1 and $\frac{1}{2}$, but it is computed for all rates of QCELP-13. As shown in Figure 5.1, CRCs of 12, 10, 8, and 6 bits are computed and attached to the bits of Rates 1, $\frac{1}{2}$, $\frac{1}{4}$, and $\frac{1}{8}$, respectively. After padding bits consisting of eight zeros are added, the total numbers of bits before convolutional coding are 288, 144, 72, and 36, respectively.

Figure 5.3 shows the bit-stream structure of each rate at this step. Figure 5.3(b) represents the case where the transmission capacity of a Rate 1 frame, 266 bits, is divided between the speech data for a Rate $\frac{1}{2}$ frame, 124 bits, and control information, 138 bits, which is signaled with the mixed mode (MM) flag. The proportion of control information can be increased by reducing the speech rate further, to Rate $\frac{1}{4}$ or $\frac{1}{8}$, or even replacing the whole capacity.

The same rate $\frac{1}{2}$ convolutional encoder used in the forward link of IS-95 for QCELP, generates 576, 288, 144, and 72 symbols, depending on the rate. Once again the symbols for rates lower than Rate 1 are repeated so that each block has 576 symbols. From interleaving to modulation, the signal processing operations are identical to those in the reverse link of IS-95 with QCELP. This means that the overall effect of using QCELP-13 is that it has higher speech bit-rate but lower error resilience, which may or may not exceed the performance of QCELP, depending on the channel conditions. For instance, higher quality may be obtained with QCELP-13 if the MS maintains line-of-sight links with the BTS over a short distance. However, lower quality may result if the MS is located at the cell edge or where the error correction capability of the rate $\frac{1}{2}$ convolutional code, which is lower than that of the rate $\frac{1}{3}$ code used with QCELP, fails to provide enough protection of the speech data.

Figure 5.4 shows the number of bits for each rate at each step in the speech and radio signal processing chain of the IS-95 forward link with QCELP-13, which is the Forward Fundamental Channel (F-FCH). Procedures identical to those of the reverse link are applied, from the QCELP-13 speech encoder to the symbol repetition block. To meet the constraints of 19.2 ksps at the input of the block interleaver and its storage of 384 elements, two out of each six symbols are punctured. After interleaving, the same procedures as used by QCELP are applied until the generated $S(t)$ is amplified and transmitted over the wireless channel. In the terminology of IS-95, speech codecs, such as QCELP, which generate all or a subset of 171, 80, 40, and 16 bits, are classified as Rate Set (RS) 1 while speech codecs generating all or a subset of 266, 124, 54, and 20 bits are classified as RS 2. Note that in Figures 5.1 and 5.4, Rate Configuration (RC) 2 is used for the reverse and forward fundamental channels, respectively.

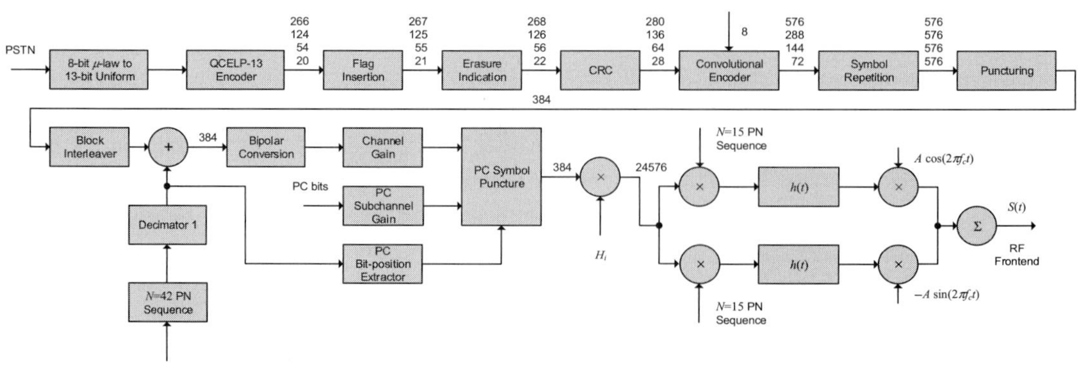

Fig. 5.4 Speech and radio signal processing operations of IS-95 forward link for QCELP-13 (RC 2).

Table 5.3 Bit-rates of rate configurations.

	Forward link bit-rate (bps)		Reverse link bit-rate (bps)	
	Voice	Data	Voice	Data
RC 1	9,600		9,600	
RC 2	14,400		14,400	
RC 3	9,600	153,600	9,600	153,600
RC 4	9,600	307,200	14,400	230,400
RC 5	14,400	230,400	9,600	614,400
RC 6	9,600	307,200	14,400	1,036,600
RC 7	9,600	614,400		
RC 8	14,400	460,800		
RC 9	14,400	1,038,600		

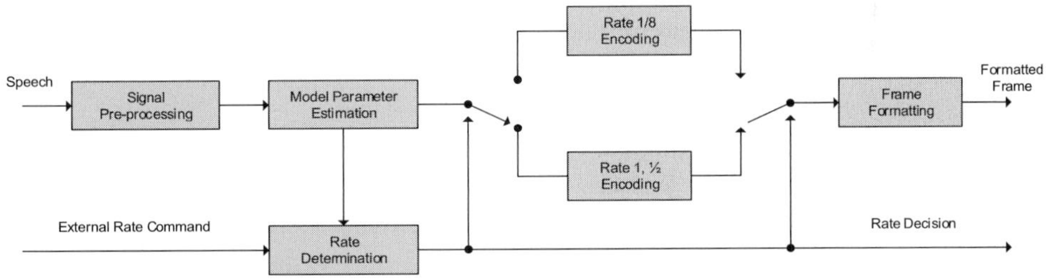

Fig. 5.5 Block diagram of EVRC speech encoder.

The rate configurations refer the sets of parameters for radio signal processing, including the frame length, channel coding rate, and modulation, which determine the achievable Quality of Service (QoS). Table 5.3 shows the maximum bit-rate of each rate configuration supported by cdma2000, for voice and data services. IS-95 supports RC 1 and RC 2, and an evolved version of IS-95, cdma2000, supports the higher configurations. Note that the high bit-rates of data services are realizable only at the sacrifice of other MSs in the cell.

5.1.2 Enhanced Variable Rate Codec

Enhanced Variable Rate Codec (EVRC) is a new speech codec for RS 1 [3GPP2 (1999)]. It generates three types of speech frames with 171, 80, and 16 bits but does not generate the 40-bit Rate $\frac{1}{4}$ frames. Figure 5.5 shows a simplified block diagram of the EVRC speech encoder. Unlike QCELP and QCELP-13, EVRC is based on a different set of compression techniques, called the Relaxed Code Excited Linear Prediction (RCELP), where the speech compression algorithms construct a time-warped version of the original residual that conforms to a simplified pitch contour, rather than that of the original speech signal. The pitch contour is computed by estimating the pitch delay per frame and linearly interpolating the pitch over frames. RCELP has the advantage that the bits saved from pitch updating can be used for stochastic excitations and parameters that can

Table 5.4 Bit allocation of EVRC speech codec.

	Rate 1	Rate $\frac{1}{2}$	Rate $\frac{1}{8}$
Spectral transition indicator	1		
LSP	28	22	8
Pitch delay	7	7	
Delta delay	5		
ACB gain	9	9	
FCB shape	105	30	
FCB gain	15	12	
Frame energy			8
Reserved	1		
Total	171	80	16

provide enhanced performance in poor channel conditions, without impacting perceived speech quality in clear channel conditions. Table 5.4 summarizes how the bits of each rate from the EVRC speech codec are allocated.

The rate is determined as in QCELP or QCELP-13 but only two thresholds are necessary since Rate $\frac{1}{4}$ is absent. Analysis of the operations of QCELP and QCELP-13 reveals that Rates 1 and $\frac{1}{8}$ constitute the majority of encoded frames while Rate $\frac{1}{4}$ is rarely used. Therefore it was determined during the design of EVRC not to use this rate. In the rate-decision procedures, energy of the input speech frame is estimated for the lower and upper frequency bands. Each estimate is compared with two thresholds which are dynamically updated. The higher of the two decisions is chosen as the rate for the frame. The neighborhood-only transition policy is also applied in rate reduction. If it is decided to encode an input frame as Rate $\frac{1}{8}$, after encoding a Rate 1 frame, the input frame is encoded at Rate $\frac{1}{2}$, to avoid an abrupt change in the speech quality.

The signaling mechanisms used by QCELP to control the average bit-rate of the reverse link with the RATE_REDUC message, which control the relative frequencies of Rates 1 and $\frac{1}{2}$ frames, are used similarly with EVRC. Although QCELP-13 and EVRC were deployed at about the same time to address the urgent quality issue of QCELP, EVRC was designed to avoid the capacity load of QCELP-13, in which the higher bit-rate necessitated a larger amount of puncturing and reduced error resilience. Therefore QCELP-13 operation required higher transmit power, which reduced network capacity. EVRC is considered to have similar speech quality as QCELP-13 but it excels in noisy environments because of its built-in noise suppressor. As the second approach to improve the speech quality and network capacity of IS-95, EVRC became the dominant speech codec of IS-95 and cdma2000, rivaling the success of AMR in GSM and W-CDMA.

5.2 cdma2000

The first attempt to improve the speech quality of IS-95, QCELP-13, was a limited success since with the modulation techniques intact, the cost for higher speech bit-rate was weaker channel coding, which had negative impact on the cell radius and the network capacity, especially in the high frequency band of PCS. Thus, even though the new speech codec had a higher bit-rate than QCELP, it could not be considered to be a clear improvement. As the number of subscribers of IS-95 increased, the new CDMA mobile communications system won the competition with D-AMPS to be the major migration path from AMPS. This put pressure on the system designers to update the speech and radio signal processing operations of IS-95 significantly for still higher quality and capacity.

cdma2000 is an updated version of IS-95 with many improvements, particularly in its radio signal processing procedures. While much of the signal processing in cdma2000, which is often referred to as cdma2000 1xRTT to differentiate it from multi-carrier (3xRTT) or data-only (EV-DO) variants of the system, re-uses the functional blocks of IS-95: cdma2000 is equipped with stronger convolutional codes and more sophisticated manipulation of symbols and PN sequences. The same network architecture as shown in Fig. 4.10 is used for cdma2000.

5.2.1 Reverse Link Signal Processing

Figure 5.6 shows the number of bits for each rate at each step in the signal processing chain of the reverse link with the EVRC speech codec, from signal digitization to interleaving. The remaining procedures are shown in Fig. 5.8. The output bit-rates of EVRC are identical to those of QCELP, except for Rate $\frac{1}{4}$ which is not supported by EVRC. Therefore, CRCs of 12, 8, and 6 bits are computed, and added to the bits of Rate 1, $\frac{1}{2}$, and $\frac{1}{8}$ frames, respectively.

CRC is a basic methodology for error detection but its bit-rate cost is not significant, usually fewer than 20 bits for an encoded speech frame. Another advantage of CRC is that its generation and verification require simple shift register circuitry that incurs negligible signal processing overhead. It is also an interesting feature of CRC to require identical levels of complexity in the generation and verification, which is not the case with speech codecs or channel codes. Therefore, adding a CRC and puncturing surplus bits are the two tools typically used to absorb changes in the bit or symbol rates, when the more complex algorithms, such as a speech codec or a channel code, are replaced in the signal processing chain.

One key difference between IS-95 and cdma2000 is that a $K = 9$, Rate $\frac{1}{4}$ convolutional code is used in cdma2000, as shown in Fig. 5.7, versus a $K = 9$, Rate $\frac{1}{3}$

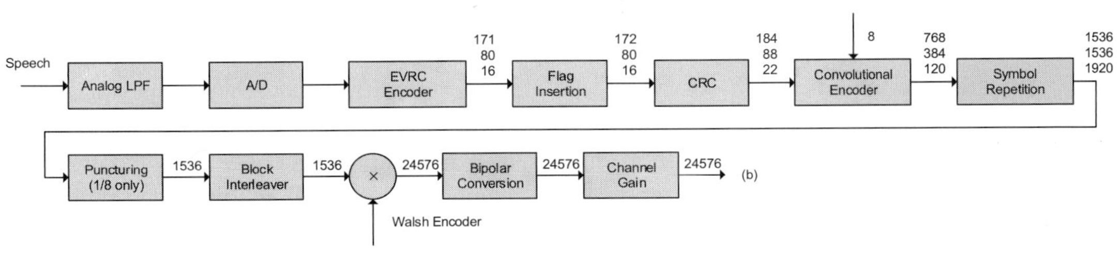

Fig. 5.6 Speech and radio signal processing operations of reverse link for EVRC (RC 3) (1).

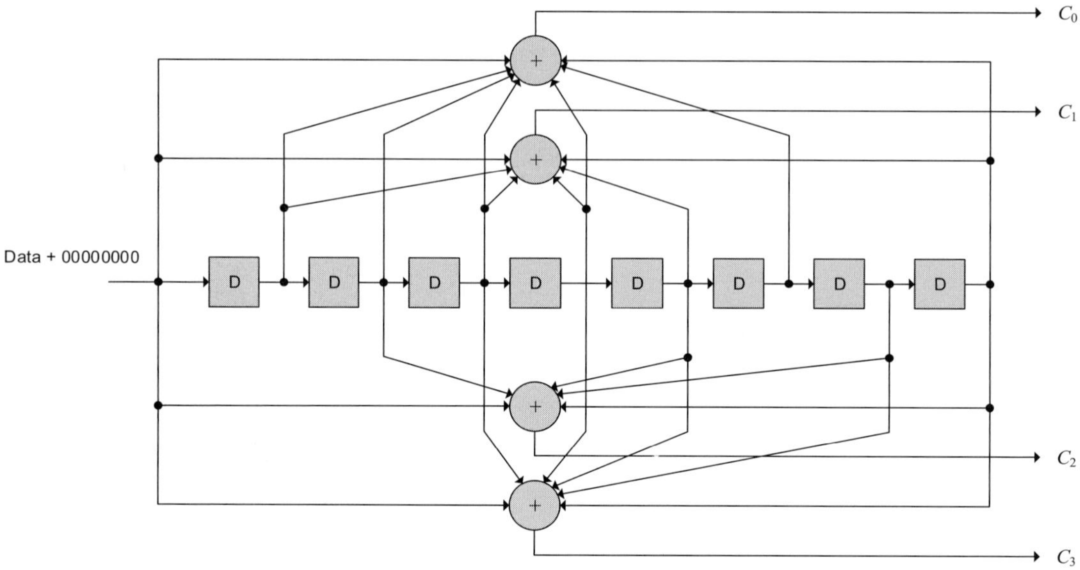

Fig. 5.7 Rate $\frac{1}{4}$ convolutional encoding.

convolutional code of IS-95. In the reverse and forward links of radio configuration 3, a Rate $\frac{1}{4}$ turbo code can be used as well. However, as turbo code exhibits noticeable performance gains only at higher bit-rates and larger codewords than those used for voice service, convolutional code is used in many applications. In the EV-DO mode of cdma2000, only turbo code is used.

The total numbers of bits before convolutional coding are 192, 96, and 30 bits for Rates 1, $\frac{1}{2}$, and $\frac{1}{8}$, respectively, which still fail to reach the familiar ratio of eight, four, and one. The Rate $\frac{1}{4}$ convolutional encoding, plus eight padding bits, generates 768, 384, and 120 bits. Then the symbols are repeated once, twice, and seven times to generate 1536, 1536, and 1920 bits. For the Rate $\frac{1}{8}$ case, one out of every five bits is periodically punctured to align the number of bits to 1536. Then the symbol rate of the reverse link is 76.8 ksps for each rate, which is about three times that of IS-95. The symbols are then input to a 64 (row) by 24 (column) interleaver that outputs the symbols in the *forward-backward bit-reversal interleaving* strategy, whose read-out order is based on different formulas than IS-95.

Multiplying each interleaved symbol by a Walsh function of length 16 spreads the information spectrum to 1.2288 MHz. This operation is similar to that used in the forward link of IS-95 which multiplexes the channels transmitted from a BTS. In the reverse link of cdma2000, this procedure is used to differentiate the channels transmitted from the MSs. Unlike in IS-95, control information, such as that used for assisting synchronization, i.e., the pilot channel, can be simultaneously transmitted with the speech channel, the reverse fundamental channel. The Walsh function used in spreading, the Walsh cover, is $-1, -1, -1, -1, 1, 1, 1, 1, -1, -1, -1, -1, 1, 1, 1, 1$. Note that the Walsh cover is used to differentiate the reverse fundamental channel from other reverse

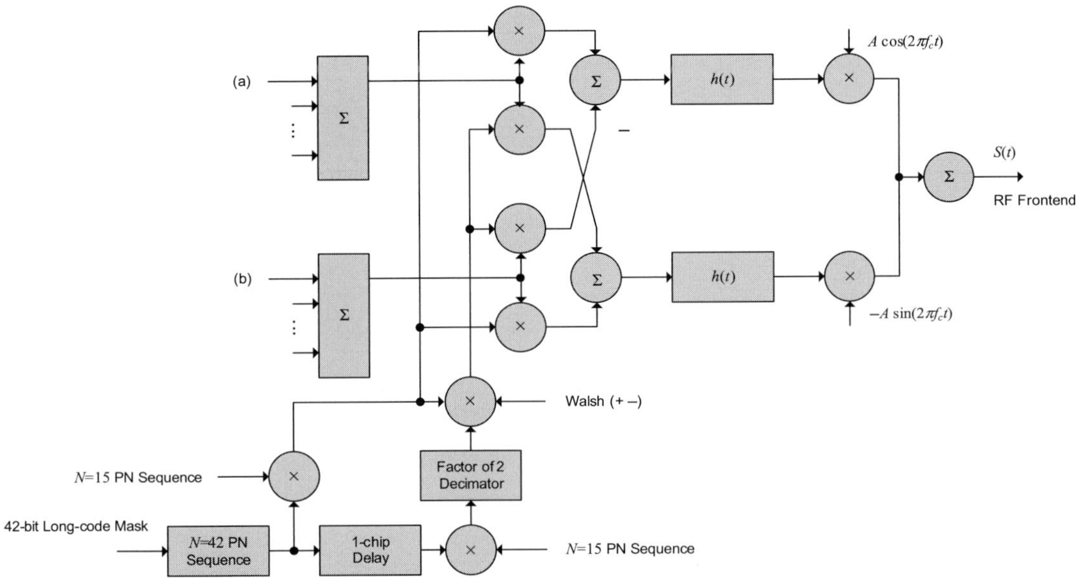

Fig. 5.8 Radio signal processing operations of reverse link (RC 3) (2).

link channels of the same MS, and is therefore used by all MSs for the same channel type.

The gain of each channel is adjusted to maintain controlled power levels among the channels of the reverse link. Figure 5.8 shows the remaining radio signal processing procedures, from PN sequence generation and manipulation to RF modulation and transmission. The gain-adjusted symbols from the signal processing chain of Fig. 5.6 are fed into the lower (quadrature) branch (b). A pilot channel should always be present when the reverse fundamental channel is used. Dedicated control and supplemental channels may also be fed to branch (a). Additional control or access channels can be fed to branch (b), each of which is multiplexed with a distinct Walsh function.

The two streams of data that enter into branches (a) and (b) are multiplied by a complex spreading sequence generated by two short PN sequences with $n = 15$ and the long PN sequence with $n = 42$ used in IS-95. The phase of the short PN sequences is used to identify the BTSs and the phase of the long PN sequence is used for data encryption. The purpose of the complex spreading is similar to that of OQPSK in the reverse link of IS-95, i.e., maintaining the signal envelope as constant as possible by avoiding zero-crossings between symbol transitions. Then the in-phase and quadrature components are entered into low-pass filters and the waveforms are modulated to RF carriers and summed to generate $S(t)$, which is amplified and transmitted over the wireless channel.

In the operations of cdma2000 outlined so far, a key responsibility of the reverse link, power control of the forward link, has not been touched upon. A separate channel fed into branch (a), the reverse pilot channel, assists the BTS to synchronize with the transmission from the MS but it also transmits power control bits for the forward link. Like the reverse fundamental channel, the reverse pilot channel consists of 16 power

Table 5.5 Gating PCGs of reverse pilot channel.

PILOT_GATING_RATE	Gating rate	Number of transmitted PCGs	Transmitted PCGs
00	1	16	0–15
01	$\frac{1}{2}$	8	1, 3, 5, 7, 9, 11, 13, 15
10	$\frac{1}{4}$	4	3, 7, 11, 15

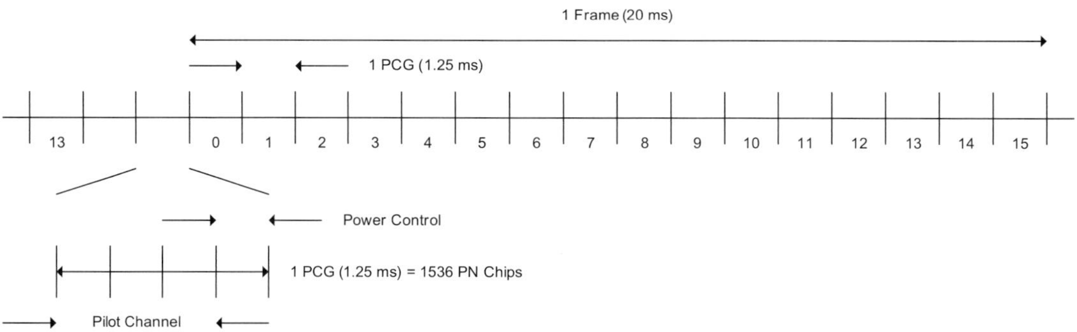

Fig. 5.9 Power control of forward link.

control groups, each of which lasts 1.25 ms with 1536 PN chips, as shown in Fig. 5.9. Three-fourths of each PCG, 1152 chips, are set to zeros, and the remaining 384 chips are used as power control bits or the EIBs, depending on the configuration signaled by the network with the FPC_MODE message. The power control chips are transmitted at the same power level as the pilot chips. The 384 chips are all set to either ones, for example, when the transmit power needs to be increased, or zeros otherwise. The reverse pilot channel is transmitted at higher power before the synchronization with BTS is established. After the fundamental channel is successfully set up, transmit power of the reverse pilot channel is reduced to a normal level and controlled at 800 Hz.

There are other reverse link channels used to initiate and maintain the link, such as the reverse dedicated control channel or the reverse supplemental channel, which are fed into branch (a). Branch (b) also accepts the reverse supplemental channel, reverse common control channel, or enhanced access channel, in addition to the reverse fundamental channel. In comparison with IS-95 where only rate set 2 speech codecs such as QCELP-13 can transmit the EIB in the reverse link, cdma2000 provides generic mechanisms to transmit the information on the status of frame reception simultaneously with speech frames, complementing the short-term procedures for link management such as power control. As in the data burst randomizing of the reverse link of IS-95, the MS gates off some PCGs to reduce the interference and power consumption. Table 5.5 shows the level of gating and the transmitted PCGs corresponding to each gating rate.

A few differences from the gating procedures of the IS-95 reverse links are noticeable. In comparison with the randomized gating pattern of IS-95, PCGs of the reverse pilot channel are transmitted periodically at any gating rate, since the pilot elements of each PCG also need to provide a timing reference to the BTS. In addition, one PCG needs to be transmitted at least once every five ms, to assist the BTS in channel detection and acquisition. Therefore, the gating rate does not fall to $\frac{1}{8}$, as in IS-95. Figures 5.6 and 5.8 correspond to radio configuration 3 of the reverse fundamental channel.

5.2.2 Forward Link Signal Processing Procedures

Figure 5.10 shows the number of bits for each rate at each step in the signal processing chain of the forward link with EVRC, from speech encoding to power control. Procedures identical to those of the reverse link are applied, from the EVRC speech encoder to the convolutional encoder. The Rate $\frac{1}{4}$ convolutional encoding, also used in the reverse link of cdma2000, generates blocks of 768, 384, and 120 bits. The bits are repeated zero, one, and three times to generate 768, 768, and 960 bits. For Rate $\frac{1}{8}$ encoding, one bit out of every five bits is periodically punctured to reduce the number of bits to 768. Then the symbols are entered into a 64 (row) by 12 (column) interleaver which outputs the symbols in the order based on the forward–backward bit-reversal interleaving strategy.

In the forward link of IS-95, the long PN sequence is sampled every 64th chip to generate a phase-shifted version of the original sequence which is then multiplied by the interleaved symbols for encryption. In the forward link of cdma2000, only the bits to encrypt the data are extracted. The long PN sequence is decimated from the masked sequence, every 32nd chip, to determine the puncturing position of the power control bits. This means that each symbol from the interleaver is covered by 32 PN chips. A pair of chips is extracted to encrypt, i.e., scramble, two symbols, which are fed into the in-phase and quadrature components. The chip scrambling the first symbol, e.g., chip $N + 31$, starts with the first symbol while the chip scrambling the second symbol, chip $N+32$, ends with the second symbol. For example, if chips $N+31$ and $N+32$ scramble symbols 5 and 6 respectively, symbol 7 is scrambled by chip $N + 95$. Therefore two consecutive chips are extracted from each set of 64 chips.

PCGs for the reverse link are built into the forward link as a *power control sub-channel*. The masked long PN sequence is decimated so that the chip rate is equal to the interleaved symbol rate, 38.4 ksps. Therefore, each 20 ms frame period is divided into 16 PCGs, each of which is spanned by 48 chips of the decimated sequence. As

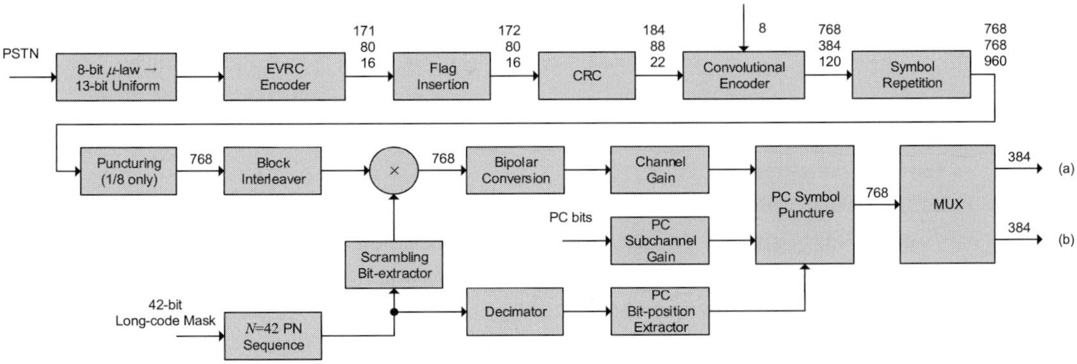

Fig. 5.10 Speech and radio signal processing operations of forward link for EVRC (RC 3) (1).

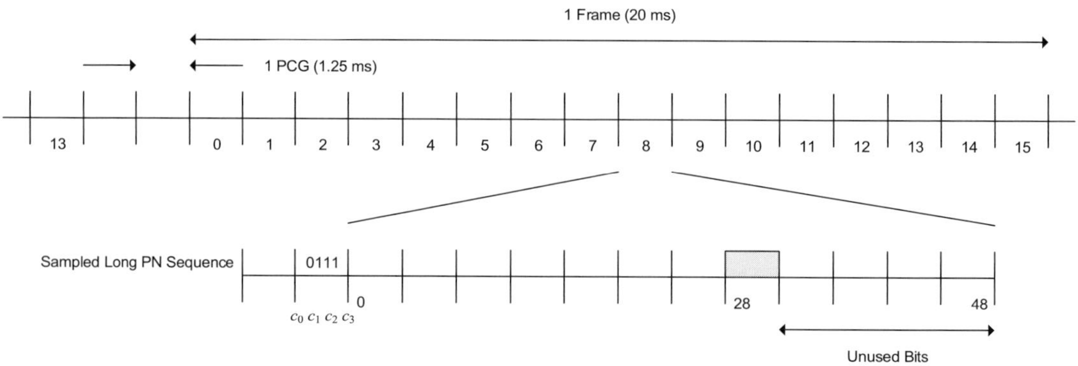

Fig. 5.11 Power control of reverse link.

shown in Fig. 5.11, let the last four PN chips of the previous PCG be $c_0c_1c_2c_3$. Then the starting location for the power control bits is computed as $2(c_32^3 + c_22^2 + c_12 + c_0)$. Notice that while the four chips represent the starting location as one of 0, 1, …, 15, in IS-95 where a PCG consists of 24 symbols, the same four chips cover a range of 0, 2, …, 30 in cdma2000. Therefore $c_0c_1c_2c_3 = 0111$ denotes the starting location of $2(2^3+2^2+2) = 28$, whose puncturing lasts for four symbols. The gain of the power control subchannel is adjusted before the power control bits are punctured over the symbols. Then the symbols are separated into two streams, (a) the in-phase and (b) the quadrature components, which consist of the odd and even symbols, respectively.

Since the forward power control subchannel transmits at higher power than the channel carrying speech data, the forward fundamental channel, the interference can be reduced by gating off the power control channel based on the speech rate. As shown in Fig. 5.12, the power control subchannels of the forward link are gated off, in line with the gating of the reverse pilot channel. The timing of the forward power control subchannel and the reverse pilot channel is arranged so that their overlap is minimized. Note that in the reverse pilot channel, the power control bits are punctured at the end of each PCG. When this is done, the reverse link power control rates of Rates 1, $\frac{1}{2}$, and $\frac{1}{4}$ are 800, 400, and 200 Hz, respectively.

Figure 5.13 shows the remaining radio signal processing procedures. The two streams of symbols entered into branches (a) and (b) are first multiplied by Walsh codes, and then multiplied by a complex spreading sequence as in the reverse link of cdma2000. QPSK is used for modulation in the forward link, as in the forward link of IS-95. The in-phase and quadrature components are entered into low-pass filters, and the waveforms are modulated to RF carriers and summed to generate $S(t)$, which is amplified and transmitted over the wireless channel. Figures 5.10 and 5.13 correspond to radio configuration 3 of the forward fundamental channel.

cdma2000 supports the transmit diversity operation, in which two antennas are used by BTS to transmit the same forward fundamental channel. Transmit diversity has the potential to achieve increased diversity gain relative to that provided by multipath only, which may be used to increase the forward link capacity. Two modes are provided: Orthogonal Transmit Diversity (OTD) and Space Time Spreading (STS),

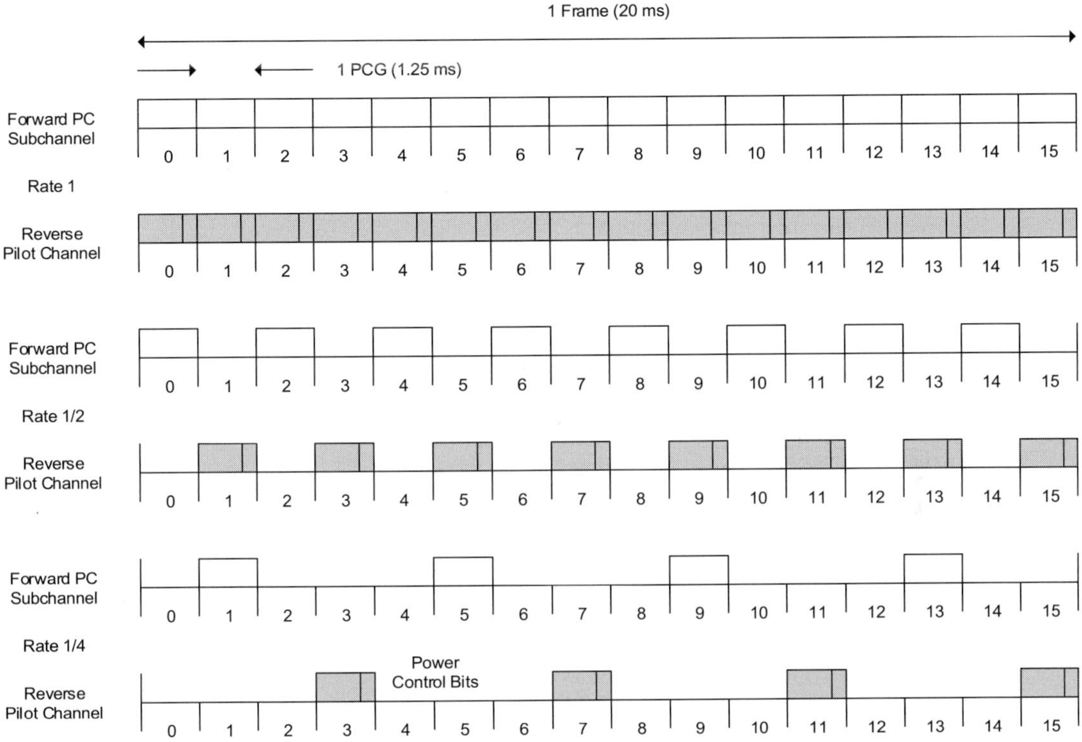

Fig. 5.12 Transmission timing of power control subchannel.

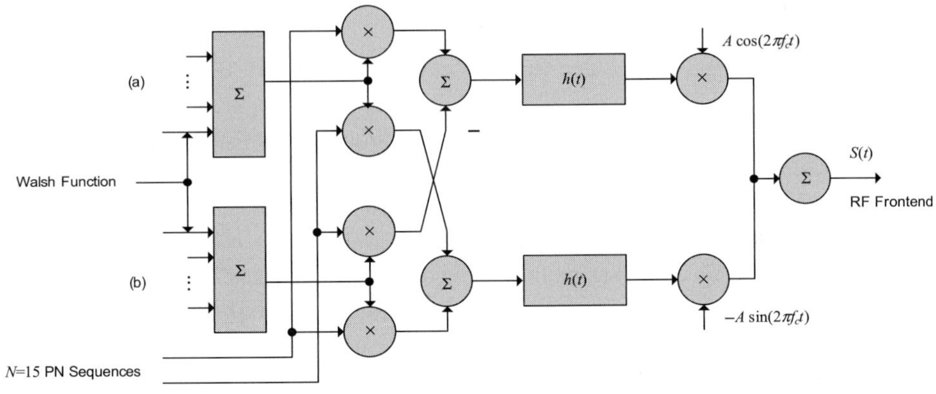

Fig. 5.13 Radio signal processing operations of forward link (RC 3) (2).

which differ in the I and Q mappings from those in Figures 5.8 and 5.13. Transmit diversity was introduced in the beginning of the third generation mobile communications systems such as cdma2000 and W-CDMA. However, they are used in many cases for the transmission of non-real-time data. Transmit diversity evolved into the Multiple-Input and Multiple-Output (MIMO), where multiple antennas are used in the transmitter and the receiver, becoming the basic operating mode of packet-switched mobile communications systems, such as the Long Term Evolution (LTE).

5.3 Enhancements in Coordination of Compression and Transmission

During the transition from IS-95 to cdma2000, enhancements were made using similar design philosophies to those used in the transition from GSM to GSM Phase 2. The basic system parameters, such as the channel bandwidth, were maintained, but the speech signal processing procedures, including the compression algorithms, were redesigned for higher speech quality and stronger error resilience. In the reverse link, separate speech and control channels are simultaneously transmitted, and the reverse pilot channel enables coherent decoding at the BTS.

Unlike GSM, where the fixed number of 456 symbols per 20 ms is maintained for many full-rate channel types, lower-rate convolutional codes result in more symbols in cdma2000 than in IS-95 but do not influence the width of the information spectrum since it is the chip rate, which is higher than the symbol rate, that finally spreads the spectrum to fit a 1.25 MHz band. In comparison with the overall improvements in GSM Phase 2+ induced by the introduction of AMR, the transition from IS-95 to cdma2000 can be regarded as an evolutionary path focused on the improvement of the radio transmission technologies. While the EVRC speech codec provided higher quality than QCELP, few innovations were made with regard to the interaction between the MS and the network that influenced the speech compression algorithms. Selectable Mode Vocoder (SMV) and the 4th Generation Vocoder (4GV) introduced several concepts not used by the speech codecs in GSM or cdma2000, providing more refined mechanisms to control the tradeoff between speech quality and network capacity.

5.3.1 Selectable Mode Vocoder

SMV is a speech codec for rate set (RS) 1, like QCELP and EVRC, whose maximum bit-rate is 171 bits per 20 ms [3GPP2 (2004b)]. Figure 5.14 shows a simplified block diagram of the SMV speech encoder, whose bit-stream composition for each rate is summarized in Table 5.6. Key improvements from QCELP and EVRC can be found in the more refined classifications of Rates 1 and $\frac{1}{2}$, which are mostly used to encode the speech. It is observed that even at the same rate, higher quality can be achieved by applying an encoding method that better matches the acoustic nature of the input speech signal. In SMV, two types, Types 0 and 1, are available for Rates 1 and $\frac{1}{2}$, respectively. The rate sanity flag, set to the result of the CRC check, indicates whether the received frame includes errors or not. If set to 1, the frame is declared a bad one and error concealment is performed.

Rates 1 and $\frac{1}{2}$ frames are encoded using the classic CELP approach where the excitation of the LPC filter is represented by entries in a fixed codebook and an adaptive codebook, which are multiplied by a fixed-codebook gain and an adaptive-codebook gain, respectively. Type 1 frames are generated from stationary voiced speech while Type 0 frames are for other speech types. In Type 0 and 1 frames, the bits are allocated for the excitation parameters and the LPC filter parameters. By allowing more than one encoding method per rate, a part of the bit-stream needs to be assigned to signal the method used to the speech decoder. A one-bit flag, Type, for Rates 1 and $\frac{1}{2}$ contains this

Table 5.6 Bit allocation of SMV speech codec.

	Rate 1 (Type 0)	Rate 1 (Type 1)	Rate $\frac{1}{2}$ (Type 0)	Rate $\frac{1}{2}$ (Type 1)	Rate $\frac{1}{4}$	Rate $\frac{1}{8}$
LSF	27	25	21	21	20	11
Type	1	1	1	1		
Adaptive codebook	26	8	14	7		
Fixed codebook	88	120	30	39		
Codebook gain	28	16	14	12		
Gain					17	
Shape					2	
Energy						5
Rate sanity flag	1	1			1	
Total	171	171	80	80	40	16

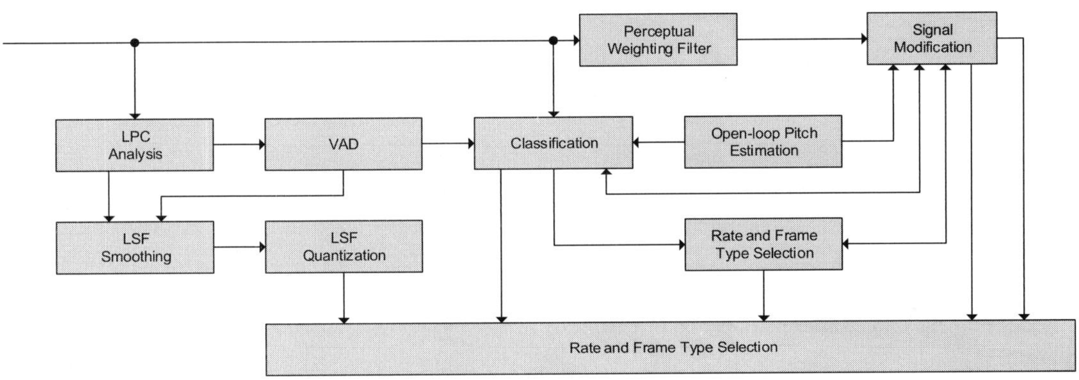

Fig. 5.14 Block diagram of SMV speech encoder.

type information. The Rate $\frac{1}{4}$ frames, once removed in the EVRC, are restored in the SMV. Therefore, there are six types of output bit-streams from the SMV speech encoder.

An interesting feature of SMV not used in the previous speech codecs is a mechanism to adjust the speech encoding method when the background of the input speech contains a considerable level of music. Under such situations, reducing the bit-rate when voice activity is absent disrupts the encoding of the background music, which is annoying to the listener. One quick remedy is to maintain the maximum bit-rate when music is detected in the background.

Note that music, in general, does not fit the acoustic model used in the linear predictive coding of speech as music is not generated through the human vocal tract. Therefore, music or audio signals are typically sampled at higher rates than are used for speech, and compressed using encoding methodologies whose bandwidth of interest is several times wider than that for speech. In addition, audio signal is often encoded with relaxed delay requirements. In the SMV speech encoder, the Music Detector (MD) recycles some parameters of the voice activity detector (VAD) and extracts additional parameters from

Table 5.7 Speech types and rates for Mode 0.

	Rate $\frac{1}{8}$	Rate $\frac{1}{4}$	Rate $\frac{1}{2}$	Rate 1
Silence	Yes		Yes	
Noise-like			Yes	Yes
Unvoiced			Yes	Yes
Onset				Yes
Non-stationary voiced				Yes
Stationary voiced				Yes

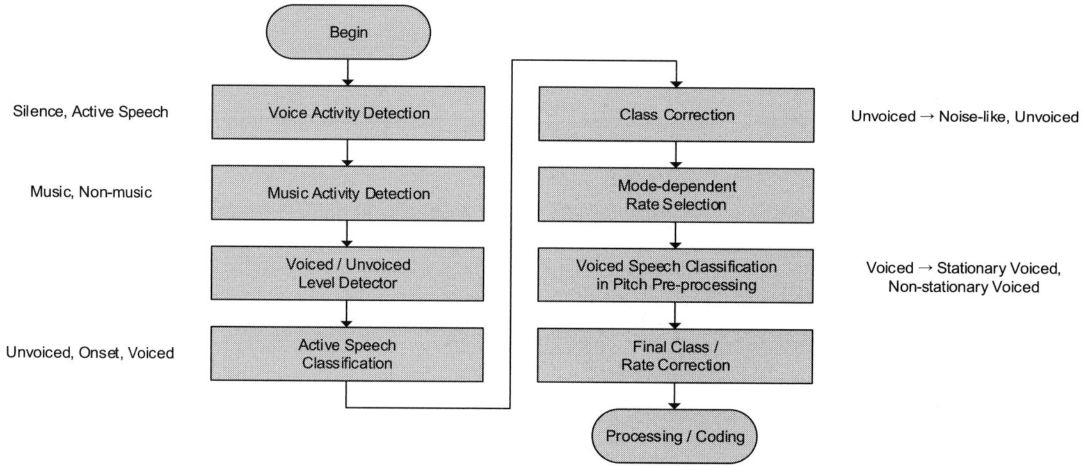

Fig. 5.15 Voice classification procedures.

the speech frame for detection of music. The rate-decision algorithm sets the rate to Rate 1 if the music detector reports a positive result, regardless of the rate indicated by other thresholds or constraints.

Figure 5.15 shows the rate-decision procedures of the SMV speech encoder. First, the VAD determines whether the input frame corresponds to silence or to active speech. In this step, the input frame is classified as either Rate $\frac{1}{8}$ or another rate. Note that the input frames diagnosed as Rate $\frac{1}{8}$ can still be upgraded in later steps. In the next step, the MD determines whether the input frames, excluding the Rate $\frac{1}{8}$ frames, contain music in the background. In the following stages, further classifications are made for those input frames not containing music. The input frames are categorized as voiced, unvoiced, and onset. The unvoiced frames are further divided into unvoiced speech and noise-like frames. Finally, the voiced frames are classified as stationary and non-stationary. Tables 5.7 and 5.8 show the mapping of each speech type to the allowed rates in four operating modes.

Table 5.9 lists a set of encoding rate transitions not allowed in the SMV speech encoder, which is more extensive than that of QCELP. These are based on the observations that an abrupt increase of the bit-rate from low rates or a steep decrease from high rates does not contribute to speech quality or network capacity noticeably. Therefore, they can be replaced with smoother adjustments.

Table 5.8 Speech types and rates for Modes 1, 2, and 3.

	Rate $\frac{1}{8}$	Rate $\frac{1}{4}$	Rate $\frac{1}{2}$	Rate 1
Silence	Yes			
Noise-like		Yes	Yes	
Unvoiced		Yes	Yes	
Onset		Yes	Yes	Yes
Non-stationary voiced			Yes	Yes
Stationary voiced			Yes	Yes

Table 5.9 Rate-transition constraints.

(a)	Rate 1 type-1 frame following a Rate $\frac{1}{8}$ frame
(b)	Rate $\frac{1}{2}$ type-1 frame following a Rate $\frac{1}{8}$ frame
(c)	Rate 1 type-1 frame following a Rate $\frac{1}{4}$ frame
(d)	Rate $\frac{1}{2}$ type-1 frame following a Rate $\frac{1}{4}$ frame
(e)	Rate $\frac{1}{8}$ frame following a Rate 1 type-1 frame
(f)	Rate $\frac{1}{8}$ frame following a Rate $\frac{1}{2}$ type-1 frame
(g)	Rate $\frac{1}{4}$ frame following a Rate 1 type-1 frame
(h)	Rate $\frac{1}{4}$ frame following a Rate $\frac{1}{2}$ type-1 frame

SMV provides four network-controlled operating modes for the reverse link: Mode 0 (Premium Mode), Mode 1 (Standard Mode), Mode 2 (Economy Mode), and Mode 3 (Capacity-saving Mode), which can be signaled to the MS using the RATE_REDUC message. A decision on the operating mode needs information on the cell loading level of the neighboring cells, as well as the link quality of the MS. It is the responsibility of the BSC to control the operating mode of each MS, as the channel type is controlled by the BSC in GSM. Mode 0 is appropriate for situations where the speech quality is considered to be most important and the network capacity is assigned a lower priority, while Mode 3 is effective in the busy-hour situations where the network capacity is regarded as critical. Modes 4 and 5 are basically Modes 0 and 1, with the additional constraint of limiting the maximum rate to Rate $\frac{1}{2}$. Notice that in the downlink, the average bit-rate can be controlled with unlimited resolution by directly adjusting the thresholds.

Each operating mode has a set of pre-defined rules to determine the rate and type of input frames but the average bit-rate depends on the nature of not only the speech but also the background noise. Tables 5.10 and 5.11 show the average bit-rate of each operating mode for clean and noisy speech. It can be seen that even with the same speech type, as the noise level increases, the proportion of Rate $\frac{1}{8}$ selections increases and the average bit-rate decreases.

5.3.2 4th Generation Vocoder

While the SMV introduced new types of tradeoffs between speech quality and network capacity, it failed to be commercially deployed, and the EVRC remained the

Table 5.10 Rate percentage for clean speech at normal level (-22 dBov).

	Mode 0 (%)	Mode 1 (%)	Mode 2 (%)	Mode 3 (%)
Rate 1	68.90	38.14	15.43	7.49
Rate $\frac{1}{2}$	6.03	15.82	38.34	46.28
Rate $\frac{1}{4}$	0.00	17.37	16.38	16.38
Rate $\frac{1}{8}$	25.07	28.67	29.85	29.85
Average bit-rate (kbps)	7.205	5.182	4.073	3.692

Table 5.11 Rate percentage for speech with street noise.

	Mode 0 (%)	Mode 1 (%)	Mode 2 (%)	Mode 3 (%)
Rate 1	55.13	29.13	16.60	6.86
Rate $\frac{1}{2}$	10.65	19.49	32.39	42.13
Rate $\frac{1}{4}$	0.00	13.17	14.93	14.93
Rate $\frac{1}{8}$	34.21	38.20	36.08	36.08
Average bit-rate (kbps)	6.214	4.507	3.940	3.472 (-15 dB)

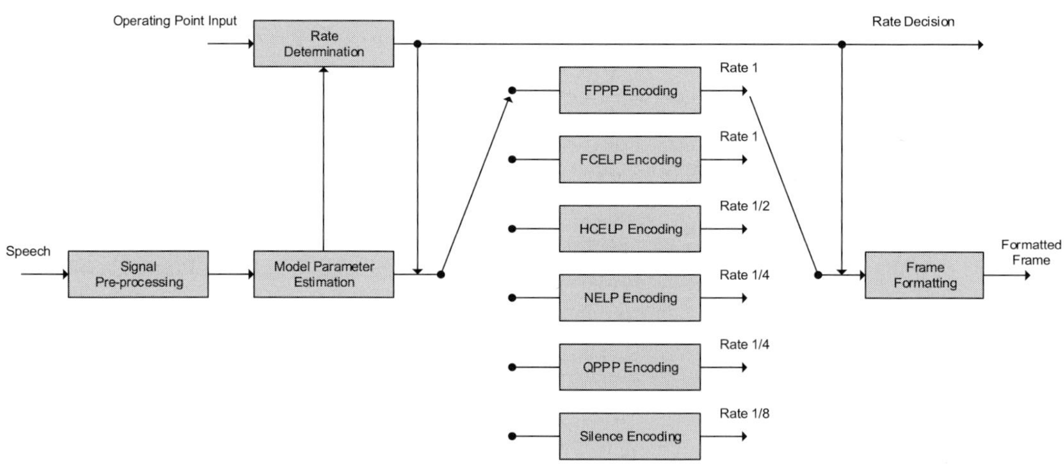

Fig. 5.16 Block diagram of 4GV speech encoder.

primary speech codec of cdma2000. When the voice services of cdma2000 were revised with the introduction of Revision E (Rev. E), EVRC became a performance bottleneck and the need to replace this aged codec became apparent. The official name of the new speech codec is EVRC-B but it is more commonly called 4GV, as the new codec can be classified as the fourth generation of a CDMA vocoder following QCELP, QCELP-13, and EVRC. In the technical specifications for EVRC and 4GV, the original EVRC is called the Speech Service Option 3 while 4GV is the Service Option 68 [3GPP2 (2010)].

4GV is a speech codec of RS 1 with four rates, like QCELP. Figure 5.16 shows a simplified block diagram of the 4GV speech encoder, which outputs every 20 ms one

Table 5.12 Coding scheme for each speech type.

Coding scheme	Speech type
FCELP (Rate 1 CELP)	Transient, bump-up, some voiced
HCELP (Rate $\frac{1}{2}$ CELP)	Ends of words, dim-and-burst signaling
Special HCELP	Packet level dimming of FCELP
FPPP/QPPP (Rate 1, $\frac{1}{4}$ PPP)	Voiced
Special HPPP	Packet level dimming of FPPP
QNELP (Rate $\frac{1}{4}$ NELP)	Unvoiced
Special HNELP	Rate $\frac{1}{2}$ containing NELP to replace Rate $\frac{1}{4}$
Rate $\frac{1}{8}$	Silence

of nine frame types depending on the input signal and the network control. In QCELP and QCELP-13, each rate is encoded with a single encoding method while SMV uses two encoding methods for Rates 1 and $\frac{1}{2}$, respectively. 4GV applies the idea of refined speech analysis and network control of average bit-rate inherited from SMV even further. 4GV offers two encoding methods for Rate 1, four encoding methods for Rate $\frac{1}{2}$, two encoding methods for Rate $\frac{1}{4}$, and a single encoding method for Rate $\frac{1}{8}$. For typical voice activity, the rate distributions of EVRC and 4GV can be roughly compared as 38%, 2%, 0%, and 60% for EVRC, and 22%, 2%, 16%, and 60% for 4GV, for Rates 1, $\frac{1}{2}$, $\frac{1}{4}$, and $\frac{1}{8}$, respectively [Jou *et al.* (2009)].

4GV incorporates three types of encoding principles: CELP, Prototype Pitch Period (PPP), and Noise-Excited Linear Prediction (NELP). CELP is used for Rates 1 and $\frac{1}{2}$, PPP is used for Rates 1 and $\frac{1}{4}$, and NELP is used for Rates $\frac{1}{4}$ and $\frac{1}{8}$. PPP is designed based on the observations that the pitch-cycle waveforms in a voiced segment change relatively slowly, which allows substituting the transmission of each pitch cycle with the transmission of a representative prototype pitch period.

Table 5.12 shows how the different coding schemes are used to match the type of input signal. Full-Rate CELP (FCELP) is used to encode the transitions, poorly voiced frames, and certain voiced frames. Half-Rate CELP (HCELP) is mainly used to encode the ends of words and low energy frames. Full-Rate PPP (FPPP) and Quarter-Rate PPP (QPPP) are used to encode strongly voiced frames. Quarter-Rate NELP (QNELP) is used for unvoiced frames. Silence is encoded by a simple single gain-shaped random noise waveform, as in EVRC. There are special versions for HCELP, HPPP, and HNELP used in special situations. The bit-stream composition of each coding scheme of 4GV is summarized in Tables 5.13 and 5.14.

The rate is determined in two steps. The first decision is invoked before the quantization and the next one is based mainly on the parameters and features of the input speech signal, including the current and previous frames. As in the other speech codecs of IS-95 and cdma2000, if an external rate command, such as RATE_REDUC, is received, it will supersede the decision of internal rate-selection mechanisms. The Anchor Operating Point (AOP), consisting of levels 0, 1, and 2, is a new concept introduced to facilitate assigning an appropriate proportion of rates while meeting a target average bit-rate.

Table 5.13 Bit allocation of 4GV speech codec for Rates 1 and $\frac{1}{2}$.

	FCELP	HCELP	Special HCELP	FPPP	Special HPPP
Mode bit	1		1	1	1
Special half-rate ID			2		2
LSP	28	22	28	28	28
Lag/special packet ID	7	7	7	7	7
Additional lag			7		7
ACB gain	9	9	9		
FCB index	105	30			
FCB gain	15	12			
Delta lag	5				
Amplitude				28	28
Global alignment					
Band alignments				99	
Energy VQ					
Filter shape					
Reserved	1		26	1	
Total	171	80	80	171	80

Table 5.14 Bit allocation of 4GV speech codec for Rates $\frac{1}{2}$, $\frac{1}{4}$, and $\frac{1}{8}$.

	QPPP	Special HNELP	QNELP	Silence
Mode bit	1		1	
Special half-rate ID		2		
LSP	16	16	16	8
Lag/special packet ID	4	7		
Additional lag				
ACB gain				
FCB index				
FCB gain				
Delta lag				
Amplitude	18			
Global alignment			7	7
Band alignments				
Energy VQ		17	17	6
Filter shape		2	2	
Reserved	1	36	4	2
Total	40	80	40	16

At the 4GV speech encoder, the input signal is roughly classified into speech, end-of-speech, and silence. If the AOP is set to 0, the three frame types are encoded with FCELP, HCELP, and Rate $\frac{1}{8}$. If the AOP is 1 or 2, the frames initially declared as non-silence are further classified into transition, voiced, unvoiced, and end-of-word. To avoid

Table 5.15 Rate control messages for reverse link.

RATE_REDUC	Encoder capacity operating point	Average speech encoding bit-rate (kbps)
000	0	8.3
001	1	7.57
010	2	6.64
011	3	6.18
100	4	5.82
101	5	5.45
110	6	5.08
111	7 (Rate $\frac{1}{2}$ max mode)	4.0

an abrupt change of the bit-rate and the encoding method, each voiced speech frame is always preceded by a transition frame, which is encoded with FCELP. The end-of-word is encoded with HCELP, unvoiced speech is encoded with QNELP, and voiced speech is mapped into the pattern sets of three frames. If the AOP is 1, the three-frame pattern set is encoded successively with FCELP, QPPP, and FCELP. If the AOP is 2, the three-frame pattern set is encoded with QPPP, QPPP and FPPP.

The concept of *frame pattern set* is a generalization of the encoding constraints that have been used in CDMA speech codecs, where a few constraints on rate transitions are enforced. The three-frame pattern sets of 4GV specify longer rate-transition paths that can accommodate the periodic nature of voiced speech better than shorter transitions. For example, a sequence consisting of Rate $\frac{1}{2}$ frames and another sequence consisting of sets of a Rate 1 frame followed by two Rate $\frac{1}{4}$ frames require the same average bit-rate but the latter encodes more Rate 1 frames and may result in higher subjective quality. Note that extending the length of a frame pattern set further, e.g., to four or five frames, may limit the capability of the speech encoder to adapt to the time-varying acoustic nature of speech.

If the changes in waveform shape cannot be represented within PPP schemes, the coding scheme is changed to another scheme. If QPPP fails either in its phase representation or amplitude quantization, the scheme is changed to FPPP if AOP$=2$, or to FCELP if AOP$=1$, and the frame pattern for the voiced speech is reset. If a Rate 1 frame is *dimmed*, i.e., downgraded, to Rate $\frac{1}{2}$, the frame pattern is re-started with FPPP if AOP$=2$, or with FCELP if AOP$=1$. The additional encoding methods are used to reduce the impact of rate-transitions or match the target bit-rate as accurately as possible. 4GV provides eight network-controlled operating modes for the reverse link, which can be signaled to MS with the RATE_REDUC message. Table 5.15 shows the average bit-rate of each operating mode. The gain of using a frame pattern set can be considered as conceptually similar to that of using vector quantization, in that the quantization of a series of speech frames is taken into account simultaneously.

In the forward link, 4GV can operate at an arbitrary average bit-rate, as commanded by the BSC. The average bit-rate for active speech, T, in bps, is computed as

Table 5.16 Partitioning of anchor operating points.

Condition	AOP
$7500 < T_{\text{ActiveAvgChRate}}$	0
$6600 < T_{\text{ActiveAvgChRate}} \leq 7500$	1
Otherwise	2

Table 5.17 Proportion of frames shifted from Rate $\frac{1}{4}$ to Rate 1.

p	AOP
$\frac{T_{\text{next}}-7.5}{(9.0-7.5)}$	0
$\frac{T_{\text{next}}-6.6}{(7.5-6.6)}$	1
$\frac{T_{\text{next}}-5.75}{(6.6-5.75)}$	2

$$T = \frac{9600N_{\text{full}} + 4800N_{\text{half}} + 2400N_{\text{quarter}}}{N_{\text{full}} + N_{\text{half}} + N_{\text{quarter}}}, \tag{5.1}$$

where N_{full}, N_{half}, and N_{quarter} are the numbers of speech frames encoded at Rates 1, $\frac{1}{2}$, and $\frac{1}{4}$ respectively, for an observation period. Let the target average rate for active speech be $T_{\text{ActiveAvgChRate}}$. Then the AOP is determined based on the formulas outlined in Table 5.16.

Once the AOP is determined, the 4GV speech encoder attempts to meet this average bit-rate target by adjusting the proportion of Rate $\frac{1}{4}$ frames, by comparing the actual average bit-rate in a window of the past 600 active speech frames to the target average bit-rate and computing a proportion, p, of frames whose rate should be converted from Rate $\frac{1}{4}$ to Rate 1. p is computed as in Table 5.17.

T_{next} and T_{prev} are the target average bit-rates for the next and previous sets of active, i.e., excluding Rate $\frac{1}{8}$, frames, which have the following relationship,

$$T_{\text{next}} = 1.98T_{\text{ActiveAvgChRate}} - T_{\text{prev}}. \tag{5.2}$$

Once p is determined, those input frames initially decided as Rate $\frac{1}{4}$ are encoded at Rate 1 in an almost uniformly distributed manner.

The Mode bit in Tables 5.13 and 5.14 indicates whether a speech frame is encoded as PPP or CELP/NELP. If a frame is encoded as PPP, then the Mode bit is set to 1. Special Half-Rate CELP (HCELP) is used when a FCELP frame is converted by removing the FCB index and gain bits, and re-packing the remaining bits. To indicate the identity of the frame to the 4GV speech decoder, the seven lag bits, the Special Half Rate CELP flag, of Rate $\frac{1}{2}$ frames are set to 0x6E, which is never produced by the 4GV encoder. Likewise, special HPPP is used when a FPPP frame is converted by removing the band alignment bits, and re-packing the remaining bits. The seven lag bits of the Rate $\frac{1}{2}$ frames are set to 0x6E.

Figure 5.17 illustrates the voice classification procedures of 4GV. The input frames diagnosed as transient are encoded as FPPP in the case of onsets, and as HCELP in the

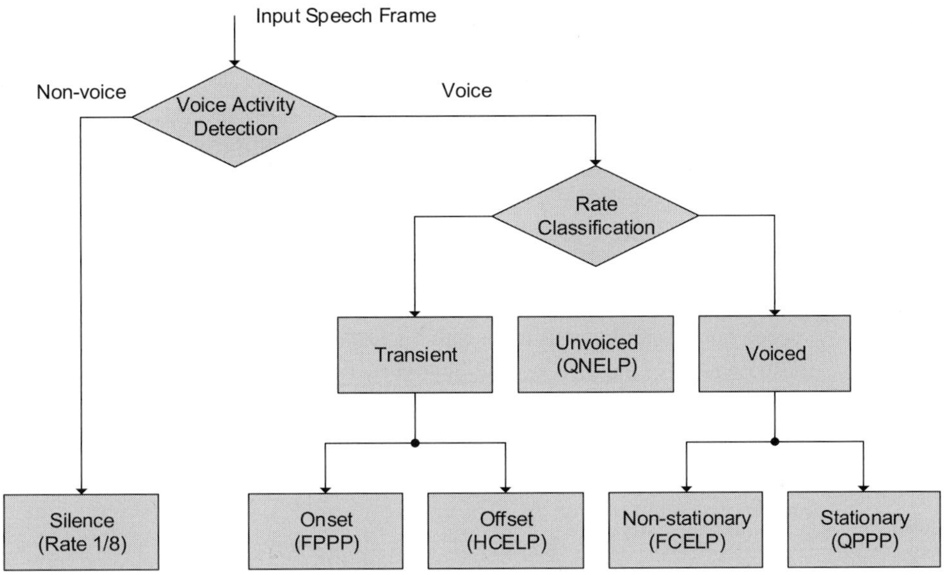

Fig. 5.17 Voice classification procedures.

case of offsets, i.e., when the speech segments correspond to the end of a word. The unvoiced speech frames are encoded as QNELP, and the voiced frames are encoded as FCELP if they are not stationary, and as QPPP otherwise. Note that three types of frames in Tables 5.13 and 5.14 are not shown in the tree. The special HCELP and special HPPP are only used when FCELP and FPPP are downgraded by network control. Finally, special Half-Rate NELP (HNELP) is only used in the Half-Rate Max (HRM) mode where the maximum rate is limited to Rate $\frac{1}{2}$. The 4GV speech decoder can identify the received frames unambiguously using the information in the 1-bit Mode bit, 2-bit special Rate $\frac{1}{2}$ ID, and the 7-bit special packet ID. Although 4GV provides a series of new encoding methods, it still shares many elements with EVRC, such as the high-pass filter, noise suppressor, VAD, and LPC analysis and quantization.

The fundamental design objective of VBR is to maintain the speech quality while reducing the average bit-rate, by replacing the rough on/off mechanisms of DTX with more soft transitions, and by transmitting a minimum amount of information on the background noise, the Rate $\frac{1}{8}$ frames, continuously when voice activity is absent. In 4GV, the DTX can be used if the bit-rate needs to be reduced further. When the DTX is enabled, Rate $\frac{1}{8}$ frames are transmitted infrequently. For example, at the end of a talk-spurt, a few Rate $\frac{1}{8}$ frames are transmitted as *eighth-rate hang-over*. However, after the hangover, the following Rate $\frac{1}{8}$ frames are not transmitted unless there is noticeable change in the characteristics of background noise. If such a change is detected, another Rate $\frac{1}{8}$ frame is transmitted. The change is detected by comparing the gain of a Rate $\frac{1}{8}$ frame with a running average of the energy of the previous Rate $\frac{1}{8}$ frames. The bit-streams with and without DTX can be converted to each other at the boundary of the networks. The typical VBR bit-stream can be converted to DTX bit-stream by discarding some Rate $\frac{1}{8}$ frames. Likewise, a DTX bit-stream can be converted to a VBR bit-stream by replicating the missing Rate $\frac{1}{8}$ frames.

5.3.3 Network Control and Voice Control of Speech Compression

Figure 5.18 conceptually compares the control strategies of the speech codecs for TDMA and CDMA mobile communications systems. The operation of FBR speech codecs such as FR or EFR is controlled only by voice activity, as the output of the speech encoder can be an encoded speech frame, an SID, or silence. Since in these codecs the speech frame can have at most one bit-rate when voice activity is present, the range of these codecs, in the direction indicated by the voice control axis, is limited to one level. The network does not influence the voice classification procedures during conversation. AMR, when operating in GSM, allows the control of speech bit-rate by the far-end, which can be either a BTS or an MS. This flexibility is reflected in the figure as the four levels, assuming an ACS of up to four modes, in the direction of the network control axis. When AMR operates in a circuit-switched CDMA network, i.e., W-CDMA, the same perspective is still valid. However, when AMR operates in packet-switched networks, such as LTE, a new axis with more levels, the peer control axis, replaces the network control axis. Note that CDMA can also be used for packet-switched networks, as in EV-DO.

Likewise, the operation of VBR speech codecs such as QCELP and EVRC is controlled mainly by voice activity, but the network can also exercise control to some extent. However, the output of the VBR speech encoder is a frame encoded at three or four non-zero bit-rates. This difference is visualized in the figure as four levels in the direction of the voice control axis. Notice that as EVRC does not encode Rate $\frac{1}{4}$ frames, its second level is absent. SMV combines the idea of network control and voice control but the output of the speech encoder is a frame at four non-zero bit-rates, and the network controls the relative proportion of each bit-rate, with four pre-defined operating modes. This approach is illustrated as the rectangle consisting of four levels on each axis respectively. 4GV drives the idea of refined network control even further so that in the reverse link, up to eight operating modes can

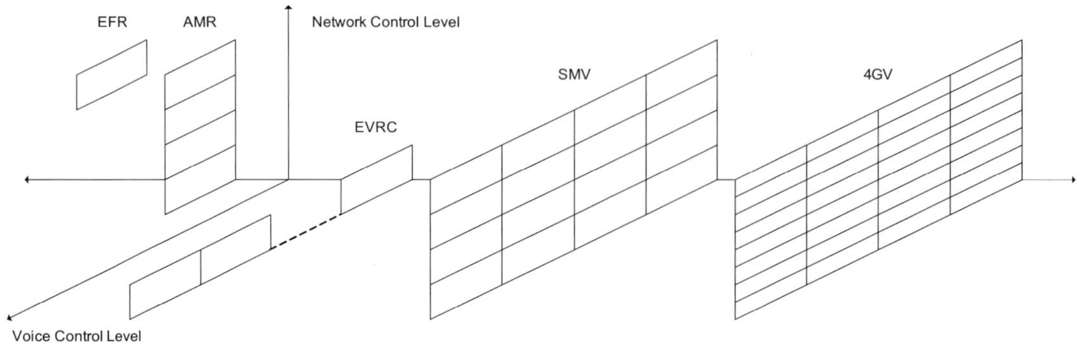

Fig. 5.18 Speech codec control of TDMA and CDMA.

be chosen while the number of operating modes in the forward link is theoretically unlimited.

It can be seen from Fig. 5.18 that to achieve higher capacity while maintaining similar quality in circuit-switched networks, control of the speech codec has to be more refined, taking both voice activity and network state into account in the speech compression. These evolutionary paths may be compared to those followed by vehicle transaxles, where developments have continuously elevated the level of control by shifting from manual to automatic transaxles, to automatic transaxles with many speeds, and to Continuously Variable Transmissions (CVT).

5.4 Enhancements in Wireless Transmission

While D-AMPS, a straightforward digital extension of AMPS, did not pose a serious threat to GSM, which was designed without consideration of the compatibility with any previous systems, IS-95, with its unprecedented signal processing operations, did become a formidable competitor to GSM. The development of AMR was a reaction to the success of CDMA. It enhanced the speech quality and network capacity of GSM significantly. As IS-95 evolved to cdma2000, AMR was selected for the next (third) generation of circuit-switched mobile communications system, Wideband CDMA (W-CDMA), which was also based on the CDMA principles but used a wider channel bandwidth, 5 MHz. The decade-long competition between cdma2000 and W-CDMA seems to have ended in the victory of the latter. The radio signal processing procedures of both GSM and cdma2000 were continuously enhanced for higher capacity while re-using the existing network infrastructure. This resulted in the development of VAMOS for GSM and Rev. E of cdma2000.

5.4.1 cdma2000 Revision E

If improving the speech codec alone does not provide enough capacity gain, the radio signal processing procedures become the natural candidates for enhancement. In the cdma2000 family of mobile communications systems, Revision E (Rev. E) of the technical specifications contains a set of improvements based on the opportunities observed in the deployed systems. Interestingly, some of these improvements were already adopted in the early phases of GSM, such as DTX or reducing the frequency of power control opportunities. Just as GSM borrowed from CDMA the concept of using orthogonality to separate the speech data from two MSs in VAMOS, cdma2000 also brought some concepts from TDMA for higher capacity. As VAMOS can operate only in MSs equipped with advanced receivers, some features of Rev. E require higher interference-cancellation capability in the receiver. Rev. E is often called 1xRTT Advanced.

5.4.2 Reverse Link Signal Processing

Figure 5.19 shows the number of bits for each rate at each step in the signal processing operations of the Rev. E reverse link, with 4GV for RC 8, from signal digitization to

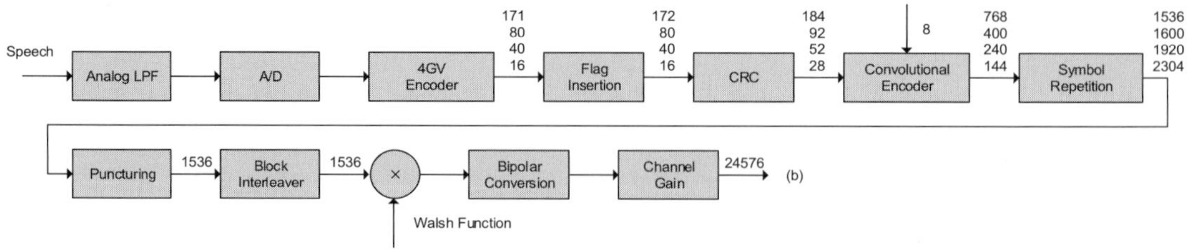

Fig. 5.19 Speech and radio signal processing operations of cdma2000 Rev. E reverse link for 4GV (RC 8).

interleaving. Since in SMV or 4GV, more than one compression method can be applied for a particular rate, blind detection of rate, as used with multiple CRC checks in QCELP or QCELP-13, cannot be applied. A new set of techniques inspired by the observations of CDMA operations includes the reduction of power control rate, blanking of Rate $\frac{1}{8}$ frames, and frame early termination, which contribute to lower interference at the sacrifice of error resilience to an acceptable extent.

First, a 12-bit CRC is computed and added to the bits of all rates. Notice that in radio configuration 3 used for EVRC, CRCs of 12, 8, and 6 bits are applied for Rates 1, $\frac{1}{2}$, and $\frac{1}{8}$, respectively. After the rate $\frac{1}{4}$ convolutional encoding is applied, frames at the four rates contain 768, 400, 240, and 144 bits, respectively. Then the symbols are repeated once, three, seven, and fifteen times to generate 1536, 1600, 1920, and 2304 bits, which necessitate puncturing for those rates other than Rate 1. Therefore, for the channel-coded symbols of Rates $\frac{1}{2}$, $\frac{1}{4}$, and $\frac{1}{8}$, one bit out of 25 bits, one bit out of five bits, and eight bits out of 24 bits are periodically punctured to reduce the total numbers of bits to 1536. After block interleaving, the symbol rate of all rates becomes 76.8 ksps.

Multiplying each interleaved bi-level symbol by a Walsh function of length 16 spreads the information spectrum to 1.2288 MHz. Finally, the gain of the traffic channel transmitting speech data is adjusted, to maintain a controlled proportion of transmit power among the channels of the reverse link. The gain-adjusted symbols from the signal processing chain of Fig. 5.19 are fed into the lower (quadrature) branch (b), of Fig. 5.8. A key difference from the procedures of RC 3 is the reverse acknowledgment channel, which signals whether a frame was successfully decoded or not. It is also fed into branch (b) while the reverse pilot channel is entered into branch (a).

The maximum rate of the reverse acknowledgment channel is one bit per PCG, i.e., 800 Hz, but the actual rate is determined by the acknowledge mask. The network controls the forward power control operating mode using the FPC_MODE_S message. If $FPC\_MODE_S = 000$, power control bits are transmitted at PCGs 1, 3, …, 13, and 15, corresponding to 400 Hz. If $FPC\_MODE_S = 011$, power control bits are transmitted at PCGs 3, 7, 11, and 15, at 200 Hz. The network can set FPC_MODE_S to match the link quality of each MS, increasing the power control rate when the link becomes unstable, and reducing it when the link stabilizes.

As shown in Fig. 5.1, the erasure indicator bit (EIB) of the RS 2 speech codec in RC 2 is used to facilitate the forward link power control, suppressing the interference by

Table 5.18 Reverse link feedback information.

RC	Speech codec	Reverse link feedback
1	QCELP	Power control
2	QCELP-13	Power control, EIB
3	EVRC	Power control, synchronization
8	4GV	Power control, synchronization, acknowledgment

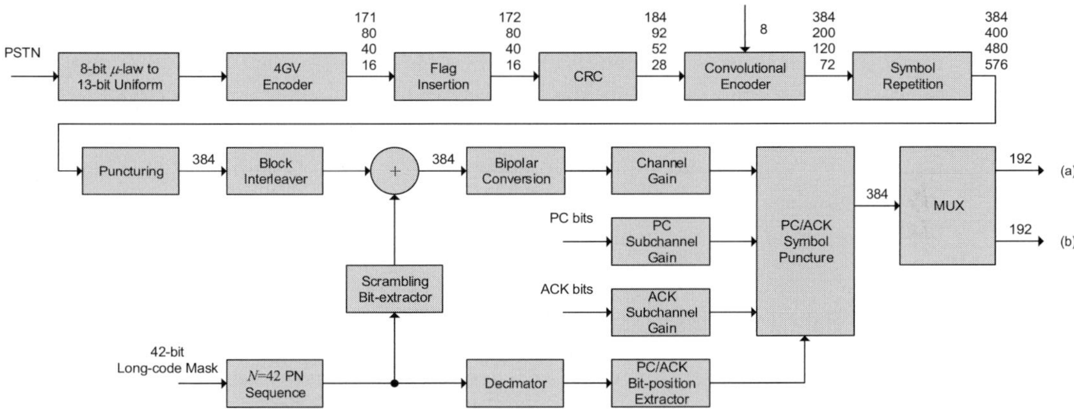

Fig. 5.20 Speech and radio signal processing operations of cdma2000 Rev. E forward link for 4GV (RC 11).

controlling the power level with a periodic supply of transmission results. On the other hand, the reverse acknowledgment channel achieves a similar objective by controlling the duration of power control at the PCG level. As outlined in Table 5.18, to achieve higher network capacity, what is required is more refined control of the interference.

It can be seen that more detailed information on transmission results or channel conditions needs to be exchanged more frequently, using lower bit-rate or transmit power, if higher quality or capacity is to be achieved. In CDMA mobile communications systems, the forward link capacity is, in general, limited by the available transmit power of the BTS, which typically reaches its maximum while the MSs in the cell can still increase the power. As a result, the forward link capacity is typically smaller than that of the reverse link. Figures 5.8 and 5.19 correspond to RC 8 of the reverse fundamental channel.

5.4.3 Forward Link Signal Processing

Figure 5.20 shows the number of bits for each rate at each step in the signal processing chain of the Rev. E forward link with 4GV, from speech encoding to power control. As with the signal processing of the reverse link, a CRC of 12 bits is computed and added to the bits of all rates. The rate $\frac{1}{2}$ convolutional encoding generates 384, 200, 120, and 72 bits, respectively. Then the symbols for rates other than Rate 1 are repeated once, three, and seven times to generate 400, 480, and 576 bits, which necessitate puncturing for rates other than Rate 1. For Rates $\frac{1}{2}$, $\frac{1}{4}$, and $\frac{1}{8}$, one bit out of 25 bits, one bit out of five, and four bits out of 12 are periodically punctured to reduce the number of bits to 384. After block interleaving, the symbol rate of all rates becomes 19.2 ksps.

The gain of the power control subchannel is adjusted before the power control bits are punctured. Then the symbols are separated into two streams, (a) the in-phase components and (b) the quadrature components, which consist of the odd and even symbols, respectively. Notice that the forward acknowledgement channel is punctured together with power control bits. Figure 5.13 shows the remaining radio signal processing procedures. The two streams of symbols entered into branches (a) and (b) are first multiplied by Walsh codes, and then multiplied by a complex spreading sequence. The in-phase and quadrature components are applied to low-pass filters and the waveforms are modulated by RF carriers and summed to generate $S(t)$, which is amplified and transmitted over the wireless channel. Figures 5.13 and 5.20 correspond to radio configuration 11 of the forward fundamental channel.

5.4.4 Blanked-Rate $\frac{1}{8}$ Frames

Unlike the DTX mechanisms of GSM where the transmitter is turned off when there is no voice activity, in the variable bit-rate speech coding of IS-95 or cdma2000, Rate $\frac{1}{8}$ frames are continuously transmitted during a silent period sending information about the background noise. When the background noise is stationary, i.e., its statistical nature does not change significantly within a short period of time, the transmit power can be reduced by *blanking*, i.e., transmitting fewer Rate $\frac{1}{8}$ frames than generated, which reduces the interference. In Rev. E, the speech codecs are allowed to produce zero-rate frames during a silence period but they are required to output at least one non-zero-rate frame every N frames, where N can be configured by the network, just as an SID is transmitted at least once per 480 ms, or 24 frames, in the FR speech codec of GSM. The background noise can be reconstructed as accurately as in the continuous transmission of Rate $\frac{1}{8}$ frames when the noise parameters do not need to be updated frequently.

Note that the Rate $\frac{1}{8}$ frames cannot be blanked when there are no other supplemental channels to maintain synchronization with the fundamental channel. In the reverse link of IS-95 where no reverse pilot channel is available, blanking can incur the loss of synchronization that cannot easily be recovered during a call. GSM does not provide such assistance in the uplink either but its TDMA frame structure and low symbol rate can be exploited by the receiver during DTX to recover synchronization. In Rev. E, the reverse pilot channel can also be gated, transmitted only in PCGs 0, 3, 4, 7, 8, 11, 12, and 15. This technique to reduce the interference can only be applied with the blanking of Rate $\frac{1}{8}$ frames [3GPP2 (2005)].

The proper operation of CDMA depends importantly on fast power control of the reverse link. Blanking frames inevitably increases the interference by decelerating the convergence of power control. Performance degradation from blanking can be avoided by controlling the step size of the closed-loop power control and the duty cycle of the

Table 5.19 Closed-loop power control step size.

$PWR\_CNTL\_STEP_S$	Power control step size (dBm)
000	1
001	0.5
010	0.25
011	1.5
100	2.0

Table 5.20 Duty cycle of blanking for reverse link (RC 8) and forward link (RC 11).

N (binary)	*N* (integer)	Description
000	1	Blanking disabled
001	4	At least 1 out of 4 frames transmitted
010	8	At least 1 out of 8 frames transmitted

blanking. The three-bit $PWR\_CNTL\_STEP_S$ message is used by the network to control the step size of the closed-loop power control when the reverse link is blanked. Table 5.19 shows the step size corresponding to each value of $PWR\_CNTL\_STEP_S$. The use of bits from 101 to 111 is reserved. With duty cycle N, which the network signals, the blanking can be executed in three levels, as summarized in Table 5.20.

5.4.5 Reduced Power Control Rate

Power-control bits for the reverse link are transmitted in each PCG on the forward fundamental channel, which allows the level of transmit power to be adjusted 800 times per second. Transmit power of the power-control bits is typically set to be the same as that for the bits of Rate 1 frames, which does not scale down when the frames are of lower rates. For example, in the Rate $\frac{1}{8}$ frames, transmit power of the power control bits would be about 9 dB higher than that of the noise. To facilitate reception by the MS during soft handover, transmit power of the power-control bits in the forward fundamental channel is further increased by the number of cells in the active set, i.e., the set of BTSs the MS is simultaneously connected to during soft handover. Figure 5.21(d) shows that in comparison with cases when there is only a single BTS in the active set, an additional 3 dB is spent for power-control bits.

Therefore, the power-control overhead in the forward link can consume a significant amount, e.g., 34% in the case of two BTSs in the active set, of the available transmit power. Reducing power-control rate from 800 times per second to 400, or to 200 reduces the interference and increases network capacity. In Rev. E, the frequency of power control for the forward and reverse links can be controlled using a two-bit flag, POWER_CONTROL_MODE. Note that the blanking of Rate $\frac{1}{8}$ frames also influences the power-control rate. POWER_CONTROL_MODE is set to 00 if the transmit power is

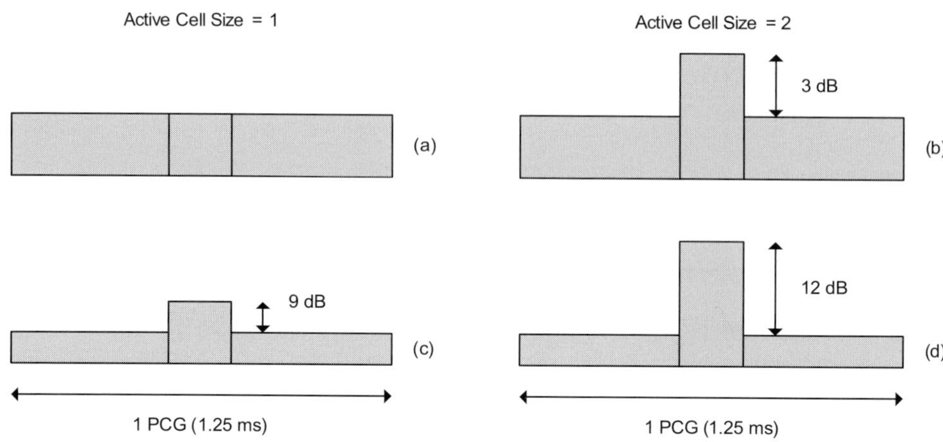

Fig. 5.21 Power control subchannel level. (a), (b) Rate 1. (c), (d) Rate $\frac{1}{8}$.

controlled at 200 or 400 Hz, and set to 01 if controlled at 200 Hz. Figure 5.21 illustrates the relative power levels of the power-control bits for Rates 1 and $\frac{1}{8}$, when an MS is connected to one and two BTSs.

5.4.6 Frame Early Termination

In the forward link of RC 3, as shown in Fig. 5.10, symbols of rates lower than Rate 1 are repeated from one to seven times and punctured to equalize the number of symbols. The repeated symbols are interleaved and transmitted at reduced power levels. At the receiver of the MS, energy of the repeated symbols is accumulated before channel decoding. However, it has been observed that channel decoding can often be completed successfully with only a subset of the symbols, i.e., using only a part of the total energy, before the entire set of symbols arrives. In the forward fundamental channel of RC 4, it was found that about 80% of Rate $\frac{1}{8}$ frames could be decoded after the first seven PCGs were received. In cdma2000, the frame error rate (FER) is controlled by power control with a typical target value of 1%. Therefore 99% of speech frames are delivered successfully under normal operating conditions. Symbols not available at the receiver at the time of decoding can be considered to be the erasures. The probability of successful channel and speech decoding, with a subset of the transmitted PCGs, depends on the rate but it can be expected that Rate $\frac{1}{8}$ frames can be decoded with a smaller number of PCGs than the frames of higher rates, as after puncturing the Rate $\frac{1}{8}$ frames contain the largest proportion of redundancy.

With Frame Early Termination (FET), the receiver attempts to decode each speech frame once a minimum number of symbols or PCGs are available. The receiver signals successful decoding to the sender at the earliest pre-defined opportunity even if there remain additional symbols or PCGs to receive. If the BTS receives a signal, acknowledgment of successful channel decoding for the frame, the BTS halts transmission of symbols belonging to the frame. The network instructs the MS about the PCGs in which

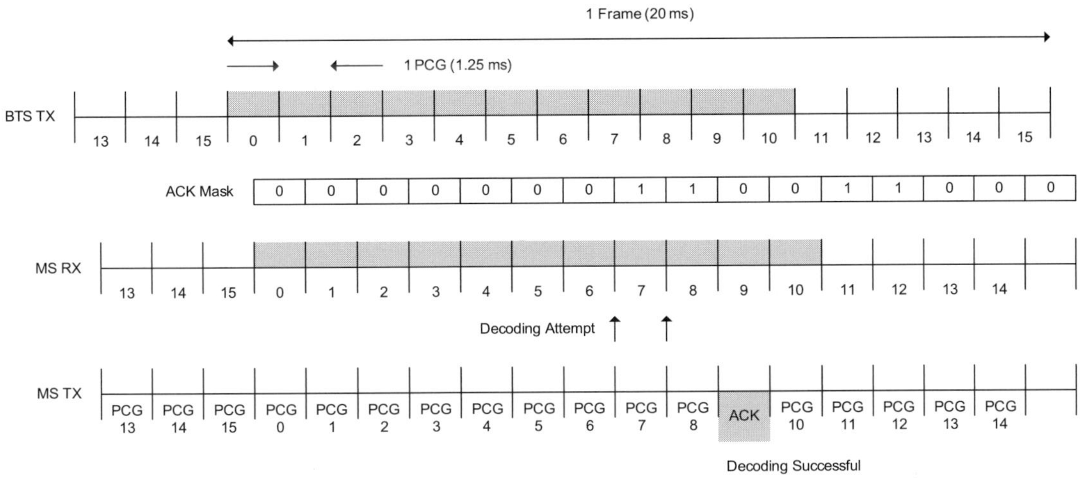

Fig. 5.22 Frame early termination in forward link.

the acknowledgment should be transmitted with a 16-bit mask, consisting of sixteen 1-bit subfields, from ACK_MASK_0 to ACK_MASK_15. When the MS successfully decodes a speech frame received on the forward fundamental channel, it returns an acknowledgment via the reverse acknowledge channel in the PCGs whose ACK_MASK bits are set to 1. For example, with a mask of 0000000110011000, the BTS may receive the acknowledgments in PCGs 7, 8, 11, and 12 of each speech frame.

Figure 5.22 illustrates the process of frame early termination. The receiver starts channel decoding and CRC check once the symbols up to PCG 6 are available. After the initial attempt fails, the receiver continues decoding and finally succeeds with symbols up to PCG 7. Then the MS transmits an acknowledgment in PCG 8 of the reverse link, which is received by the BTS and the remaining PCGs are not transmitted. Although successful reconstruction of the speech frame can only be confirmed subjectively after speech decoding, the probability of an erroneous speech frame passing the CRC check successfully is negligible. To reduce such a probability of false CRC passes, a 12-bit CRC is used for all rates in the forward fundamental channel of RC 11, as shown in Fig. 5.20.

5.4.7 Interference Cancellation

As in the downlink advanced receiver performance (DARP) of GSM, the network capacity can also be increased by using higher-complexity receivers that can recover the transmitted signal in lower SNR environments. Sophisticated receivers that cancel the interference between signals received from multiple paths, using complex iterative algorithms, may achieve higher capacity, but the end-to-end delay constraints of voice services limit the use of such exhaustive approaches to higher bit-rate, non-real-time services such as data access. Because of the computational complexity and power consumption limitations of the MS, interference cancellation techniques can be more easily

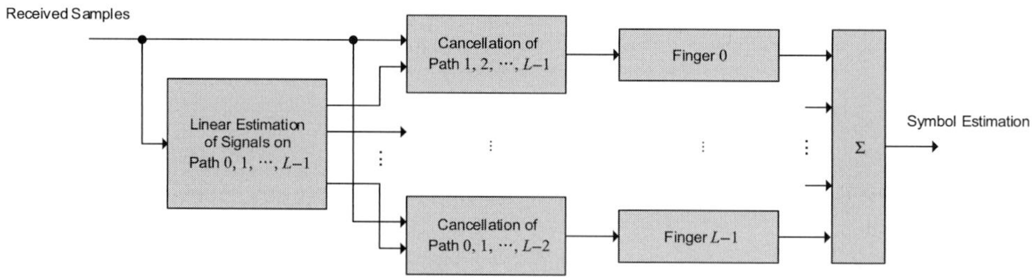

Fig. 5.23 Interference cancellation with rake receiver.

applied in the receivers at the BTS so that the interference of the reverse link is removed from the received signals. However, some gain from advanced receivers can also be obtained in the forward link.

Figure 5.23 shows the structure of a typical linear interference cancellation receiver for CDMA mobile communications systems. The correlated signal components of other paths are linearly estimated and subtracted from the input to each finger. This increases the SNR at the combiner. The receiver may also take the spreading codes used into account. Unlike in the DARP of GSM, the Rev. E of cdma2000 does not specify any performance requirements for interference cancellation. However, linear interference cancellation is a simple but efficient technique implemented in many receivers. Additional antennas of the MS will increase the performance of interference cancellation significantly. If the proportion of MSs capable of interference cancellation increases beyond a certain level, the forward link interference will be reduced noticeably. Therefore, in contrast to the conventional understanding of CDMA capacity where the forward link is the bottleneck, due to the limitations of the power amplifiers in the BTS, the forward link may have a larger capacity than the reverse link if enough MSs in the cell are equipped with receivers capable of interference cancellation.

5.5 Performance Evaluation

In this section, the mean opinion score (MOS) performances of the speech codecs are evaluated under error-free and frame-loss conditions. It is shown that the complexity of the variable bit-rate speech codecs for CDMA is, in general, higher than that of the fixed bit-rate speech codecs for TDMA. In addition, VBR provides a similar level of quality at lower average bit-rate. It is also shown that the tradeoff between speech quality and network capacity can be controlled by adjusting the average bit-rate of the speech codecs. This is a more elegant control than halving or doubling the bit-rate by switching between the full-rate and half-rate channels of TDMA. The results from a live cdma2000 call, measured from commercially operational networks, are introduced to visualize the dynamic operation of the signal processing procedures. Finally it is shown that rough bounds for voice capacity can be derived for the reverse and forward links.

Table 5.21 Complexity and quality of speech codecs.

	Maximum MIPS	MOS (clean channel)
VSELP	14	3.5
FR	6	3.5
EFR	14	3.8
QCELP	15	3.3
EVRC	20	3.8

Table 5.22 Complexity and quality of speech codecs.

	QCELP-13	EVRC	AMR
WMOPS	17.8	20	16.8
Memory (kWords)	27.7	42.3	38.8
MOS (clean channel)	3.832	3.852	3.932

5.5.1 Speech Compression and Transmission Performance

Table 5.21 compares the complexity and MOS performance of the speech codecs for IS-54, GSM, and IS-95 under error-free conditions [Gibson (2000)]. Notice that the two criteria should not be considered as more than a very basic set of design requirements. VSELP achieves the same MOS as FR at a lower maximum bit-rate, at the cost of higher complexity. It is apparent that the VBR speech codecs used in IS-95 and cdma2000 require higher complexity than that of the speech codecs for TDMA. During the initial deployments of IS-95, QCELP was often criticized for poor quality, as shown in its MOS value of 3.3, which accelerated the development of QCELP-13 and EVRC.

Table 5.22 compares the complexity and MOS values of EVRC, QCELP-13, and AMR under comparable conditions [3GPP2 (2003b)]. It is not easy to compare the performance of speech codecs designed for different radio access technologies but in this test, the SNR is set to -22 dB to equalize the operating environments. The Weighted Million Operations Per Second (WMOPS) is a measure of computational complexity used in the speech and audio coding community to evaluate the efficiency of compression algorithms. For typical Digital Signal Processors (DSP), WMOPS generates metrics similar to the values of million instructions per second (MIPS) [Salami *et al.* (2006)].

To measure the complexity of different codec implementations, a standardized set of signal processing routines, mathematical functions, and bit-manipulation tools sufficient to develop typical speech codecs [ITU-T (2009)] was provided. In the test, the average bit-rates of EVRC and QCELP-13 are computed as 7.246 and 10.868 kbps at a voice activity factor of 75%, which is computed as

$$T = \frac{9600N_{\text{full}} + 4800N_{\text{half}} + 2400N_{\text{quarter}} + 1200N_{\text{eighth}}}{N_{\text{full}} + N_{\text{half}} + N_{\text{quarter}} + N_{\text{eighth}}}. \tag{5.3}$$

N_{full}, N_{half}, N_{quarter}, and N_{eighth} are the numbers of speech frames encoded at Rates 1, $\frac{1}{2}$, $\frac{1}{4}$, and $\frac{1}{8}$, respectively, during an observation period.

Although the estimated complexity depends on how the speech codecs are implemented and cannot be considered as a unique metric to represent the performance, it can still be used to assess the cost for quality. In this regard, from comparing the performance of QCELP and QCELP-13 shown in Tables 5.21 and 5.22, using that of EVRC as a reference, it can be seen that QCELP-13 achieves higher quality than QCELP more by increasing the bit-rate than by increasing the complexity. With similar MOS values, speech codecs of RS 1 are preferred to those of RS 2, because of the higher level of error protection and corresponding benefits in the coverage and capacity.

EVRC requires an even higher level of complexity than QCELP or QCELP-13. However, the signal processing capability and the storage for data and programs are basic resources of mobile communications systems that can be continuously increased with little additional cost thanks to advances in semiconductor technology. This may not be the situation for other resources such as transmit power or frequency spectrum. QCELP-13 and EVRC were deployed at similar times to replace QCELP but the superiority of EVRC was soon recognized. Therefore EVRC became the dominant speech codec of CDMA, which continued as IS-95 evolved into cdma2000. While different units are used in Tables 5.21 and 5.22, the MOS and WMOPS values of EVRC and AMR are comparable to the MOS and maximum MIPS of EVRC and EFR. However, the MOS values depend significantly on the testing conditions and should be considered only as a measure to evaluate relative performance in the same test.

The official testing of speech codecs consists of time-consuming and costly procedures whose criteria may differ, depending on the design objectives. Therefore, the test cases, which specify the testing methods and the target values or ranges, are usually chosen to minimally satisfy the objectives. As a result, when the performances of the candidates for a new standard codec are compared with those of earlier speech codecs, it is unlikely that speech codecs spanning as long as three generations of mobile communications systems would be evaluated together. Therefore in the official testing where SMV or 4GV was evaluated, QCELP was considered as already phased-out from commercially operational CDMA networks, and therefore excluded from testing. Note that in Table 3.24 and Fig. 3.25, FR was compared with the codec modes of AMR in the half-rate channel mode.

Figure 5.24 compares the MOS performance of PCM, EVRC, AMR, and SMV under error-free conditions [3GPP2 (2003a)]. The MOS of 64 kbps PCM format, G.711 μ-law, is provided as a reference for the uncompressed speech quality. It is noticeable that the qualities of SMV Mode 0, EVRC, and AMR 12.2 kbps are almost identical but exceeded by that of PCM with MOS differences of 0.17–0.18, which is statistically significant, and cannot easily be reduced by refining the speech compression algorithms. To leap over such a large performance gap using similar bit-rates, one of the available measures is to double or quadruple the sampling rate so that the information bandwidth is significantly extended, and design speech compression algorithms that exploit the higher information bandwidth. The mobile communications systems of later generations than IS-95 or cdma2000 do provide higher-quality voice services based on this approach, i.e., using the *wideband* or *super-wideband*.

Table 5.23 Performance comparison of QCELP-13, EVRC, and SMV in error-free channel.

	QCELP-13	EVRC	SMV Mode 0	SMV Mode 1	SMV Mode 2	SMV Mode 3
MOS	3.83	3.85	3.86	3.78	3.63	3.59

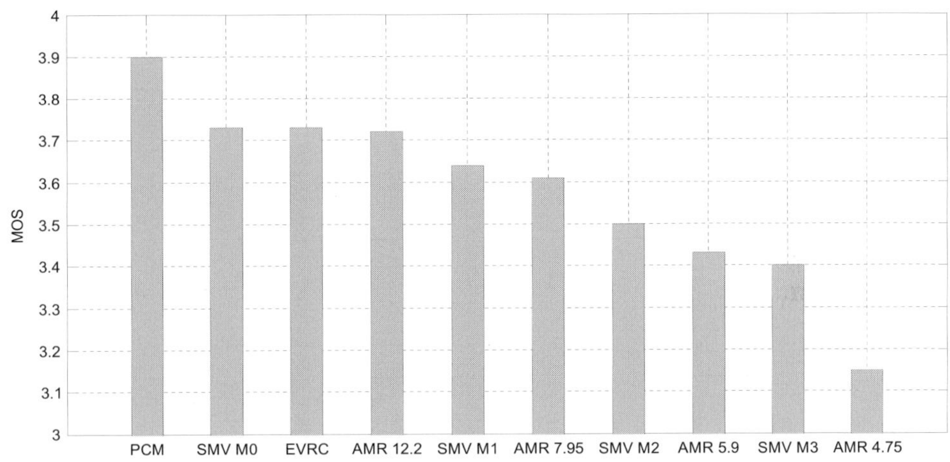

Fig. 5.24 Performance comparison of PCM, EVRC, SMV, and AMR.

In Fig. 5.24, as the bit-rates of AMR and SMV are decreased, the speech quality is correspondingly reduced. Note that while the mode of SMV running on cdma2000 can be used to control the average bit-rate, the codec mode of AMR running on GSM cannot be used for this purpose. In GSM, reducing the speech bit-rate does not necessarily decrease the speech quality as a reduction of the speech bit-rate is translated into an increase of the error resilience, which can improve the overall quality in error-prone channel conditions. It will be shown that the codec mode of AMR running on W-CDMA, can be used for a similar purpose, i.e., to control the tradeoff between speech quality and network capacity.

Table 5.23 compares the MOS performance of QCELP-13, EVRC, and SMV under error-free conditions, which is consistent with the results of Table 5.22. EVRC achieves almost identical quality as QCELP-13, and with the results of Fig. 5.24, SMV Mode 0 provides comparable performance to EVRC. Table 5.24 summarizes the capacity gains from deploying SMV at operating modes other than Mode 0 [Greer and DeJaco (2001), 3GPP2 (2001)]. By reducing the average bit-rate, the transmit power is reduced and the network capacity is increased, while sacrificing the speech quality to some extent. It is notable that the gains are greater in the forward link than in the reverse link. This has a positive impact on the overall capacity of CDMA systems since the forward link in general has a smaller capacity than that of the reverse link.

In GSM running AMR, some codec mode limitations are enforced in the uplink and downlink. For example, at any instant different bit-rate allocations between encoded speech and channel code can be used in the uplink and downlink, but the channel

Table 5.24 Capacity gain of each SMV mode.

Mode	Forward link (%)	Reverse link (%)
0	0	0
1	27	16
2	49	29
3	60	35

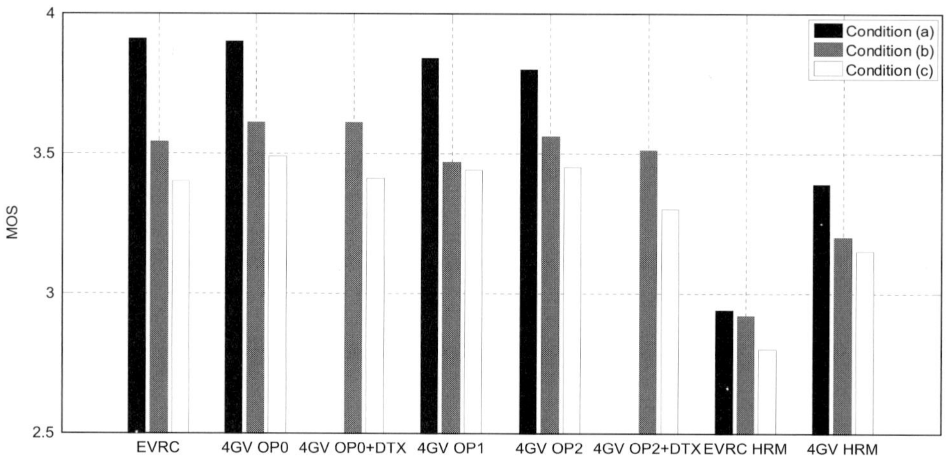

Fig. 5.25 Performance comparison of EVRC and 4GV.

type and the codec mode set must be identical to simplify the channel allocation, slow frequency hopping, and interworking procedures. By way of contrast, no such constraints are enforced in the reverse and forward links of CDMA. A recommended deployment scenario is to deploy SMV Mode 0 in the reverse link and another mode with a lower average bit-rate in the forward link. Deploying Mode 0 over the higher-capacity reverse link provides higher speech quality without impacting network capacity significantly while modes with lower average bit-rates provide lower quality but higher capacity in the power-limited forward link.

In Fig. 5.25, performance of EVRC and 4GV are compared over a wide range of channel conditions. Condition (a) is set to error-free and a nominal input level. In condition (b), car noise at 15 dB is applied as the background to the input signal. Condition (3) is made up of street noise at 15 dB and 1% FER. It is shown that 4GV Operating Point 0 (OP0) does not perform better than EVRC in error-free conditions. However, the gain from the stronger error resilience of 4GV becomes noticeable as the background noise level or the FER increases.

Table 5.23 and Fig. 5.24 show that the operating points of 4GV, as the modes of SMV, provide an elegant control of the tradeoff between speech quality and network capacity. Turning on DTX at 4GV OP0 and OP1 has a noticeably negative effect on the speech quality. Notice that unlike in GSM where the training sequences for synchronization are built into the TDMA frame structure, DTX in CDMA, whose symbol rate is much

Table 5.25 Performance comparison of EVRC and 4GV for music signals in error-free channel.

	MOS		
	Classic	Pop	Rock
EVRC	1.88	1.75	2.19
4GV OP 0	2.00	1.75	1.88

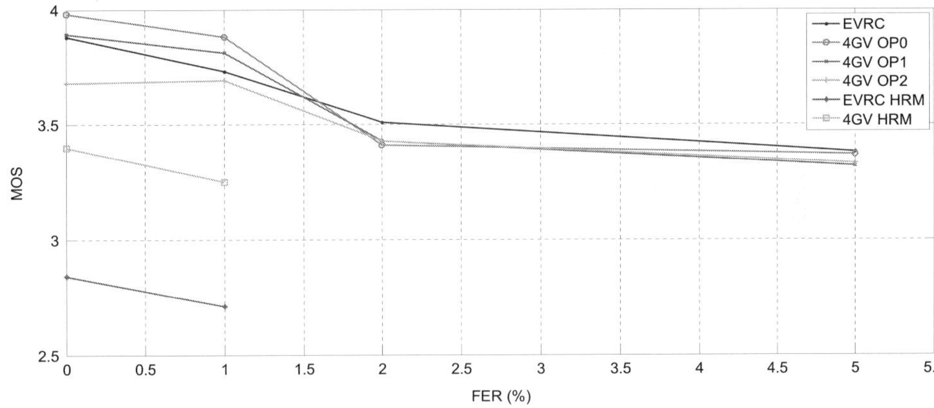

Fig. 5.26 Performance of EVRC and 4GV in error-prone channel conditions.

higher, requires advanced radio reception capabilities such as those provided in Rev. E, as well as the pilot channels, to compensate for the shortcomings. Significant quality differences between EVRC and 4GV are observed in the half-rate max (HRM) mode where only rates lower than Rate 1 are allowed. In contrast to EVRC which has to use either Rate $\frac{1}{2}$ or $\frac{1}{8}$, 4GV can also use Rate $\frac{1}{4}$ to encode the speech more efficiently at low average bit-rates.

As mentioned earlier in the discussion of SMV, the EVRC and 4GV were not designed with music detection capability. Table 5.25 compares the performances of the two speech codecs in the handling of a few music types: classic, popular, and rock [3GPP2 (2006)]. As expected, neither codec exhibits notable advantages in handling these input signals. The performance will degrade further if background noise or channel error is also present.

Figure 5.26 shows the performance of the codec configurations in conditions where the speech frames can be lost during transmission. The envelope of the curves representing the MOS-FER performances of 4GV in several operating modes stays above the curve of EVRC. It may be concluded that with appropriate switching of the operating modes by the network, 4GV can consistently provide quality superior to that of EVRC, just as AMR outperforms FR or EFR in GSM over a wide range of channel conditions.

While the gains in MOS values from enhancing the speech compression algorithms can be interpreted as improving the speech quality, the gains in network capacity can

Table 5.26 Voice capacity of forward link [Jou *et al.* (2009)].

Receiver type	EVRC (RC 3)	4GV (RC 11, PC 400 Hz)	4GV (RC 11, PC 200 Hz)
1-antenna	28	66	74
1-antenna (with IC)	33	79	88
2-antenna (with IC)	41	99	114

be estimated from system-level simulations in which the activity of multiple MSs and BTSs in voice services is modeled. Table 5.26 outlines the simulation results over 19 cells (57 sectors). The network capacity is defined here as the maximum number of MSs whose average FER does not exceed 2% while the probability of rejecting service requests, the *outage rate*, is maintained below 5%, for each 120-degree sector of a 1.25 MHz channel. The forward link capacity of cdma2000 is, when the MSs are evenly distributed over the cells, limited by the transmit power of the BTS whose power amplifier typically produces 20 watts of output power. Assuming that the MSs are equipped with a RAKE receiver with an antenna, 28 MSs can be served by a basic cdma2000 system with EVRC. The capacity can be increased to 33 MSs if each MS cancels the interference from other MSs in the same or neighboring cells. A capacity of 41 MSs can be achieved if each MS is equipped with two antennas and the interference-cancelation capability.

Replacing EVRC with 4GV alone may not produce a significant gain in the network capacity. However, if it is combined with the new radio signal processing procedures of Rev. E, the combined enhancements more than double the capacity. At a power control rate of 400 times per second, reduced from 800 Hz, 66, 79, and 99 MSs can be served with the receiver configuration of an antenna, an antenna with interference cancelation, and two antennas with interference cancelation, respectively. At 200 times per second, further gains, up to 74, 88, and 114, can be achieved. However, it is not easy to quantify the relative contributions from frame early termination, reduced power control rate, or DTX, since their operations are interrelated and also depend on the internal functions of 4GV. The total improvements from 4GV and Rev. E can be regarded as a package of well-known techniques to reduce the average bit-rate, transmit power, and the interference, thereby increasing the SNR at the receiver, whose performances and limitations were proven in the TDMA and CDMA mobile communication systems.

5.5.2 Live Call Analysis

Figure 5.27, consisting of eight plots, shows the time-varying characteristics of the key parameters measured in a live cdma2000 call using EVRC, for 30 frames or 0.6 seconds. The first and second plots show the power levels of the received and transmitted signals, whose average values for the period are -80.2 and -2.8 dBm, respectively. The next plot shows how the MS adjusts the transmit power, following the power control command from the BTS. Notice that the SNR at the chip level, $\frac{E_c}{I_o}$, roughly tracks the power.

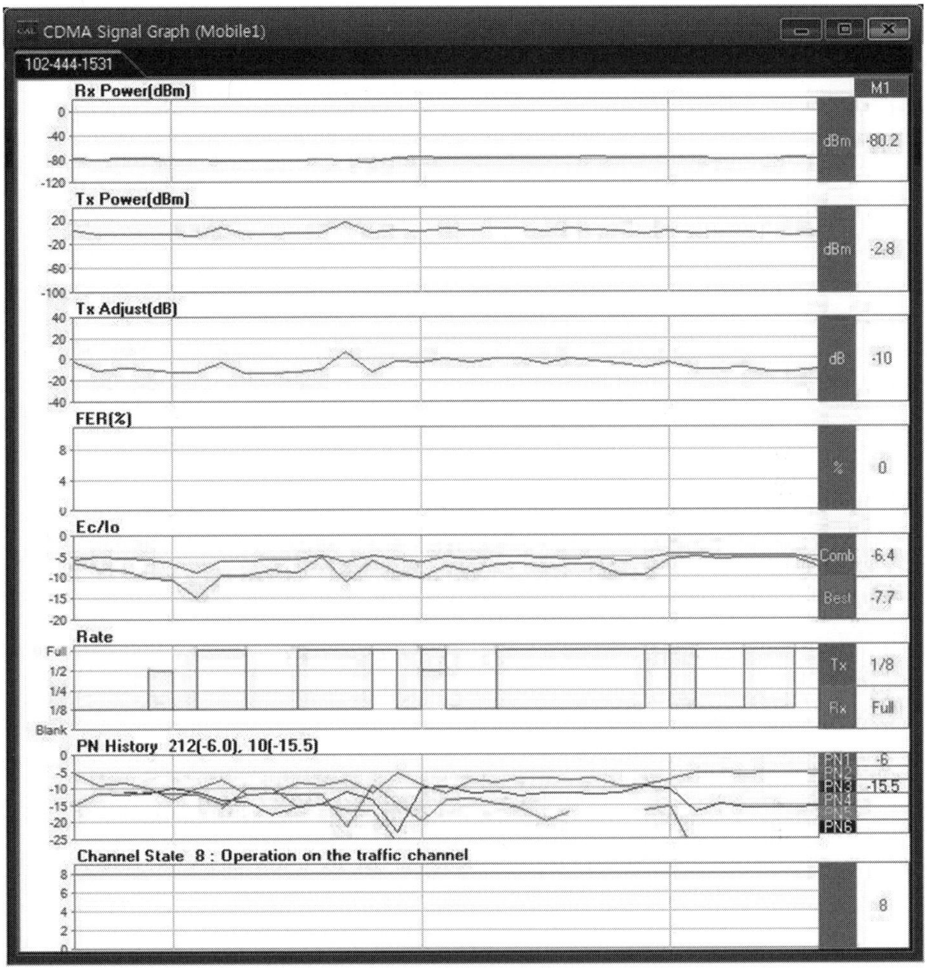

Fig. 5.27 cdma2000 call analysis (courtesy of Innowireless Co., Ltd.).

The cell used in the analysis is loaded with a small number of MSs, and the FER of the forward link is maintained at a negligible level in spite of the rainy weather during the measurements. It is shown that the frequency of encoding the speech at Rate $\frac{1}{2}$ is lower than those for other rates.

5.5.3 Derivation of CDMA Voice Capacity

The voice capacity of FDMA and TDMA mobile communications systems can be upper-bounded relatively easily, as the difference between the total number of traffic channels/slots and the total number of control channels/slots for the available bandwidth. In analog FDMA systems such as the AMPS, two channels carrying voice traffic for the reverse and forward links are assigned to each MS during a call, and the shared control channels are not used for voice services. Therefore, the voice capacity of FDMA is simply the maximum number of traffic channels assigned to each cell. In AMPS, from a bandwidth of 12.5 MHz assigned to each service provider, approximately 18 channels can be allocated to each sector. In other words, a BS can support approximately 54 simultaneous calls if the MSs are uniformly scattered over the cell.

In digital TDMA systems such as GSM, all channels and slots can be used for voice services and control purposes, but the voice capacity can still be computed from the number of channels assigned to each cell. In practice, the interference from the MSs in the same cell or neighboring cells cannot be avoided completely, and channels or slots whose SNR does not exceed the required level cannot be used for conversation. Therefore, crowded cells can be split into smaller ones when the expected gain in network capacity is more important than the additional cost.

Similar approaches can be considered for CDMA, with codes replacing channels or slots. However, the asymmetric structure of the reverse and forward links, and the higher complexity of CDMA make such an analysis less practical. To simplify the analysis, assume that the entire capacity of each cell, except some for controlling purposes, is used for voice services. In CDMA, because of the absence of guard band or guard time, it is more complicated to estimate an upper-bound of voice capacity. From the signal processing procedures of IS-95 and cdma2000, a few logical places that may be related to the limitation of voice capacity can be identified, in addition to the limited transmit power of the BTS.

For example, the Walsh function of length 64 that limits the number of forward channels is one of such locations. In IS-95, from the 64 forward channels differentiated with Walsh functions, only 55 channels can be assigned for voice services. As a result, approximately 18 MSs can be served with 1.25 MHz if the channels are divided for three 120-degree sectors, which is almost a ten-fold increase from the capacity of AMPS. Since the required SNR of IS-95 is far lower than that of AMPS, the capacity increase is even larger.

In IS-95, which was designed mainly for voice services, the length of the Walsh function is fixed at 64 in the forward link. In cdma2000, which supports data as well as voice, the length of the Walsh functions ranges from 4 to 128 in the Spreading Rate (SR) 1 and 4 to 256 in SR 3. With SR 1, both the reverse and forward links use a carrier whose chip rate is 1.2288 Mcps. This operation is called cdma2000 1xRTT, which is the most commonly deployed configuration. With SR 3, the reverse link operates on a broader channel whose chip rate is 3.6864 Mcps while the forward link combines three contiguous carriers each of which is spread at 1.2288 Mcps. Walsh functions are generated from a tree structure where each node corresponds to a function.

For the Walsh functions of different lengths to be orthogonal, all nodes located under a higher node of the same tree cannot be used for other channels until the Walsh function corresponding to the higher node is released. Therefore, the high-speed data services using short Walsh functions can drain the available Walsh functions rapidly. The limited space of Walsh functions can be expanded with Quasi-Orthogonal Functions (QOF) when the code space is exhausted. QOFs are generated by multiplying special vectors to Walsh functions so that each vector generates a set of orthogonal functions.

The orthogonality is maintained only within the same set but not between the QOFs of different sets [3GPP2 (2009)].

An upper-bound for the reverse link voice capacity of cdma2000 can be straightforwardly derived. The objective of this derivation is not to obtain a numerically accurate estimate but to provide an insight into the relationship of the signal processing procedures and the key parameters involved. We follow the system model and theoretical development from [Vanghi *et al.* (2004)]. We assume a cell where the power received from each MS is perfectly controlled so that the power from each of K MSs are all S (watts). Let E_b be the energy per bit including the reverse pilot channel and N_t be the total noise power over the channel bandwidth W (Hz). The effects of VBR speech coding are not considered yet and the bit-rate is assumed to be fixed at its maximum value, R_b (kbps). Then the SNR at the BTS is

$$\frac{E_b}{N_t} = \frac{S/R_b}{N_o + (K-1)S/W} = \frac{W}{R_b}\frac{S/(N_oW)}{1+(K-1)S/(N_oW)}, \tag{5.4}$$

where N_o is the thermal noise power. It is assumed that the interference is generated only by the other MSs in the same cell. Since the reverse link is not synchronized, the RAKE receiver of the BTS assigned for each MS considers the received signals from all the other MSs as noise. Let the receiver sensitivity be the ratio of received signal power to the background noise power,

$$\frac{S}{N_oW} = \frac{\frac{R_b}{W}\frac{E_b}{N_t}}{1-\frac{R_b}{W}\frac{E_b}{N_t}(K-1)}. \tag{5.5}$$

For the receiver sensitivity to be positive, which implies that the reverse link signals transmitted at finite transmit power can be detected and recovered by the BTS, the following inequality should be met,

$$K < 1 + \frac{W/R_b}{E_b/N_t}. \tag{5.6}$$

We replace N_t with I_o, the total received power spectral density, and ignore the constant 1. Since $I_o > N_t$, this approximation has the effect of relaxing the upper bound to some extent,

$$K < \frac{W/R_b}{E_b/I_o}. \tag{5.7}$$

The resulting bound on K is called the *pole capacity* in that it represents the maximum capacity which cannot actually be achieved in the stable operation of CDMA systems. In reality, the interference is also generated by the neighboring cells and the transmit power of the MS cannot be controlled perfectly. To tighten the bound, it is necessary to limit I_o to a level comparable to N_o. We define $\eta = N_o/I_o$, and call its reciprocal the Rise over Thermal (RoT). Then the total received signal power can be expressed as

$$KE_bR_b = (I_o - N_o)W = (1-\eta)I_oW. \tag{5.8}$$

To assure a stable operation distanced from the pole capacity, we limit the total received signal power to be less than the right-hand side of the equation, which generates the following upper-bound for K,

$$K < \frac{W/R_b(1-\eta)}{E_b/I_o}, \tag{5.9}$$

where the upper bound of K has been reduced by a factor of $1-\eta$.

To this point we have considered an isolated cell that is served by a single BTS. To model realistic deployment scenarios where the interference from neighboring cells is also taken into account, we define f to be the ratio of interference from the other cells to the interference from the given cell. Then the total noise power N_t and its approximation I_o are increased by a factor of $1+f$, which tightens the upper bound further. On the other hand, by ignoring the effects of VBR speech coding, the capacity has been underestimated to some extent.

Let ν be the voice activity factor, the proportion of a call during which voice activity is present. In IS-95 or cdma2000, ν corresponds to the proportion of speech frames whose rate is other than Rate $\frac{1}{8}$. Then the average bit-rate is reduced by a factor of ν from the maximum bit-rate, which relaxes the bound. The upper bound of the reverse link voice capacity considering the two factors is therefore

$$K < \frac{W/R_b(1-\eta)}{\nu E_b/I_o(1+f)}, \tag{5.10}$$

where typical values for ν and f are 0.4 and 0.6 respectively. A series of simplifying assumptions are made to derive the capacity model but nevertheless the model for the upper bound clarifies the relationship of the key parameters to the quality and capacity of CDMA systems.

Deriving an upper bound for the forward link voice capacity is more complex since the forward link includes shared control channels with fixed transmit power and power-controlled dedicated fundamental channels. Let o be the proportion of maximum transmit power assigned for the shared channels and ϕ_k be the ratio of the chip energy of the kth fundamental channel to the total transmitted power spectral density. Defining ν_k to be the voice activity factor of the kth fundamental channel, a constraint that the BTS should meet for stable operation of CDMA systems is to maintain the proportion of total transmit power for K fundamental channels less than $1-o$, which can be expressed as

$$\sum_{k=1}^{K} \nu_k \phi_k < 1 - o. \tag{5.11}$$

Since ν_k and ϕ_k are independent, by averaging both sides of the inequality, K can be upper-bounded as

$$K < \frac{1-o}{E\{\nu\}E\{\phi\}}, \tag{5.12}$$

where o and $E\{\nu\}$ are constant.

Let the SNR of the kth MS be $(E_b/N_t)_k$, then

$$\left(\frac{E_b}{N_t}\right)_k = \frac{E_{b,k}}{N_{t,k}} = \frac{I_{c,k}W/R_b}{N_{t,k}} = \frac{I_{c,k}}{I_{or,k}}\frac{W}{R_b}\frac{I_{or,k}}{N_{t,k}} = \phi_k\frac{W}{R_b}CINR_k, \tag{5.13}$$

where $I_{c,k}$, $I_{or,k}$, and $CINR_k$ are the chip energy of the kth fundamental channel, the transmitted power spectral density, and the carrier-to-interference-and-noise ratio, measured at the kth MS. Then ϕ_k is expressed as

$$\phi_k = \frac{R_b}{W}\left(\frac{E_b}{N_t}\right)_k\left(\frac{1}{CINR_k}\right). \tag{5.14}$$

In CDMA systems where an MS can be simultaneously connected to multiple BTSs during soft handover, ϕ_k depends on the status of the handover since the signal energies from the BTSs are combined at the RAKE receiver. Therefore the average of ϕ_k can be expressed as a weighted sum of its conditional expectations on h, the number of BTSs connected to the MS,

$$E\{\phi_k\} = \sum_h E\{\phi_k|h\}P_h. \tag{5.15}$$

After dropping k, the expectation of ϕ when the MS is connected to h BTSs becomes

$$E\{\phi|h\} = \frac{R_b}{W}E\{(E_b/N_t|h)(1/CINR|h)\}. \tag{5.16}$$

Then the forward link voice capacity is upper-bounded as

$$K < \frac{1-o}{E\{\nu\}\sum_h \frac{R_b}{W}E\{(E_b/N_t|h)(1/CINR|h)\}P_h}. \tag{5.17}$$

In [Vanghi *et al.* (2004)], it is shown that further approximations can be made with additional parameters such as the excess transmit power or the outage probability.

5.6 References

3GPP2. 1999. C.S0014-0 V1.0 Enhanced Variable Rate Codec (EVRC). December.

3GPP2. 2001. C11-20010326-003 SMV Post-Collaboration Subjective Test – Final Host and Listening Lab Report. March.

3GPP2. 2003a. C10-20030317-020B Considerations for Choosing Vocoders in 3GPP2 MMS. March.

3GPP2. 2003b. C10-20030317-022 MMS Speech Codec Selection Criteria Analysis. March.

3GPP2. 2004a. C.S0020-A V1.0 High Rate Speech Service Option 17 for Wide Band Spread Spectrum Communication Systems. May.

3GPP2. 2004b. C.S0030-0 V3.0 Selectable Mode Vocoder (SMV) Service Option for Wideband Spread Spectrum Communication Systems. March.

3GPP2. 2005. C.S0076-0 V1.0 Discontinuous Transmission (DTX) of Speech in cdma2000 Systems. December.

3GPP2. 2006. C11-20060327-015 Characterization Test Report for EVRC-Release B. March.

3GPP2. 2009. C.S0002-E V1.0 Physical Layer Standard for cdma2000 Spread Spectrum Systems. September.

3GPP2. 2010. C.S0014-D V2.0 Enhanced Variable Rate Codec, Speech Service Options 3, 68, 70, and 73 for Wideband Spread Spectrum Digital Systems. January.

Gibson, J. D. 2000. Source Representations, Networks, and Applications. In: Gibson, J. D. (ed), *Multimedia Communications: Directions and Innovations*. Academic Press.

Greer, S. C., and DeJaco, A. 2001. Standardization of the Selectable Mode Vocoder. *IEEE International Conference on Acoustics, Speech, and Signal Processing*, **2**(May).

ITU-T. 2009. Recommendation G.191 STL-2009 Manual. November.

Jou, Y., Black, P., Wu, Q., *et al.* 2009. cdma2000 1X Rev. E Forward Link Voice Capacity. *IEEE International Conference on Communications*, June.

Salami, R., Lefebvre, R., Lakaniemi, A., *et al.* 2006. Extended AMR-WB for High-Quality Audio on Mobile Devices. *IEEE Communications Magazine*, May.

Vanghi, V., Damnjanovic, A., and Vojcic, B. 2004. *The cdma2000 System for Mobile Communications: 3G Wireless Evolution*. Prentice Hall.

6 Signal Processing in W-CDMA Systems

As GSM evolved, its speech quality and network capacity improved significantly. This improvement saw the introduction of new modulation schemes such as 8-PSK. Furthermore the number of channels per carrier was increased by HR and VAMOS. However, multiple access schemes based on the FDMA/TDMA principles and a fixed channel bandwidth of 200 kHz placed fundamental limitations on the achievable performance, blocking further progress. The Universal Mobile Telecommunications System (UMTS) was the official successor to GSM. It was designed with a completely new set of system parameters. The Radio Access Technology (RAT), i.e., the radio signal processing procedures between the mobile and the network, of UMTS is called Wideband CDMA (W-CDMA) [Tachikawa (2002)], which is often used as an alternative name for the UMTS.

Formally, cdma2000 and UMTS, both of which are based on the CDMA principles, are classified as the third generation (3G) mobile communications systems, since they can theoretically meet the performance requirements of 3G, the International Mobile Telecommunications (IMT)-2000, set by ITU-T: 384 kbps at vehicular speed and 2 Mbps in a fixed position [ITU-R (2014)]. However, achieving the maximum transmission speed requires very favorable transmission conditions, such as short LOS paths, and leaves little capacity for other mobiles in the cell.

The radio access technologies of W-CDMA are closer to those of cdma2000 than GSM, but the core networks of W-CDMA share many technical features with GSM to reduce the deployment cost and thus to ease the migration from the earlier networks. A key feature of UMTS is the introduction of mobile video telephony, whose development was simplified by combining multimedia telephony technologies proven in circuit-switched fixed networks, such as PSTN or the Integrated Services Digital Network (ISDN), with newly developed W-CDMA radio access technologies. Nevertheless, support of video necessitates significant changes in the signal processing operations, as video data is much more fragile than speech data, because of its higher compression ratio. Video codecs also cannot be built into the signal processing procedures as tightly as speech codecs.

We begin this chapter by discussing the speech and radio signal processing of W-CDMA. Before we move to the topics of 3G-324M mobile video telephony that operates over W-CDMA, we outline the basic idea of video signal processing and compare the fundamental differences between it and speech signal processing, originating from the multi-dimensional nature of the video signal itself. In this chapter, no new speech codecs are introduced but the compression principles of two video codecs are briefly discussed with regard to 3G-324M.

6.1 W-CDMA Release 99

Table 6.1 summarizes the key parameters of W-CDMA Release 99 Frequency Division Duplex (FDD). Release 99, often called Rel. 99 or Rel-99, is named for the year when the first set of W-CDMA technical specifications was completed. The naming system was changed later to use an integer, as Release 5, 6, ... Currently the term Release is used also to represent the period of technology development and standardization in other mobile communications systems such as GSM and LTE. In addition to FDD, W-CDMA also supports the Time Division Duplex (TDD) mode, which uses a single frequency band for both the uplink and downlink, even for full-duplex services such as voice and video telephony.

Since control signalling is required to switch between transmission and reception, TDD cannot match FDD in terms of spectral efficiency; it often exhibits smaller coverage per base transceiver station or less stable link quality. However, TDD can still be economical as the frequency spectrum for FDD, consisting of at least a pair of appropriately separated bands for the uplink and downlink, is much more costly. TDD can reduce the burden of the most expensive spectrum of FDD, e.g., that used for voice services and roaming, by handling traffic without tight delay constraints or uni-directional traffic. W-CDMA TDD uses a similar set of parameters as FDD but its power control rate is significantly lower. In this chapter, we focus on the analysis of W-CDMA FDD, and its media and radio signal processing operations.

The channel bandwidth of W-CDMA is 5 MHz, which is larger than the 1.25 MHz bandwidth of cdma2000. However, its chip rate is 3.84 Mcps while the chip rate of cdma2000, 1.2288 Mcps, is high enough to spread the information spectrum to fill up the channel bandwidth. Therefore in W-CDMA, the channel bandwidth divided by the chip rate is 1.302 compared to 1.017 for cdma2000. This means that W-CDMA uses a larger proportion of the bandwidth for its guard band. However, the higher chip rate of W-CDMA differentiates the multi-path components in a more refined fashion. This enables more robust transmission when fading is present over a wide spectrum and reduces the CIR required for proper operation.

Since the difference in distance corresponding to one chip delay is about 300 meters in cdma2000 and is about 78 meters in W-CDMA, position of an MS can be estimated more precisely with radiolocation when the GPS information is not available. For the uplink, Binary Phase Shift Keying (BPSK) is used to modulate the signal to maximize the CIR at the cost of spectral efficiency. For the downlink, Quadrature Phase Shift Keying (QPSK) is used. The rate of power control adjustment is higher than for cdma2000. The duration of transmission was not controllable in the earlier mobile communications systems but W-CDMA allows data to be transmitted over one of four time intervals. This can be used to reduce the interference by spreading transmission in the time domain as widely as possible when the requirement for delay is not stringent. On the other hand, urgent

Table 6.1 System parameters of W-CDMA (FDD).

Multiple access	FDMA/CDMA
Modulation	BPSK, QPSK (uplink), QPSK (downlink)
Channel bandwidth (MHz)	5
Frame length (ms)	10
Chip rate (Mcps)	3.84
Power control rate (Hz)	1500
Transmission time interval (ms)	10, 20, 40, 80

Table 6.2 W-CDMA FDD frequency bands.

Operating band	Uplink (MHz)	Downlink (MHz)
I	1920–1980	2110–2170
II	1850–1910	1930–1990
III	1710–1785	1805–1880
IV	1710–1755	2110–2155
V	824–849	869–894
VI	830–840	875–885
VII	2500–2570	2620–2690
VIII	880–915	925–960
IX	1750–1785	1845–1880
X	1710–1770	2110–2170

Table 6.3 W-CDMA cell coverage.

Band (MHz)	900	1800	2100	2600
Coverage (km^2)	17.6	7.6	6.2	3.7

data can be transmitted in a shorter time interval. Table 6.2 shows the frequency bands of W-CDMA, which include the Primary GSM-900 band as band VIII, and the AMPS and IS-95 bands as bands V and II, respectively. The channel frequency can be adjusted in increments of 200 kHz.

Table 6.3 shows the estimated coverage for the major W-CDMA bands, in which the cell radius decreases rapidly as the channel frequency increases [Holma and Toskala (2010)]. It is clear that the lower frequency bands require fewer base transceiver stations to serve the same area. Unlike the gradual evolution of GSM from TCH/FS to VAMOS, or the slow transition of CDMA from IS-95 to cdma2000 Rev. E, even the latest GSM and the earliest W-CDMA share so few technical elements that they cannot be operated on the same frequency bands. A typical strategy to migrate the network from GSM to W-CDMA is to deploy W-CDMA in the same area but in different, often higher, frequency bands so that W-CDMA provides video call and data service, and GSM, whose coverage is extensive, handles the majority of voice traffic and messaging.

Notice that with the same speech codec operating at the same bit-rate, e.g., AMR 12.2 kbps, there is little audible quality difference between GSM and W-CDMA as the networks are configured to achieve similar quality levels. As the coverage of W-CDMA gets wider and more intensive, more services will shift from GSM to the new network. After sufficient time has passed, depending on the pattern of traffic and cost for the equipment, GSM will finally enter the sunset stage, in which the obsolete networks are phased out, new ones are installed, and the frequency spectrum is re-assigned for W-CDMA or later radio access technologies. In rural areas with low traffic, the life span of the network is longer and it may be that GSM outlives W-CDMA, migrating directly toward LTE.

6.1.1 Network Architecture

Figure 6.1 shows the network architecture of W-CDMA. It is quite similar to that of the previous circuit-switched mobile communications systems. The terminology of W-CDMA is different; the MS, BTS, and BSC of the earlier systems are changed to the User Equipment (UE), the NodeB, and the Radio Network Controller (RNC), respectively. Like the BTS and BSC of IS-95, the NodeB and RNC are equipped with various measures to control the tradeoff between speech quality and network capacity, or to spread the network load over neighbouring cells, using techniques such as power control, handover, or admission control. The U_u interface connects the UE and NodeB, and the I_u interface connects the RNC and MSC. A major proportion of the radio signal processing occurs in U_u. The I_{ub} and I_{ur} interfaces, which connect the NodeB and RNC, and the RNC and another RNC respectively, initially run on the Asynchronous Transfer Mode (ATM) Adaptation Layer Type 2 (AL2). The transport technology provides the

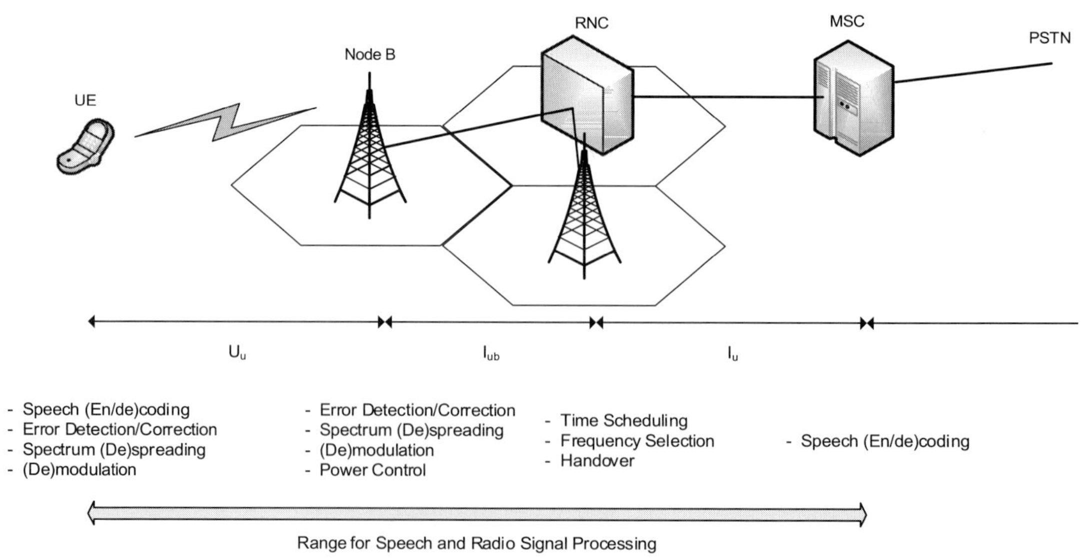

Fig. 6.1 W-CDMA network architecture.

required Quality of Service (QoS) for the traffic, but the Ethernet can also be used for the interfaces from Release 5.

As in cdma2000, soft handover is supported in W-CDMA, under the control of the RNC, which also controls the speech bit-rate to and from each UE in the cell. However, it will be shown that neither the RNC nor NodeB, or any other network node, can control the bit-rate of speech or video in 3G-324M. A key difference between W-CDMA and cdma2000 is that the NodeBs are not synchronized to a common timing signal while in cdma2000, the BTSs are synchronized to the GPS signal. cdma2000 has the advantage of faster synchronization but W-CDMA also has the advantage of providing synchronization in locations where the GPS signal is not available, such as underground or inside buildings. Another difference is that in earlier circuit-switched mobile communications systems, a set of speech codecs is installed at the TRAU located somewhere between the BTS and BSC to serve the MSs in the cells. In W-CDMA, the speech codecs are located in the MSC, which sits at a higher level in the network architecture than the RNC controlling the codecs in both the UE and MSC. The speech codecs are elevated in the hierarchy to maximize the distance over which the speech is transported in its most compressed format, since the cost for the I_{ub} and I_u interfaces, such as those for installing and running ATM AL2, is significant. Note that an MSC typically hosts a large number of speech codecs that can be shared by RNCs.

6.1.2 Protocol Stack Principles

In the previous chapters, speech and radio signal processing operations involved in TDMA and CDMA mobile communications systems were presented. In these systems, it was possible to specify the bit-rate at each step precisely, following the transmission paths from the input to the speech encoder to the input to the RF front-end. Little information other than the compressed speech, redundancy for error resilience, and the signaling to control the bit-rate or transmit power is transported over these paths. However, such a detailed analysis is no longer feasible in W-CDMA and later mobile communications systems. The radio signal processing in some steps becomes too complex to track as precisely as before, and the transmission paths have to be shared by the bit-streams for other network operations at all times, even when voice activity is present.

Although cdma2000 and W-CDMA are based on similar network architectures and multiple access schemes, W-CDMA was designed from the very beginning to support multimedia services other than voice, such as data, video telephony, and multimedia broadcasting. Hence the radio signal processing procedures of W-CDMA are not optimized only for the limited bit-rates of some speech codecs. Instead, the procedures are designed to be as versatile as possible so that an arbitrary set of QoS parameters, such as the bit-rate, bit or packet error rate, or delay, can be finely controlled to meet the diverse needs of those services. Therefore in W-CDMA, a new concept, *Protocol Stack*, is employed to ease the description of the media and radio signal processing operations.

In this and the following chapters, the concept of stacked or layered communications protocols is used to describe the flow of media and control signaling between the UE and the NodeB. Each protocol, or layer, is assigned a set of responsibilities and capabilities that are clearly defined so that each protocol can interact with those located above or below, or with its counterpart on the other side of the channel. The interfaces of the protocols must be designed so that neighboring protocols developed by different manufacturers, or protocols on the same level but separated by the channel find no difficulty in communicating. This principle of designing a protocol stack enables each protocol to be separately updated as long as the interoperability between the interfaces is maintained.

A block of data input to a protocol is called a Service Data Unit (SDU), and a block of data output from a protocol is called a Payload Data Unit (PDU). When an SDU passes through a communications protocol, a header, a small amount of overhead used to indicate the nature of the input data or the status of certain parameters, or to request certain actions from the counterpart protocol over the channel, can be attached to or detached from the SDU. As each protocol typically has limitations on the size of data block it can handle, an SDU can be merged with another, or split into multiple units inside the protocol and output as several PDUs.

Figure 6.2 shows the protocol stack architecture of a W-CDMA baseband modem, which consists of five major protocols: PHY, MAC, RLC, RRC, and PDCP. These are dynamically configured depending on the status of various connections and services. In the figure, functionality of the W-CDMA protocol stack is vertically separated into two groups: the Control Plane (C-Plane) and the User Plane (U-Plane). The C-Plane manages the control signaling used to set up and manage the wireless links between the

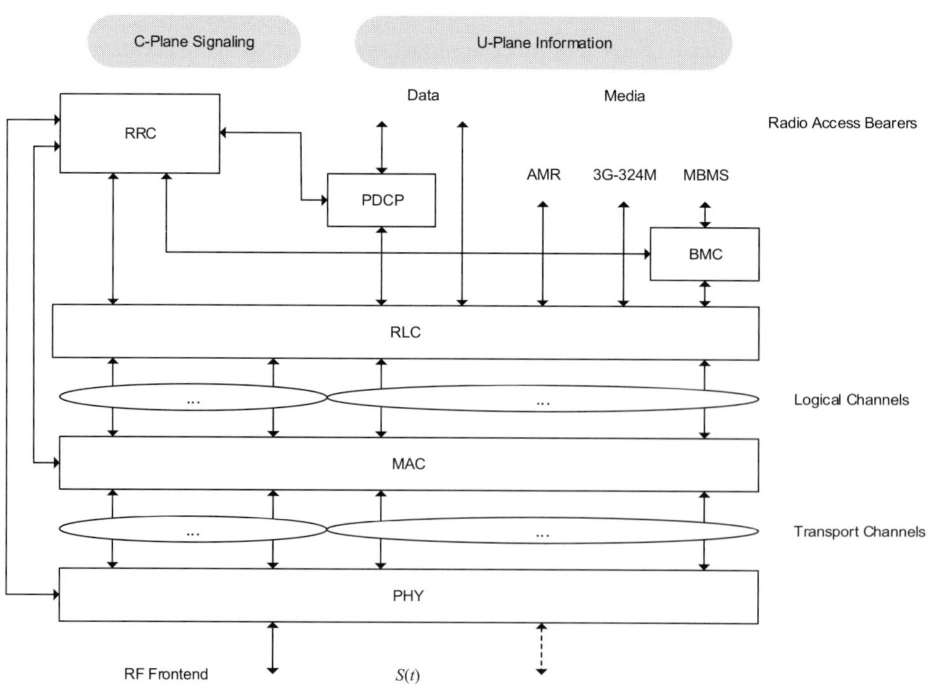

Fig. 6.2 W-CDMA modem protocol architecture.

UE and the NodeB; the U-Plane manages the transmission and reception of applications data or multimedia services. Each protocol may belong to the C-Plane or the U-Plane, or both. Although, in principle, one can assemble the stack using protocols from different manufacturers, typically all of the modem protocols are developed by the same manufacturer to optimize the interaction between protocols and minimize the latency in the propagation of control signals and data through the stack. The architecture of the protocol stack shown in Fig. 6.2 is applicable to both the UE and the NodeB.

At the Packet Data Convergence Protocol (PDCP) [3GPP (2017h)], data packets to and from the applications have their IP headers compressed to minimize the bit-rate before the packets are transmitted over the wireless channel. To compress the headers using the Robust Header Compression (ROHC), the statistical nature of each field in the headers is analysed and appropriate data compression techniques are applied [Bormann *et al.* (2001)]. Once the packets cross over the wireless channel safely, the compressed headers are restored to their full sizes.

When the packets are generated frequently and each packet carries a small amount of data, as at the output of the speech encoder, the overhead from RTP/UDP/IP headers required for VoIP can outweigh the speech data itself. However, since in W-CDMA, the compressed speech and video frames are transmitted over circuit-switched networks, the PDCP is bypassed. When the packets are generated infrequently or each packet carries a large amount of data, as at the output of the video encoder, compression of the IP headers does not return a significant gain in spectral efficiency. Since the scope of the control signalling is limited to the UEs, NodeBs, or nearby core network nodes, the IP headers are not attached to these messages. The messages carrying control information can be exchanged between them without the IP address. Therefore PDCP belongs only to the U-Plane.

Radio Link Control (RLC) [3GPP (2017g)] assumes a set of functions to facilitate the transmission of data between the UE and the NodeB. Depending on the required levels of overhead, error resilience, and delay, the RLC has three operational modes: Transparent Mode (TM), Unacknowledged Mode (UM), and Acknowledged Mode (AM). The paths between the RLC and the layers above it are called the Radio Access Bearers (RAB). The RAB for carrying the control information is called the Signaling Radio Bearer (SRB). SRBs typically have requirements different from those of RABs, such as larger delay but lower bit and error rate.

Medium Access Control (MAC) is assigned the responsibility for reliable transmission of data [3GPP (2016b)]. The paths between the RLC and the MAC are called the Logical Channels. A key functionality of the MAC is priority handling, in which the MAC multiplexes the SDUs received from the RLC by applying different priorities to the logical channels. It also de-multiplexes the PDUs received from a lower layer, the Physical Layer (PHY), where the conventional radio signal processing of CDMA, including the CRC computation, channel coding, interleaving, power control, spreading, and modulation, is performed. The paths between the MAC and the PHY are called the Transport Channels (TrCH). Note that the MAC at the NodeB has larger responsibilities than its counterpart at the UE, as it also handles the priorities of UEs in the cell as well as the priorities of their SDUs.

Radio Resource Control (RRC), present only in the C-Plane, is responsible for setting up and managing the radio links between the UE and the NodeB [3GPP (2017j)]. RRC makes key decisions including admission control and handover. The Broadcast/Multicast Control (BMC) manages the control signalling and media flow of the Multimedia Broadcast Multicast Service (MBMS) [3GPP (2017i)], so that the MBMS is targeted to a subset of the UEs, i.e., as multicasting, or to the entire set of UEs in the cell, as broadcasting. In Release 99, the BMC supports only the Short Message Service (SMS) but later versions also support multimedia streaming. As the voice service and video telephony of W-CDMA use the TM or UM of the RLC, most media and radio signal processing operations are taken in the media codecs and the physical layer, bypassing the intermediate protocols, RLC and MAC.

6.2 Radio Signal Processing

We approach the radio signal processing of W-CDMA in three steps: analysis of the RLC and MAC protocols commonly applicable to the uplink and downlink, and analysis of the radio signal processing operations in the physical layer for each link. The radio signal processing operations are further separated into the transport channel multiplexing, and the spreading and modulation. In the seven layer classification of the Open Systems Interconnection (OSI) [ISO/IEC (1994)] reference model, PHY is regarded as layer 1 (L1) while MAC, RLC, and PDCP are considered as layer 2 (L2).

In GSM, where all the transport channels on the uplink and downlink are synchronized, identical speech and radio signal processing is used in the uplink and downlink. In cdma2000, on the other hand, the reverse link and the forward link employ significantly different radio signal processing procedures, since the channels from different MSs on the reverse link are not synchronized while in the forward link they are. In W-CDMA, the uplink and downlink use similar procedures for the RLC and MAC but apply significantly different ones for the PHY, although the same protocol architecture is applied to the uplink and downlink.

AMR is the basic speech codec of Release 99, which has to be supported by any UE. It is not impossible to use other speech codecs of GSM in W-CDMA but there is little reason not to use AMR, with its superiority in speech quality and error resilience. However, as the underlying radio access technologies are now based on the CDMA principles, AMR needs to operate in a different fashion, which involves neither the joint source-channel coding of GSM nor the variable bit-rate coding of cdma2000 to control the tradeoff between speech quality and network capacity.

6.2.1 Radio Link Control

Figure 6.3 shows the PDU formats of the RLC for the TM, UM, and AM. As shown in the protocol architecture of W-CDMA, the speech codec directly interfaces with the RLC, which is typically configured in the transparent mode (TM). In TM, as the name transparent suggests, no header field is attached to the SDU, whose length is not

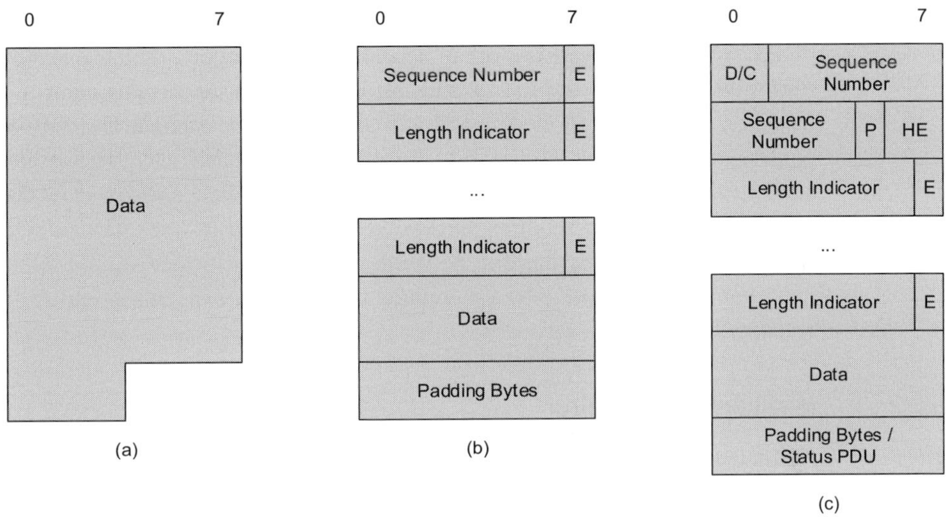

Fig. 6.3 RLC PDU formats. (a) TM. (b) UM. (c) AM.

constrained to be a multiple of bytes. However, the SDU may be segmented if it exceeds the maximum sizes allowed for transmission. TM, which does not provide any means for dealing with a loss of data or an out-of-order delivery, is typically employed for delay-constrained applications such as voice call. Figure 6.3(a) illustrates the case when the SDU, as with a frame of AMR 12.2 kbps, is not byte-aligned, i.e., not made up of an integer-number of bytes.

UM provides an elevated level of error resilience compared to TM. In UM, length of an SDU is an integer multiple of bytes. The PDU of UM starts with a 7-bit sequence number that indicates the order of each PDU, which is followed by one or more 7-bit length indicators representing the length of each SDU. The interpretation of the Extension (E) bit depends on the mode and configuration of the higher layer. In normal operation, $E = 0$ indicates that the next field is data, a PDU including the status information, or padding bits. $E = 1$ indicates that the next field is another length indicator and an E bit. With the information in the header, a large amount of data can be transmitted in multiple PDUs and re-assembled at the receiver even when the fields arrive out of order. In addition, lost PDUs can be identified easily using this information. Since UM attaches only a small amount of overhead to the SDU, and neither acknowledges received PDUs nor retransmits lost PDUs, UM can still be used for delay-constraint applications, such as 3G-324M mobile video telephony, which requires a higher level of error resilience than voice service. If necessary, the data is padded with an integer number of bytes, for example, to fill up the assigned transmission capacity.

In the RLC mode with the highest error resilience, AM, the PDU starts with a 2-bit D/C field that indicates whether the PDU includes data or control signal. A Polling (P) bit is used to request a status report from the receiver. A 2-bit Header Extension (HE) type field is used to indicate whether the next byte is data ($HE = 00$), or a length indicator and an E bit ($HE = 01$). $HE = 10$ indicates that the next byte is data, and the last byte of the PDU is the last byte of an SDU. The usage of HE = 11 is reserved for other needs. A status report, which can be periodically transmitted, includes the list of missing PDUs. With the re-transmission mechanisms based on the status report, the AM is mainly used for applications with relaxed delay constraints. The necessity of AM might be questioned as re-transmission schemes such as the Transport Control Protocol (TCP) are often employed in the application layer. However, such schemes necessitate end-to-end re-transmission which can take more time than re-transmission between two RLC entities separated by a single wireless channel. Therefore AM can reduce the frequency of end-to-end re-transmissions significantly.

6.2.2 Medium Access Control

PDUs from the RLC are entered into the MAC as SDUs, using logical channels such as the Dedicated Traffic Channel (DTCH), which is the channel used to transport speech and video. In the MAC, a series of fields is attached to facilitate the prioritization, concatenation, and segmentation of the SDUs. Figure 6.4 shows the format of a MAC PDU when an SDU is entered using DTCH. Depending on the transport channels to which the PDUs are output and whether the dedicated channels are multiplexed or not, some of the fields can be omitted.

The Target Channel Type Field (TCTF), whose length is two to eight bits in the FDD, identifies the logical channel from which the SDU comes. The two-bit UE-Id Type field is an identifier of the UE using common transport channels. These channels do not carry the speech or video data of conversational services. This field ensures a correct decoding of the following 16- or 32-bit UE-Id field. Finally, a four-bit Channel Type (CT) field can be used to identify each logical channel when multiple logical channels are mapped onto the same transport channel. For a single stream of SDUs, a CT field may not be attached.

When a MAC PDU is passed to the physical layer using a transport channel, it is denoted as a transport block. The set of MAC PDUs transported within a time unit, known as the Transmission Time Interval (TTI), is defined as a Transport Block Set (TBS). All transport blocks in a TBS are of the same size. When the MAC multiplexes the RLC PDUs from multiple logical channels onto a single transport channel, which is possible only when the logical channels require the same QoS level, the order of transport blocks from the same logical channel is maintained. MAC also prioritizes the order of logical channels in a transport block set. Figure 6.5 shows the processing of an AMR frame in the RLC and MAC, as the first step of the uplink. If the UED or UEP is not used, the logical channels would be merged in the MAC.

2-8	2	16/32	4	
TCTF	UE-Id Type	UE-Id	C/T	MAC SDU

Fig. 6.4 MAC PDU format for DTCH.

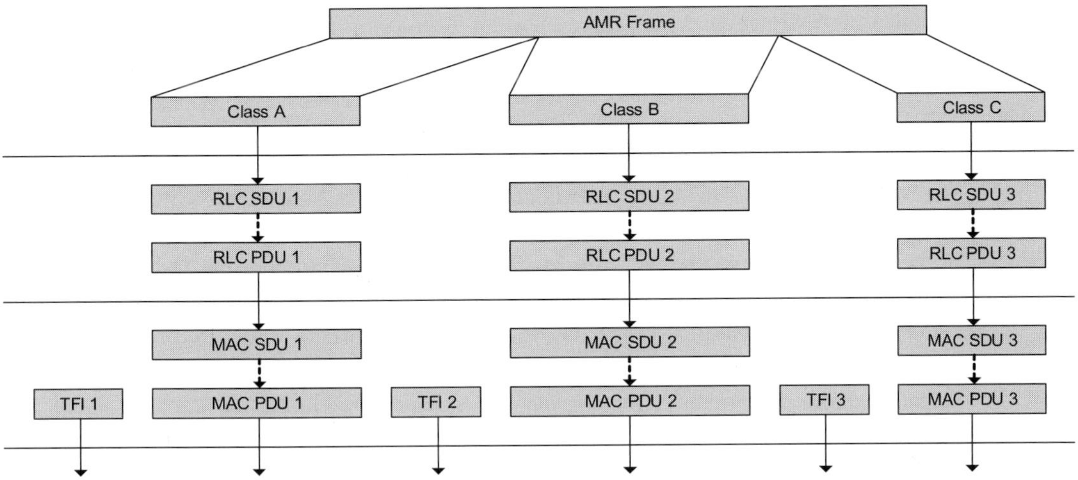

Fig. 6.5 Step 1: L2 header processing for speech with UED/UEP.

Table 6.4 Bit-stream classification of AMR for W-CDMA.

Codec mode (kbps)	Class			Total (bits)
	A	B	C	
12.2	81	103	60	244
10.2	65	99	40	204
7.95	75	84	0	159
7.40	61	87	0	148
6.70	58	76	0	134
5.90	55	63	0	118
5.15	49	54	0	103
4.75	42	53	0	95

In TDMA mobile communications systems such as IS-54 and GSM, to apply the unequal error detection (UED) and protection (UEP) to the speech bit-stream, output bits from the speech encoder are categorized into multiple classes, depending on the relative influence of each bit on the quality. Notice that the three-level classification of AMR bit-stream, as shown in Table 6.4, is slightly different from that of the same codec for GSM. To match the nature of a wireless channel as closely as possible by adjusting the assignment of CRC and channel code, such a classification depends more on the characteristics of radio signal processing than on those of media signal processing. Note that the bit-stream of AMR is classified differently even for TCH/AFS or TCH/AHS of GSM.

Depending on the codec mode of AMR, two or three classes of bit-streams are sent to the radio bearers, each of which has a different set of QoS parameters such as the SDU error ratio and the residual bit error ratio that can be tolerated. As only the 12.2 and 10.2 kbps have three protection classes, only two bearers are necessary for the codec modes of AMR equal to or lower than the 7.95 kbps. Then the bit-streams propagate through the RLC to the logical channels and further propagate through the MAC to the transport channels. In this procedure, no header field is attached to each bit-stream as unequal error detection and protection are realized in the lowest layer, PHY. Unlike in GSM where the SID of the AMR speech codec is padded with bits to fill the time slots completely, in W-CDMA, the SID is classified as Class A, and transmitted without the bits of other bearers.

For each transport channel, the MAC assigns a Transport Format Indicator (TFI), a label indicating a specific transport format that is transferred to the physical layer in the C-Plane. The Transport Format (TF) includes the information required for the delivery of a transport block, and consists of two parts: a dynamic part and a semi-static part. The dynamic part includes the sizes of the transport block and the transport block set, which can be used to infer the number of transport blocks. The semi-static part includes the error detection and correction schemes, such as the type of error correction coding, coding rate, parameters for repetition or puncturing to make the size of data match the available bit-rate, and the size of the CRC.

An example of the dynamic part for Class A of the 12.2 kbps includes {35 bits, 35 bits} and {81 bits, 81 bits} for the SID and speech frame, and an example of its semi-static part includes {20 ms, convolutional coding only, state rate matching parameter = 1, CRC = 8 bits}. Therefore the transport block set includes only one transport block. Otherwise, transmission of a speech frame has to be delayed to be transmitted with the next frames. On the other hand, an example of the dynamic part for Class B of the 12.2 kbps includes {103 bits, 103 bits}, and an example of its semi-static part can be {20 ms, convolutional coding only, state rate matching parameter = 1, CRC = 0 bits}, as no CRC is applied to Class B or C bits. This concludes the first step of W-CDMA radio signal processing operations, which is applicable to both the uplink and downlink. The meaning of parameters in the TF will be clarified in what follows.

As illustrated in Fig. 6.5, a negligible amount of signal processing is taken in the layer 2 of W-CDMA, as data passes through the protocols with a minimum or no overhead attached, and complex operations are concentrated in the lower layer, PHY. In packet-switched networks such as LTE, the signal processing operation of PDCP to compress the IP headers can be substantial.

6.2.3 Physical Layer

Uplink Signal Processing Procedures

In the physical layer, a series of radio signal processing operations is applied to the transport blocks from the MAC, based on the accompanying TFI, before the transport blocks are multiplexed into a single bit-stream [3GPP (2017f)]. Thus, reinforcement of the transport blocks for transmission over the wireless channel, in the form of unequal error detection and protection are realized in this layer. Figure 6.6 shows the first part of the radio signal processing taken in the PHY, as the second step of the uplink. The bit-streams from the transport channels are coded and multiplexed into the Coded Composite Transport Channel (CCTrCH), whose bit-stream is de-multiplexed to fill in the slots of the physical channel.

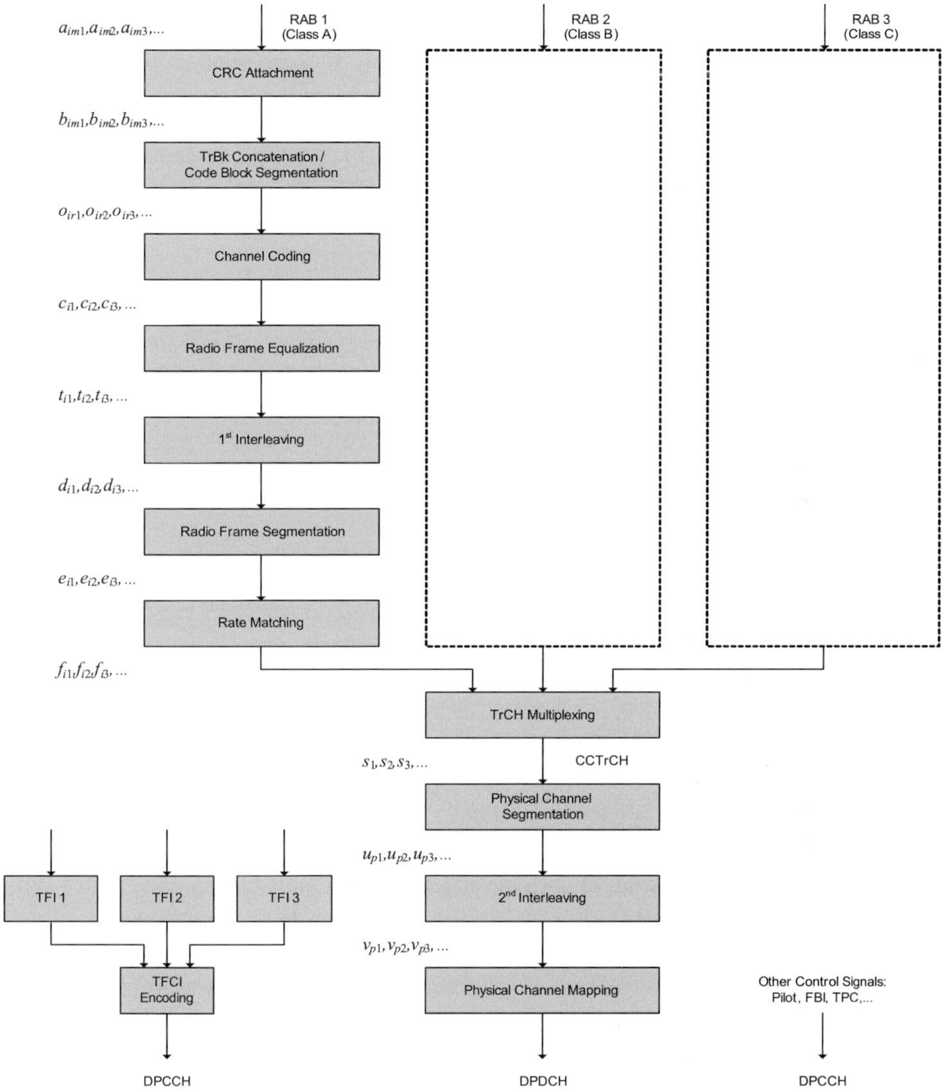

Fig. 6.6 Step 2: transport channel multiplexing for uplink.

To support the variations in voice activity or codec mode, each transport channel can have multiple transport formats. Suppose that four transport channels, TrCH1, TrCH2, TrCH3, and TrCH4, are assigned for voice service using AMR 12.2 kbps. In Table 6.5, TrCH1, TrCH2, and TrCH3 are defined for the three classes of AMR bit-streams. Notice that TrCH1 can carry no data during silence, an SID, or Class A bits from the speech codec. Likewise, TrCH2 can carry the silence or Class B bits. The last transport channel, TrCH4, supports a 148-bit transport block every 40 ms for signaling purpose.

The maximum transport block sizes for the first three transport channels are the same as the numbers of Class A, B, and C bits. However, the last transport channel, operating in the AM of the RLC, is assumed to be made up of 128 bits of control data, a 16-bit RLC header, and a four-bit MAC header. As shown in Table 6.5, TFI of the four

Table 6.5 TFI for voice service.

TFI	AMR (bits)			Signaling radio bearer (bits)
	TrCH1	TrCH2	TrCH3	TrCH4
0	0	0	0	0
1	39	103	60	148
2	81			

Table 6.6 TFCI for voice service.

TFCI	TFI				Application
	TrCH1	TrCH2	TrCH3	TrCH4	
0	0	0	0	0	DTX
1	1	0	0	0	SID
2	2	1	1	0	AMR 12.2 kbps
3	0	0	0	1	signaling
4	1	0	0	1	SID, signaling
5	2	1	1	1	AMR 12.2 kbps, signaling

transport channels can have 3, 2, 2, and 2 levels, respectively, which can represent the status of the transport channels when the output of the AMR speech encoder is silence, an SID, or a 244-bit frame (12.2 kbps), and when the control signaling is present in the fourth transport channel or absent.

The Transport Format Combination Indicator (TFCI) is constructed from the received TFIs, and transmitted with the processed transport blocks and control signaling including power control or feedback information, using the Dedicated Physical Data Channel (DPDCH). Table 6.6 shows how each TFCI can be generated from the TFIs for the permitted cases. When the radio access bearer is set up, the RNC informs the UE and the NodeB of the allowed combinations of transport formats. Using this information, the receiver is able to decode and separate the transport blocks of each transport channel. However, data and control information are not mixed as they require different treatment during transmission. The control information is transported using the Dedicated Physical Control Channel (DPCCH), which is a separate physical channel that accompanies another physical channel that carries the data. In cdma2000, the reverse pilot channel is similarly paired with the reverse fundamental channel to assist synchronization but the DPCCH can also carry various types of control information simultaneously.

As the second stage of the uplink signal processing, Fig. 6.6 shows the transport channel multiplexing procedures, a combination of error detection, error correction, rate matching, interleaving, and mapping of the transport channels onto the physical channels. The procedures feed the three radio access bearers corresponding to the three classes of AMR bit-streams with three different QoS targets to the third stage. Once every TTI, data for each transport channel, denoted in the figure by $a_{im1}, a_{im2}, \ldots, a_{imA_i}$, arrives in the form of a transport block set. The TTI, which depends on the required QoS of the transport channel, has a value from the set {10 ms, 20 ms, 40 ms, 80 ms}. For transport channels requiring error detection, e.g., those carrying Class A bits, a CRC is computed and attached to each transport block. The size of the CRC can be 8, 12, 16, or 24 bits, depending on the importance of the data and the transport channel assigned.

Figure 6.7 illustrates how transport blocks with different sizes and TTIs are carried by the transport channels before being multiplexed into a single CCTrCH. For each TTI, all transport blocks in a transport channel must be of the same size. However, even in the same transport channel, transport blocks of different sizes can be used in different TTIs. Let A_i and L_i be the size of transport block TrCH i and the number of its parity bits, respectively. Then, the bit-stream of the transport block after CRC attachment, the code block, will be denoted by $b_{im1}, b_{im2}, \ldots, b_{imB_i}$, where $B_i = A_i + L_i$. In the next step, if the number of bits to be transmitted within a TTI is larger than the maximum size of a code block, which depends on the type of channel coding to be used (see below), then the code block is segmented after the transport blocks are concatenated, acknowledging the expected overhead from the channel coding. W-CDMA provides two channel coding schemes, conventionally used convolutional code for two rates and new turbo code that requires higher decoder complexity and higher bit-rate. The choice for the channel code can be made to match the nature of the services. Figure 6.8 shows the $K = 9$, Rates $\frac{1}{2}$ and $\frac{1}{3}$ convolutional encoders.

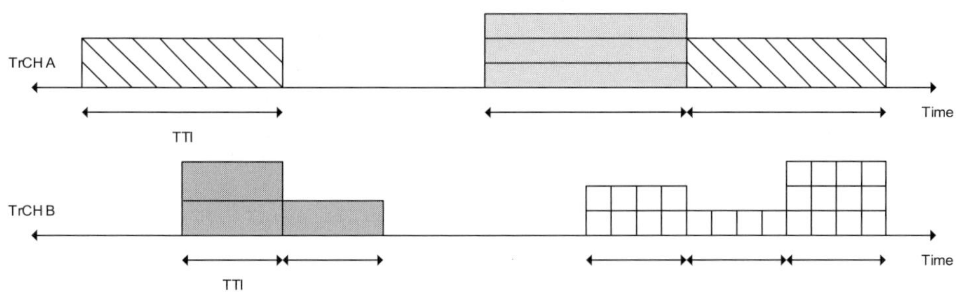

Fig. 6.7 Transport block set for CCTrCH.

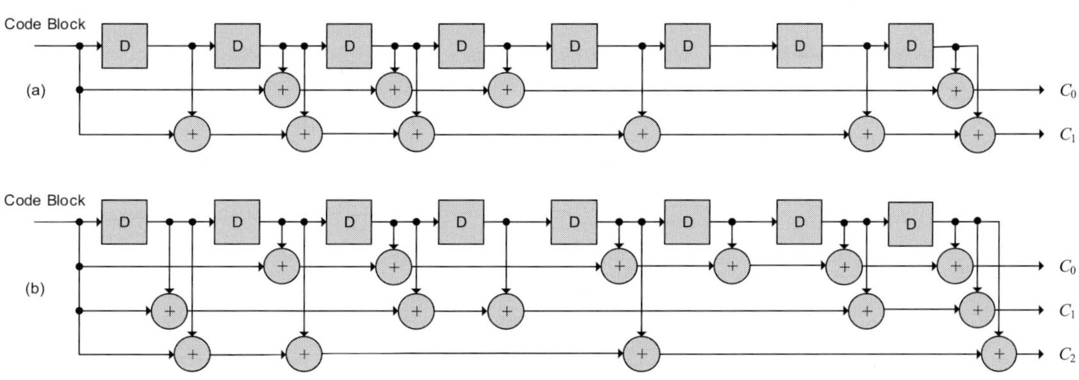

Fig. 6.8 Convolutional encoding. (a) Rate $\frac{1}{2}$. (b) Rate $\frac{1}{3}$.

Table 6.7 Channel codes for W-CDMA.

Channel code	Number of output bits (for k input bits)	Maximum code block size (bits)
Rate $\frac{1}{2}$ convolutional	$2(k+8)$	504
Rate $\frac{1}{3}$ convolutional	$3(k+8)$	504
Rate $\frac{1}{3}$ turbo	$3k+12$	5114

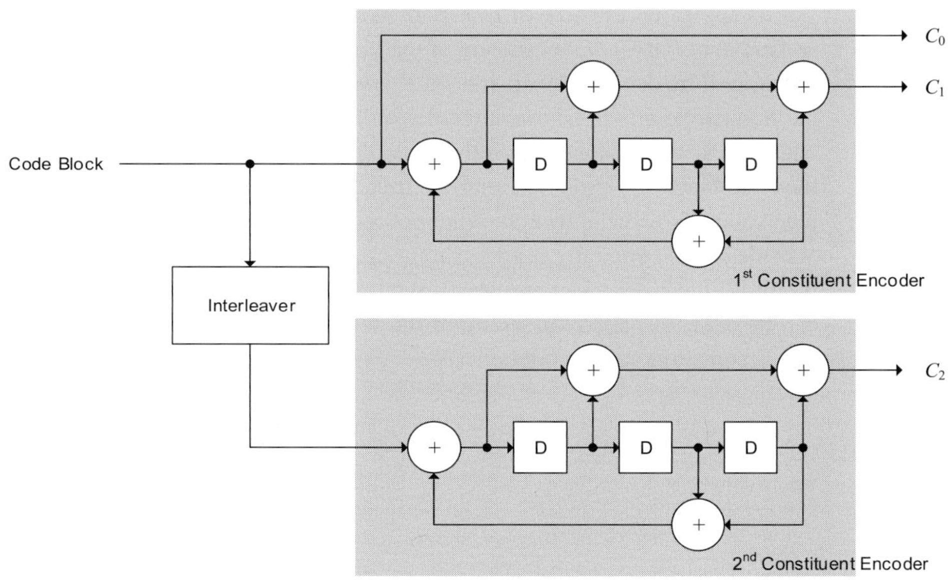

Fig. 6.9 Rate $\frac{1}{3}$ turbo encoding.

The turbo code is a new type of channel code whose performance can be increased by iterative processing with soft-decision metrics [Berrou *et al.* (1993)]. The code can approach the theoretical limit of channel capacity. The performance increase depends on the number of iterations but the gain per iteration becomes smaller once the majority of achievable gain is reached, making it usable in both real-time and non-real-time applications. Figure 6.9 shows the Rate $\frac{1}{3}$ turbo encoder of W-CDMA, which is a Parallel Concatenated Convolutional Code (PCCC) with two eight-state constituent encoders and an interleaver. Whether to use the convolutional or turbo code depends on the amount of data and the required Quality of Service (QoS). In general turbo codes outperform convolutional codes significantly for code block sizes that exceed hundreds or thousands of bits, while convolutional codes are typically used to protect relatively small code blocks such as those of the speech bit-stream or control signaling.

Table 6.7 outlines the relationships between the numbers of input and output bits, and the maximum code block sizes for the three channel codes. With a turbo encoder the bits used to complete a code block are not all zeros, unlike the zero-padding bits of a convolutional encoder that flush the values of the shift-registers to zeros. Instead, bits internally generated during turbo encoding are used to complete the code block. From the design constraints, the two convolutional codes of W-CDMA can encode up to 504 bits, which can cover the bit-rates of most speech codecs. Since the turbo code of W-CDMA can encode up to 5114 bits, it is typically used to take care of the data of services that operate at bit-rates higher than 32 kbps, such as video call. After channel encoding, the coded blocks are serially concatenated so that they are output in the order of arrival.

The next step is denoted as radio frame equalization, which only appears in the uplink. Let i denote the transport channel number. Then the input bit-stream is padded with zeros so that the output bit-stream can be divided into multiple segments of the same size, F_i, when the TTI is longer than 10 ms. The first interleaver receives an integer multiple of F_i bits and outputs the permuted bits, which are segmented into the radio frames. In the first interleaving, the bits can be shuffled for a period longer than 10 ms, the length of a W-CDMA radio frame. Therefore, each 20 ms speech frame of AMR is carried in two radio frames. In the following step, denoted as rate matching, bits of a transport channel are repeated or punctured as necessary, as the number of bits of a transport channel can vary for each TTI. For example, when the numbers of bits in two TTIs are different, the bits are repeated or punctured to ensure that the total bit-rate after the transport channels are multiplexed into a single bit-stream is equal to or less than the total bit-rate of the allocated dedicated physical channel. Figure 6.10 illustrates the separation, processing, and collection of bits for rate matching in the uplink. In the case of puncturing, it is necessary to avoid the reduction of error resilience caused by removing key bits, such as the systematic bits from the turbo encoder, as shown in Fig. 6.10(a) where the bits can be eliminated in the bit-processing modules. On the other hand, convolutionally-coded transport channels or turbo-coded transport channels with repetition are rate-matched as in Fig. 6.10(b).

The rate-matching procedures can be configured with a set of parameters whose values are based on the type of channel coding, number of available physical channels, spreading factor, and whether the transport channels have fixed or flexible positions in the physical channels. Each 10 ms, one radio frame from each transport channel is multiplexed with other radio frames to form a single bit-stream, the coded composite transport channel (CCTrCH). Thus, throughout the rate-matching procedures, bit-streams of transport channels with higher QoS are transmitted with more redundancy than those with lower QoS.

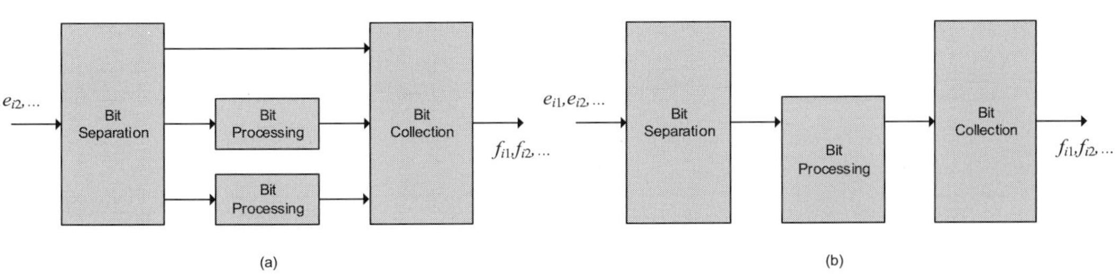

Fig. 6.10 (a) Rate matching of turbo-coded transport channels. (b) Rate matching of convolutional-coded transport channels and turbo-coded transport channels with repetition.

If the size of a CCTrCH requires the use of multiple physical channels, bits of the CCTrCH are divided among the physical channels although for voice service, such a segmentation is unlikely. After another round of interleaving, whose scope is limited to a radio frame, the bits are finally mapped onto the physical channels. In the operations of the transport channel multiplexing shown in Fig. 6.6, the radio signal processing from CRC attachment to radio frame segmentation is done to each transport channel of a TTI, e.g., 20 ms for voice service. Further radio signal processing from rate matching to physical channel segmentation and the second interleaving is done to all transport channels for each radio frame, 10 ms.

Figure 6.11 shows the structures of two physical channels, the dedicated physical data channel (DPDCH), and the dedicated physical control channel (DPCCH) where the basic W-CDMA time unit is divided into 15 slots or five subframes. For each 0.667 ms slot, the uplink DPDCH transports the data from the transport channels, and the uplink DPCCH carries the control information generated at the physical layer, which can include the Pilot bits to support coherent detection at the NodeB, the Feedback Information (FBI) bits, and the Transmit Power Control (TPC) commands. Such a clear separation of data and control information can also be seen in the reverse link of cdma2000. In the DPCCH, the TFCI informs the receiver of the instantaneous transport format combination of transport channels mapped to the simultaneously transmitted uplink DPDCH radio frame. Thus, the data of each slot needs to be buffered until the accompanying TFCI is decoded. Depending on the processing capability, a UE may operate up to six DPDCHs simultaneously but can use only one DPCCH. In practice, more than one DPDCH is rarely supported due to the elevated complexity of the NodeB.

Figure 6.12 shows the third radio signal processing step in the uplink. The two bit-streams, DPDCH and DPCCH, are fed into the two branches, converted into a bipolar format, and multiplied by the channel gain and the channelization code, C_d and C_c, respectively. β_d and β_c are the scaling factors used to control the relative power levels. When more than one DPDCH is required, which rarely happens in practice, the

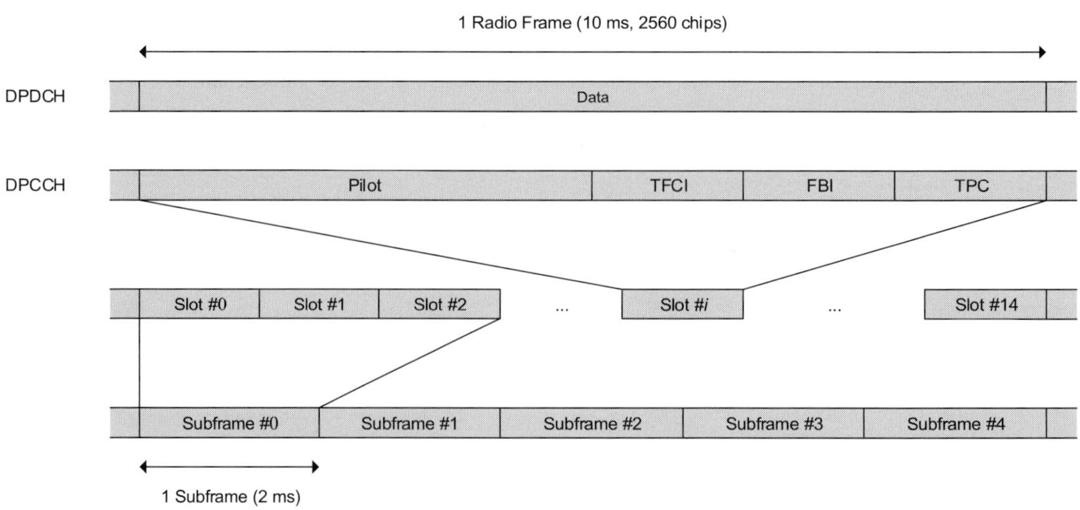

Fig. 6.11 Uplink dedicated channel structure.

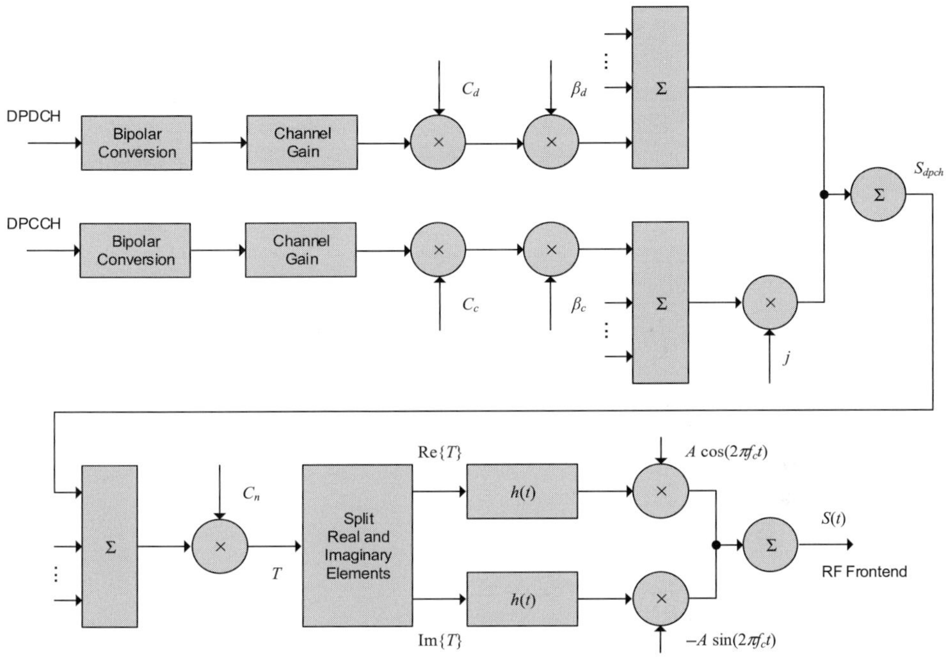

Fig. 6.12 Step 3: uplink spreading and modulation.

additional DPDCHs are split between the two branches as evenly as possible so that the variation of the signal envelope is minimized. For example, the second DPDCH is fed into the lower branch while the third is fed into the upper branch. To differentiate each physical channel, a channelization code similar to the Walsh code of cdma2000, the Orthogonal Variable Spreading Factor (OVSF) code, is multiplied by each symbol.

The OVSF code, which maintains the orthogonality between physical channels, is defined using a tree-structure shown in Fig. 6.13 as $C_{ch,SF,k}$, where SF is the spreading factor, a parameter indicating the level of spectrum spreading with the OVSF code, and k is the code number. The tree-structure is basically another formula to generate a family of Walsh codes as used in cdma2000 but lists them in a different order. The spreading factor and code number of the DPDCH are dynamically changed based on the amount of data and the required QoS, while in the DPCCH only $C_{ch,256,0}$ is used. For fixed bit-rate services, such as video call, there is no need to change the spreading factor and the code number.

It is possible that a shortage of the OVSF codes might limit the network operation even when there is enough power headroom in the transmitter of the NodeB, and the interference is within an acceptable level. The network, i.e., the RNC, needs to pay attention to the ongoing consumption of the OVSF codes. To reserve as many codes as possible, it is important for the codes to be as long as possible as the orthogonality between two OVSF codes is guaranteed regardless of their lengths. When the NodeB serves higher bit-rate services such as video calls, which require shorter OVSF codes than voice services, there will remain fewer codes.

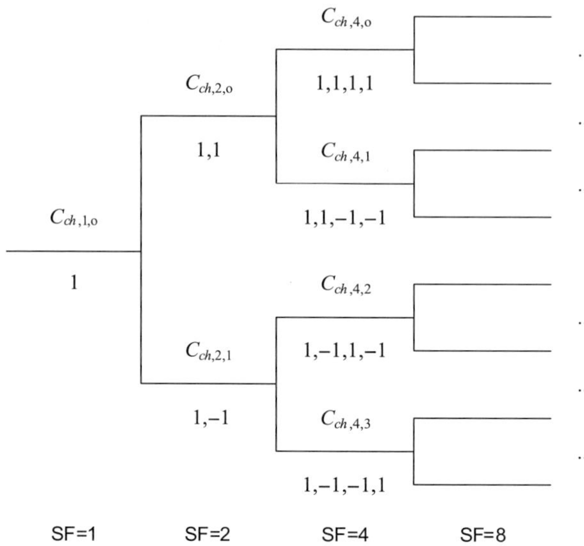

Fig. 6.13 Code-tree for OVSF codes.

One channel in the upper branch and another in the lower branch of Fig. 6.12 can share the same channelization code because the two channels can be separated in the lower branch by using a phase shift, indicated as a multiplication by j. When only one DPDCH is transmitted, $C_{ch,SF,k}$ is used, where $k = \frac{SF}{4}$. When more than one DPDCH is transmitted, all DPDCHs have spreading factors equal to 4. The nth DPDCH is spread by $C_{ch,4,k}$, where $k = 1$ if n is 1 or 2, $k = 3$ if n is 3 or 4, and $k = 2$ if n is 5 or 6. The network sends a four-bit integer, representing values 0–15, to the UE, which indicates the power ratio of DPCCH to DPDCH from 0 to 1, in steps of $\frac{1}{15}$.

The power ratio, set by the network, is applied to all DPDCHs if there are more than one. After the spectrum spreading and the power balancing, the symbol streams of the lower branches are summed and phase-shifted by $\frac{\pi}{2}$, to ensure they can be separated from those of the upper branches at the receiver. Then the signals in the upper and lower branches are summed to form S_{dpch}, which is combined with the symbol-streams of other physical channels for the uplink.

The basic approach for higher bit-rates is transmission using multiple DPDCHs, i.e., multi-code transmission, but using a single higher bit-rate DPDCH simplifies the implementation of the receiver. In addition, the spectral efficiency can be reduced during the time the DPDCHs are assigned and restored. To overcome these limitations, W-CDMA evolved to the High Speech Packet Access, where in the downlink the UE can use the entire spectrum for a shorter time than the duration of a W-CDMA radio frame [Holma and Toskala (2006)]. In the uplink of HSPA, dedicated channels are still used but they are complemented with the Hybrid Adaptive Repeat and Request (HARQ), a PHY-level re-transmission scheme that operates faster than the acknowledged mode of RLC. In the HSPA, the NodeB allocates the opportunity for transmission and reception to the UEs using scheduling algorithms designed to meet the criteria for service quality and network capacity. The NodeB can provide both W-CDMA and HSPA on the same channel simultaneously. Note that the BTS of cdma2000 cannot offer 1xRTT and EV-DO as such.

Since W-CDMA is basically a CDMA system, all of the uplink physical channels need to be scrambled so that the combined signals looks like noise. However, unlike the case in cdma2000, a complex-valued scrambling code is used in W-CDMA. Figure 6.14 shows the configuration of shift registers used to generate the long scrambling sequence for the uplink. It is based on a combination of two sequences, $C_{1,n}$ and $C_{2,n}$ where $C_{2,n}$ is generated by shifting $C_{1,n}$ by 16,777,232 chips. The two sequences are constructed by element-by-element modulo-2 addition of the 38400 chip segments of two binary sequences constructed using two primitive polynomials, $x^{25} + x^3 + 1$ and $x^{25} + x^3 + x^2 + x + 1$. The resulting sequences are the segments of a set of Gold sequences [3GPP (2017d)]. Then the complex-valued scrambling sequence is defined as

$$C_n(i) = C_{1,n}(i)(1 + j(-1)^i C_{2,n}(2\lfloor \frac{1}{2} \rfloor)), \tag{6.1}$$

where $i = 0, 1, \ldots, 2^{25} - 2$, and the floor function $\lfloor\ \rfloor$ rounds the value of its argument down to the nearest integer.

Unlike the PN sequences used in cdma2000, the length of the scrambling sequence is limited to that of a radio frame, 10 ms. However, the auto-correlation and cross-correlation properties of the scrambling sequence are sufficient for synchronization. The complex scrambling sequence is designed to minimize the zero-crossings in order to reduce the peak-to-average ratio of the signal, as in the OQPSK of the IS-95 reverse link. A shorter scrambling code of length 255 can be used if the NodeB is equipped with advanced receivers such as the Multi-User Detector (MUD) [Verdú (1998)] although typical NodeBs are equipped with the RAKE receivers. When the symbols are scrambled, the chip rate can be as high as 3.84 Mcps and the signal spectrum is fully spread over the 5 MHz channel bandwidth. Each UE is assigned a unique scrambling code by the network.

Finally, the real and imaginary components of the scrambled symbols are separated, filtered for shaping the spectrum, QPSK-modulated, and summed before transmission over the wireless channel. The pulse-shaping filter, shown as $h(t)$ in Fig. 6.12, is a

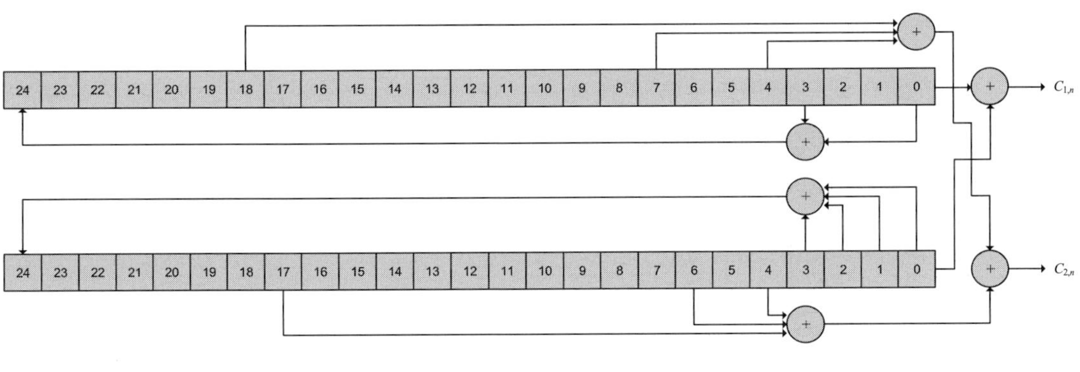

Fig. 6.14 Uplink long scrambling sequence.

root-raised cosine (RRC) filter with a roll-off factor $\alpha = 0.22$ [3GPP (2016a)]. The transmitted RF signals are received by a RAKE receiver running several fingers, which decodes them into the baseband signal. The level of overall transmit power is determined as the product of the channel gain and A, the magnitude of the RF carriers. The conversion from baseband signal to RF signal has conventionally been achieved in two steps, by modulating to an Intermediate Frequency (IF) somewhere below the carrier frequency, and then to the final frequency. With the Direct Conversion (DC), the baseband signal is modulated in one step to the carrier frequency, which can save the cost for an oscillator that tunes to the IF. However, the IF-based conversion has an important advantage that the channel gain can be controlled at the IF [Tanner and Woodard (2004)]. A theoretical analysis of the modulation and spreading procedures of W-CDMA and its scrambling sequences can be found in [Glisic (2003)].

Downlink Signal Processing Procedures

The first step of the uplink processing, the L2 header processing shown in Fig. 6.5, is also the first step in the downlink processing. Figure 6.15 shows the second step, which includes a combination of error detection and correction, rate matching, interleaving, and mapping of transport channels onto physical channels. This is slightly different from what is done for the uplink.

After attachment of the CRC, segmentation of the code block, and channel coding, the codeword bits undergo rate matching. The next step, which is unique to the downlink, is the first insertion of the discontinuous transmission (DTX) indication. The DTX bits indicate when the transmitter should be turned off, and the signal bits are not transmitted. This step of inserting DTX bits is used only when the positions of the transport channels in the radio frame are fixed. There is another possibility where the positions are neither fixed nor flexible, but *pseudo-flexible* when an advanced transmission scheme for the dedicated channel (DCH), set by the DL_DCH_FET_Config parameter, is used. DCH refers to the channels dedicated to each UE, including DPDCH and DPCCH. The pseudo-flexible rate matching is similar to the fixed-position rate matching but applied to each TTI independently. Therefore, the DTX indication bits need to be tri-valued: 0, 1, and a third state that only needs to be internally recognizable. After that, the radio frame segmentation and the transport channel multiplexing are applied to the bits.

Then another set of DTX indication bits are inserted, which are placed at the end of the radio frame to signal that the assigned capacity is incompletely used. There remains another step of interleaving after which the DTX indication bits will be distributed over all slots. The remaining three procedures, the physical channel segmentation, the second interleaving, and the physical channel mapping, are the same as in the uplink. The TFCI and other control signals that will accompany data are then generated. Figure 6.16 shows the channel structures of DPDCH and DPCCH for the downlink, whose slot formats are summarized in Table 6.12. Note that in the downlink, the DPDCH and DPCCH are multiplexed into a single bit-stream, and synchronization at the UE is assisted by a channel broadcast at high power that is shared by all UEs in the cell.

Figure 6.17 shows the third step of the downlink processing. This includes the spreading and modulation procedures, where each pair of consecutive symbols is converted

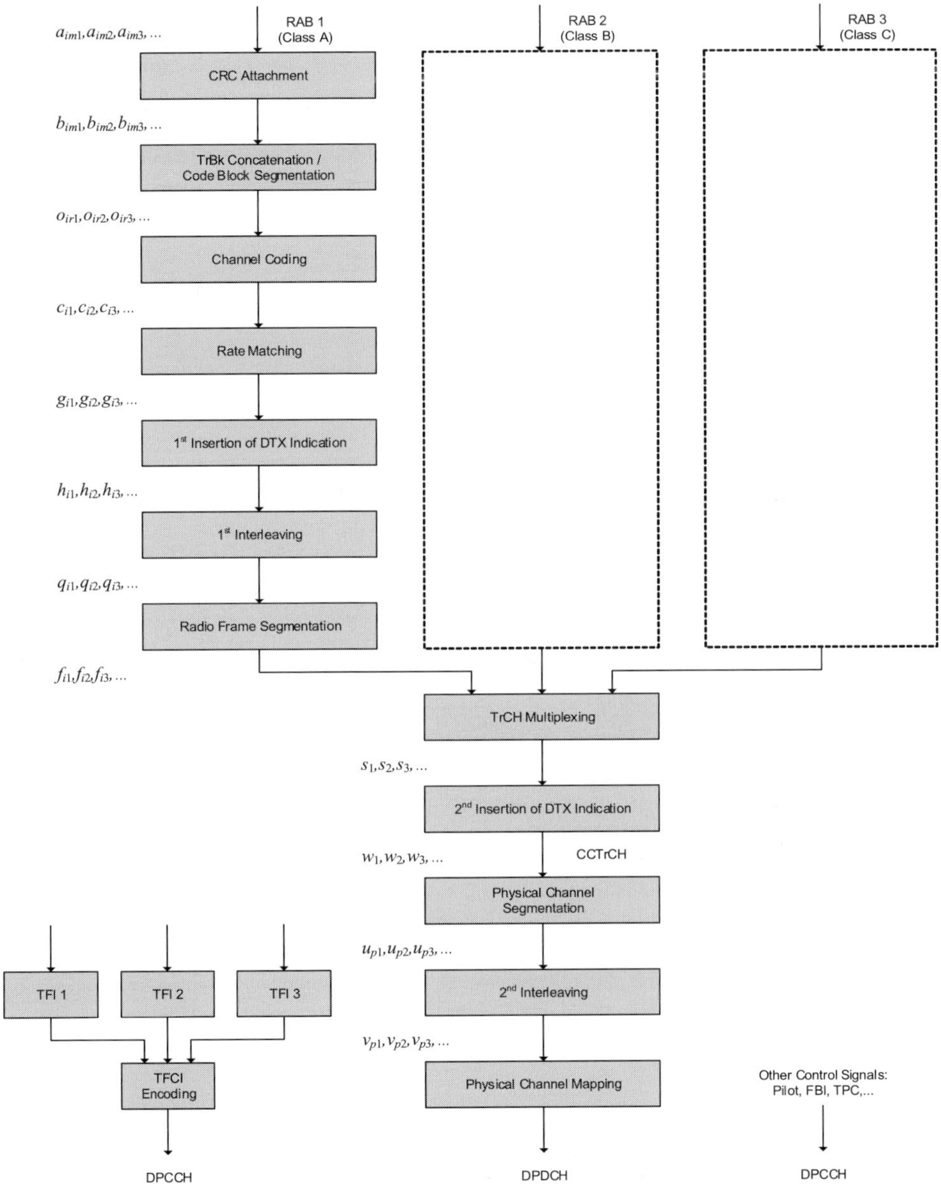

Fig. 6.15 Step 2: transport channel multiplexing for downlink.

from a serial to a parallel formation and mapped to the I and Q branches. At the modulation mapper, 0 is mapped to +1, 1 is mapped to −1, and the DTX indication bit is mapped to 0 before they are all multiplied by the channel gain. Then the spectra of the symbols on the I and Q branches are spread by a real channelization code, C_{ch}, as shown in Fig. 6.13. After the phase of the symbols on the Q branch is shifted by $\frac{\pi}{2}$ radians, the two sequences of symbols on the two branches are summed on a symbol-by-symbol basis. Then the new sequence is scrambled using a complex scrambling code generated

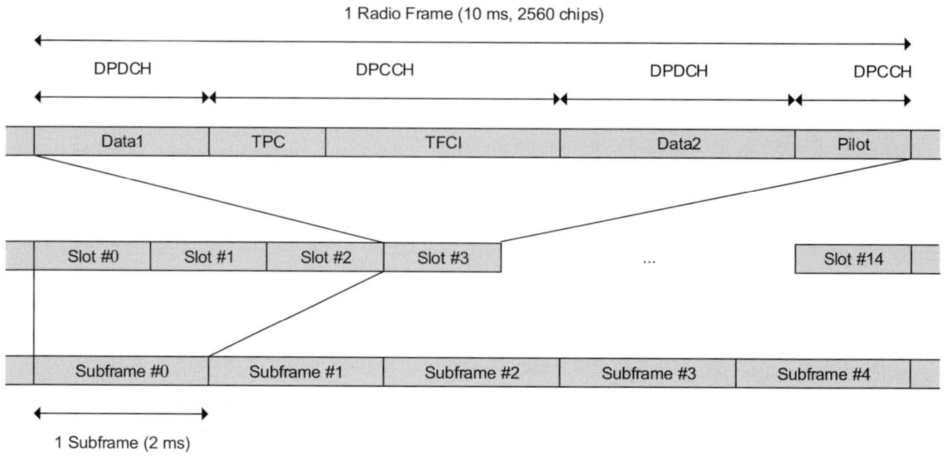

Fig. 6.16 Downlink dedicated channel structure.

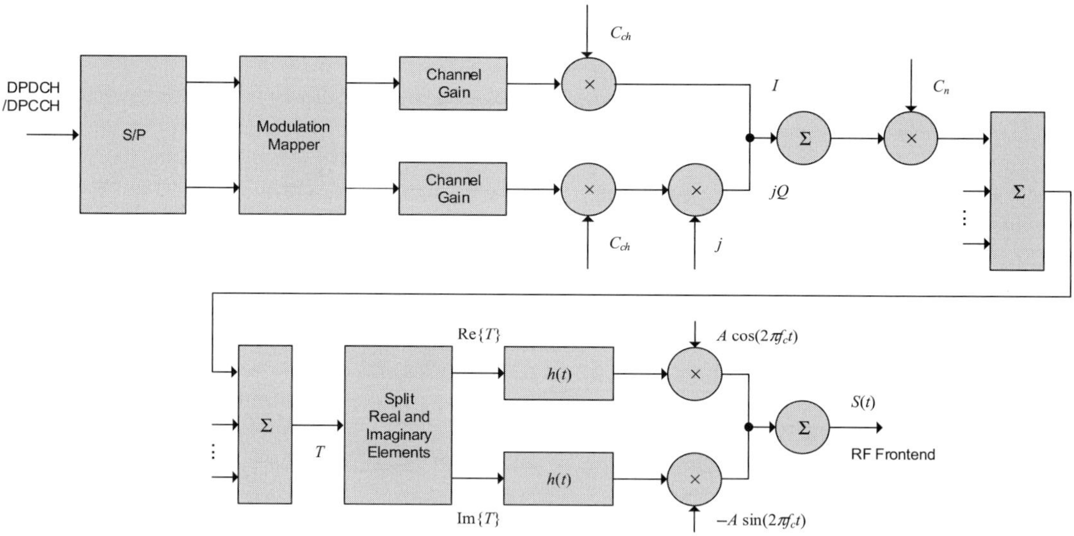

Fig. 6.17 Step 3: downlink spreading and modulation.

with the configuration of shift registers shown in Fig. 6.18, which is different from the configuration of shift registers used in the uplink.

Theoretically $2^{18} - 1$ scrambling codes can be generated from the circuitry but only a subset of those codes is actually used for scrambling. This reduces the required complexity for the code acquisition, synchronization, and cell search process. The scrambling codes are classified into 512 primary scrambling codes, each of which is related to 15 secondary codes. The 512 primary scrambling codes are divided further into 64 code groups, and each code in the code group is associated with two other codes for use in the *compressed mode*, an operation for monitoring the channel status of neighboring networks such as GSM or cdma2000. Each NodeB is assigned a scrambling code by the

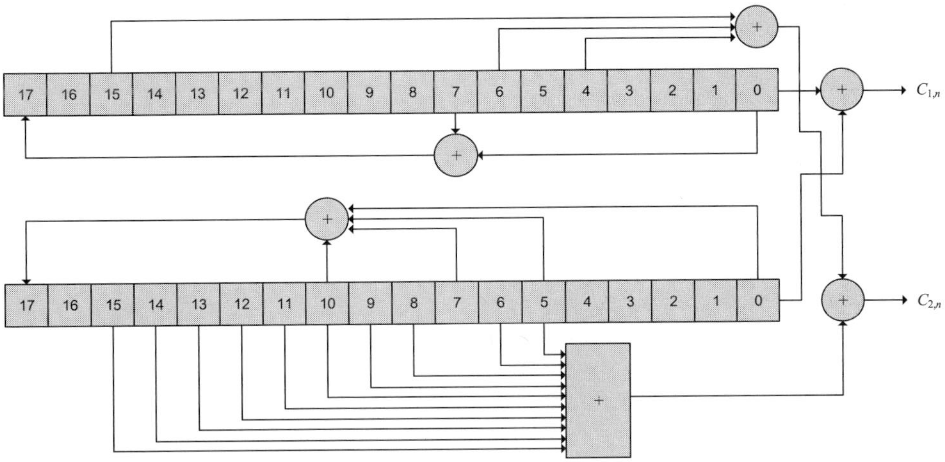

Fig. 6.18 Downlink scrambling sequence.

network, which is broadcast over the Synchronization Channel (SCH) at high transmit power for one tenth of each slot, 256 chips. In the downlink, a unique channelization code is assigned to each physical channel but a single scrambling code is used for all physical channels.

Then the downlink physical channels carrying data and control signal are summed, and also combined with other physical channels used for network management, such as the synchronization and pilot channels. Finally, the real and imaginary components of the scrambled symbols are separated into the real and imaginary elements, QPSK-modulated, summed, and filtered for pulse-shaping using the same RRC filter as in the uplink before being transmitted over the wireless channel.

Figure 6.19 shows the downlink radio signal processing operations from the CRC attachment to the physical channel mapping, focusing on the variations of bit-rate from the overhead attached in each layer. It is assumed that a 244-bit speech frame output from the AMR speech encoder, encoded at 12.2 kbps, and a 100-bit control signal are to be transmitted over the downlink. The TTI for the speech frame is 20 ms while the control signal needs to be transmitted within a larger TTI of 40 ms. The three classes of bit-streams from the speech encoder are input to the physical layer with no overhead attached at the RLC and MAC, as outlined in Table 6.8. It is also assumed that slot format 11, whose DPDCH can transport 420 bits per frame as shown in Table 6.12, is used.

Three radio access bearers (RAB) are established for Class A, B, and C bit-streams. A 12-bit CRC is attached only to Class A bits, and nine padding bits are attached to all classes. Then the bit-streams are inserted into the channel encoder, which applies rate $\frac{1}{3}$, $\frac{1}{3}$, and $\frac{1}{2}$ convolutional encoding, respectively. A total of 772 bits is generated after the channel coding, which is reduced to 686 bits after symbol repetition and puncturing to match the available bit-rate that has to be shared with the control signal. It can be expected that some bits from Classes B and C will be removed in the process. Then the rate-matched bits are interleaved and segmented for two radio frames, which assigns 343 data bits for each frame.

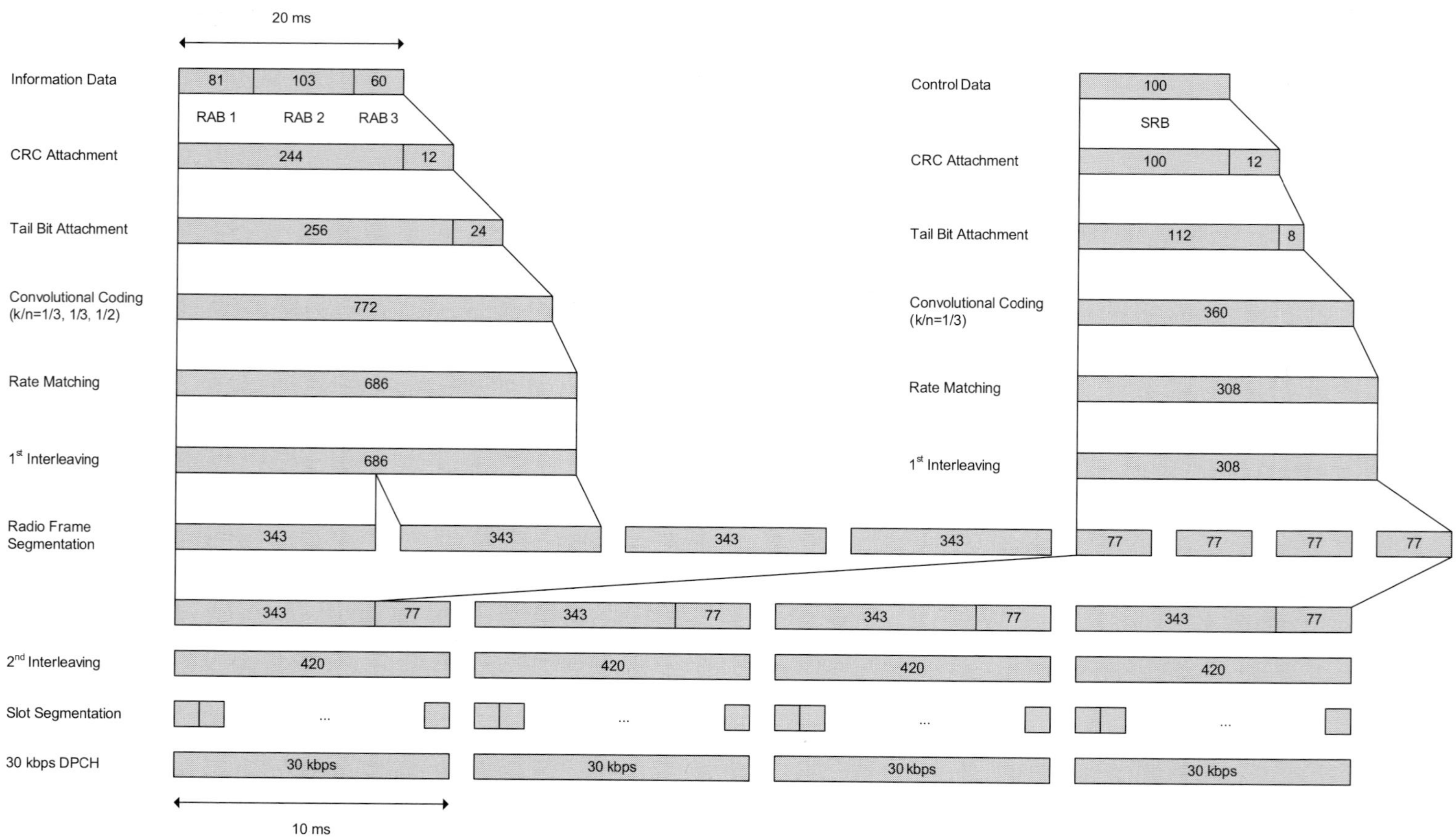

Fig. 6.19 Radio signal processing operations of W-CDMA downlink for AMR.

Table 6.8 W-CDMA protocol parameters for voice service.

Protocol	Parameter	RAB 1 (Class A)	RAB 2 (Class B)	RAB 3 (Class C)	SRB
	Mode	TM	TM	TM	AM
RLC	Payload size (bits)	0, 39, 81	103	60	128
	Max bit-rate (bps)		12200		3200
	Header size (bits)		0		16
MAC	Multiplexing		None		
	Header size (bits)		0		4
	TB size (bits)	0, 39, 81	103	60	148
	TTI (ms)		20		40
	CRC size (bits)	12	None	None	16
	Coding type, rate	Convol., $\frac{1}{3}$	Convol., $\frac{1}{3}$	Convol., $\frac{1}{2}$	Convol., $\frac{1}{3}$
PHY	Max bits / TTI (after channel coding)	303	333	136	516
	Max bits / frame (before rate matching)	152	167	68	
	RM attribute	180–220	170–210	215–256	

The control signal from the signaling radio bearer consumes the remaining 77 bits per frame. To the control signal, a 12-bit CRC is attached, which is shorter than the 16-bit CRC used in Table 6.8. The bits are zero-padded, and rate $\frac{1}{3}$ convolutional-coded, which generates 360 bits. The rate-matching reduces the number of bits to fit the available capacity of DPCCH, which corresponds to four radio frames each with 77-bit capacity. Then the bits from the three RABs and an SRB are concatenated and interleaved before being mapped to the respective locations of each slot. In practice, the radio frames are used for either RABs or SRBs, as frame stealing for transporting urgent control signals happens rarely in cdma2000. When there is a large amount of control information to transport, or when it is unnecessary to transport control information immediately, the UE or NodeB can simply wait for a silent period during which the radio frames can be used solely for the control signal, with less or no puncturing of bits that have to be received error-free.

Table 6.8 summarizes the key parameters used in each protocol layer of W-CDMA, when three radio access bearers (RAB) for the three classes of AMR bit-stream, and a signaling radio bearer (SRB) for control information, such as the handover message from RRC, are configured. Compared with the radio signal processing operations of GSM that use the same speech codec, the generic framework of unequal error detection and protection is maintained, but the introduction of protocol stack and rate matching complicates the analysis. In addition, a second opportunity for UEP can be provided by rate matching, i.e., by controlling the amount of redundancy depending on the relative importance of the transport channels, with the Rate Matching (RM) attribute [3GPP (2017c)]. Notice that unlike in the cdma2000, where some control information such as power control is punctured over data, each type of control information is assigned a dedicated field in the frame structure of a physical channel.

Such practices in the frame structure of physical channels are consistent with the policy of designing the traffic channel and control signaling in GSM. In the RLC, the 3200 bps for SRB is computed by assuming a maximum data size of 128 bits for each TTI of 40 ms. No overhead is attached in the RLC in the speech bit-stream but a four-bit MAC header is attached to control data. In the PHY, the maximum number of bits per TTI is computed using the formulas of Table 6.7. Then the maximum number of bits per radio frame is half the maximum bits per TTI rounded up to the nearest integer.

The rate matching (RM) attribute, which can have an integer value between 1 and 256, is related to the amount of puncturing applied to each transport channel in a CCTrCH. The larger the value of the RM attribute, the more repetition, or the less puncturing, of the bits. The RM attributes are adjusted to maintain the same number of total bits for the physical channel, and absorb the changes of bit-rate from voice activity or control signal. In Table 6.8, the target residual bit-error rate (BER) of each radio bearer, i.e., the BER measured after channel decoding, is assumed to be 5×10^{-4}, 10^{-3}, and 5×10^{-3}, respectively. Since the SRB is used to transport critical, but less delay-constrained, data, a TTI of 40 ms is set for this radio bearer, and the acknowledged mode (AM) of the RLC is used. Note that although the same channel coding rate is used as for Class A bits, control data of the SRB is more strongly protected as it is retransmitted by the RLC if not successfully received.

Transmit Diversity

Receiver diversity is a relatively cost-effective technique that increases the SNR at the receiver by using additional antennas. With receiver diversity, more signal energy can be gathered from independent channels between the transmitter and the receiver without using additional radio resources. In general, with N antennas at the receiver, the combined SNR can be as much as N times that of a single antenna. Moreover, variance of the SNR can be reduced, which can also reduce the required SNR for a given QoS level. For these reasons, receiver diversity is supported by most NodeB implementations.

However, as in cdma2000, W-CDMA does not require a second antenna in the UE. It is often pointed out in the literature that it is not feasible to realize the theoretical performance of receiver diversity, because it is difficult to separate the antennas within a UE by more than half a wavelength of the lowest frequency supported. However, non-negligible gains can be observed even when the antennas are separated by only 1–2 centimeters. The cost for receiver diversity is the increased complexity for the additional receiver circuitry. The gains from receiver diversity can be compromised if the received signals are not separated in the time or spatial domain sufficiently to be resolved. Such situations can happen when the distance between the UE and the NodeB is small, strong LOS paths exist from the NodeB to the UE, or the speed of UE is low.

Transmit diversity is a way to compensate for the limitations of receiver diversity. It is a set of techniques to artificially increase the spatio-temporal distance of multi-path signals. Like power control, transmit diversity can be classified into Open-Loop

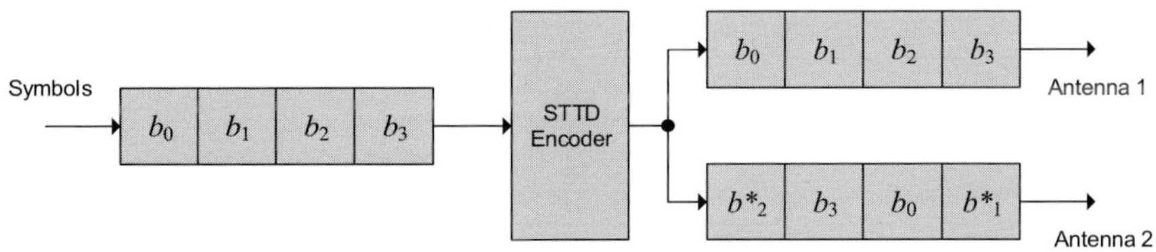

Fig. 6.20 STTD encoder for QPSK.

Transmit Diversity (OLTD) and Closed-Loop Transmit Diversity (CLTD). Figure 6.20 shows an open-loop transmit diversity scheme of W-CDMA for QPSK, which is called Space-Time Block-coding-based Transmit Diversity (STTD). In the STTD, a set of four complex symbols is transmitted from an antenna, and the same set is re-ordered, partially complex-conjugated, and transmitted from another antenna. Re-ordering of the symbols and physical separation of the antennas generate the necessary distance for transmit diversity in the temporal and spatial domains. In Fig. 6.20, b_2^* represents the complex conjugate of b_2.

Since the STTD is basically an OLTD technique, the relationship of the two sets of symbols transmitted in parallel does not change, and no control information is fed back from the receiver to the transmitter. Let four symbols to transmit be b_0, b_1, b_2, b_3. The first and third slots of the sets include only the original and complex conjugates of b_0 and b_2, and the second and fourth slots include the original and complex conjugates of b_1 and b_3. Thus, the performance of STTD can be analysed like classic space-time coding [Paulraj *et al.* (2008)]. STTD is an efficient transmission technique to improve the SNR at a relatively small cost for complexity when the gains of receive diversity are not sufficient. However, STTD complicates the operation of the receiver in soft hand-over when some NodeBs transmit to the UE in STTD while others do not. cdma2000 also provides similar OLTD techniques, Space-Time Spreading (STS) and Orthogonal Transmit Diversity (OTD) [Vanghi *et al.* (2004)].

Figure 6.21 shows a block diagram of a CLTD transmitter for DPCH in which the spread and scrambled symbols are transmitted from two antennas, after weighting by antenna-specific factors. To simplify the channel estimation, w_1 is set to a constant scalar, and w_2 is complex valued and dynamically adjusted by feedback information transmitted via the FBI field of DPDCH. Two orthogonal pilot fields of DPDCH, CPICH_1 and CPICH_2, are transmitted from two antennas to assist the receiver in estimating each channel. For each slot, the UE computes, ϕ, the phase adjustment required to maximize the SNR at the receiver, for the antenna in the lower branch of Fig. 6.21. The phase adjustment is crudely quantized into two levels, 0 and π, which are signaled as 0 and 1, respectively. To compensate for the rotation of the signal constellation, the NodeB interprets the phase adjustment value in a different fashion for each slot, as shown in Table 6.9.

Note that the gain is not adjusted in CLTD, even though the format of the FBI field includes bit space for both gain and phase. w_1 is set to $\frac{1}{\sqrt{2}}$, and w_2 is computed as

Table 6.9 Phase adjustment, ϕ_i, for slot i by feedback command.

Slot		0	1	2	3	4	5	6	7	8	9	10	11	12	13	14
Phase interpretation	0	0	$\frac{\pi}{2}$	0	$\frac{\pi}{2}$	0	$\frac{\pi}{2}$	0	$\frac{\pi}{2}$	0	$\frac{\pi}{2}$	0	$\frac{\pi}{2}$	0	$\frac{\pi}{2}$	0
	1	π	$-\frac{\pi}{2}$	π	$-\frac{\pi}{2}$	π	$-\frac{\pi}{2}$	π	$-\frac{\pi}{2}$	π	$-\frac{\pi}{2}$	π	$-\frac{\pi}{2}$	π	$-\frac{\pi}{2}$	π

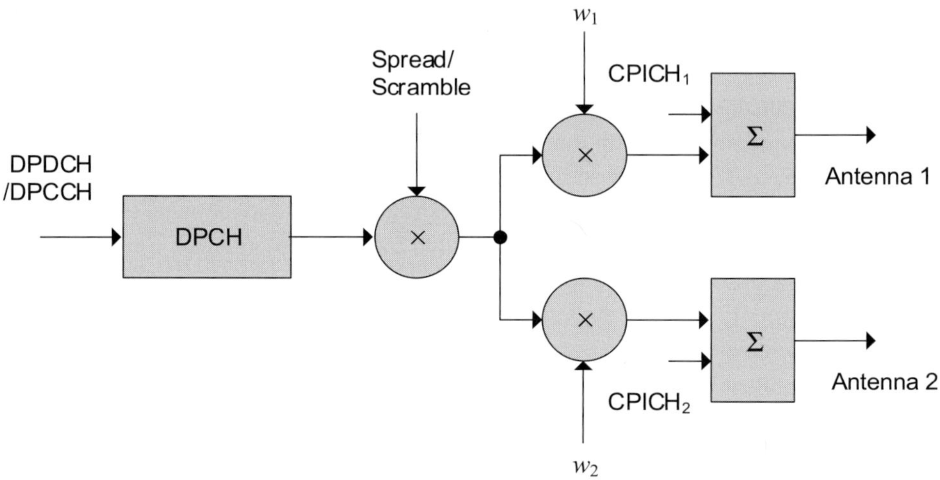

Fig. 6.21 CLTD transmitter for DPCH.

$$w_2 = \frac{\sum_{i=n-1}^{n} \cos(\phi_i)}{2} + j\frac{\sum_{i=n-1}^{n} \sin(\phi_i)}{2}, \tag{6.2}$$

where $\phi_i \in \{0, \pi, \frac{\pi}{2}, -\frac{\pi}{2}\}$. CLTD may not exhibit its maximum performance when the transmission is bursty, as in voice service, but still can complement the performance of power control significantly when a fixed bit-rate is maintained, as in a video call. Figure 6.2 illustrates the flow of transmit diversity with a dashed path to the RF frontend.

6.2.4 Link Management

Uplink Control Information

Figure 6.11 shows that various types of uplink control information, including the Pilot, transport format combination indicator (TFCI), feedback information (FBI), and transport power control (TPC) fields, are transmitted in the DPCCH. TFCI informs the receiver of the transport format combination used in the CCTrCH. As the TFCI plays an indispensable role in decoding the received data, the ten-bit TFCI is strongly protected with a (32,20) second-order Reed–Muller code. The channel encoding generates a 32-bit code word whose two bits at the end are punctured. TFCIs shorter than ten bits are complemented with padding bits before the channel coding. However, transmission of the TFCI can be avoided if certain conditions related to the transport formats and the

Table 6.10 Uplink DPCCH formats.

Slot format	Bit-rate (kbps)	Symbol-rate (ksps)	Spreading factor	Bits per frame	Bits per slot	Pilot	TFCI	FBI	TPC
0	15	15	256	150	10	6	2	0	2
1	15	15	256	150	10	8	0	0	2
2	15	15	256	150	10	5	2	1	2
3	15	15	256	150	10	7	0	1	2
4	15	15	256	150	10	6	0	2	2
5	15	15	256	150	10	5	2	2	1

transport channels are satisfied, and the receiver can identify the transport format used in the radio frame *blindly* by some trial-and-error approaches.

With an opportunity for power control in each slot, the transmit power of the downlink can be controlled as frequently as 1500 times per second, which is higher than the 800 Hz of IS-95 or cdma2000. As with other CDMA systems, the UE measures the link quality in terms of the SNR or SIR, and compares the value of the measured metric with a threshold. If the measured value is larger than the threshold, then the UE sets the TPC bit of the corresponding slot to 0, which asks the NodeB to decrease transmit power by one step. Likewise, if the threshold is larger than the metric, the UE transmits a TPC bit of 1, which asks the NodeB to increase the power by one step. As there is only one DPCCH in the uplink, a TPC bit on the uplink controls multiple DPDCHs of the downlink. The step size for the downlink power control can be 0.5, 1, 1.5, or 2 dB but the NodeB must support the 1 dB step size in any case. Depending on the slot format used, as outlined in Table 6.10, the TPC bit is repeated twice to generate the TPC field, such as 00 and 11.

The FBI field is used to signal the weight factor in closed-loop transmission diversity [3GPP (2017e)] when transmit diversity is used. If the FBI field is unnecessary, slot formats that do not include the channel information, such as slot format 0 or 1, can be used for DPDCH in the uplink. The conventional advantages of CDMA, resolving signals that arrive as multipaths and combining their energies to operate in low SNR environments, are not guaranteed if the length of the impulse response is not long enough, as is the case in small cells. Transmit diversity is a technique that can be used in the downlink to artificially increase the length of the impulse response in either an open-loop or closed-loop fashion, by transmitting the signal from multiple antennas. As shown in Table 6.11, ten bits are available for each slot of DPCCH, which need to be distributed to the four fields. Therefore, the TFCI and FBI fields can be omitted in the slot. When there is a space of six bits for the Pilot field, as in slot format 0 or 4, 111110 is used to indicate the start of each slot. For AMR 12.2 kbps, slot format 1 of DPDCH can match the bit-rate requirements for the service, providing transmission capacity for a speech frame encoded at the bit-rate and the overhead generated during the radio signal processing operations illustrated in Figures 6.6 and 6.12.

The physical layer (PHY) selects the slot format for the uplink, and may change the format for each frame to match the situation. Although Table 6.11 includes a wide range

Table 6.11 Uplink DPDCH formats.

Slot format	Bit-rate (kbps)	Symbol-rate (ksps)	Spreading factor	Bits per frame	Bits per slot
0	15	15	256	150	10
1	30	30	128	300	20
2	60	60	64	600	40
3	120	120	32	1200	80
4	240	240	16	2400	160
5	480	480	8	4800	320
6	960	960	4	9600	640

Table 6.12 Downlink DPDCH/DPCCH formats.

Slot format	Bit-rate (kbps)	Symbol-rate (ksps)	Spreading factor	Bits per frame DPDCH	Bits per frame DPCCH	Bits per slot	Data1	Data2	TPC	TFCI	Pilot
0	15	7.5	512	60	90	10	0	4	2	0	4
1	15	7.5	512	30	120	10	0	2	2	2	4
2	30	15	256	240	60	20	2	14	2	0	2
3	30	15	256	210	90	20	2	12	2	2	2
4	30	15	256	210	90	20	2	12	2	0	4
5	30	15	256	180	120	20	2	10	2	2	4
6	30	15	256	150	150	20	2	8	2	0	8
7	30	15	256	120	180	20	2	6	2	2	8
8	30	60	128	510	90	40	6	28	2	0	4
9	60	30	128	480	120	40	6	26	2	2	4
10	60	30	128	450	150	40	6	24	2	0	8
11	60	30	128	420	180	40	6	22	2	2	8
12	120	60	64	900	300	80	12	48	4	8	8
13	240	120	32	2100	300	160	28	112	4	8	8
14	480	240	16	4320	480	320	56	232	8	8	16
15	960	480	8	9120	480	640	120	488	8	8	16
16	1920	960	4	18720	480	1280	248	1000	8	8	16

of bit-rates, the highest bit-rates such as 480 or 960 kbps are only available when the UE is in unrealistically favorable channel conditions, or there remain few UEs competing for radio resources in the cell. Practically, the highest bit-rates in Release 99 are about 384 kbps. In W-CDMA, a UE is classified according to the bit-rate and supported capabilities for transmission and reception.

Downlink Control Information

Table 6.12 summarizes the parameters of the downlink slot formats. Unlike in the uplink, the slot formats are assigned by the protocol layers of the NodeB higher than the PHY to balance radio resources such as transmit power among the downlink channels. Comparison of the DPDCH and DPCCH slot formats for the uplink and downlink shows that, in general, it takes a one-level higher spreading factor for the downlink to provide throughput capability similar to that of the uplink. In the case of voice service, the radio network control (RNC) assigns a set of bit-rates to each UE. Different bit-rates can be used for the uplink and downlink at any time but the same set of codec modes must be used for both links. However, since the range of speech bit-rates is relatively small, using different bit-rates does not change the spreading factor significantly. In the downlink, the TPC bit can be repeated up to eight times to generate the TPC field.

Timing Relationship

In the uplink, the DPCCH and DPDCHs transmitted from a UE share the same frame timing, i.e., starting or ending radio frames simultaneously. In the downlink, the DPCCH and all DPDCHs carrying CCTrCHs of a dedicated type transmitted to a UE have the same frame timing. At the UE, transmission of the uplink DPCCH/DPDCHs takes place in approximately T_O chips, where T_O is a constant defined to be 1024, after the reception of the first detected path (in time) of the corresponding downlink DPCCH/DPDCHs measured at the antenna. In Fig. 6.22, the time difference between the Data1 field of the downlink DPDCH and the Pilot field of the uplink DPCCH is shown to be T_O chips.

The 1024 chip delay, which is used regardless of the slot format, is chosen to ensure sufficient opportunity in the time domain for SIR measurement and TPC configuration. Figure 6.22 illustrates the timing relationship between the TPC and Pilot fields. Once a TPC field is received, the UE must change the transmit power of the uplink DPCCH/DPDCH at the beginning of the first Pilot field after the reception of the TPC field. Likewise, upon the reception of a TPC field using the uplink DPCCH, the NodeB must change the transmit power of the downlink DPCCH/DPDCH at the beginning of the next, which may not be the first, Pilot field after the reception of the TPC field. Therefore, the NodeB may accept or postpone the execution of the TPC field from the UE, depending on factors such as the power headroom at the transmitter. This asymmetric relationship of the UE and the NodeB in the handling of received TPC field is similar to that of the MS and the BTS in GSM in the handling of received codec mode request/command. The TPC field is typically set based on the estimated SNR or CIR,

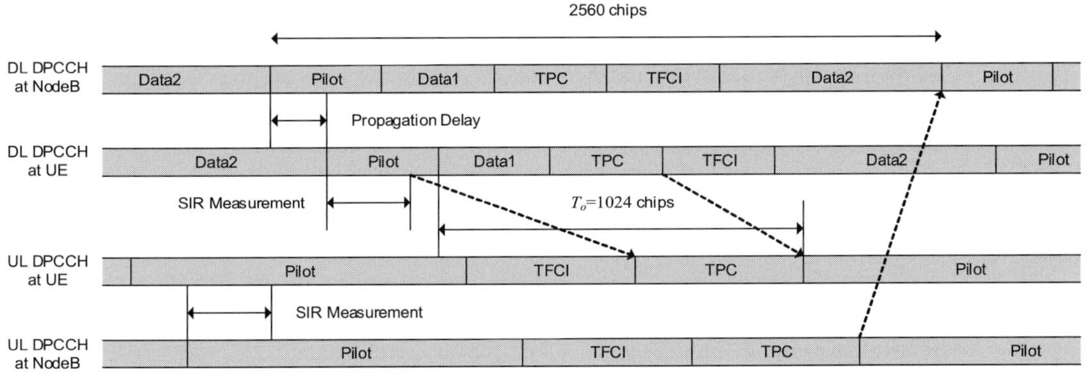

Fig. 6.22 Power control timing relationship.

which is computed by correlating the Pilot field. However, the period for SIR measurement is not specified for the UE or the NodeB, and is left to the discretion of the implementation.

6.2.5 Operational Strategy

In cdma2000, there are mechanisms to control the tradeoff between speech quality and network capacity. In the uplink of cdma2000, the average bit-rate of the speech encoder such as EVRC can be controlled among eight levels by the BSC using the RATE_REDUC message. In the downlink, the average bit-rate can be controlled in a more refined fashion by directly controlling internal codec parameters such as the thresholds for rate-classification. In W-CDMA, the RNC can exercise similar control over the speech bit-rate. A key advantage of the RNC is that it can control the precise bit-rate of the speech codecs located in the UE and the MSC. In contrast, the BSC of cdma2000 can control only the average bit-rate since the VBR speech codecs are used there.

Unlike in GSM where a fixed bit-rate is dynamically allocated between the speech bit-stream, CRC, and channel code, AMR can be operated in W-CDMA using several strategies in which the total bit-rates of the uplink or downlink are neither identical nor constant. For example, in a scheme, a fixed codec mode may be allocated to all UEs in a cell or in an area, and the codec mode is maintained throughout voice service. In another scheme, each UE might be assigned a unique codec mode that helps the network maintain a balance between speech quality and network capacity, based on the channel condition or the cell loading level. In this *per-call allocation* approach, higher codec modes can be assigned when there is more network capacity or the interference level is low.

If the request for service is admitted and a codec mode is allocated, the mode is maintained even when the channel condition or the cell loading level changes. Note that a codec mode cannot be assigned in the call admission based on channel condition or link quality, as the control channels used to set up services are transmitted at higher power levels than those of the voice channels to be assigned. Therefore, the quality of control channel does not necessarily represent the quality of transport channel in actual services. Moreover, it is not feasible to measure the link quality when the transmissions are bursty. Therefore, the cell loading level is the key factor in assigning codec modes in the per-call assignment. Finally, in a more complex scheme, a codec mode is assigned in the call admission process depending on the network load but can be dynamically switched to another assignment when the situations change. Various strategies for call admission and codec mode control of W-CDMA can be found in [Karlsson *et al.* (2002)].

Figure 6.23(a) illustrates the modeling of speech quality as a function of the codec mode and the Block Error Rate (BLER). The quality is assumed to be a monotone-increasing function of the codec mode but also a monotone-decreasing function of the BLER. This kind of model can be taken into account by the RNC when it admits call requests and adjusts codec modes. Although the shape of quality plane is shown in the figure as concave, the slope or height of the plane will depend on quality criteria and

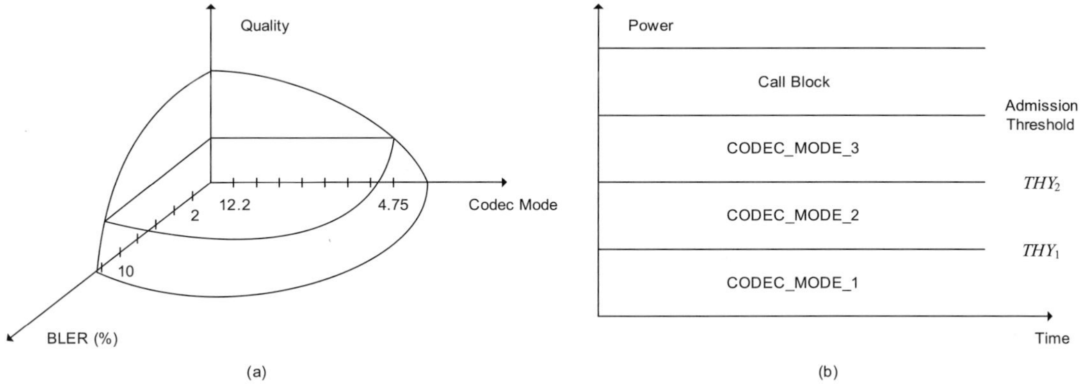

Fig. 6.23 (a) Quality plane as a function of codec mode and BLER. (b) Load-driven admission control.

service policies. Figure 6.23(b) illustrates the decision process for admission control and codec-mode assignment, which also depend on the network load. The power axis indicates the transmit power of the NodeB, which can be considered as a metric for the cell loading level. As in cdma2000, capacity of the downlink is smaller than that of the uplink and the traffic of voice services influences both links. Thus the transmit power of the NodeB is an appropriate indicator for the network load.

The call admission algorithms based on the load-driven admission control can be outlined as follows: if the power is below THY_1, e.g., when the UE is closer to the NodeB or the link quality is good, CODEC_MODE_1, usually the highest codec mode as CIR, is the metric used there. In Fig. 6.23(b), the one with the highest bit-rate, in the active codec-mode set (ACS), is assigned. Note that the CODEC_MODE_4 of GSM, shown in Fig. 3.12, represents the highest codec mode. A lower codec mode, CODEC_MODE_2, is assigned if the power is measured to be between THY_1 and THY_2. If the power exceeds the Admission Threshold, the requests for call setup are turned down temporarily until the power falls below the threshold. If the power is between the Admission Threshold and THY_3, CODEC_MODE_3 is assigned. The use of thresholds and hysteresis parameters may also be considered as in Fig. 3.12 if the radio resource is to be allocated as fairly as possible. However, the relative magnitude of power fluctuation is smaller compared with the cases of GSM where the SNR of channels to and from a UE is compared with the thresholds.

Other metrics such as the frame error rate, call blocking rate, or call dropping rate may be considered in the construction of the quality plane or in the design of the admission control algorithms. Intuitively, the simplest admission control scheme using the allocation of a fixed codec mode does not have any measure for adapting to the situations and underperforms more advanced schemes. The static allocation of a fixed codec mode is known to perform well in lightly-loaded network conditions but the per-call or dynamic allocation outperforms the static one in the majority of network and channel conditions. In addition to codec mode control, inter-cell or intra-cell handover is another measure the RNC can take to manage the loading level over multiple cells or large geographic areas, by spreading heavy traffic concentrated in some areas over less crowded neighboring areas. The handover for balancing network load can be indirectly triggered by controlling the power level of the Common Pilot Channel (CPICH) as the UE is likely to be connected to the NodeB with the highest link quality.

Compressed Mode

Figure 2.37 illustrates that during the slow frequency hopping (SFH) of GSM, the MS tunes to a downlink frequency used by the current or an adjoining cell to measure its signal strength, before moving to the next channel frequency used by the current cell. Information on the link quality of neighboring cells can be taken into account for deciding on the timing or direction of handover, or for balancing the network load among the cells. In W-CDMA, the situation is quite different, as the majority of the network coverage over which W-CDMA is to be deployed is already served by GSM or other previous systems. W-CDMA uses fundamentally different technologies for radio access, but shares many elements of the core networks with GSM. Therefore the actions of UE required at the network boundary were taken into account from the design phase so that the UE can switch between the coverages of W-CDMA and other systems as quickly as possible.

In the *compressed mode* of W-CDMA, the transmission time is temporarily reduced to secure an opportunity for monitoring the link quality of a downlink frequency that has been assigned to the current or an adjoining cell, or to a cell with a different radio access technology (RAT) such as GSM or cdma2000. In the mode, the same number of bits is transmitted in a shorter time. In W-CDMA, the compressed mode is used mainly in the downlink but in some cases it can also be used in the uplink. Figure 6.24 illustrates the concept of compressed mode where a transmission gap of $M+N-1$ slots is secured in the middle of a radio frame. The main saving from compressed mode is not in the complexity, although a single RF circuit with an oscillator typically serves both GSM and W-CDMA modems, but in the power consumption, as it is not necessary for the two pairs of transmitters and receivers to operate simultaneously at the network boundary.

The reduction of transmission opportunities caused by compressed mode can be compensated using two approaches. If lower bit-rates can be temporarily tolerated, as would be the case for transport channels carrying data for services with relaxed delay constraints, the transmission gap can be generated simply by puncturing more bits at rate matching. For example, the channel coding rate can be increased from Rate $\frac{1}{3}$ to $\frac{1}{2}$, or the transmission can be halted and the data can be temporarily held at a higher protocol layer. On the other hand, if the bit-rate cannot be reduced, such as in the middle of a conversation, the transmission gap can be generated by temporarily increasing the

Fig. 6.24 Downlink compressed mode transmission.

Table 6.13 Downlink compressed DPDCH/DPCCH formats.

Slot format	Bit-rate (kbps)	Symbol-rate (ksps)	Spreading factor	Bits per frame DPDCH	DPCCH	Bits per slot	Data1	Data2	TPC	TFCI	Pilot
2A	30	15	256			20	2	14	2	0	2
2B	60	30	128			40	4	28	4	0	4

bit-rate of the slots adjacent to the gap, which would require the spreading factor of the slots to be reduced. There are two slot format types in the compressed mode. Slot format A is used when the radio frames are compressed by higher-layer scheduling, and slot format B is used when the radio frames are compressed by reducing the spreading factor. Table 6.13 shows the two versions of downlink slot format 2 defined for the compressed mode. Note that the number of slots per radio frame can range from 8 to 14 in the compressed mode, corresponding to a transmission gap of one to seven slots.

In either case, the slot format needs to be changed so that a minimum set of DPCCH bits is transmitted. Although the compressed mode can only be used for the dedicated channels, if used in the uplink, transmission over other channel types may have to be prohibited to assist the measurement in the downlink. The minimum duration of the transmission gap is three slots, and the maximum duration is seven slots. The compressed mode does reduce the network capacity, and so by under-utilizing transmission opportunities in the time domain or by increasing the interference if the transmit power is increased to compensate for the reduced transmission time, the network needs to manage the proportion of UEs in the compressed mode, and the frequency and duration of transmission gaps.

Transcoder Free Operation

As shown in Fig. 6.1, in W-CDMA, the logical distance between where the speech is encoded and where it is decoded is greater than in GSM. This is because the speech codec is located at a higher location than the RNC in the network architecture. Such an architectural shift saves the network capacity since more of the speech transmission is in its most compressed format. If the codecs and bit-rates of the far-end UE or MS supports are known and also supported by the near-end UE, there is no need to transcode to and from PCM. This will not only save the network capacity on the fixed side but also increase the speech quality. The transmission delay is also reduced as the conversion between speech formats is unnecessary. To enable Transcoder-Free Operation (TrFO), where two MSCs negotiate the codec and bit-rates to use in voice services so that the compressed speech can be transmitted end-to-end between two UEs managed by the MSCs [3GPP (2017b)], W-CDMA provides Out-of-Band Transcoder Control (OoBTC), a set of signaling mechanisms for the operation. The media paths between two MSCs, which may belong to the same or two different service providers, can be connected by the Asynchronous Transfer Mode (ATM) or Internet Protocol (IP) networks, using the Nb interface [3GPP (2016c)].

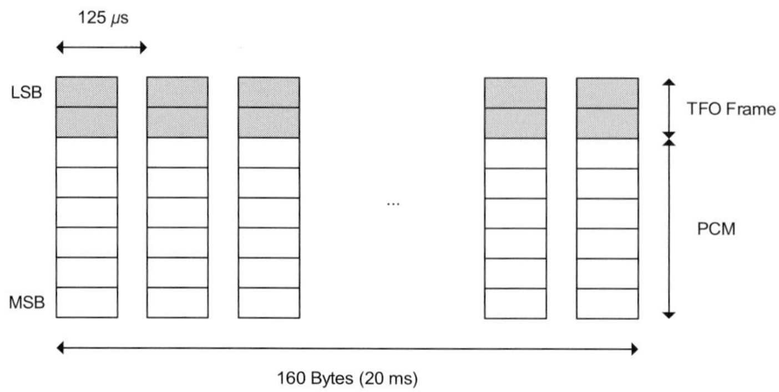

Fig. 6.25 TFO frame format.

When TrFO support is not provided by the core networks of W-CDMA, for example, when only 64 kbps PCM channels are available between MSCs, Tandem-Free Operation (TFO) can be used instead for similar objectives [3GPP (2017n)]. In TFO, the compressed speech is transmitted using the unused two Least Significant Bits (LSB) of PCM. Figure 6.25 illustrates the format of PCM used in TFO, in which 320 bits per 20 ms are reserved for a speech frame and control signaling without infringing on the bits assigned to the PCM speech samples. If a speech frame at a bit-rate higher than 16 kbps is to be transported, additional LSBs can be assigned to TFO. As the wired links between MSCs in general do not depend on the sampling rate of speech codecs, TrFO and TFO can be used for wideband speech codecs such as AMR-WB [3GPP (2017m)] where speech samples are digitized at 16 kHz.

The price paid for the benefits of TrFO or TFO is the reduced capability of the RNC to control the speech bit-rate of each UE connected to the NodeBs under its management, because for an RNC to adjust the speech bit-rate, the signaling messages need to propagate not only to the near-end MSC but also to the far-end MSC. Moreover, because of the limitation in the number of bits available for control, the signaling has to be simplified. For example, not all codec modes of AMR can be used in TrFo or TFO. As the measures to control the cell loading level that the RNC can exercise are slower and less accurate, the capability of optimizing network capacity is also compromised when TrFO or TFO is used. Note that even if the bandwidth for control signaling is sufficient, tracking the link quality or managing the cell loading level would be more complicated in TrFO and TFO as the end-to-end paths can include two independent wireless channels that might use two different radio access technologies such as GSM and W-CDMA.

Soft Handover

Figure 6.26 illustrates the soft handover procedures of W-CDMA, which are similar to those used in cdma2000. As the measured signal strength to and from the UE, which is reported by the source NodeB, decreases, the RNC asks several candidate NodeBs around the source NodeB to transmit the same downlink signal as the source NodeB. The candidates are selected based on the link quality measured by the UE during compressed mode. The RNC also asks the NodeBs to receive the uplink signal from the UE,

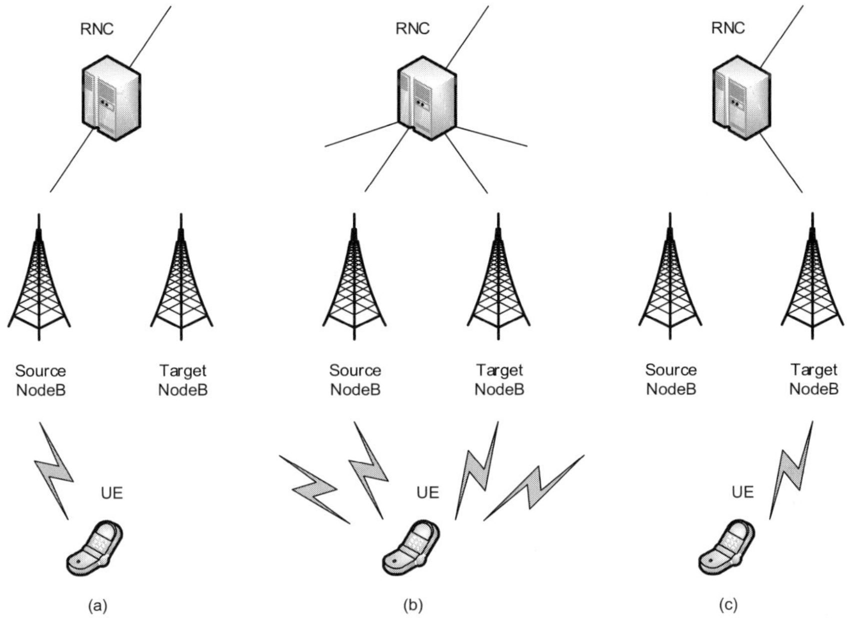

Fig. 6.26 Soft handover procedures.

and combines the signals which undergo different gains and delays in their respective channels, realizing a form of macro diversity. When the link quality of a NodeB outperforms those of the other candidates and exceeds certain thresholds, the RNC selects that NodeB as the destination of handover and removes the connections with the other NodeBs. This completes the soft handover procedures. In the case of hard handover used in the TDMA mobile communications systems, the BSC drops connection of the MS with the source BTS before connecting it to the destination BTS. During the transition, a short call-break is often generated that may be audible.

Note that although the candidate NodeBs transmit the same downlink signal to the UE in soft handover, the TPC fields in the DPCCHs are likely to be different. NodeBs located in the direction of the UE might ask the UE to reduce transmit power while NodeBs behind the moving UE might ask it to increase the power. Therefore, in the power control of the uplink, the UE follows the TPC fields from the NodeB with the highest link quality to minimize the interference. Therefore the UE increases transmit power only when all NodeBs ask it. In CDMA, soft handover is one of the key characteristics that contribute to the speech quality being higher than with TDMA, especially during handover. However, soft handover reduces the network capacity significantly as the UEs at cell edge require high transmit power from multiple cells.

To manage the level of interference, the RNC can update the SIR target of each NodeB, and set the initial level of transmit power correspondingly. A more sophisticated but common type of handover, *softer handover*, refers to the intra-cell handover from one sector of a NodeB to another sector of the same NodeB, which is not perceived by the UE. Figure 6.26(b) illustrates the procedures of a 4-way soft handover where the UE communicates with a maximum four NodeBs, which may belong to a single or multiple RNCs. At the network boundary between W-CDMA and GSM, the UE may be handed over to GSM at a call-break time of less than 80 ms, which may be audible but would not reduce the speech quality significantly.

6.3 Video Signal Processing

The speech and radio signal processing operations that are a part of GSM, cdma2000, and W-CDMA, and have been discussed in the previous and current chapters do not provide enough theoretical background for understanding the principles underlying 3G-324M. The transmission of video over a wireless channel is a technically mature area, considering the omnipresence of digital broadcasting systems and televisions. However, the complexity and power constraints inherent in mobile communications, and the delay requirements of real-time conversations render many of the established video and radio signal processing techniques of broadcasting less useful and necessitate new approaches.

As a signal, video has two more dimensions than speech, the horizontal and vertical coordinates as well as time. It also requires much higher bandwidth and stronger signal processing capability. However, the theoretical background for video compression is relatively straightforward, and in many respects conceptually relatable to the principles of adaptive differential pulse coded modulation (ADPCM). Just as modeling human aural and voicing systems plays a fundamental role in the design of signal processing algorithms for compressing speech, characteristics of the human visual system are taken into account in the areas of video capture, compression, and display.

Compressing video is a very computationally-demanding signal processing process that reduces the required bandwidth for dynamic visual information considerably, without significantly hurting its quality as perceived by the human eye. Conceptually, video compression is an extension of image compression to the time domain, where the difference between two successive video frames, instead of each frame, is compressed to reduce the bit-rate. With the strong correlation between two successive video frames, if the scenes do not change abruptly, this approach of compressing video using *motion estimation* and *compensation* can achieve very high compression efficiency.

Figure 6.27 illustrates the encoding of Independent (I) and Predicted (P) frames, which are the two basic types of encoded video frames. If the frame rate of the captured video is high enough, the difference between captured frames will be an array of *pixels*, i.e., picture samples, with small values. The video encoder generates an I-frame as a static image, i.e., one encoded without regard to a previous frame, or a P-frame, which is generated from the difference between the current frame and a previous motion compensated frame. Depending on the practices used in each video codec, different terms, such as the Instantaneous Decoding Refresh (IDR) or intra-frame, and P picture or inter-frame, are used to represent the I- and P-frame, respectively. When the requirements for encoding delay are not as strict as real-time, a third frame type, Bi-directional (B) frame, can be considered to improve the compression efficiency further. A B-frame is generated by prediction from both a previous frame and a subsequent frame. B-frames can

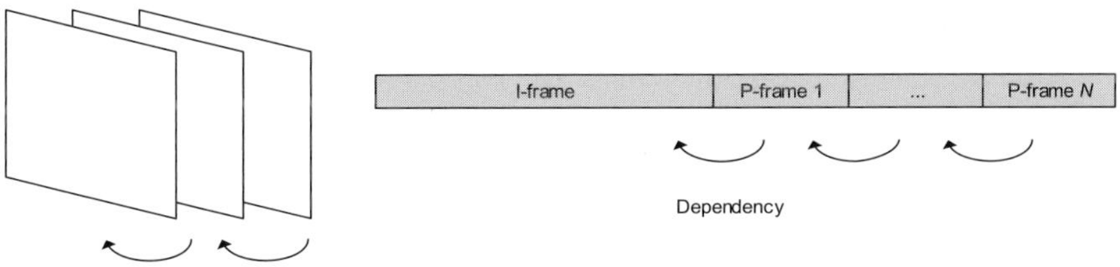

Fig. 6.27 Dependence between independent and predicted frames.

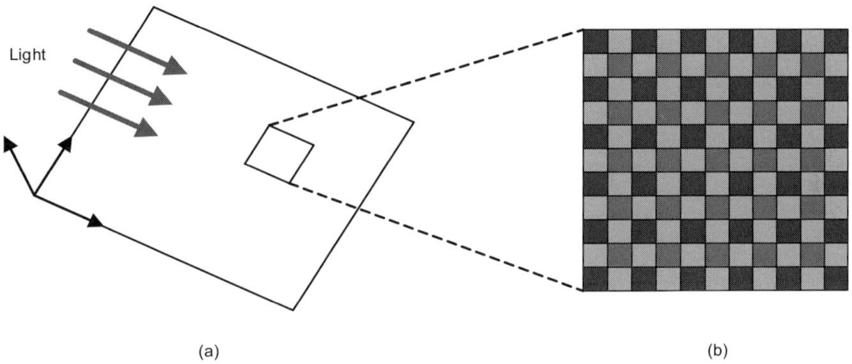

Fig. 6.28 (a) Image sensor. (b) Bayer filter array for red (R), green (G), and blue (B).

also reduce the need for encoding an I-frame when the video contents change abruptly from rapid motion activity or the appearance of new objects, as they can bridge the gap between two different frames. In the compression of video, extensive use of classical source coding techniques such as the Variable Length Codes (VLC) increases the spectral efficiency but reduces the resilience against transmission errors as the effects of errors would propagate in the codewords.

6.3.1 A/D Conversion

Although many technical elements of image and video compression are theoretically based on multi-dimensional extensions of classic signal processing theory [Dudgeon and Mersereau (1983)], there are many unique procedures in the capturing, compression, and displaying of video for which one-dimensional counterparts do not exist. Figure 6.28(a) illustrates the collision of photons on the surface of an image sensor, which converts the photons to electrons, and further to the variation of voltages for three color components, Red (R), Green (G), and Blue (B). Figure 6.28(b) shows a typical pixel structure of the Complementary Metal Oxide Semiconductor (CMOS) image sensor often used in the cameras of a UE [Nakamura (2005), Kuroda (2014)].

The image sensor is typically fabricated by covering an array of photodiodes with a sequential pattern of color filters so that the total number of green filters in the array is equal to the numbers of red and blue filters combined. This pattern is called the Bayer filter array. A larger space is assigned to green filters since the human visual system reaches its maximum sensitivity in the wavelength region of the visible spectrum, around 550 nm, represented by this color. The R, G, and B components of each sampled pixel are converted to an alternative color space, the YCbCr-space, using

$$\begin{bmatrix} Y \\ Cb \\ Cr \end{bmatrix} = \begin{bmatrix} 16 \\ 128 \\ 128 \end{bmatrix} + \begin{bmatrix} 0.1826 & 0.6142 & 0.0620 \\ -0.101 & -0.3391 & 0.4392 \\ 0.4392 & -0.3988 & -0.0404 \end{bmatrix} \begin{bmatrix} R \\ G \\ B \end{bmatrix}. \tag{6.3}$$

The Y component is called the luminance component; it represents the brightness of the sample. Cb and Cr are the chrominance components that incorporate the color information. YCbCr is one of the color spaces designed to transport color information over the transmission channels of black and white television systems. Thus, RGB and YCbCr are alternative representations of the uncompressed digital image, conceptually corresponding to the 13-bit Uniform format of speech. One benefit of YCbCr is the reduced dynamic range that can be exploited by the image or video encoder to reduce the bit-rate. The dynamic range of R, G, and B for a 24-bit image is 0–255 for all colors but those for Y, and Cb and Cr are 16–235, and 16–240 respectively [Berns (2000), Wyszecki and Stiles (2000), ITU-R (2002)].

Exploiting the fact that the human eye is more sensitive to the changes of luminance than to the changes of chrominance, the Cb and Cr arrays are subsampled so that the ratio of Y:Cb:Cr components is reduced from 4:4:4 to 4:2:2 or 4:2:0, which represents the first stage of video compression. Then the frame made up of each component is separately encoded by the video encoder, taking the previously compressed and reconstructed frames of the same component into account. Figure 6.29 shows the sampling pattern of YCbCr, where the circle, box, and triangle represent the Y, Cb, and Cr samples, respectively.

In the 4:2:0 sampling, four Y components are matched with one Cb component and one Cr component by decimating three Cb components and three Cr components from four YCbCr pixels in the hypothetical color space. For each block of $M \times N$ samples where eight bits are assigned to each component, i.e., each sub-pixel, of YCbCr, the 4:2:0 sampling reduces the number of sub-pixels to $1.5MN$. If we assume that the resolution and frame rate of video is 176×144 pixels at 15 frames per second, the bit-rate

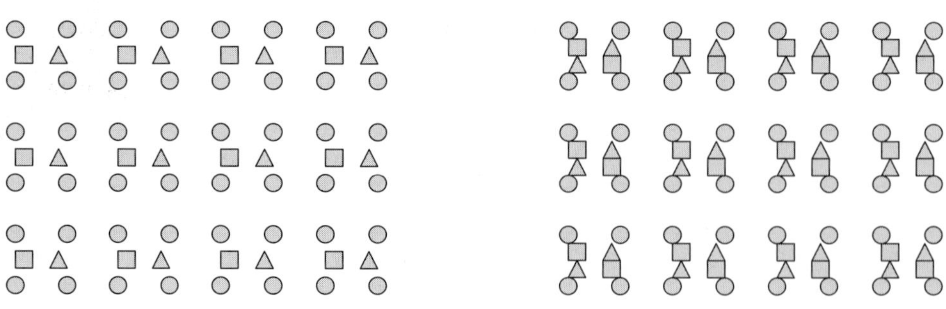

(a) (b)

Fig. 6.29 YCbCr sampling patterns. (a) 4:2:0. (b) 4:2:2.

for YCbCr 4:2:0 format would be as high as 4.56 Mbps. Therefore, even after the decimation, the bit-rate for such a low-quality video still cannot be supported by any slot format of W-CDMA. The video encoder takes YCbCr as the input, and reduces the bit-rate by removing the redundant information in the spatial and temporal domain as much as possible.

6.3.2 Motion Estimation and Compensation

At the video encoder, each frame is divided into 16×16 rectangular arrays, called macroblocks. Figure 6.30(a) shows a set of macroblocks in the 4:2:0 format. For each macroblock in a frame, a similar macroblock in the previously encoded and reconstructed frame is identified using a search. A vector is determined that represents the displacement of the current macroblock from the matching macroblock in the previous frame, which is called a *motion vector*. This search process, *motion estimation*, is one of the most computation-intensive operations in video compression. Figure 6.30(b) illustrates the motion estimation process between a reference frame and a predicted frame, in which a motion vector, $(x_2 - x_1, y_2 - y_1)$, that minimizes a metric that represents the magnitude of the macroblock difference, is computed.

To limit the amount of computation and delay, the range over which potential motion vectors are searched is limited. The metric used to identify the best matching macroblock also influences the complexity required. For example, minimizing the Mean Square Error (MSE) of the macroblock difference takes far more computation than minimizing its Sum of Absolute Difference (SAD). To reflect the fine movement of objects between consecutive frames, the value of the motion vector is not confined to be an integer. Figure 6.31 illustrates a three-level motion estimation process in which a matching macroblock is found first at the integer level. From 16 candidate pixels of the rectangular search area, the center of the first match is designated as the dark circle. In the next step, by interpolating pixels, the video encoder constructs eight hypothetical macroblocks whose centers are located between the pixels of the first search area. By comparing the metric for each macroblock, a match that is closer than the one determined in the

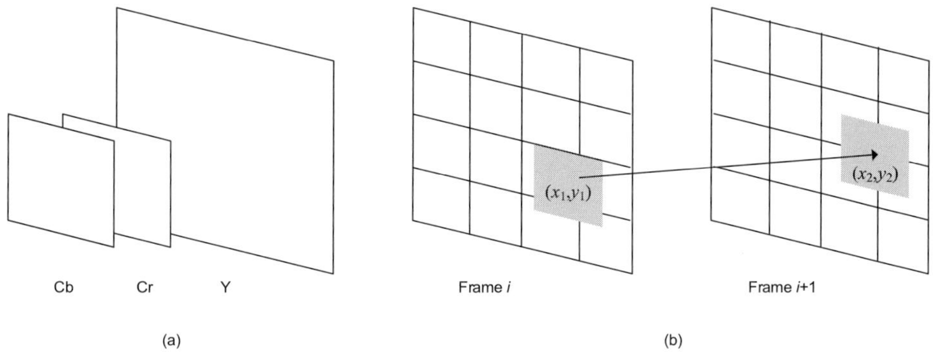

Fig. 6.30 (a) 4:2:0 YCbCr. (b) Motion estimation between successive frames.

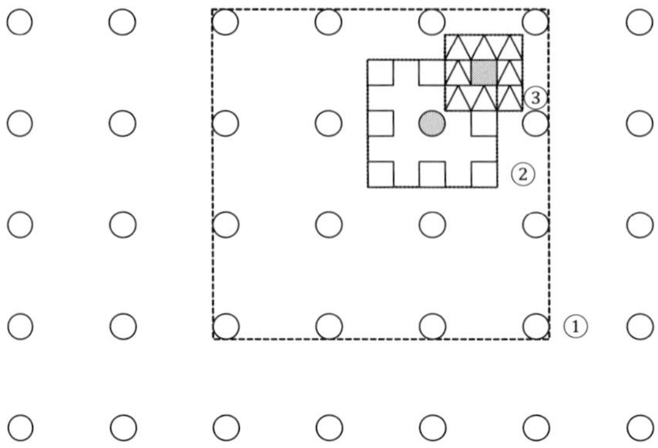

Fig. 6.31 3-level motion estimation.

first step is found, whose center is the upper-right position of the rectangle. Depending on the type of video codec, and the available complexity, the video encoder may search for an even better match at the quarter-pixel level, but it may stop there if better ones are not found. As the motion of objects tends to be consistent between consecutive frames, which results in similar values for the motion vectors, the motion vectors are also differentially encoded to reduce the bit-rate.

After a best matching macroblock has been identified, the difference between the two macroblocks is transformed onto the frequency domain where the spectrum is represented by an ordered set of coefficients that enables easy analysis and modification. These coefficients are then divided by appropriate quantization steps given by a quantization table and rounded up to the nearest integer values. The low-frequency coefficients, which are visually most significant, are kept while the high-frequency coefficients are reduced or zeroed. The compression ratio can be adjusted by scaling the values in the quantization table; larger quantization steps reduce the bit-rate, and smaller ones increase it. Note that the spectrum can also be crudely shaped by adjusting the relative values in the quantization table.

The residuals from the quantizer and the computed motion vectors are encoded with variable length codes, and combined with additional elements, such as headers and markers, to construct the video bit-stream. If there is no macroblock in the previous frame that is a good match to the current macroblock, the current macroblock itself is transformed. Even with sophisticated algorithms, the motion estimation and compensation cannot handle video containing high motion activity or newly-appearing objects. It is also possible that an entire frame needs to be encoded without taking previous frames into account, for example, after a complete change of scenes.

Figure 6.32 shows the bit-stream structure of typical video codecs. At the highest level, the bit-stream starts with an I-frame, which is followed by a series of N P-frames. It is assumed that after the Nth P-frame, the same frame pattern is repeated. Due to the origin of the frames, frame P_1 cannot be decoded unless frame I_0 is successfully decoded, and decoding of frame P_2 depends on the decoding of frame P_1, and so on. From this sequential relationship, loss of a frame during transmission or decoding affects

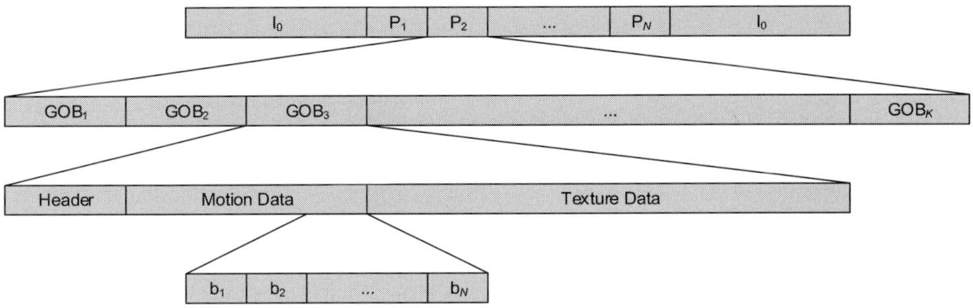

Fig. 6.32 Hierarchical frame structure of compressed video.

the decoding of all following frames until the next I-frame is received and successfully decoded. At the next lower level, each frame consists of a number of Groups of Blocks (GOB), each of which has three functional elements: header, motion data, and textual data. At the lowest (bit) level, each element is made up of hundreds of bits.

For each frame, the three bit-streams for YCbCr are separately constructed and concatenated. At the beginning of the video bit-stream, the basic information to initialize the video decoder, such as its resolution, is included. Therefore, the bit-stream structure of video is fundamentally different from that of speech, where a structural dependence between encoded frames does not exist. Decomposing each frame into smaller blocks and applying motion estimation and compensation have been the conventional approaches for video compression, called *hybrid video coding*. Targeting specific applications, a few different approaches have been proposed, in which the frame is decomposed into subbands to enable more flexible control over the information spectrum or where only the I-frames are encoded for applications requiring very high error resilience [Said and Pearlman (1996), Yang and Ramchandran (1998), Goyal *et al.* (1998)].

6.3.3 Multi-Dimensional Signal Processing

The effect of transformation and quantization for video compression can be explained using the framework of low-pass filtering. Figure 6.33 compares the concepts of low-pass filtering in the one-dimensional and two-dimensional frequency domains. Figure 6.33(a) illustrates the process of reducing the bit-rate by removing signal components above f_1 Hz, which can be implemented either in the time- or frequency-domain. In the figure, the darker regions on the left represent the visually important lower-frequency signal components with higher energy density. When the signal is transformed into the frequency domain, the majority of the signal energy is seen to lie at low frequencies. After quantization of the frequency components most of the higher frequency components are removed, resulting in an effect similar to low-pass filtering.

If there is a transform for video that can efficiently represent a macroblock in the frequency domain, then such a representation would require two axes, since two-dimensional sinusoids have both horizontal and vertical wavelength components. Suppose the grayed rectangle in Fig. 6.33(b) represents the magnitude of the frequency

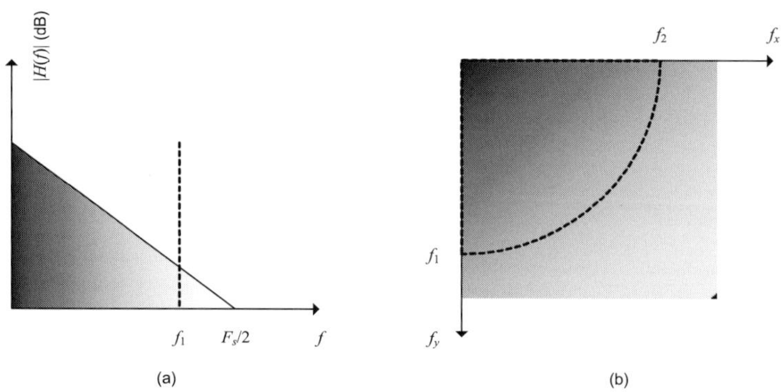

Fig. 6.33 Concept of low-pass filtering. (a) One-dimensional. (2) Two-dimensional.

domain representation of a macroblock, where the darker areas represent signal components with higher energy density. For images these are located in the low frequency areas of the plane. The origin would indicate the DC component, and the spectral components for the horizontal and vertical edges of video contents are located in the lower left and upper right areas respectively [Rao and Hwang (1996)]. Then the effects of two-dimensional low-pass filtering can be visualized as suppressing the signal components outside the dotted arc while keeping the internal signal components intact.

The two-dimensional (2D) Discrete Cosine Transform (DCT) is a transform that can be used to compact the visual energy into low-frequency components [Lim (1990)]. The 2D DCT of Type II is defined as

$$C_x(k_1, k_2) = \sum_{n_1=0}^{N_1-1} \sum_{n_2=0}^{N_2-1} 4x(n_1, n_2) \cos\left[\frac{k_1(2n_1+1)\pi}{2N_1}\right] \cos\left[\frac{k_2(2n_2+1)\pi}{2N_2}\right], \qquad (6.4)$$

for $0 \le k_1 \le N_1 - 1$ and $0 \le k_2 \le N_2 - 1$. The 2D DCT is an invertible transform that converts an array of real values, such as the subsampled frames of YCbCr, into another array of real values. Since the procedures for computing the DCT are similar to those for the Discrete Fourier Transform (DFT), the complexity of DCT computation can be reduced significantly using Fast Fourier Transform (FFT) algorithms. For the luminance component of images, $C_x(k_1, k_2)$ has its maximum at the DC component, i.e., when $k_1 = k_2 = 0$. Since the DCT is a real transform, less computation and smaller storage are required than with complex transforms such as the DFT.

Figure 6.34 visualizes the 2D DCT of a 16×16 macroblock, through which the uneven plane of positive values shown in (a) is mapped to a relatively flat plane with peak values around DC as shown in (b). The factor 4 in Eq. 6.4 is replaced by a scaling factor of $\frac{2}{\sqrt{N_1 N_2}}$ in the computation. The visual effects of energy compaction are clear. In the frequency domain, the values of $C_x(k_1, k_2)$ are negligible when k_1 or k_2 is larger than about five. By dividing each value of $C_x(k_1, k_2)$ by a quantization step size, and rounding the quotient down, the higher-frequency components can be suppressed. The I-frames typically result in many more non-zero quantized DCT coefficients than the P-frames.

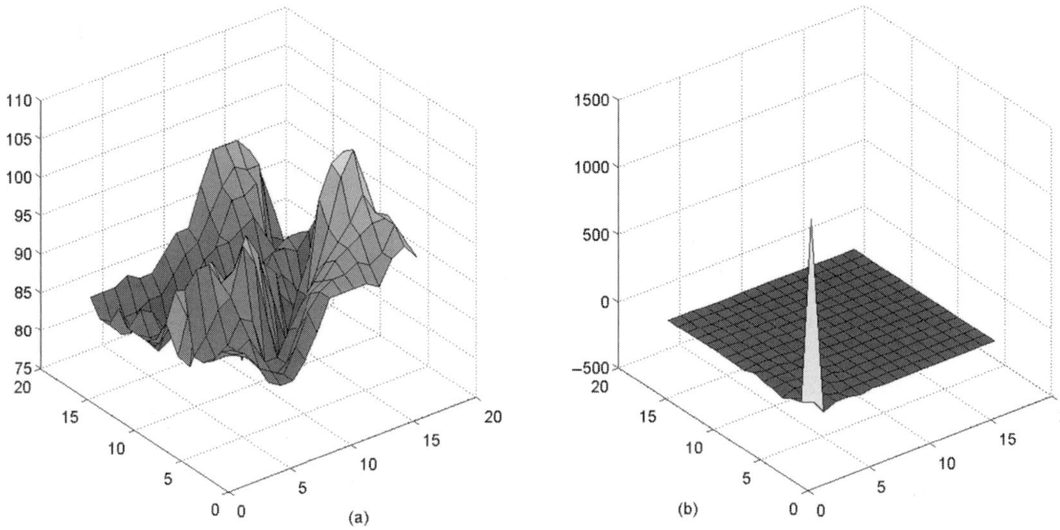

Fig. 6.34 Two-dimensional DCT. (a) Spatial domain. (b) Frequency domain.

The compression efficiency can also be increased by controlling the block size used for motion estimation and compensation. If a satisfactory match using the default macroblock size is not found in the previous frame, the area can be searched again for matching blocks of a smaller size. The size can be successively reduced until an acceptable candidate is found or until a pre-defined number of attempts have been tried. However, smaller block sizes for motion estimation and compensation increase the bit-rate as more, and potentially larger, motion vectors need to be transmitted. Therefore using the smallest block size does not always result in the highest compression efficiency.

The video decoder reverses the compression process to reconstruct video frames. When the received video frames cannot be decoded or are found to contain errors that cannot be corrected, the negative effects on the image quality can be reduced by interpolating the previously decoded frames. However, unlike the case for speech frames, the errors in video frames are likely to propagate to the following frames, because of the motion estimation and compensation. The error propagation can be stopped if an I-frame is received without error. These can be periodically generated at the video encoder or explicitly requested by the receiver via signaling. The I-frames have the capability of flushing errors and restoring video quality although they provide similar quality as the P-frames in error-free channels. To enhance the quality, additional filtering may be applied to decoded video frames. Such post-processing does not consume additional bandwidth but increases the delay and power consumption. Note that in many standards for video codecs, only the bit-stream structures and procedures of the decoder are specified, leaving the design of the encoder to the discretion of the implementation as long as the bit-stream can be interpreted by a standard-complying video decoder.

6.3.4 D/A Conversion

As in analog-to-digital conversion of video, the digital-to-analog conversion process includes signal processing procedures for which no clear counterparts exist in speech. The reconstructed frames from the video decoder are in the YCbCr or RGB formats. Figure 6.35 shows the pixel structures of the displays often used in the UE. Figure 6.35(a) is the RGB stripe, the conventional pixel format used in the Liquid Crystal Display (LCD) where the red, green, and blue sub-pixels are of the same size and proportion. The output from the video decoder can be entered to the input of the display, either directly or after super-sampling and format conversion to modulate the sub-pixels.

Figures 6.35(b) and 6.35(c) show the RGBG and RGBW structures, which are often called the *PenTile matrix* family of sub-pixel layouts. RGBG, which exploits the fact that the human visual system is most sensitive to green, is often used when sub-pixels of certain colors are more difficult to generate, as in the case of the Organic Light Emitting Diode (OLED) [Tsujimura (2012)]. As blue sub-pixels contribute less than the ones with other colors in resolving and recognizing images, their frequency on the layout can be reduced without impacting the quality noticeably. RGBG realizes a color space providing luminance comparable to that of RGB, with only two-thirds of the sub-pixels that RGB uses. RGBG realizes a color space providing luminance comparable to that of RGB, with only two-thirds of the sub-pixels that RGB uses. This approach is effective when the resolution of display is high. In RGBG, there are twice as many green sub-pixels as there are blue and red, but the green are typically smaller. To match the physical nature of display materials as closely as possible, non-rectangular sub-pixel layouts are also used, as illustrated in a more realistic fashion in Fig. 6.35(c) where each sub-pixel is separated from its neighbors. The diamond shape structure increases the achievable density of sub-pixels.

6.3.5 Combined Distortion from Compression and Transmission

Another characteristic of video for which no speech counterpart exists is the controllable tradeoff between the spatial and temporal quality. Figure 6.37 shows the SNR values of

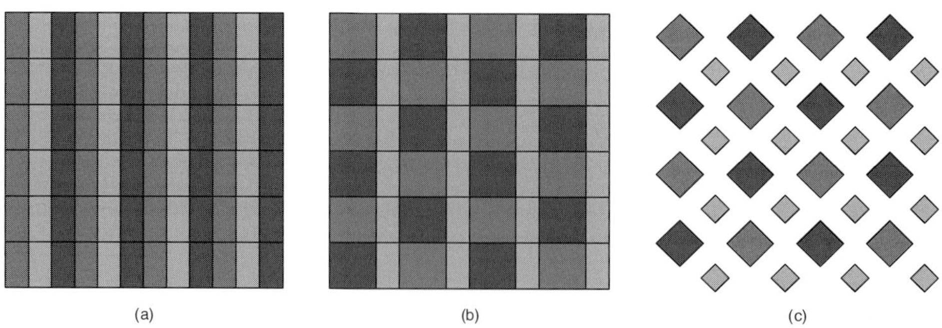

Fig. 6.35 Display pixel structures. (a) RGB. (b) RGBG. (c) RGBG (Diamond).

Fig. 6.36 Video sequences. (a) *Carphone*. (b) *Akiyo*. (c) *Mother and daughter*.

a 900-frame synthetic video sequence, generated by concatenating the standard *Carphone*, *Akiyo*, and *Mother and Daughter* sequences. Figure 6.36 shows the images of their first frames. They are typical head-and-shoulder video sequences that differ significantly in the level of motion activity. The synthetic sequence was constructed to capture the wide variety of input video sequences that can be expected in mobile communications, depending on the location or velocity of the UE.

To evaluate the efficiency of video compression in a sequence-wise fashion, an average of the Peak Signal-to-Noise Ratio (PSNR), is commonly used. It is related to the frame-by-frame MSE between an uncompressed video sequence and its compressed and decoded version. PSNR is defined as

$$\frac{1}{L}\sum_{l=0}^{L-1} 10\log\left\{\frac{1.5 * M * N * (255)^2}{\sum_{i=0}^{M-1}\sum_{j=0}^{N-1}\left(Y_l(i,j)-\hat{Y}_l(i,j)\right)^2 + \sum_{i=0}^{M/2-1}\sum_{j=0}^{N/2-1}\left(Cr_l(i,j)-\hat{Cr}_l(i,j)\right)^2 + \sum_{i=0}^{M/2-1}\sum_{j=0}^{N/2-1}\left(Cb_l(i,j)-\hat{Cb}_l(i,j)\right)^2}\right\}, \tag{6.5}$$

where the differences between Y, Cb, and Cr of each uncompressed frame and their counterparts in the compressed and reconstructed frame are computed, respectively. Although the above formula, which is often called the PSNR for brevity, is extensively used in comparisons of video compression algorithms and codec implementations, limitations of this metric in which the results of subjective evaluations are not aligned with the PSNR values have often been reported in the literature [Richardson (2003)]. Information on the objective and subjective methods designed to evaluate low bit-rate video quality for mobile communications systems can be found in [3GPP (2001), 3GPP2 (2007)].

In this experiment, $M = 176$, $N = 144$. L is set to 450, 225, and 150, which correspond to the frame rates of 15, 7.5, and 5 frames per second (fps), respectively, as the video sequences are assumed to be originally captured at 30 fps. Following a typical configuration of 3G-324M, the mobile video telephony system of W-CDMA, the bit-rate of the MPEG-4 video encoder is maintained at 48 kbps, and an I-frame is generated every 20 s for all three frame rates. As the PSNR is defined assuming that each uncompressed frame of a video sequence is matched by its compressed and reconstructed version, it can be difficult to apply PSNR to situations where some frames are lost during compression or transmission, and the frames are not synchronized. The possibility of dropping frames, i.e., intentionally giving up encoding frames when the bit-rate or processing time is insufficient, is another characteristic of video compression to which there is no counterpart in speech compression where the frame rate is fixed, e.g., to 50 fps if each frame covers 20 ms of speech, as long as voice activity is present.

Figure 6.37 shows that the PSNR can have a large dynamic range, from 28.00 to 43.57 dB. Although the PSNR values do not represent the subjective quality directly, an image with a PSNR lower than 30 dB is in general too blurred to recognize its contents clearly. On the other hand, increasing the PSNR beyond 40 dB contributes little to further image clarity. The dynamic range of the PSNR would decrease significantly if the bit-rates were increased, and the high-motion areas in frames 100–300 were encoded with more bits. The variations in the image quality are often misunderstood as being caused by defective focusing mechanisms of the camera. Mobile video telephony systems have to cope with such a highly fluctuating image quality, which depends on the input video sequence and the encoding method. In classical video optimization techniques, the quantization step size or the frame type is dynamically selected during compression to satisfy a rate-distortion criterion, while maintaining a fixed frame rate and an image size [Ortega and Ramchandran (1998)]. The selection is typically done to maximize the PSNR or, equivalently, to minimize the reconstruction error. With a fixed bit-rate as in W-CDMA, a more straightforward approach for increasing PSNR is to reduce the frame rate, which conceptually corresponds to shifting some of the bits spent in the temporal domain to the spatial domain.

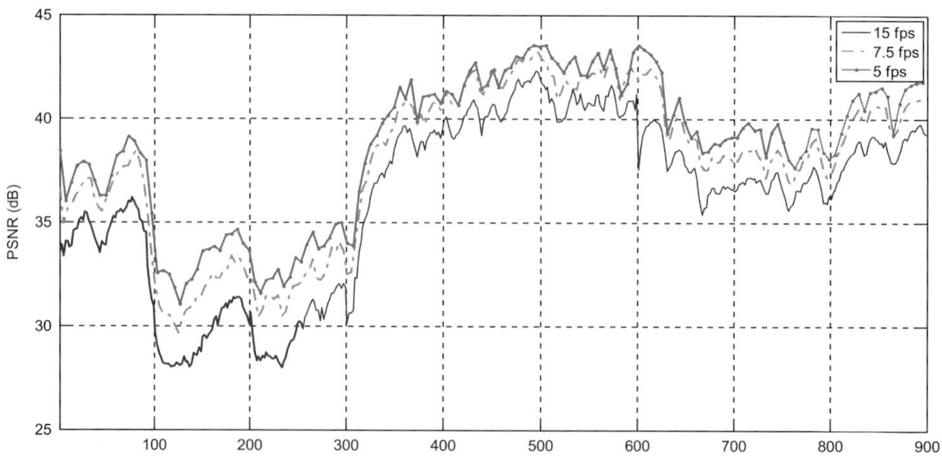

Fig. 6.37 PSNR of concatenated video sequences.

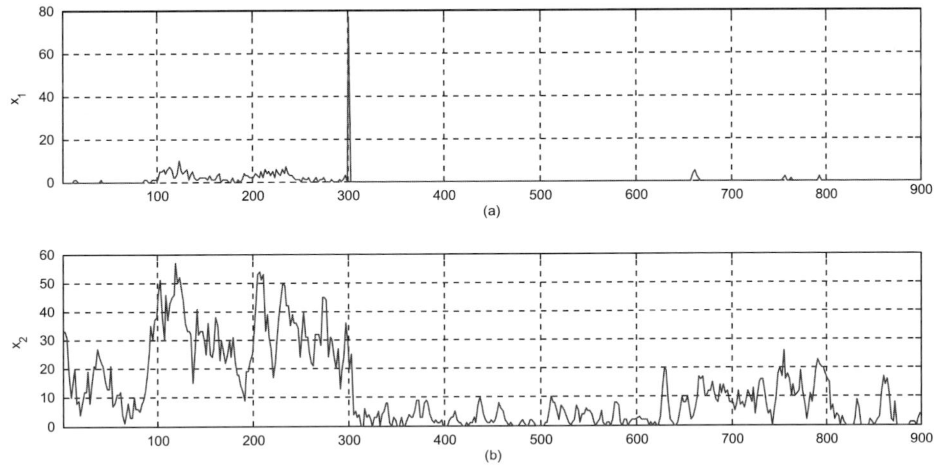

Fig. 6.38 Number of macroblocks. (a) Intra-coded. (b) 8 × 8 inter-coded.

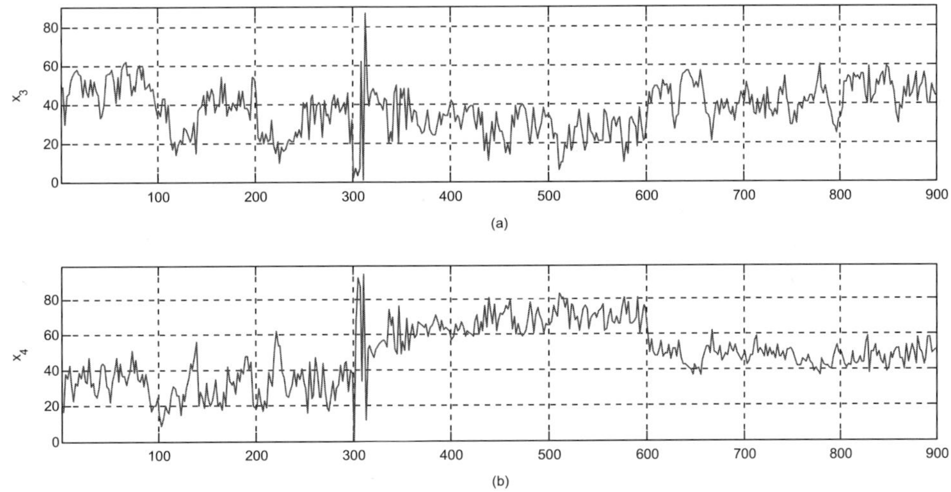

Fig. 6.39 Number of macroblocks. (a) 16 × 16 inter-coded. (b) Not-coded.

Figures 6.38 and 6.39 show the statistics of the macroblock encoding mode, for the same sequence as in Fig. 6.37. x_1, x_2, x_3, and x_4 denote the number of 16 × 16 macroblocks of a P-frame that are intra-coded, 8 × 8 inter-coded, 16 × 16 inter-coded, and not-coded, respectively. The intra-coding refers to compressing macroblocks without motion estimation and compensation, and the inter-coding refers to the use of both techniques as in the encoding of P-frames. If the difference between the current macroblock and the previously encoded and reconstructed macroblock at the same location is a zero matrix, the macroblock need not be encoded at all. For this resolution, $x_1 + x_2 + x_3 + x_4 = 99$. The following parameters are computed inside the reference MPEG-4 video encoder in order to determine the macroblock encoding mode:

$$MB\_mean = \frac{1}{N_c} \sum_{i=1,j=1}^{16,16} y_{\text{orig}}[i,j], \tag{6.6}$$

$$A = \sum_{i=1,j=1}^{16,16} |y_{\text{orig}}[i,j] - MB\_mean|, \tag{6.7}$$

$$SAD_N(x,y) = \sum_{i=1,j=1}^{N,N} |y_{\text{orig}}[i,j] - y_{\text{pred}}[i,j]|, \tag{6.8}$$

$$SAD_{Kx8} = \sum_{1}^{K} SAD_8(x,y), \tag{6.9}$$

$$SAD_{\text{inter}} = \min(SAD_{16}(x,y), SAD_{Kx8}). \tag{6.10}$$

Where y_{orig} is a pixel value from the luminance component of the macroblock being encoded, N_c is the number of pixels inside a macroblock, $N = 8, 16$, and $SAD_N(x,y)$ is the Sum of Absolute Differences (SAD) between the pixels in that block and their predicted values based on the earlier frames. Other than a division by N_c for MB_mean, no multiplication is necessary in the computation of the parameters. As a metric to indicate the level of motion activity, SAD can be considered as a computationally lower-cost alternative to PSNR. The motion vectors used in making those predictions are integer-valued, and constant over 16×16 macroblocks. The four motion vectors that produce the lowest values of SAD_8 are used to select the integer-valued motion vectors for the four 8×8 subblocks. The ultimate motion vectors are obtained by further refining these integer-valued vectors to have a half-pixel resolution. SAD_{Kx8} is the SAD for a whole macroblock using the 8×8 displacement vectors. $K = 4$ is selected for the simple profile of the MPEG-4 video codec, which represents the set of encoding features supported.

The decision to use intra- or inter-coding for a macroblock is based on a test involving these parameters. The intra-coding is selected if

$$A < (SAD_{\text{inter}} - 2 * N_B), \tag{6.11}$$

where N_B is the number of pixels inside a video frame multiplied by $2^{(\text{bits\_per\_pixel}-8)}$. If this test fails, the macroblock is inter-coded. Once the inter-coding has been decided upon for the macroblock, the next decision to be made is whether it is to be encoded at 16×16 or decomposed into four 8×8 blocks. The 8×8 blocks are chosen if

$$SAD_{Kx8} < SAD_{16} - (N_B/2 + 1). \tag{6.12}$$

Otherwise, a 16×16 block is used. Finally a 16×16 inter-coded macroblock is not-coded if its motion vector is zero and there is no data for the DCT coefficients. Detailed descriptions of the decision procedures can be found in [ISO (2001)]. Notice that other procedures can also be used as long as they comply with the required bit-stream structures. Following the encoding procedures, the bit-rate for 176×144 video at 15 fps in YCbCr 4:2:0 format is reduced to only 1–2% of its original value. Considering that speech encoders such as AMR or EVRC reduce the bit-rate of sampled speech from the

104 kbps of 13-bit Uniform to 12.2 or 8.55 kbps respectively, video encoders in general achieve much higher compression efficiency than speech encoders.

From the decision logic used to select the macroblock encoding mode, it can be expected that smaller block sizes would be more frequently used in video sequences with high mobility levels or in complex scenes. When Figures 6.37, 6.38, and 6.39 are compared, two parameters, x_2 and x_4, seem to be directly related to the PSNR, albeit in the opposite direction. x_2, the number of 8×8 inter-coded blocks in each frame, looks inversely related to the PSNR, which is understandable since this encoding mode is used when the correlation between successive frames is low, as in abrupt changes of scene. Even when the motion of objects in the scene is slow, the correlation between frames can be low if the scenes contain complex objects. The macroblocks are intra-coded when the correlation is low, but the video encoder uses this mode only when it is absolutely necessary, as the intra-coding significantly reduces the compression efficiency.

At frame 300, a peak value of x_1 appears, when the video sequence changes from *Carphone* to *Akiyo*, which triggers a complete change in the scenes. x_2 reaches its highest values during the frames corresponding to *Carphone*, and lowest for *Akiyo*. On the other hand, x_4 is positively related to PSNR, reaching its highest values in *Akiyo* and lowest in *Carphone*. x_3 seems to have a low correlation with PSNR. The relationships of the measured parameters to the PSNR are likely to change to some extent, depending on video sequences and codec configurations such as the bit-rate, frame rate, or resolution, but they can still be used to assess the status of encoding in progress without computing PSNR. For example, if the statistical distribution of the macroblock encoding modes is similar to that of *Carphone*, reducing the frame rate may increase the image clarity to an acceptable level. Likewise, when the input video sequence contains little motion activity, as in *Akiyo*, the frame rate can be increased without compromising the image clarity.

Figure 6.40 shows the PSNR values for the *Carphone* sequence compressed in the configuration of Fig. 6.37, i.e., at 48 kbps and 15 fps, and transmitted for two seconds over a channel whose bit error rate (BER) is 0.001. This low BER corresponds to no more than a few bits decoded in error spread over the two-second period. The video sequence is encoded with a lower-complexity MPEG-4 video encoder than the one used in Fig. 6.37, resulting in lower PSNR values. Typically simple compression algorithms not only reduce the PSNR as the motion estimation becomes less accurate but also increase the fluctuation of bit-rate around a target value. The PSNR, which started at around 28 dB, drops abruptly after frame 41, in which some key fields of the bit-stream have been corrupted by errors and the difference between the original frame and the compressed, transmitted, and reconstructed frame increases abruptly. However, the video decoder may not be aware that the incoming bit-stream is corrupted by errors if the received bit-stream complies with the grammar. Once the PSNR drops, it does not rebound even when there are no further errors but starts generating noticeable visual artifacts that propagate to the following frames. The image quality degrades rapidly as the effects of the errors propagate and contaminate the decoding of clean frames, reaching an unrecognizable state once the quality loss exceeds a certain level. Until one or more clean I-frames arrive at the decoder, the PSNR will remain low. In Fig. 6.40, the

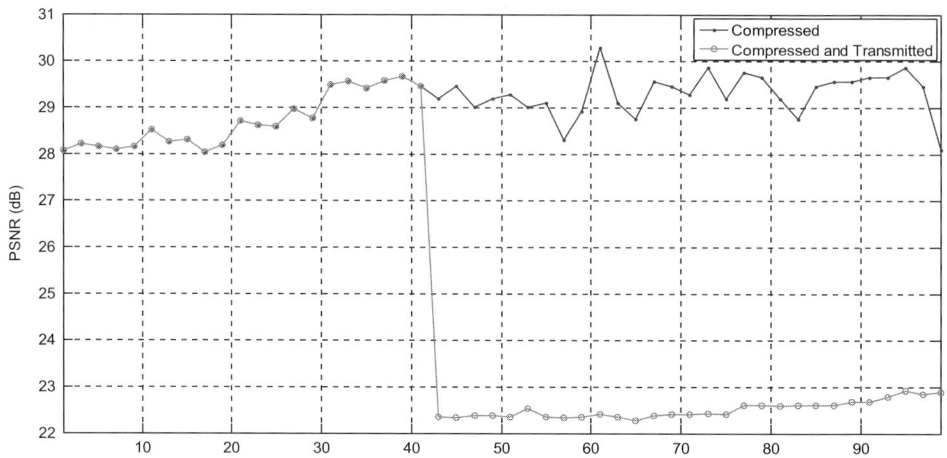

Fig. 6.40 PSNR of compressed and transmitted video sequence.

Fig. 6.41 Video frames (*Carphone*). (a) Uncompressed. (b) Compressed. (c) Compressed and transmitted.

PSNR drops more than 6 dB in a frame, and the gap between the PSNR after compression and the PSNR after transmission is maintained. Figure 6.41 shows the last frames of *Carphone* whose variations of PSNR are shown in Fig. 6.40, when the sequence is (a) uncompressed, (b) compressed, and (c) compressed and transmitted over a channel with BER = 0.001. It can be observed that the image is blurred first from compression, and loses intelligibility after transmission. The loss of quality from erroneous transmission is visually larger than that from compression. However, it is also likely that other types of bit errors at the same BER may result in little noticeable degradation if the corrupted areas are visually not important or if the video decoder corrects the compromised bit-stream. If the video decoder identifies that certain areas of the bit-stream have been damaged, it may invoke error concealment, replacing lost macroblocks with the corresponding areas in the previously decoded video frames.

Note that a speech encoder typically reduces the bit-rate to 10–15% of its original value, as shown in Fig. 1.2, and the loss from BER = 0.001 would be virtually unnoticeable, considering the testing conditions and results of GSM and cdma2000. The price for the higher compression efficiency, e.g., up to 1–1.5% of its original bit-rate, of video compression is that the structural responsibility of each bit is increased radically so that the loss of only a small number of bits can render the quality of the reconstructed video practically useless, as can be seen in Fig. 6.41(c).

Because the propagation of bit errors can have immediate and detrimental effects on the visual quality of all succeeding frames, PSNR, which is an efficient metric in the analysis and comparison of video compression efficiency, may not be able to represent the combined results of compression and transmission. This phenomenon of an abrupt drop in PSNR when some key bits are lost is caused by the inter-frame dependence from motion estimation and compensation, the elevated structural importance of each bit from the low bit-rate, and the low encoding frequency of I-frames. Note that encoding I-frames more frequently at a fixed low bit-rate can result in a lower frame-rate which is perceived as jerky motions.

The Percentage Degraded Video Duration (pDVD) is a new metric that was defined to evaluate the quality of video communications over error-prone channels. pDVD is defined as

$$pDVD(x) = \frac{1}{L}\sum_{l=0}^{L-1} 1(PSNR_l^c - PSNR_l^e > x), \tag{6.13}$$

where $1(y)$ is defined to be 1 if condition y is true, and 0 otherwise. L is the number of compressed and transmitted video frames.

pDVD represents the percentage of the video period during which the PSNR falls below a threshold x. $PSNR_l^c$ and $PSNR_l^e$ represent the PSNR values of the lth frame under error-free and error-prone conditions, respectively. Therefore a lower pDVD represents higher transmission performance. In Fig. 6.40, $PSNR_l^c - PSNR_l^e > x$ from frame 42, assuming a reasonable value for x, as the PSNR drops significantly after this frame. Even when the video decoder fails to reconstruct corrupt frames, or some frames are not received, the pDVD can still be computed by setting the corresponding $PSNR_l^e$ to pre-defined low values. pDVD is an efficient metric to measure the performance of video compression and transmission at low bit-rates but becomes less effective when the bit-rate becomes higher and the I-frames can be inserted more frequently.

6.3.6 Rate Control

Figure 6.37 showed that the video quality strongly depends on the contents of the scene and the motion activity. Vastly different PSNR values can result even at the same bit and frame rates. As seen in the voice transmission with GSM, cdma2000, and W-CDMA, an upper bound for the bit-rate has to be determined between the UE and the NodeB, or between the MS and the BTS, before starting the services. The allowable bit-rates of speech codecs are precisely defined, and the output bit-rate of the speech encoder does not exceed the upper bound in any case, regardless of the input signal or channel quality. In contrast, in typical video codecs using motion estimation and compensation, it is not possible to control the number of bits for each frame as precisely as for speech codecs.

To transport real-time video over mobile communications networks, the output bit-rate of the video encoder needs to be kept strictly below an upper bound. The rate control, which is in general not an integral part of the video codec standards, lies in a class of implementation-specific techniques to maintain the bit-rate of a video encoder about a target value as closely as possible. It does this by controlling various encoding parameters, such as the frame type, frame rate, quantization step size, or the macroblock encoding mode. The fullness of the buffer for storing encoded frames before transmission also has to be taken into account in the selection of the encoding parameters. Figure 6.42 illustrates the concept of video rate control.

Figure 6.42(a) shows that a fixed set of encoding parameters generates a video bit-stream whose bit-rate fluctuates over a large range, influenced by the input signal. Frames with rapid motion or complex scenery require more bits while stationary, simple frames require fewer encoding bits. To stabilize the bit-rate, the output bit-stream of the video encoder can be temporarily stored in a buffer, whose level can be fed back to affect the rate control. The output of the buffer connects to the interface of the communications network or storage, which periodically fetches bits from the buffer. Based on the loading level of the buffer, the rate controller adjusts the parameters of the video encoder so that the buffer neither overflows nor underflows. For example, if video sequences with high motion activity or complex scenes are encoded at a fixed bit-rate, the quantization step size can be increased, or the frame rate can be decreased to suppress the increase of bit-rate that might cause the buffer to overflow.

Figure 6.42(b) shows the output bit-rate of a rate-controlled video encoder with a buffer size of N bits. The bit-rate still fluctuates around a target value, albeit over a smaller range. The rate controller may choose not to encode some frames if the buffer is filled beyond a threshold. In the case of underflow, when the bit-rate falls below the

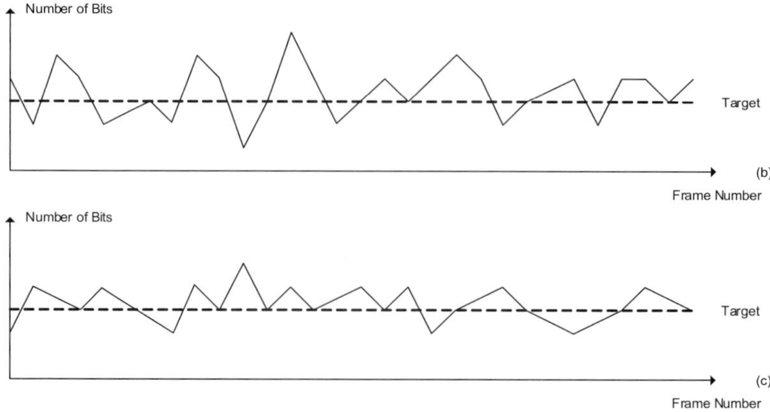

Fig. 6.42 (a) Uncontrolled. (b) Buffer size = N. (c) Buffer size = $2N$.

target value, the network may leave the unused capacity as such or fill it with padding bits, depending on the transport technology or service policy. In the case of overflow, the surplus bits have to be removed by either the video encoder or the network. Since the network in general cannot be aware of the visual impact from dropping some bits of the bit-stream, the bit-rate needs to be conservatively controlled by the video encoder. Therefore the upper bound on the video bit-rate needs to be smaller than the transport bit-rate supported by the network that includes other overhead required to manage the transmission. Figure 6.42(c) shows the output bit-rate of a rate-controlled video encoder with twice the size of the buffer used in (b). Due to the higher smoothing effects, variation of the bit-rate is reduced further, at the cost of a larger encoding delay. Note that although the bit-rate is smoothed with rate control, the PSNR may still fluctuate significantly.

Depending on the available delay budget, rate control of video can be considered as a dynamic optimization problem. Maximizing quality at a given range of bit-rates is of major concern when a long encoding delay can be tolerated, as in video storage applications. In this case the encoding can be controlled over an observation window of several frames, and decisions about the encoding parameters for the current frame can be made in the context of the decisions of multiple previous and future frames. The B-frames can play important roles in this case. Such problems are often formulated using rate-distortion models, and numerous solutions based on iterative or exhaustive searches are described in the literature [Ding and Liu (1996), Lee and Dickinson (1997)]. On the other hand, when both the bit-rate and the encoding delay are tightly limited, such as in 3G-324M, the algorithms for rate control have to be compromised. In this case, the observation window and the frame buffer have to be smaller, and encoding decisions based on future frames, i.e., the use of B-frames, is not possible. Located somewhere between these two extreme cases is live streaming, as used in the Internet broadcasting of business or entertainment events, where the combined delay of several seconds, for compression, transmission, and reconstruction of video, can be tolerated.

Figure 6.43 shows the number of bits a rate control algorithm assigns to each frame of the *Carphone* sequences in Fig. 6.40. The sequences are encoded at 48 kbps, and 5, 7.5, and 15 fps respectively. Only the first frame is encoded as an I-frame, and the remaining frames are encoded as P-frames. It is clear that as an I-frame is significantly larger than P-frames, encoding an I-frame would consume the transmission capacity for at least 5–6 frames in these encoding conditions. As a result, when an I-frame is decoded and played, a short period of a frozen image can be perceived at a low bit-rate. Since the bit-stream of an I-frame does not depend on either a previous or a future frame, its size does not directly depend on the frame rate. However, it can be seen that the lower the frame rate, the smaller the deviation of bits assigned to each frame.

In contrast to the periodic fluctuations at 15 and 7.5 fps, the numbers of bits at 5 fps are similar, regardless of the scenes. This can be understood since a lower frame rate provides more time for rate control and encoding parameters such as the quantization step size can be adjusted in a longer-term fashion. Although the reduced frame rate increases the differences between frames, the overall compression efficiency is increased

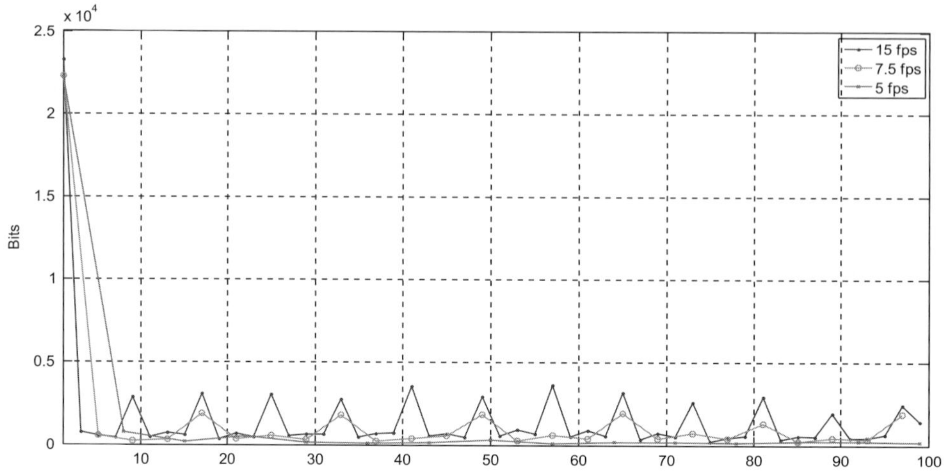

Fig. 6.43 Number of bits per frame (*Carphone*).

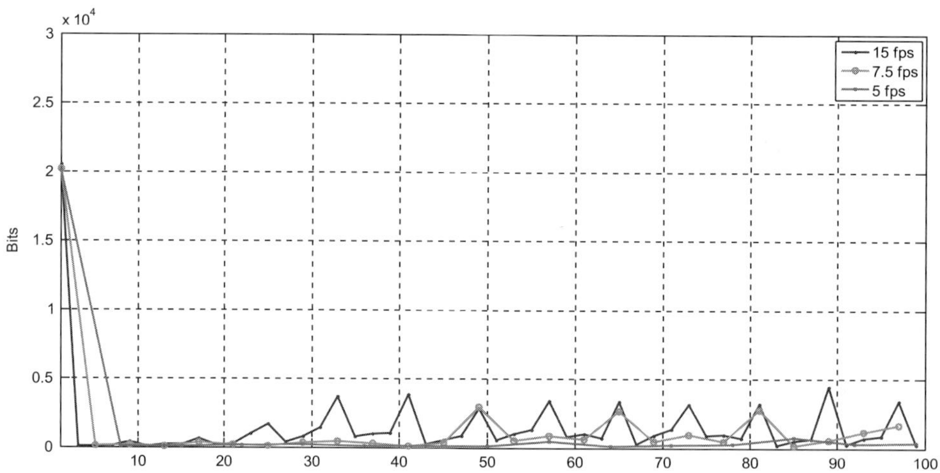

Fig. 6.44 Number of bits per frame (*Akiyo*).

so that fewer bits are used per frame at low frame rates. However, this reduction of bit-rate cannot be immediately exploited as in the DTX of GSM or the VBR of cdma2000, since the unused capacity has to be filled with overhead to maintain a constant bit-rate over the wireless channel.

Recall that an efficient operation of DTX or VBR requires a complete integration of speech codecs into radio signal processing procedures, which cannot be realized for video codecs whose control over the bit-rate is not as precise as that of speech codecs. Figure 6.44 shows the number of bits assigned to each frame of the *Akiyo* sequences, which are encoded identically. As shown in Fig. 6.37, *Akiyo* includes lower motion activity and achieves higher PSNR than *Carphone*, although the rate control equalizes the bit allocations for both sequences to some extent.

6.4 Video Codecs

From the very beginning of digital mobile communications systems, speech codecs have been designed as a key component whose performance directly influences speech quality, network capacity, and user satisfaction. The characteristics of the underlying radio signal processing operations were taken into account in the design of the speech codecs. However, such benefits are difficult to realize even if video codecs are designed as an element of mobile communications systems. As the video is a high bit-rate, three-dimensional signal whose values are decomposed into three color components over the spatial and temporal domains, video codecs operate in too complex a fashion to be efficiently controlled by the network. For example, the network cannot completely control the tradeoff between video quality and network capacity, since unlike in the case of speech codecs, the operating points of video codecs are not limited by the standards. In most video codecs, only the decoder is standardized, which allows the performance of the video encoder to be differentiated among implementations with algorithmic or computational innovations. There are no standardized messages, such as the CMC of GSM or the RATE_REDUC of cdma2000, for the network to control the behavior, i.e., the bit-rate, of the video encoder. In fact, the network is aware that video, which requires a higher QoS than that required for speech, is transmitted but unaware of the type of video codec or its operating conditions used in the services. Only the UE is aware of the type and detailed configuration of the video codec.

There have been numerous approaches for enhancing the video compression for mobile communications introduced in the literature. Many of these do not employ the conventional framework of block-based partitioning, followed by transformation and quantization. To achieve higher compression efficiency or error resilience than the conventional methods, they resort to new signal processing techniques such as wavelet-based or multiple description codecs. However, as video codecs are typically implemented in hardware to handle the large amount of data and reduce power consumption, and also used in applications other than video calls, for example, in streaming or recording, standard video codecs that can cover a wide variety of applications are used for the video telephony services of W-CDMA.

6.4.1 H.263 Video Codec

H.263 can be classified as one of the second generation video codecs, developed after H.261 and MPEG-1 [ITU-T (2005)]. Unlike the Motion Picture Expert Group (MPEG) video codecs developed by the International Organization for Standardization (ISO) for various applications including home entertainment and multimedia broadcasting, H.263 was developed by the International Telecommunication Union Telecommunication Standardization Sector (ITU-T) as a video component of H.32x multimedia communications terminal standards. These were designed for fixed circuit-switched and packet-switched networks. For example, H.320 is for ISDN, H.323 is for Ethernet, and H.324 is for PSTN. Among this family of standards, H.324 became the parent technology for the video telephony system of W-CDMA, 3G-324M.

Table 6.14 Supported resolutions of H.263.

Format	Resolution
Sub Quarter Common Intermediate Format (SQCIF)	128 × 96
Quarter Common Intermediate Format (QCIF)	176 × 144
Common Intermediate Format (CIF)	352 × 288
4 Common Intermediate Format (4CIF)	704 × 576
16 Common Intermediate Format (16CIF)	1408 × 1152

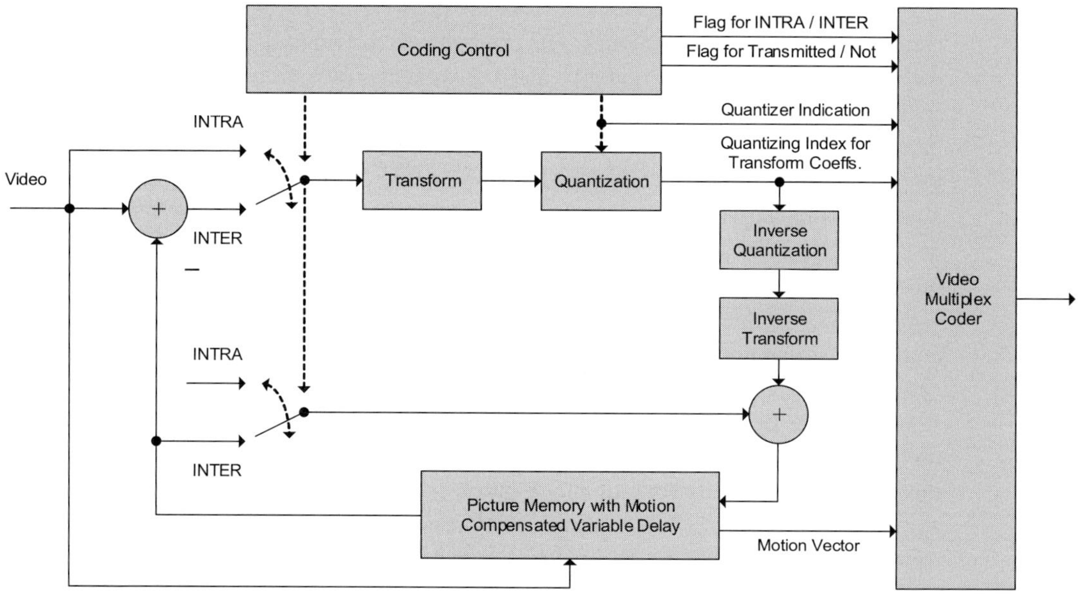

Fig. 6.45 Block diagram of H.263 video encoder.

Figure 6.45 shows a block diagram of the H.263 video encoder, which supports five resolutions at a fixed frame rate of about 29.97 fps, as outlined in Table 6.14. At a particular resolution, the frame rate can be adjusted depending on the bit-rate and the input signal. The architecture of the H.263 encoder follows the basic flow of video compression algorithms using block-based decompositions of frames, motion-estimation and compensation, followed by transformation and quantization of coefficients. The encoded frame is internally decoded and stored in the picture memory to be used as the reference for motion estimation. The key information to decode each frame, quantization step sizes and motion vectors, is multiplexed with the transform coefficients to construct the bit-stream.

As new features to increase compression efficiency and error resilience were added to the initially standardized H.263 in the form of annexes, the constraints of encoding video at only pre-defined resolutions were removed in the enhanced versions of the codec, called H.263+ and H.263++. The concepts of profile and level were introduced to

Table 6.15 Annexes of H.263 recommended for 3G-324M.

Annex	Name	Effects
I	Advanced intra-coding mode	Use coefficients of neighboring macroblocks to intra-code current macroblock
J	Deblocking filter mode	Apply filtering to block edges before encoding to reduce blocking effects
K	Slice structured mode	Manage macroblocks in slice, non-rectangular group of macroblocks
T	Modified quantization mode	Apply different quantization step to luminance and chrominance

facilitate the representation of feature sets and the processing capability of video codec. As a minimum requirement, 3G-324M requires the baseline profile, which supports a set of basic encoding features and resolutions up to QCIF. 3G-324M also recommends the implementation of Annexes I, J, K, and T. See Table 6.15. Annexes J and T improve compression efficiency while Annex K enhances error resilience. Annex I contributes to both aspects.

6.4.2 MPEG-4 Video Codec

The MPEG-4 video codec, also called MPEG-4 Visual or MPEG-4 Part 2 [ISO (2001)], was developed by the ISO as the successor to the MPEG-2 video codec, which has been successfully employed in various applications including digital televisions and digital video disks. Figure 6.46 shows a block diagram of the MPEG-4 video encoder. Although the architectures of H.263 and MPEG-4 are slightly different, both codecs use similar block-based motion estimation and compensation, and DCT-based transformation and quantization schemes. With 3G-324M, where H.263 was the default video codec envisioned, MPEG-4 was designed so that the bit-stream of H.263 supporting the baseline profile could be decoded by the MPEG-4 video decoder. However, maintaining such a minimum compatibility in the bit-stream structure with a previous codec limits the compression efficiency of a new codec significantly. Therefore the quality difference between H.263 and MPEG-4 in basic video compression is not significant but more advanced compression techniques can be used in MPEG-4 for other profiles. In practice, H.263 has the advantage of more mature implementations than the new codec. As a minimum requirement, 3G-324M requires the simple profile of MPEG-4 at level 0. As 3G-324M operates at a single total bit-rate including both speech and video, a higher codec capability does not bring any benefit.

The design of MPEG-4 focused on higher error resilience so that its bit-stream could still be decoded in situations when the bit-streams of other codecs could not because of the loss or corruption of bits. A class of source encoding techniques, called *error-resilient encoding*, is used in the bit-stream construction process of the MPEG-4 video encoder. In error-resilient encoding, the video encoder sacrifices compression efficiency in a controlled manner in order to enhance the robustness of its bit-stream against errors.

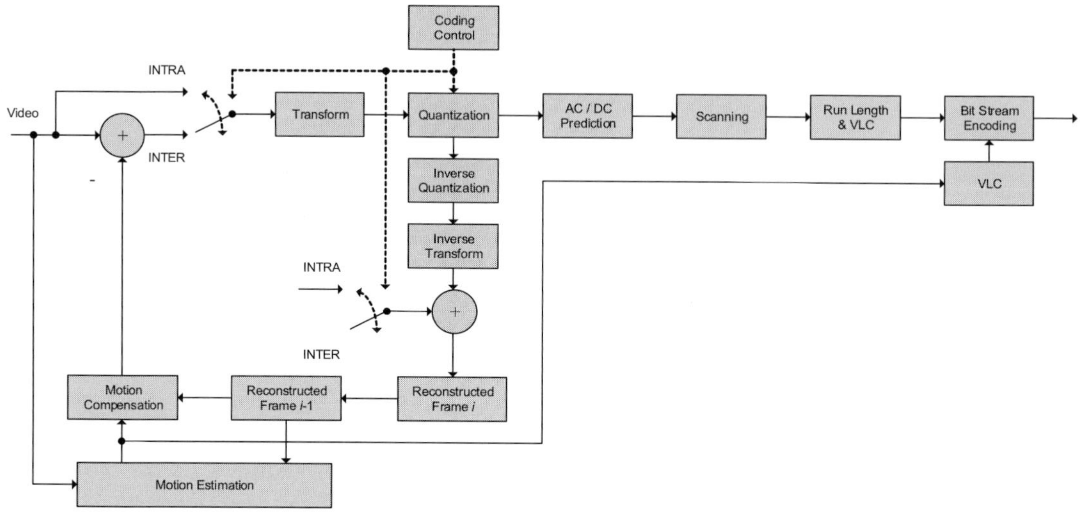

Fig. 6.46 Block diagram of MPEG-4 video encoder.

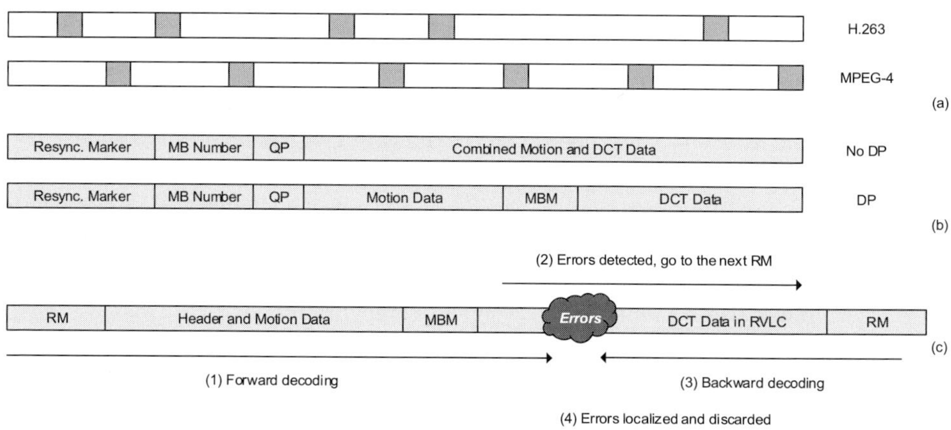

Fig. 6.47 MPEG-4 error resilience tools. (a) Resynchronization. (b) Data partitioning. (c) Reverse variable length code.

In MPEG-4 terminology, a set of such techniques are classified as the Error Resilience Tools (ERT), each of which is a relatively simple procedure to modify the structure of the bit-stream so that the error resilience is enhanced at minimum overhead [Talluri (1998)]. The ERTs are optional features that need not be employed on benign channels. Figure 6.47 illustrates the modified bit-stream and decoding process when the ERTs of MPEG-4 are activated.

A major source of errors for the video decoder is a loss of synchronization with the video encoder. Periodic insertion of Resynchronization Markers (RM) into the bit-stream can be a simple but effective remedy for this problem. RM is a series of bits,

0000000000000001, that cannot appear in other parts of the bit-stream. When the video decoder encounters an RM, its location in the bit-stream is reconfirmed and the decoding process is allowed to continue. The fields before and after the RM are also identified. In H.263, an RM can be inserted only at certain locations in the bit-stream, e.g., at the beginning of each Group of Blocks (GOB). A direct disadvantage of this constraint is that the interval between two RMs can vary significantly, depending on the motion activity or video contents, as illustrated in Fig. 6.47(a). For example, GOBs with high motion activity might receive RMs less frequently, while GOBs with low motion activity receive more. As a means to address the problem of fluctuating error resilience, MPEG-4 allows the RMs to be inserted at approximately equally-spaced intervals. In the terminology of MPEG-4, the bit-stream between two RMs is called a video packet. A video packet contains the bit-stream corresponding to an integral number of macroblocks, which can be decoded separately from other packets. Once the length of the bit-stream after an RM exceeds a predefined value, a new video packet is generated at the next macroblock. Naturally there is a tradeoff between error resilience and compression efficiency which can be controlled by choosing the size of a video packet.

Data Partitioning (DP) is a low-overhead technique for improving error resilience based on the observation that the motion data typically has a higher priority than the DCT data in the reconstruction of video sequence. If a similar number of bits is lost from the bit-stream, the damage to video quality is likely to be higher if the lost bits belong to the motion data. In MPEG-4, DP refers to the separation of motion data and DCT data in a video packet, with a marker. Motion Boundary Marker (MBM) is a unique 17-bit word that cannot appear in the DCT data. The structure of the video bit-stream modified for DP is shown in Fig. 6.47(b). Conventionally, the motion data and DCT data for each macroblock are located together in the bit-stream. Thus errors in either data can result in the loss of both types of data. With data partitioning, if the DCT data is lost but the motion data is intact, decoder error concealment techniques can be effectively used to minimize the visual impact of losing the DCT data. On the other hand, if the motion data is lost but the DCT data is intact, the video packet can be lost and the influence of errors propagate to the following frames. If the header is lost, the damage will be even higher than in the loss of motion data. Header Extension Code (HEC) is a 1-bit flag that signals, if set to 1, the presence of a duplicated header of the frame for higher error resilience.

Reversible Variable Length Codes (RVLC) is a class of variable-length codes (VLC) that can be decoded in either direction. The source decoding can start from the first bit to the last, or vice versa. This property makes it possible to salvage some of the corrupt data that otherwise has to be jettisoned, as illustrated in Fig. 6.47(c). In MPEG-4, an RVLC can replace the VLC used to encode the DCT data at the cost of some compression efficiency. It is observed that the average length of the codewords coded with an RVLC is longer by only about 5% than those with a VLC [Takishima *et al.* (1995)]. Table 6.16 shows a simple approach to transform a VLC into an RVLC, where each codeword of the RVLC, with a constant Hamming-weight of 3, is formed by adding a prefix 1 and a suffix 1 to each VLC codeword, which has a constant Hamming-weight of 1.

Table 6.16 Generation of RVLC from VLC.

VLC	RVLC
1	111
01	1011
001	10011
0001	100011

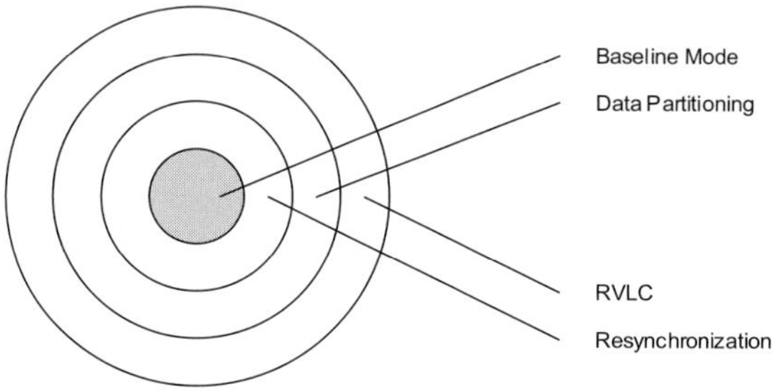

Fig. 6.48 Layers of MPEG-4 error resilience tools.

These error resilience techniques cannot be independently employed since the use of RVLC requires the use of DP, which further requires the RM. Figure 6.48 illustrates the conceptual structure of MPEG-4 ERT layers where the innermost core is the baseline mode which is compatible with the baseline profile of H.263. Among the tools, the RM and DP are commonly implemented in 3G-324M as they provide relatively high error resilience at small cost for bit-rate and complexity.

6.5 3G-324M

In a strict sense, 3G-324M was not exactly the first mobile video telephony. A system based on the Personal Handy-Phone System (PHS), a Japanese cordless telephony based on TDMA [ARIB (2011)], preceded 3G-324M by a couple of years. In the system, about two images in the Motion Joint Photographic Experts Group (JPEG) format at 220×254 resolution were transmitted per second over a 32 kbps channel, with speech in 32 kbps ADPCM format. The intra-only compression, i.e., the Motion JPEG, was used to avoid the high complexity of video codecs such as MPEG-4. The Motion JPEG also compensated for the low error resilience and limited mobility of the PHS. In contrast, 3G-324M used fully-fledged video compression using motion estimation and compensation, at a lower resolution of 176×144, and supported video services at up to vehicular speed using the robust transport capability of W-CDMA.

6.5.1 System Architecture

Figure 6.49 shows the protocol architecture of 3G-324M, in which the UE is separated into a media client and a W-CDMA modem. The media client consists of the speech and

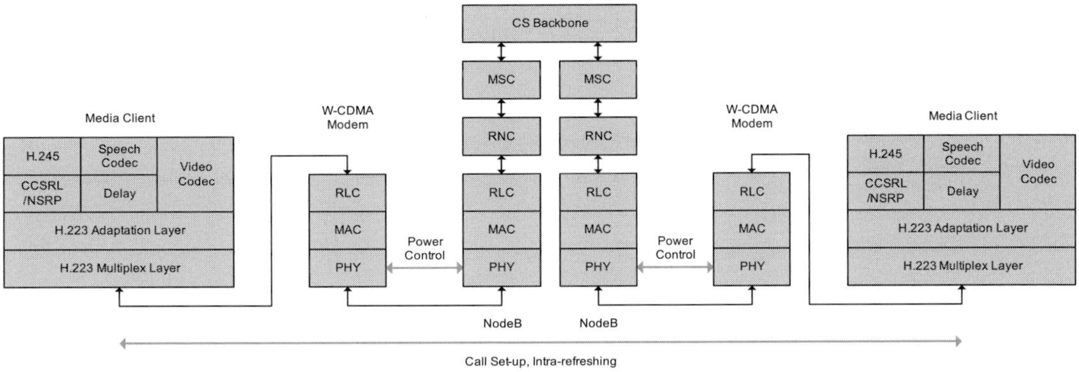

Fig. 6.49 End-to-end media compression and transmission paths of 3G-324M.

video codecs, H.245 call control protocol, and H.223 multiplexing protocol that mixes the bit-streams from the media and control into a single fixed bit-rate bit-stream. The call control protocol directly communicates with the counterpart in the far-end UE to determine the media configurations and recover the media quality if a loss of data is identified. 3G-324M is not originally for W-CDMA but has evolved from H.324 [ITU-T (2002), 3GPP (2017k), 3GPP (2017l)], which was designed for PSTN. To cope with the harsher mobile environments, where the error rate and transmission cost are both higher, 3G-324M can employ more recently developed speech and video codecs with higher compression efficiency and error resilience than those of the parent standard [3GPP (2017a)].

6.5.2 Media Adaptation and Multiplexing Procedures

This set of signal processing procedures takes place only at the UE for the uplink, and the intermediate network nodes between two UEs, including the NodeB, RNC, and MSC, do not modify the bit-stream generated in this step. The H.223 multiplexing protocol consists of two stacked layers: an adaptation layer and a multiplex layer. The adaptation layer provides only primitive services in H.324 where the channel is assumed to be nearly error-free, e.g., having a BER of about 10^{-6}. The unequal error protection (UEP) is applied to the bit-streams of the logical channels that enter the adaptation layer, depending on the nature of the data. The H.245 call control protocol, which is used to set up a video call, is interfaced with H.223 via the Control Channel Segmentation and Reassembly Layer (CCSRL) and the Numbered Simple Re-transmission Protocol (NSRP). CCSRL and NSRP are used to segment large control messages and re-transmit lost segments for reliable delivery. In H.324 a weaker protection by the Simple Re-transmission Protocol (SRP) is provided for control messages against transmission errors. The *intra-refreshing* refers to a series of actions between two media clients for recovering video quality. Table 6.17 summarizes the key parameters of 3G-324M, which runs on W-CDMA Release 99.

Table 6.17 System parameters of 3G-324M.

Speech codec	AMR
Video codec	H.263, MPEG-4
Call control protocol	H.245
Multiplexing protocol	H.223
Speech bit-rate (kbps)	12.2
Video bit-rate (kbps)	48
Video frame-rate (fps)	5–15
Video resolution	176 × 144
Total bit-rate (kbps)	64

When H.324 was upgraded to 3G-324M, MPEG-4 was included as an optional video codec, and AMR replaced G.723.1 [ITU-T (2006)] as the default speech codec. Although G.723.1 was designed with H.324 in mind, its low quality at 5.3 and 6.3 kbps, and its 30 ms frame length made it difficult to integrate the codec into typical radio signal processing procedures and limited its use in mobile communications systems. The adaptation and multiplex layers of H.223 have been made more robust, to compensate for the differences in the channels considered for H.324 and 3G-324M. H.324 with a new annex for enhanced error resilience, Annex C, was named H.324M, and finally became 3G-324M after additional customizations are made for the characteristics of W-CDMA.

In the adaptation layer of H.223, three methods for adding redundancy to the payload, AL1, AL2, and AL3, are defined to match the characteristics of the control signal, speech, and video, respectively [ITU-T (2001)]. In practice, AL2 was found to be appropriate also for video, since the higher error resilience of AL3 does not always result in higher video quality even with the higher overhead. In the multiplex layer of H.223, four methods for multiplexing the logical channels from the adaptation layer are defined. Level 0 was developed originally for H.324 over PSTN, and offers low-level error protection. The other higher levels, levels 1, 2, and 3, were developed for increasingly more error-prone channels. The higher levels enable higher protection but leave lower bit-rates for the payload. Level 2 is a good tradeoff that is typically employed in 3G-324M.

In typical 3G-324M operations, a 64 kbps bearer of W-CDMA transports a 1280-bit block every 20 ms, via two radio frames. This includes either one or no speech frame, a portion of a video frame, and the overhead bits from the H.223 and H.245 protocols. As in a voice call, negligible or no overhead is attached in the radio link control (RLC) and the media access control (MAC). In typical mobile-to-mobile video calls, the assembled H.223 multiplex (MUX) protocol data unit (PDU) from the multiplex layer is sent to the W-CDMA modem and transported over the uplink. The PDU passes the radio network control (RNC) and the mobile switching center (MSC), and enters the circuit-switched backbone networks connecting the networks of the service providers. Finally the PDU is transmitted over the downlink and disassembled at the far-end multiplex layer so that each payload of the logical channels is fed to the appropriate source decoder or call control entity.

A type of quality degradation that is found only in the multimedia communications, which can occur even when no data is lost, is the loss of synchronization between the speech and video frames, where the contents of the audio channel do not match those of the video channel. Compared with the speech frame rate of 50 fps, the frequency of generating video frames is lower, and also less regular. In addition, each video frame is typically fragmented into smaller chunks that are transported in multiple MUX PDUs. Therefore the speech frames almost always arrive before their corresponding video frames, even when both media types are digitized at the same time. As a result, the end-to-end delay constraints for 3G-324M are somewhat relaxed, e.g., 400–500 ms, compared with the strictly less than 280 ms of voice call. To help the media client synchronize the playback times, a delay can be inserted in the receive path for the speech. This delay is set by the sender with an H.245 message, H223SkewIndication. 3G-324M differs from other conversational services of mobile communications in that there is no capacity gain from voice activity detection. When voice activity is reduced and no speech frames are generated, or only the SIDs are transmitted, the unused capacity is filled up with the multiplexing overhead.

Multiplexing Protocol

Figure 6.50 shows the format of H.223 MUX-PDU, which is the output bit-stream of the H.223 multiplex protocol at level 0. The 8-bit header is followed by an information field, which is made up of an integer number of bytes containing the data from various logical channels. The source of the logical channels can be the speech codec, the video codec, or the H.245 control protocol. Note that in W-CDMA, the logical channel is a term also used to indicate the signaling and data path between the RLC and the MAC.

The header is made up of three fields: a three-bit Header Error Control (HEC), a four-bit Multiplex Control (MC), and a one-bit Packet Marker (PM). By referencing an entry from the MultiplexEntryDescriptor table that can have up to 15 entries, the MC indicates the structure of logical channels to which the bytes of the MUX-PDU information field belong. HEC is computed from the MC as a CRC with polynomial $x^3 + x + 1$, and the PM indicates the end of the segmented logical channels. Table 6.18 shows an example with eight entries, where it is shown that a MUX-PDU can transport the data and control information from multiple logical channels. Entry 0 is permanently assigned to the call control.

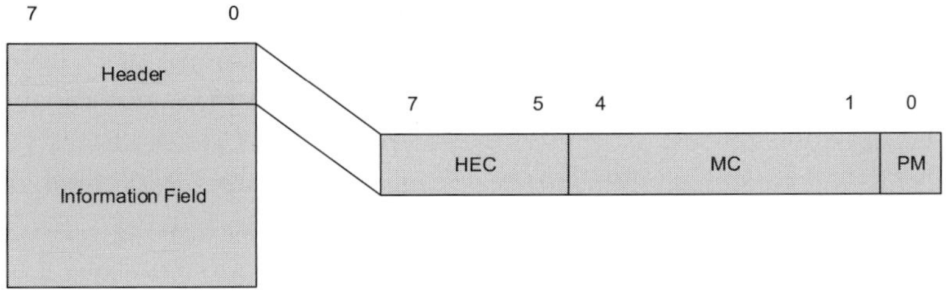

Fig. 6.50 H.223 MUX-PDU format for level 0.

Table 6.18 MultiplexEntryDescriptor examples.

Row	MultiplexEntryDescriptor	Element ListSize	Nesting Depth	Subelement ListSize	Application
1	{LCN1,RC UCF}	1	0	0	All audio
2	{LCN3,RC UCF}	1	0	0	All video
3	{LCN1,RC21},{LCN3,RC UCF}	2	0	0	Audio, all video
4	{{LCN2,RC1},{LCN3,RC3},RC UCF}	1	1	2	1:3 data video
5	{LCN1,RC4},{{LCN2,RC1},{LCN3,RC2}, RC UCF}	2	1	2	Audio, 1:2 data video
6	{LCN1,RC21},{{LCN2,RC2},{LCN3,RC6}, {LCN0,RC1},RC UCF}	2	1	3	Audio, 2:6:1 data video control
7	{LCN1,RC21},{LCN4,RC25},{{LCN2,RC1}, {LCN3,RC1},RC UCF}	3	1	2	Audio I, Audio II, 1:1 data video
8	{{LCN1,RC25},{{LCN2,RC1},{LCN3,RC1}, RC5}, RC UCF}	1	2	2	2-level nesting

In the second column of Table 6.18, the entry of MultiplexEntryDescriptor specifies the pattern of logical channel multiplexing for constructing the information field of the MUX-PDU. LCN0, LCN1, LCN2, LCN3, and LCN4 represent the logical channels for control, audio, data, video, and second audio, respectively. The Repeat Count (RC), which is associated with each logical channel, is used to represent the number of bytes from the logical channel. RC is either followed by the number of bytes or the Until Closing Flag (UCF). Therefore entries 1 and 2 represent the multiplexing methods where the information field is filled with audio and video bytes until the next flag. On the other hand, in entry 3, 21 bytes of audio are followed by video bytes until the next flag. Entry 4 includes two sub-entries such that a chunk of data consisting of one data byte and three video bytes is repeated until the next flag, i.e., the end of the information field. The following entries represent more complex multiplexing methods that include multi-level nested descriptions. The MultiplexEntryDescriptor of entry 0 is {LCN0,RC UCF}, which signifies multiplexing call control bytes until the next flag. In the call setup stage, entries of the multiplex table for the send direction are proposed by a media client to the far-end client using the H.245 MultiplexEntrySend message, which is usually accepted. The entries are exchanged for each direction independently.

Figure 6.51 illustrates the construction of an H.223 information field with three bytes from LCN1, one byte from LCN2, two bytes from LCN3, two bytes from LCN2, and finally one byte from LCN3. One entry of MultiplexEntryDescriptor that can describe this construction process is {LCN1,RC3},{LCN2,RC1},{LCN3,RC2},{LCN2,RC2}, {LCN3,RC1}. Note that other entry types can generate the same format, depending on the MultiplexEntryDescriptor tables and logical channels. In level 0, two MUX-PDUs are separated by a flag, 01111110, which is grammatically guaranteed not to appear inside a MUX-PDU since a 0 is inserted after five consecutive ones are observed in the bit-streams of logical channels.

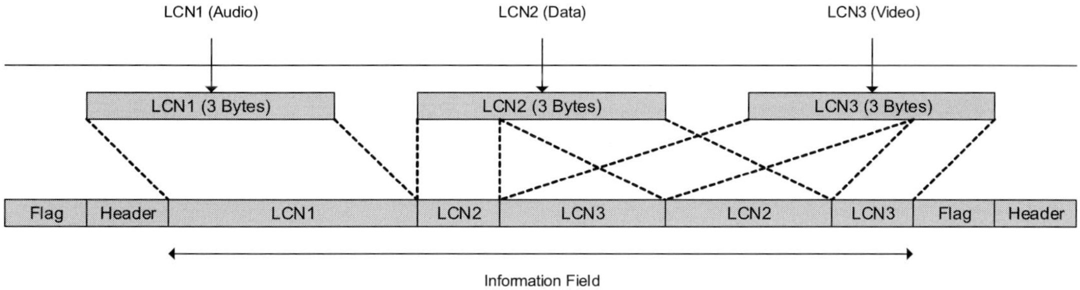

Fig. 6.51 H.223 information field construction.

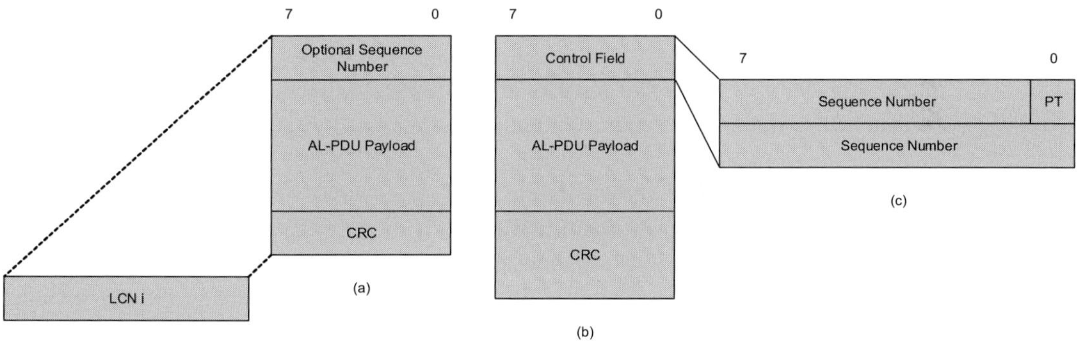

Fig. 6.52 AL-PDU formats. (a) AL2. (b) AL3. (c) Control field of AL3.

Before the multiplexing process, the bytes from each logical channel undergo the adaptation process, which depends on the nature of the payload but not on the level of the multiplex layer until level 2. Figure 6.52 shows two types of Adaptation Layer (AL) PDU, Adaptation Layer Type 2 (AL2) and Type 3 (AL3), which are the output bit-streams of the adaptation layer. AL2 and AL3 were originally defined with speech and video in mind but AL2 is often used for video. The re-transmission scheme supported by AL3 leaves insufficient time for transmission, and also complicates the synchronization with AL2 carrying real-time media such as speech. AL1 includes neither a sequence number nor a CRC.

In AL2, an eight-bit CRC is computed from the input payload of the adaptation layer, AL-SDU, and attached for error detection. In addition, an eight-bit sequence number can be attached before the AL-SDU to represent the order, although in circuit-switched networks, it is very unlikely that AL-PDUs will arrive in a different order than transmitted. Therefore the number does not add significant error resilience. AL3 supports a higher capability for error detection, using a 16-bit CRC, 1110000101001101, which is statistically more difficult to emulate in the presence of bit errors than the eight-bit CRC of AL2. There can be a zero-, one-, or two-byte control field in front of the AL-SDU of AL3. When the control field is present, at least a one-bit Payload Type (PT) is present in the field. When PT is set to 0, the AL-SDU includes the control information used for re-transmission. Otherwise, AL-SDU includes the media or data payload. When a two-byte control field is used, a 15-bit sequence number is used, which can be essential in the re-ordering of retransmitted payloads.

A combination of a level 0 multiplex layer and an adaptation layer provides barely enough error resilience for benign channels such as PSTN, whose BER is about 10^{-6}. Such a low error rate comes mainly from thermal noise. It is unlikely that level 0 MUX-PDUs would survive the transmission over a wireless channel without a loss of information. Even with a low error rate, it is not theoretically impossible that the eight-bit flag 01111110 is emulated by bit errors and the de-multiplexer would then fail to detect the boundary of the MUX-PDUs. Level 1 of the multiplex layer is defined for higher error resilience than level 0 but the only difference between levels 0 and 1 is that level 1 uses a longer, 16-bit flag which is almost impossible to be emulated by bit errors. Since an extended flag, without reinforcing the structure of MUX-PDU, is not enough to cope with the challenges of a wireless channel, a higher level, level 2, was introduced. Figure 6.53 shows the formats of MUX-PDU at level 2, which include a basic format with a three-byte header field, and an optional format with a four-byte header field.

The three-byte header shown in Fig. 6.54(a) consists of a four-bit MC field, an eight-bit Multiplex Payload Length (MPL) field, and a twelve-bit Parity (P) field. MPL indicates the length of the information field, 0–254, in bytes, leaving 255 reserved for another use. The P field is the parity in the codeword of a systematic Extended Golay (24,12) code whose twelve input bits are the MC and MPL fields. By the definition of this block code, the minimum Hamming distance between two codewords is eight. Therefore any patterns of bit errors on the header up to four bits can be corrected.

The four-byte header consists of a three-byte header of level 2, which is followed by a one-byte header of the previous MUX-PDU in the format of level 0. If a four-byte header is used, the de-multiplexer does not throw away a MUX-PDU whose header is corrupt immediately but waits for the next one. If the next MUX-PDU includes a valid eight-bit header for the previous MUX-PDU whose processing is held, at least the information field may be salvaged and sent to the respective logical channels. Even if

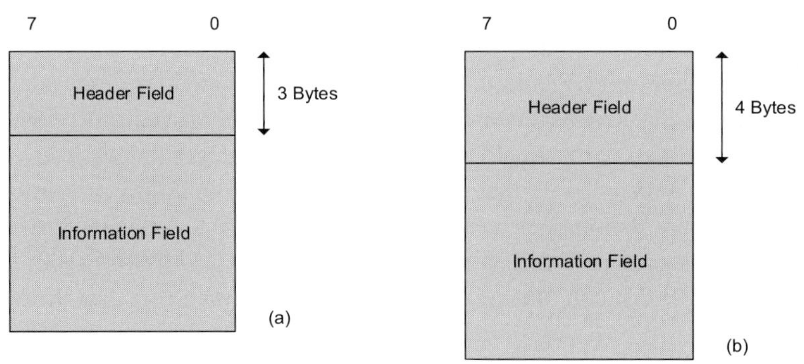

Fig. 6.53 (a) H.223 MUX-PDU format for level 2. (b) Optional H.223 MUX-PDU format for level 2.

Table 6.19 Error resilience capability of H.223.

Level	Multiplex layer	Adaptation layer	Target channel type
0	1-byte flag, 1-byte header	1-byte sequence number, 1-byte CRC (AL2), 15-bit sequence number, 2-byte CRC (AL3)	Almost error-free
1	Same as level 0, 2-byte flag	Same as level 0	Low error-prone
2	3-, 4-byte header	Same as level 0	Moderate error-prone

(a)

7							0
MPL4	MPL3	MPL2	MPL1	MC4	MC3	MC2	MC1
P4	P3	P2	P1	MPL8	MPL7	MPL6	MPL5
P12	P11	P10	P9	P8	P7	P6	P5

(b)

7							0
MPL4	MPL3	MPL2	MPL1	MC4	MC3	MC2	MC1
P4	P3	P2	P1	MPL8	MPL7	MPL6	MPL5
P12	P11	P10	P9	P8	P7	P6	P5
HEC3	HEC2	HEC1	MC4	MC3	MC2	MC1	PM

Fig. 6.54 Headers of H.223 MUX-PDU formats for level 2. (a) 3-byte header. (b) 4-byte header.

the information field includes bit errors, it is still better than no reception and the impact on media quality may be reduced by error concealment.

The same two-byte flag used in level 1 is used but unlike level 1, the flag cannot be transmitted consecutively. If there is no MUX-PDU to send, the transmitter combines a flag with a three- or four-byte header field to indicate the level currently used, which is called the *stuffing mode* in H.324 terminology. In the header of a stuffed MUX-PDU, the MPL and MC fields are set to 00000000 and 0000 respectively. Therefore, in 3G-324M, the cost for higher error resilience, by elevating the level of the multiplex layer, is a corresponding reduction of the bit-rate that can otherwise be used for media and control, when the total bit-rate is fixed. Higher levels also require larger processing delays.

There is a fourth level that provides even higher error resilience. Level 3 uses the same multiplex layer as level 2 but its adaptation layer provides the capability of retransmitting lost AL-PDUs and acknowledging whether an AL-PDU has been successfully received or not. However, considering the round-trip and processing delays, re-transmission of a lost AL-PDU is not considered useful for real-time conversational services. Therefore, although level 3 was introduced into H.324M, in 3G-324M, AL2 at level 2 is the typical configuration for both speech and video. Table 6.19 summarizes the key features for error resilience at each level of H.223.

Error Control

From error-resilient source coding in the video codec to turbo coding in the PHY, there are schemes built into many layers of the 3G-324M protocol stack for error control. As 3G-324M typically operates at the transparent or unacknowledged mode of the RLC, if the MUX-PDUs are lost either in the uplink or downlink, re-transmission of lost data by the RLC is not used, because of its large delay, although the re-transmission of acknowledged mode by the RLC will be at least faster than the end-to-end re-transmission of AL3.

As unequal error detection (UED) and protection (UEP) are applied in the speech and radio signal processing of voice calls in GSM and W-CDMA, at the adaptation layer of 3G-234M, they are applied to speech, video, and control information in different levels with the AL-PDU formats. Some control signals, which are less sensitive to delay but should be delivered to the far-end media client absolutely error-free, are acknowledged and retransmitted if not received properly. From the eight-bit CRC and the sequential number, it is possible to confirm the integrity of the payload and identify AL-PDUs lost during transmission. When the payload is speech or video, error concealment at the source decoder can be invoked to reduce the audiovisual impact of data loss. The error concealment techniques range from the simple replacement of a corrupt frame by a previous uncorrupted one to sophisticated algorithms that estimate the lost data using some form of interpolation. Using error concealment, higher quality can be obtained without increasing the bit-rate, but this is at the cost of additional decoder delay and complexity.

The effectiveness of these schemes is influenced by the pattern of errors imposed on the bit-stream. For example, it might be difficult to use the techniques designed to detect bit-level errors in handling packet-level errors, where data packets containing hundreds or thousands of bits are completely lost. In some cases, error concealment is not sufficient to preserve quality. The transmission of motion-estimated and compensated video requires remedial actions also in the video encoder to limit ongoing error propagation.

While the loss of an encoded speech frame is hardly perceivable, the loss of an I- or P-frame influences all succeeding P-frames until a new and unharmed I-frame arrives at the video decoder. As a remedy for this structural problem, 3G-324M has inherited the videoFastUpdatePicture message of H.245 from H.324, which asks the far-end video encoder to encode an I-frame at its earliest opportunity [ITU-T (2003)]. As can be seen in the rate control of Figures 6.43 and 6.44, to encode an I-frame, multiple P-frames need to be sacrificed, i.e., not encoded to make enough space and time for the I-frame in the transmission buffer. The size of the videoFastUpdatePicture message is small enough not to reduce the bit-rate for video or speech noticeably.

Figure 6.55 illustrates the *intra-refreshing* process in which the adaptation layer at the receiver identifies an error in the MUX-PDUs for P-frame l but decides that the error is manageable and it is not necessary yet to inform the sender. However, in the MUX-PDUs for P-frame m, the adaptation layer finds another error that can reduce the quality significantly. Therefore the receiver transmits a videoFastUpdatePicture message

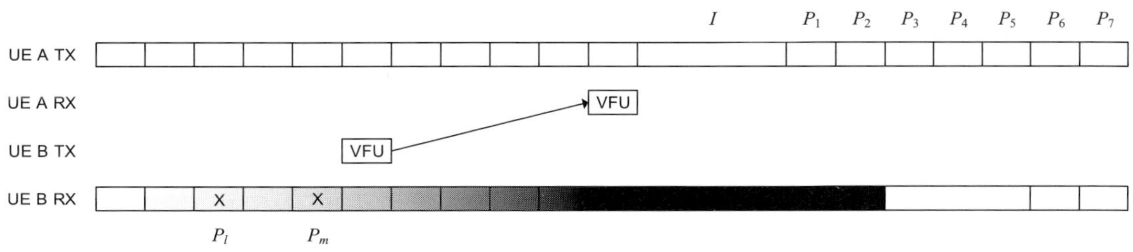

Fig. 6.55 Recovery of video quality with intra-refreshing.

to the far-end, which is accepted by the sender. In Fig. 6.55, the color darkening is a graphic visualization of the image quality loss induced by the propagation of errors to the following P-frames and the rapid decrease of PSNR. As the color of frames reaches complete black, it is difficult to recognize the contents of video once the PSNR falls significantly, as was seen in Figures 6.40 and 6.41. Eventually the error propagation and damaged images are stopped by the I-frame the sender transmits immediately after accepting the videoFastUpdatePicture message. The I-frame needs to be received without any error to have such cleaning effects. H.245 also provides similar commands, such as the videoFastUpdateGOB and videoFastUpdateMB messages, which ask for intra-refreshing of areas smaller than a frame. However, they are considered insufficient against the pattern of data loss in 3G-324M, where corrupt MUX-PDUs are not delivered since the RLC typically drops corrupted PDUs rather than passes them with error marks. With a fixed total bit-rate, neither the codec mode control of speech nor the rate control of video helps when the channel condition deteriorates.

6.5.3 Radio Signal Processing

At the end of the multiplex layer, the constructed MUX-PDUs leave the media client and enter the W-CDMA modem, where they are fed into the RLC for the uplink radio signal processing. With the complex shuffling of media and control data at the adaptation and multiplex layers, it is not possible to classify the incoming bit-streams and apply UED and UEP. Therefore, the MUX-PDUs are entered into a single radio access bearer and processed as a single bit-stream until transmission over the wireless channel. However, information on the QoS required for 3G-324M is transferred to the modem to configure each layer of the protocol stack for the send and receive directions.

Table 6.20 outlines the set of parameters for the RLC, MAC, and PHY where a radio access bearer for the MUX-PDU and a signalling radio bearer for the control signal are set up. At the RLC, transparent mode is typically used for 3G-324M but unacknowledged mode is also used as its header overhead is not significant. As the sizes of transport blocks for 3G-324M are larger than those for the voice call, and a constant bit-rate has to be maintained throughout the service, turbo code is almost always used. Because of the delay constraints, the maximum number of iterations in the turbo decoder needs to be set at an acceptable value. In the uplink, the receiver at the NodeB

Table 6.20 W-CDMA protocol parameters for 3G-324M.

Protocol	Parameter	RAB	SRB
RLC	Mode	TM	AM
	Payload size (bits)	640	128
	Maximum bit-rate (bps)	64000	3200
	Header size (bits)	0	16
MAC	Multiplexing	N/A	
	Header size (bits)	0	4
PHY	TB size (bits)	640	148
	TTI (ms)	20	40
	CRC size (bits)	16	16
	Coding type, rate	Turbo, $\frac{1}{3}$	Convol., $\frac{1}{3}$
	Max bits / TTI (after channel coding)	3948	516
	Max bits / frame (before rate matching)	1974	
	RM attribute	150–195	

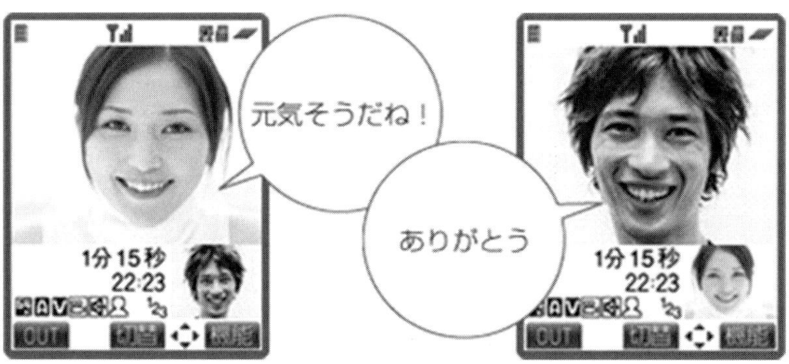

Fig. 6.56 User interfaces of 3G-324M (courtesy of NTT DOCOMO, Inc.).

decodes the received bit-stream and sequentially removes the redundancy the transmitter attached at the UE. Then the reconstructed MUX-PDUs are sent via a 64 kbps channel of the circuit-switched backbone network to the far-end, which might be another UE supporting 3G-324M or other devices types such as a computer. Regardless of the channel conditions, parameters of the W-CDMA modem protocol, outlined in Table 6.20, are maintained throughout the service, leaving the transmit power to be the only controllable parameter available to adapt to time-varying channel conditions.

3G-324M requires several features that need not be considered in the voice call. Figure 6.56 shows the User Interfaces (UI) of the UEs engaging in 3G-324M conversations. On the display of the UE, a rectangular area for the received video is prepared, which does not necessarily have to be 176 × 144 as the decoded video is often resized for a different image size or for a different aspect ratio. In addition, a smaller rectangular area, Picture-In-Picture (PIP), is necessary for the video currently being captured and encoded to help the user adjust the direction and distance between her/himself and the camera.

6.6 References

3GPP. 2001. TR 26.912 V4.0.0 Codec for Circuit Switched Multimedia Telephony Service; Quantitative Performance Evaluation of H.324 Annex C over 3G. April.

3GPP. 2016a. TS 25.101 V14.0.0 User Equipment (UE) Radio Transmission and Reception (FDD). July.

3GPP. 2016b. TS 25.321 V14.0.0 Medium Access Control (MAC) Protocol Specification. December.

3GPP. 2016c. TS 26.102 V14.0.0 Mandatory Speech Codec; Adaptive Multi-Rate (AMR) Speech Codec; Interface to Iu, Uu and Nb. September.

3GPP. 2017a. TR 26.911 V14.0.0 Codec(s) for Circuit Switched Multimedia Telephony Service; Terminal Implementor's Guide. March.

3GPP. 2017b. TS 23.153 V14.0.0 Out of Band Transcoder Control; Stage 2. March.

3GPP. 2017c. TS 25.212 V14.0.0 Multiplexing and Channel Coding (FDD). January.

3GPP. 2017d. TS 25.213 V14.0.0 Spreading and Modulation (FDD). March.

3GPP. 2017e. TS 25.214 V14.1.0 Physical Layer Procedures (FDD). March.

3GPP. 2017f. TS 25.302 V14.0.0 Services Provided by the Physical Layer. March.

3GPP. 2017g. TS 25.322 V14.0.0 Radio Link Control (RLC) Protocol Specification. March.

3GPP. 2017h. TS 25.323 V14.0.0 Packet Data Convergence Protocol (PDCP) Specification. March.

3GPP. 2017i. TS 25.324 V14.0.0 Broadcast/Multicast Control (BMC). March.

3GPP. 2017j. TS 25.331 V14.2.0 Radio Resource Control (RRC) Protocol Specification. March.

3GPP. 2017k. TS 26.110 V14.0.0 Codec for Circuit Switched Multimedia Telephony Service; General Description. March.

3GPP. 2017l. TS 26.111 V14.0.0 Codec for Circuit Switched Multimedia Telephony Service; Modifications to H.324. March.

3GPP. 2017m. TS 26.190 V14.0.0 Speech Codec Speech Processing Functions; Adaptive Multi-Rate - Wideband (AMR-WB) Speech Codec; Transcoding Functions. March.

3GPP. 2017n. TS 28.062 V14.0.0 Inband Tandem Free Operation (TFO) of Speech Codecs; Service Description; Stage 3. March.

3GPP2. 2007. C.R1008-0 V1.0 cdma2000 Multimedia Services Evaluation Methodology. February.

ARIB. 2011. RCR SDT-28 V6.0 Personal Handy Phone System. March.

Berns, R. S. 2000. *Billmeyer and Saltzman's Principles of Color Technology*. 3rd edn. Wiley-Interscience.

Berrou, C., Glavieux, A., and Thitimajshima, P. 1993. Near Shannon Limit Error-Correcting Coding and Decoding: Turbo Codes. *IEEE International Conference on Communications*, May.

Bormann, C., Burmeister, C., Degermark, M., *et al.* 2001. IETF RFC 3095 RObust Header Compression (ROHC): Framework and Four Profiles: RTP, UDP, ESP, and Uncompressed. July.

Ding, W., and Liu, B. 1996. Rate Control of MPEG Video Coding and Recording by Rate-Quantization Modeling. *IEEE Transactions on Circuits and Systems for Video Technology*, **6**(February).

Dudgeon, D. E., and Mersereau, R. M. 1983. *Multidimensional Digital Signal Processing*. Prentice Hall.

Glisic, S. G. 2003. *Adaptive WCDMA: Theory and Practice*. Wiley.

Goyal, V. K., Kovacevic, J., Arean, R., and Vetterli, M. 1998. Multiple Description Transform Coding of Images. *International Conference on Image Processing*, **1**(October).

Holma, H., and Toskala, A. 2006. *HSDPA/HSUPA for UMTS: High Speed Radio Access for Mobile Communications*. Wiley.

Holma, H., and Toskala, A. 2010. *WCDMA for UMTS: HSPA Evolution and LTE*. 5th edn. Wiley.

ISO. 2001. MPEG-4 Information Technology – Coding of Audio-Visual Objects – Part 2: Visual. December.

ISO/IEC. 1994. Open Systems Interconnection – Basic Reference Model: The Basic Model. November.

ITU-R. 2002. Recommendation BT.709-5: Parameter Values for the HDTV Standards for Production and International Programme Exchange. April.

ITU-R. 2014. M.1457 Detailed Specifications of the Terrestrial Radio Interfaces of IMT-2000 (IMT-2000). November.

ITU-T. 2001. H.223 Multiplexing Protocol for Low Bit Rate Multimedia Communication. July.

ITU-T. 2002. H.324 Terminal for Low Bit-Rate Multimedia Communication. March.

ITU-T. 2003. H.245 Control Protocol for Multimedia Communication. July.

ITU-T. 2005. H.263 Video Coding for Low Bit Rate Communication. January.

ITU-T. 2006. G.723.1 Dual Rate Speech Coder for Multimedia Communications Transmitting at 5.3 and 6.3 kbit/s. May.

Karlsson, M., Almgren, M., Bruhn, S., Larsson, K., and Sundelin, M. 2002. Joint Capacity and Quality Evaluation for AMR Telephony Speech in WCDMA Systems. *IEEE Vehicular Technology Conference*, September.

Kuroda, T. 2014. *Essential Principles of Image Sensors*. CRC Press.

Lee, J., and Dickinson, B. W. 1997. Rate-Distortion Optimized Frame Type Selection for MPEG Encoding. *IEEE Transactions on Circuits and Systems for Video Technology*, **7**(June).

Lim, J. S. 1990. *Two-Dimensional Signal and Image Processing*. Prentice Hall.

Nakamura, J. (ed). 2005. *Image Sensors and Signal Processing for Digital Still Cameras*. CRC Press.

Ortega, A., and Ramchandran, K. 1998. Rate-Distortion Methods for Image and Video Compression. *IEEE Signal Processing Magazine*, **15**(November).

Paulraj, A., Nabar, R., and Gore, D. 2008. *Introduction to Space-Time Wireless Communications*. Cambridge University Press.

Rao, K. R., and Hwang, J. J. 1996. *Techniques and Standards for Image, Video, and Audio Coding*. Prentice Hall.

Richardson, I. E. G. 2003. *H.264 and MPEG-4 Video Compression: Video Coding for Next-generation Multimedia*. Wiley.

Said, A., and Pearlman, W. 1996. A New, Fast, and Efficient Image Codec Based on Set Partitioning in Hierarchical Trees. *IEEE Transactions on Circuits and Systems for Video Technology*, **6**(3).

Tachikawa, K. 2002. *W-CDMA Mobile Communications System*. Wiley.

Takishima, Y., Wada, M., and Murakami, H. 1995. Reversible Variable Length Codes. *IEEE Transactions on Communications.*, **43**(February).

Talluri, R. 1998. Error-Resilient Video Coding in the ISO MPEG-4 Standard. *IEEE Communications Magazine*, June.

Tanner, R., and Woodard, J. 2004. *WCDMA: Requirements and Practical Design*. Wiley.

Tsujimura, T. 2012. *OLED Display Fundamentals and Applications*. Wiley.

Vanghi, V., Damnjanovic, A., and Vojcic, B. 2004. *The cdma2000 System for Mobile Communications: 3G Wireless Evolution*. Prentice Hall.

Verdú, S. 1998. *Multiuser Detection*. Cambridge University Press.

Wyszecki, G., and Stiles, W. S. 2000. *Color Science: Concepts and Methods, Quantitative Data and Formulae*. 2nd edn. Wiley-Interscience.

Yang, X., and Ramchandran, K. 1998. Optimal Multiple Description Subband Coding. *International Conference on Image Processing*, **1**(October).

7 Evolution of W-CDMA Systems

Although W-CDMA achieved an unrivaled success in 3G mobile communications systems, introducing mobile video telephony as an affordable service, efforts to improve its performance while maintaining the key system structure continued. The first mobile communications system to offer multimedia telephony, 3G-324M, revealed new opportunities to improve media quality and network capacity. This chapter discusses how the radio signal processing of W-CDMA was systematically improved by using lessons learned from network operations. Then it introduces a new type of quality loss inherent only in multimedia communications, long call setup delay, and the measures taken to address it. Finally, the performance of W-CDMA and 3G-324M is evaluated with experimental and field results.

7.1 Enhancements in Wireless Transmission

The evolution of GSM and cdma2000, which involved the introduction of new speech codecs and the enhancement of radio signal processing operations, was driven by the rapid expansion of mobile communications at a time when the major application was voice service. After the launch of W-CDMA, data access became one of the dominant services and even exceeded video telephony. This shifted the direction of network evolution toward packet-switched mobile communications systems, such as High Speed Packet Access (HSPA) and Long Term Evolution (LTE). Therefore the speech and radio signal processing algorithms of W-CDMA did not receive the level of attention that had been applied to GSM and cdma2000 for several releases.

Release 12 made significant efforts to improve the circuit-switched voice services of W-CDMA. A set of techniques from these efforts, which included many of the approaches that had been introduced in the Rev. E of cdma2000, became known as W-CDMA Plus (W-CDMA+). Like Rev. E, WCDMA+ can increase the network capacity for voice services significantly without incurring excessive cost. It generates the gains by avoiding redundant transmissions and optimizing signaling procedures. Since W-CDMA+ can co-exist in the same channel with HSPA, it can reduce the capacity gap between W-CDMA and LTE, especially when the large chunks of bandwidth required to maximize the performance of LTE are not available. WCDMA+ can also release more radio resources for data services.

Table 7.1 Downlink pilot-free DPDCH/DPCCH formats.

Slot format	Bit-rate (kbps)	Symbol-rate (ksps)	Spreading factor	Bits per frame DPDCH	DPCCH	Bits per slot	Data1	Data2	TPC	TFCI	Pilot
17	30	15	256	270	30	20	18	0	2	0	0
18	60	30	128	570	30	40	38	0	2	0	0

7.1.1 Pilot-Free Slot Format

Comparing the fundamental channels of cdma2000 and the physical channels of W-CDMA shows that both systems use a pilot channel dedicated to each MS or UE in the reverse link and the uplink to assist the synchronization at the receiver. In W-CDMA, a dedicated pilot channel, embedded in DPCCH, is also used in the downlink. If the signal strength of the shared pilot channel is high enough for a UE, the pilot channel dedicated to the UE can be removed without reducing the performance of the receiver significantly. This can reduce the interference and increase network capacity. In addition, the bits freed from of the pilot field in DPCCH can be recycled for data transmission.

In the pseudo-flexible rate matching of Release 99 where the TFCI is not transmitted, bit-positions of the DPDCH are allocated to the transport channels based on the maximum transport block sizes. This can simplify the channel decoding but wastes the radio resource for signaling radio bearer (SRB) if there is no control information to transmit. In W-CDMA+, bit-positions of the DPDCH are not assigned to the SRB if the SRB does not carry any data. If the rate matching attribute of the DPCCH is set to 0, bits for the DPDCH can be transmitted in the bit locations for DPCCH. Table 7.1 shows two downlink slot formats designed for the pilot-free transmission of W-CDMA+. Compared with slot formats 2 and 9 of Table 6.12, slot formats 17 and 18 can carry 30 and 90 additional bits in the DPDCH, which can be used for higher codec modes.

7.1.2 SRB Power Boost

In the power control of the downlink, transmit powers of the DPDCH and DPCCH are adjusted by the same amount upon the reception of a TPC bit, so that the difference between the powers of the two physical channels, *power offset*, is maintained. The transmit powers of the TFCI, TPC, and Pilot fields of the DPCCH are offset relative to the transmit power of the DPDCH by PO1, PO2, and PO3 dB, respectively. When the SRBs are transmitted in the data, the channel coding rate has to be increased to make room for the additional bits, leaving the real data for the encoded speech or applications less protected. Therefore in W-CDMA+, to compensate for the reduction of error resilience from transmitting the SRBs with data, the transmit power of the DPDCH is increased by PO_{SRB}.

Figure 7.1 illustrates the power control procedures in which the transmit powers of DPDCH and DPCCH in slot N are reduced by the same amount, following the

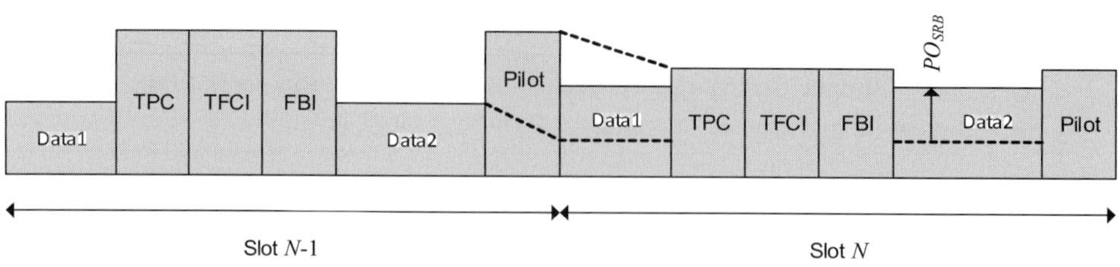

Fig. 7.1 Downlink power boost.

information in the TPC field of slot $N - 1$. However, the transmit power of the DPDCH is increased by PO_{SRB} to compensate for the SRB bits. It is not necessary that the TPC, TFCI, FBI, and Pilot fields be transmitted at the same power, i.e., PO1 = PO2 = PO3, although it is drawn as such in the figure. Note that in cdma2000 where the bit-rate of the variable bit-rate (VBR) speech codec can be reduced to make temporary room for data, it is unnecessary to boost the transmit power, as the bits of the encoded speech and control data are treated together, with the same channel coding rate and CRC.

7.1.3 Compressed DPDCH

Since W-CDMA Release 99 follows many of the design practices of the TDMA and CDMA circuit-switched mobile communications systems, the combined length of the radio frames transporting each speech frame is identical to the length of the time period the speech frame spans. Therefore a TTI of 20 ms is used to transmit a speech frame encoded in AMR. However, since the flexible radio signal processing procedures of W-CDMA enable the transmission of multiple transport channels with different QoS or TTI values, untying the two periods when it is possible or necessary may provide another opportunity to increase the spectral efficiency.

As can be seen in the protocol parameters for the SRB, shown in Table 6.8, a larger TTI can reduce the bit-rate but provide a higher level of error resilience. This can reduce the interference and increase network capacity. On the other hand, a smaller TTI can increase the time during which the transmitter (DTX) or the receiver (DRX) can be turned off, which can also reduce the interference and increase network capacity. However, just as the compressed mode requires increasing the transmit power, a smaller TTI also requires increasing the transmit power. Therefore the TTI needs to be carefully selected taking such conflicting factors into account. Notice that in GSM, the use of the half-rate channel, e.g., TCH/HS or TCH/AHS, does not reduce the time period over which the bits are transmitted. As a result, the time-domain diversity is not compromised when the half-rate channel is used. Instead, only the number of time slots is halved. Figure 7.2(a) illustrates the frame construction process of W-CDMA for the uplink when the speech is encoded at AMR 12.2 kbps.

If the DPDCH is compressed in W-CDMA+, a 10 ms radio frame can be used in the uplink to transport a 20 ms speech frame. During the period corresponding to the second radio frame, the transmitter can be turned off as illustrated in Fig. 7.2(b). To compact the bits of a speech frame into a half of the space, the transmit power needs to be increased

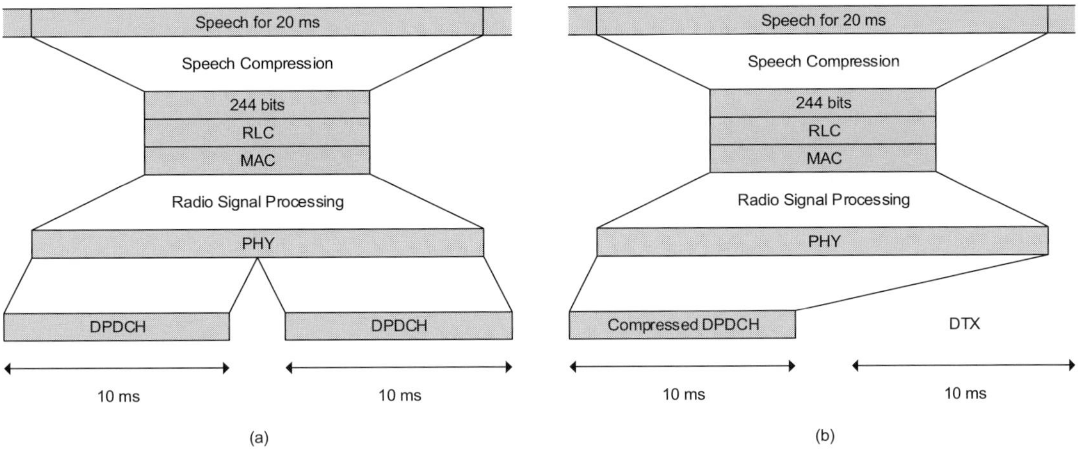

Fig. 7.2 Uplink frame length switching. (a) 20 ms mode. (b) 10 ms mode.

to maintain a similar value of SNR per bit. In addition, a smaller spreading factor needs to be used to accommodate the additional bits. Since increasing the transmit power may not be possible in some situations, for example, when the UE is located near the cell edge and little power headroom remains in the amplifier of the transmitter, it may be necessary for the UE to switch from a compressed to a normal DPDCH.

The decision for switching the DPDCH type has to be made in the UE to track the rapid variation of uplink channel conditions. For each TTI, the UE estimates the transmit power necessary for a 10 ms period, and enters the compressed DPDCH mode if the required increase of power is within a tolerable level. The UE informs the NodeB of the DPDCH type used with the TFCI. When the compressed DPDCH is configured, the TFCI information bits correspond to the unsigned binary representation of the sum of TFC index and the size of the TFCS. Therefore, protocol layers above the PHY are not aware of the switching. It is not possible to assign a TTI smaller than the fixed 10 ms length of a radio frame. However, modifying the signaling of W-CDMA such that the channel type can be controlled at the subframe or slot level might not necessarily improve the network capacity, considering that key parameters such as the modulation scheme, channel coding rate, and the channel bandwidth are still fixed.

7.1.4 Frame Early Termination

The frame early termination (FET) of the W-CDMA+ downlink operates under very similar principles to those of the technique with the same name used in the forward fundamental channel of cdma2000 Rev. E. FET enables the NodeB to halt transmission in the middle of sending a speech frame when the UE reports that the frame was successfully decoded before all of the slots are transmitted. The downlink transmission can be terminated earlier if a subset of transport blocks received contains sufficient bits for the speech decoder to reconstruct the frame. This would be feasible if the remaining transport blocks correspond to the redundant bits generated from the repetition and

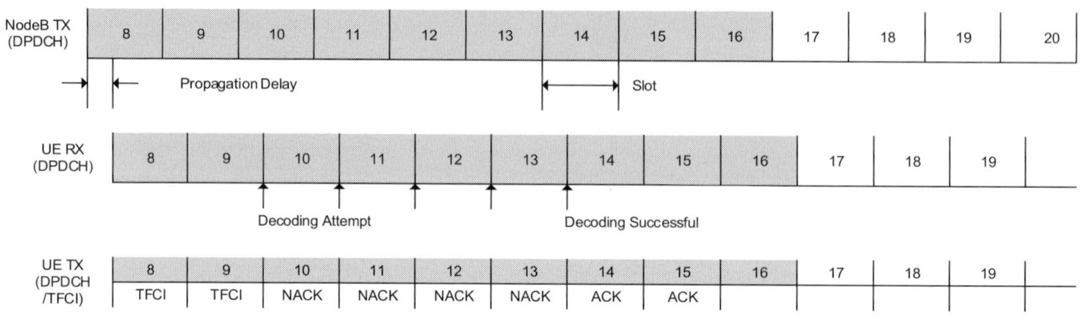

Fig. 7.3 Downlink FET operation.

puncturing of rate matching. Whether the transmission can be terminated is checked for every slot received but the result is sent to the NodeB via the ACK/NACK indicator bits only when certain conditions are met. For example, it is very likely that the speech decoder fails to reconstruct the frame with the slots initially received. Once the FET is declared for a speech frame, the transmitters in the UE and the NodeB can be turned off, which would reduce the interference. The same principle can be applied to the uplink transmission.

When the downlink FET is configured, slot format 5 of DPCCH is used for the uplink, where two bits per slot are available for TFCI. Figure 7.3 illustrates the interaction between the UE and the NodeB in the downlink FET operation. In each 20 ms TTI, the first ten slots of the uplink DPCCH include two bits for the TFCI, which is replaced with the repeated ACK/NACK bit. It is assumed that in the first ten slots, enough bits of the speech frame are included, albeit in channel-coded form. Note that the convolutional encoders of W-CDMA do not generate systematic bits. The remaining 20 slots can contain downlink FET ACK/NACK indicator bits unless the slots are in a transmission gap of the compressed mode. The FET ACK/NACK indicator bits can only be transmitted in a pair, i.e., over two consecutive slots, to reduce the probability of mistakenly concluding a successful decoding.

7.2 Enhancements in Media Negotiation

7.2.1 Media Configuration Delay

The commercial deployments of the first mobile video telephony system revealed some unsatisfactory aspects of media quality that were not anticipated during the design stages of 3G-324M. The response of the general public to the new service, which was billed as a premium, was not anticipated by the system designers. In fact, the transmission of highly compressed video bit-streams turned out to be acceptable, as the 1500 Hz power control of W-CDMA, with a lower block error rate (BLER) target than required for speech, keeps the channel quality good enough. There were few issues on the transmission side of 3G-324M while on the compression side, the low quality issue from encoding video

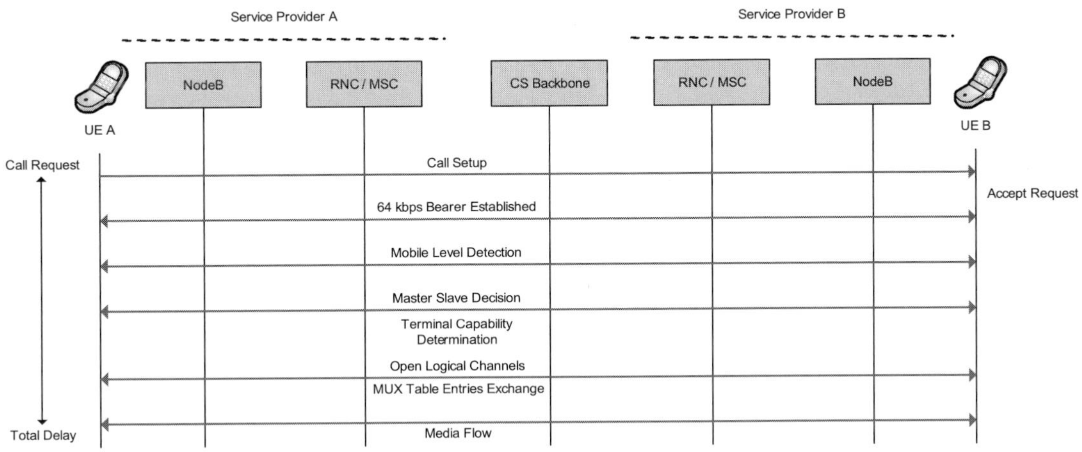

Fig. 7.4 Call setup procedures of 3G-324M.

at 48 kbps was inevitable. However, an aspect of the overall quality that was noticeably worse than for the conventional voice call was in the delay measured from the moment the video call is placed by a UE to the moment the video of the far-end appears on the display of the UE, i.e., the call setup delay. In 3G-324M, the delay was significantly longer than the delay measured from the moment the voice call is placed by a UE to the moment the voice of the far-end is heard from the speaker of the UE, assuming that the networks are not overloaded and that the call requests are not rejected by the admission control. It is also assumed that the far-end accepts the requests immediately. Figure 7.4 shows the call setup procedures of 3G-324M.

In the analysis of the signal processing procedures of 3G-324M, it was shown that the media client and the W-CDMA modem do not interact with each other, at least with officially standardized control messages, which is in contrast to the speech and radio signal processing operations where immediate measures are taken if the channel quality drops abruptly or a handover occurs. Since the total bit-rate as well as the bit-rate of each component medium is fixed in 3G-324M, there is nothing the W-CDMA modem needs to ask of the media client, or vice versa, during a video call. More importantly, they do not share the control messages to set a video call up. Once the user of UE A pushes a SEND button or an icon on the UI to initiate 3G-324M, control messages are sent to UE B to establish two-way 64 kbps paths, i.e., bearers, between the UEs. During this stage of the signaling process, the messages inform many network nodes en route of the needs for preparing the service, such as the resource reservation but no information on the detailed configuration of the media codecs is exchanged. If a voice call is initiated, then by the time the first answer from UE B arrives at UE A, most information required to initiate the voice call, including the codec type and bit-rates, are exchanged. It is also assumed that the user of UE B pushed an ANSWER button or an icon to confirm the acceptance of the 3G-324M request. The first two lines of Fig. 7.4 show the message exchange of this stage.

In the next stage, the parameters of 3G-324M are negotiated between two media clients under the control of H.245. To agree upon the level of the H.223 multiplex layer between two media clients, each media client transmits a series of flags and also inspects the received flags to compare the levels for the two directions. The transmission of flags starts at the highest level each media client supports, which would typically be level 2 in 3G-324M. Note that the media client can differentiate level 2 from level 1, which uses the same two-byte flag, with the header for the stuffing mode where the flag cannot be transmitted consecutively. In the generic call setup procedures of H.324, the media client reduces the level by one if the flag it is receiving is in a level lower than the transmitted one. In 3G-324M, the level detection process typically stops at level 2, when the two media clients learn that they are transmitting flags in the same level, after confirming at least five consecutive flags of the same kind.

Although the bit-stream structure of 3G-324 is reinforced for use in harsher mobile environments, the elevated error resilience is not fully usable in the interworking situations. When a UE supporting 3G-324M is communicating with an H.324 device connected to the fixed network, it is likely that the multiplex level will be negotiated to be lower than level 2, to the maximum level that the H.324 device supports, such as level 1 or 0, unless a media gateway is inserted in the network border for equalizing the configurations. For the direction from 3G-324M to H.324, the gateway would negotiate with the UE as another UE supporting 3G-324M and negotiate to level 2 but re-multiplex the received MUX-PDUs into another level the H.324 device can handle, and vice versa.

The shifting down of the multiplex level continues until a mutually-agreeable level is reached as the simultaneous use of two different levels is not allowed. Once a level has been agreed upon, the elements of the media clients are configured. The TerminalCapabilitySet message includes the information on the supported codecs, the maximum sizes for AL2 and AL3 SDUs or MUX-PDUs, or the multiplexing features at the negotiated level, such as the support of optional flags or headers.

Figure 7.5 shows several examples of the message source code, which is encoded in the Abstract Syntax Notation One (ASN.1) format to reduce the bit-rate. By transmitting a series of messages describing the features supported by a media client, a subset of which is selected by the other client, the conditions for media compression and multiplexing are mutually negotiated. Although a common multiplex level has to be agreed upon, it is not necessary for two communicating media clients to be configured identically in all aspects. In the H223Capability message, an appropriate AL is selected for each medium by setting its Boolean value to 1. Note that it is not impossible to transmit multiple streams of the same medium as long as the multiplex table is configured as such, but an AL has to be selected for each medium. Other detailed parameters such as the maximum sizes of the SDU to the adaptation layer can also be negotiated.

In the H263VideoCapability message, as shown in Fig. 7.6, qcifMPI indicates the maximum frame interval supported for the resolution, QCIF, as an integer multiple of $\frac{1}{29.7}$ seconds. Therefore qcifMPI=3 corresponds to a frame rate of about 10 fps. maxBitRate = 480 represents the maximum bit-rate of 48 kbps, as an integer multiple

```
TerminalCapabilitySet ::= SEQUENCE
{
        sequenceNumber                SequenceNumber,
        protocolIdentifier            0.0.8.245.0.16
        multiplexCapability           MultiplexCapability OPTIONAL,
        capabilityTable               SET SIZE (1..256) OF CapabilityTableEntry OPTIONAL,
        capabilityDescriptors         SET SIZE (1..256) OF CapabilityDescriptor OPTIONAL,
        ...,
        genericInformation            SEQUENCE OF GenericInformation OPTIONAL
                                      -- generic information associated with the message
}

H223Capability ::= SEQUENCE
{
       TransportWithI-frames          BOOLEAN,                    -- I-frame transport of ITU-T H.245

        videoWithAL1                  BOOLEAN,
        videoWithAL2                  BOOLEAN,
        videoWithAL3                  BOOLEAN,
        audioWithAL1                  BOOLEAN,
        audioWithAL2                  BOOLEAN,
        audioWithAL3                  BOOLEAN,
        dataWithAL1                   BOOLEAN,
        dataWithAL2                   BOOLEAN,
        dataWithAL3                   BOOLEAN,

        maximumAl2SDUSize             INTEGER (0..65535),         -- units octets
        maximumAl3SDUSize             INTEGER (0..65535),         -- units octets

        maximumDelayJitter            INTEGER (0..1023),          -- units milliseconds

        h223MultiplexTableCapability  CHOICE
        {
                Basic                 NULL,
                enhanced              SEQUENCE
                {
                        maximumNestingDepth                       INTEGER (1..15),
                        maximumElementListSize                    INTEGER (2..255),
                        maximumSubElementListSize                 INTEGER (2..255),
                        ...
                }
        },
        ...,
        maxMUXPDUSizeCapability               BOOLEAN,
        nsrpSupport                           BOOLEAN,
        mobileOperationTransmitCapability     SEQUENCE
        {
                modeChangeCapability                  BOOLEAN,
                h223AnnexA                            BOOLEAN,
                h223AnnexADoubleFlag                  BOOLEAN,
                h223AnnexB                            BOOLEAN,
                h223AnnexBwithHeader                  BOOLEAN,
                ...
        } OPTIONAL,
        h223AnnexCCapability          H223AnnexCCapability OPTIONAL,

        bitrate                       640
        mobileMultilinkFrameCapability SEQUENCE
        {
                maximumSampleSize                     INTEGER (1..255),       -- units octets
                maximumPayloadLength                  INTEGER (1..65025),     -- units octets
        ...
        } OPTIONAL
}
```

Fig. 7.5 TerminalCapabilitySet message format.

```
GenericCapability ::= SEQUENCE
{
        capabilityIdentifier        0.0.8.245.1.1.1              -- GSM AMR
        maxBitRate                  122                          -- Units 100 bit/s
        ...
}

H263VideoCapability ::= SEQUENCE
{
        qcifMPI                     3                            -- units 1/29.97 Hz
        maxBitRate                  480                          -- units 100 bit/s
        unrestrictedVector          BOOLEAN,
        arithmeticCoding            BOOLEAN,
        advancedPrediction          BOOLEAN,
        pbFrames                    BOOLEAN,
        temporalSpatialTradeOffCapability  BOOLEAN,
        hrd-B                       INTEGER (0..524287) OPTIONAL,   -- units 128 bits
        bppMaxKb                    INTEGER (0..65535) OPTIONAL,    -- units 1024 bits
        ...,
        errorCompensation           BOOLEAN,
        h263Options                 H263Options OPTIONAL
}
```

Fig. 7.6 Audio and video capability messages.

of 100 bps. In the message, encoding of the video codec can be configured with Boolean parameters. Note that the total bit-rate at the output of H.223 is 64 kbps, as the bit-rate field is set to 640 in the H223Capability message. The difference of 16 kbps is used by AMR 12.2 kbps and the overhead added in the two layers of H.223. Messages including codecs or features not supported by the receiving media client are ignored.

In the next stage, the communicating media clients decide who will be the *master* and who will be the *slave*, which is necessary to resolve conflicts between them. Although 3G-324M is a new type of service which does not require any direct interactions between the UE and the NodeB, there can be situations when a decision needs to be made by a UE. For example, when the outcomes of message exchanges are not clear and it is necessary to start again, or when both media clients transmit messages to open bi-directional logical channels for speech or video at similar times, the master can settle the stand-off situations by sending a message that rejects the channel proposal, which must be accepted by the slave. Compared with the strictly vertical master–slave relationships between the MS and the BTS in GSM or cdma2000, or between the UE and the RNC in W-CDMA, the master in 3G-324M has only limited capability for settling ambiguous situations.

Then the media clients open logical channels for each medium. The logical channels can be opened as uni-directional or bi-directional, the latter of which is most common in 3G-324M. In the last stage of the call setup, entries of the multiplexer tables, e.g., as shown in Table 6.18, are exchanged and set. Then the media can be compressed and transmitted in two directions, and shortly the display and speaker of each UE begins playing the far-end video and audio.

The call setup delay, the total delay measured from pushing the SEND button or icon to media flowing, can vary depending on the network conditions and reactions of the far-end but more than ten seconds, much larger than those for typical voice calls, were frequently experienced in the early deployments of W-CDMA. The delay can be even longer if the channel condition is unstable and the messages have to be retransmitted. This reduced the perceived quality of the new service significantly. Opportunities to reduce the call setup delay can be found in two areas: (1) the delay to set up the radio bearer, and (2) the delay to configure the media client. A joint reduction of the two delays is not feasible as the first step has to be completed before the second can begin.

Such a limitation was inevitable since the key building blocks of 3G-324M, H.223 and H.245, were not designed with transmission over W-CDMA in mind. Therefore, the signaling of 3G-324M is rarely integrated into those of W-CDMA completely. If low bit-rate, end-to-end transmission channels are available between two UEs, at the time 64 kbps radio bearers are being established, it would be possible to simultaneously configure two media clients. However, such an approach would complicate the design of signaling and the management of compatibility, as it would make it more difficult to replace 3G-324M with a new type of, e.g., packet-switched, multimedia communications systems.

7.2.2 Accelerated Media Negotiation

Media Oriented Negotiation Acceleration (MONA) is a set of techniques designed to reduce the latter part of the call setup delay [ITU-T (2006)]. Once the end-to-end bearers have been established, media clients capable of MONA start transmitting the preference messages that indicate the preferred configuration of each MONA technique, as shown in Fig. 7.7. Each preference message is preceded and followed by a two-byte synchronization flag, 1010001100110101, which is different from the two-byte flag used in level 1 or higher multiplex layers. Therefore, MONA-capable media clients can identify the received messages while conventional media clients disregard the messages and flags.

The first byte of a preference message is the Frame Information (FI) field, which consists of a three-bit Segment Sequence Number (SSN), a one-bit Last Segment (LS),

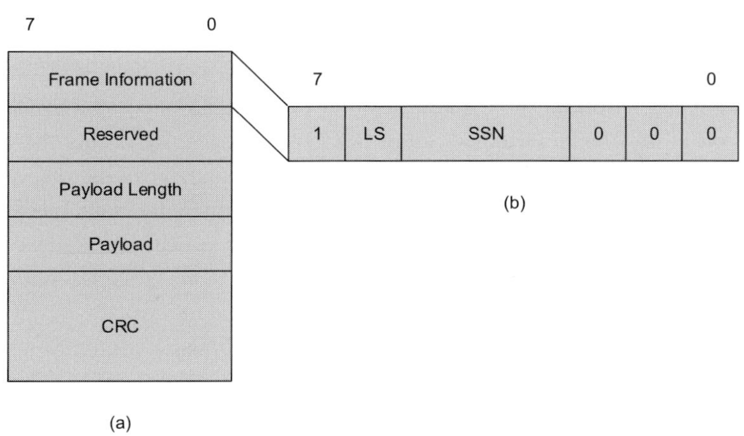

Fig. 7.7 (a) MONA preference message format. (b) Frame Information field.

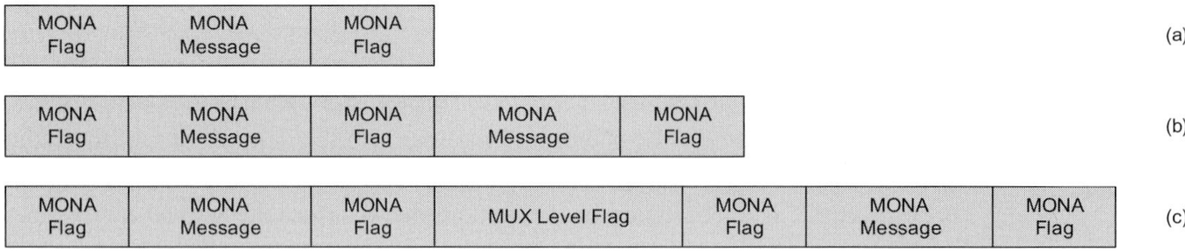

Fig. 7.8 Acceptable patterns of MONA flag, MONA message, and MUX level flag.

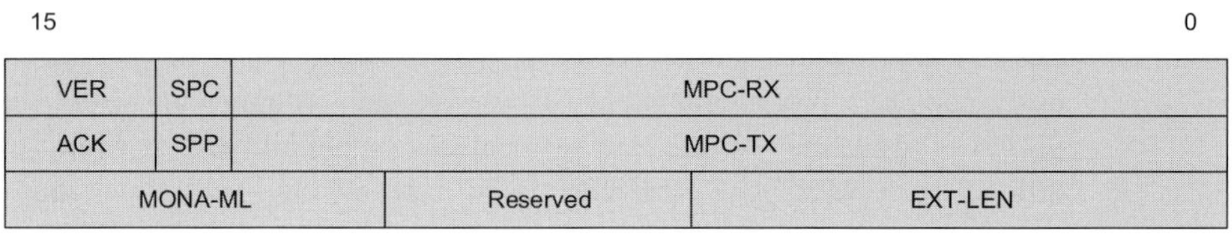

Fig. 7.9 Payload format of MONA message.

and a one and three zero bits. The 16-bit CRC is computed from the entire frame except for the synchronization flags. For media clients incapable of MONA, conventional flags can be transmitted between MONA flags but the pattern of message transmission needs to be restricted to enable the bit-correlator of the media client to synchronize with the bit-stream as soon as possible. Figure 7.8 shows the acceptable transmission patterns.

Figure 7.9 shows the payload format of a MONA preference message including the configurations for two techniques of accelerating the 3G-324M call setup, Signaling of Pre-configured Channels (SPC) and Media Pre-configured Channels (MPC). The SPC and SPP fields are used to indicate the support and preference of SPC, and the MPC-TX and MPX-RX fields are used to indicate the configurations of MPC for the send and receive directions. Table 7.2 summarizes the fields of the MONA preference message.

The MONA-ML field indicates the preferred level of the multiplex layer, bypassing the lengthy procedures of mutually converging to a level. The ACK field includes the reception status of the MONA preference messages. If the SPC is supported, all necessary information on the capabilities and preferences is compactly organized and transmitted in the Media Oriented Setup (MOS) messages, bypassing the conventional message exchange of H.245. The compressed media can arrive immediately after a two-way delay when the MOS Request messages transmitted by both media clients are answered with the MOS requestAck messages, if the processing delays are ignored. Therefore, from the message exchanges shown in Fig. 7.4, the delay until media flow can be reduced by two two-way delays.

In MPC, the media client selects a constrained set of media configuration parameters and includes them in the MUX-TX field of the preference message. Then the media client compresses the media based on the configuration and transmits it to the far-end with the preference message, without confirming the capability or preference of the far-end media client. If the initial guess was correct, the compressed media can arrive after

Table 7.2 MONA preference message fields.

Capability	Length (bits)	Description
VER	2	MONA version number (set to 0)
SPC	1	Set to 1 if media client supports SPC. Otherwise, set to 0.
MPC-RX	13	Describes configurations of MPC for receive direction
ACK	2	Set in outgoing preference messages to 00 if not successfully received any preference messages. 01 if acknowledges receiving at least one preference message whose ACK=00 10 if acknowledges receiving at least one preference message whose ACK=01 11 (reserved)
SPP	2	Set to 1 if media client prefers SPC. Otherwise, set to 0.
MPC-TX	13	Describe configurations of MPC for send direction
MONA-ML	5	Preferred multiplexer level
Reserved	4	
EXT-LEN	7	Length of additional capability information (in bytes)

Table 7.3 Bit-fields of MONA preference message.

Codec	Mux code	LCN	Multiplex table entry
Reserved (for H.245)	0		
AMR	1	1	1 ucf
AMR-WB	2	2	2 ucf
H.264	3	3	3 ucf
MPEG-4	4	4	4 ucf
H.263	5	5	5 ucf
Reserved	6–11		
Unspecified	12, 13		
Signaling Preconfigured Channel (SPC)	14	14	14 ucf
Reserved (for WNSRP)	15		

a one-way delay. Therefore the MPC can set up a 3G-324M call theoretically faster than the SPC but is less flexible. If the initial guess turns out to be wrong, i.e., if the media and multiplex configurations used for the initial transmission are not acknowledged in the received preference message, the media client can switch to the configuration supported by the far-end. Therefore, even when the initial guess was wrong, the compressed media can arrive after a two-way delay, which can still match the delay of SPC. However, it is not possible to introduce new media and multiplexing configurations into the fixed MPC-RX and MPC-TX fields. Table 7.3 shows the standard configurations supported by MPC, which also include speech and video codecs, such as AMR-WB and H.264, used in the mobile communications systems following W-CDMA.

Figure 7.10 shows the accelerated session negotiation of UEs supporting MONA, in which several exchanges of messages shown in Fig. 7.4 are not present. Depending on

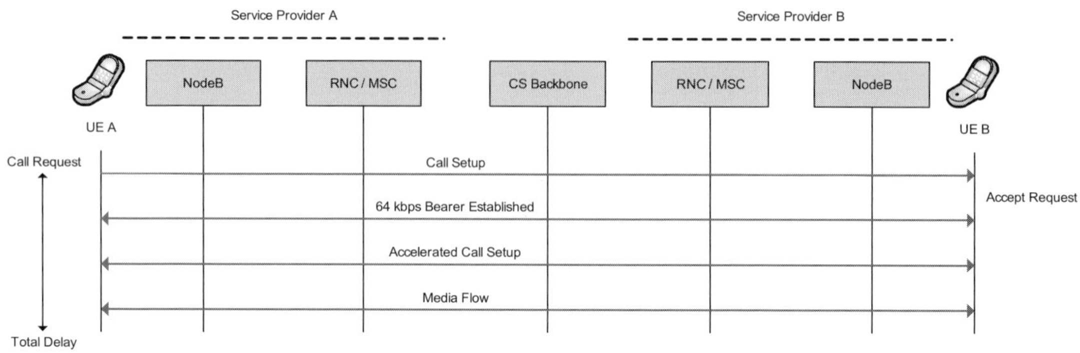

Fig. 7.10 Accelerated call setup procedures of 3G-324M.

the supported features, a media client capable of MONA is classified into one of three groups, Class I, II, or III. The introduction of MONA to 3G-324M, several years after the initial deployments, exemplifies that an unexpected source of quality loss can emerge with the introduction of new media or service types. In the design of the media and radio signal processing procedures after W-CDMA, the call setup delay became an important quality metric that needed to be minimized.

7.3 Performance Evaluation

Assuming that AMR 12.2 kbps is used as a common codec mode, speech quality of 3G-324M should be theoretically higher than that of the voice services in GSM or W-CDMA. One reason supporting this claim is that the target error rate of 3G-324M is lower than that of voice call in W-CDMA, which would be comparable to that of GSM, because of the need for protecting video more carefully. Unlike GSM or W-CDMA, the codec mode of 3G-324M is typically fixed at 12.2 kbps regardless of the channel condition or network load, as there is no signaling provided for the radio access or core network to control the configuration of 3G-324M. In addition, DTX can be turned off and the additional bit-rate can be used for higher quality by transmitting the background sound continuously. Notice that whether to use DTX or not is not negotiated in H.245 or MONA.

However, the end-to-end delay is significantly larger in 3G-324M than in voice call. In fact, the delay is sometimes longer than the longest allowable delay in the toll-quality voice services, 280 ms, because of the larger encoding delay of video, processing delay in the adaptation and multiplex layers, and the delay for audiovisual synchronization. In practice, an end-to-end delay as high as 400–500 ms is commonly observed in commercially operating W-CDMA networks. Thus, quality loss of 3G-324M comes from delay as well as from compression or transmission.

Unlike in speech codecs, video codec standards leave the structure of the encoder to the discretion of the implementation, allowing many alternative algorithms especially in the computationally intensive operations such as motion estimation and compensation.

Table 7.4 Video codec configurations.

Configuration	Video codec information
H.263 (a)	H.263 Annexes D, F, I, J, N, T, videoFastUpdatePicture
H.263 (b)	H.263 Annexes D, F, N, R
MPEG-4 (a)	MPEG-4 simple profile, resynchronization, data partitioning, RVLC
MPEG-4 (b)	MPEG-4 simple profile, resynchronization, data partitioning, RVLC, HEC
MPEG-4 (c)	MPEG-4 simple profile, resynchronization, data partitioning, RVLC, videoFastUpdatePicture
MPEG-4 (d)	MPEG-4 simple profile, resynchronization, data partitioning, RVLC, HEC

Table 7.5 Additional annexes of H.263.

Annex	Name	Effects
D	Unrestricted motion vector	Allow motion vectors to point outside picture
F	Advanced prediction	Allow four motion vectors per macroblock, and overlapped motion compensation for luminance
N	Reference picture selection	Select reference picture from buffer to stop error propagation
R	Independently segmented decoding	Allow picture to be decoded across slice or GOB boundaries

Due to the limitation in complexity, video codecs in UEs supporting 3G-324M are constructed with less hardware or simpler algorithms. Therefore, the PSNR values of video codecs in UEs, even with the same encoding parameters, will generally be lower than those of codecs in computers or video cameras. Therefore, applying new video compression algorithms does not necessarily result in higher quality as previous codecs with more hardware and refinement can perform as well.

7.3.1 Video Compression and Transmission Performance

Table 7.4 outlines six configurations of 3G-324M, in which four configurations use the MPEG-4 video codec, and the other two use H.263. In the configurations for H.263, some annexes of H.263 not recommended in 3G-324M are used in the tests. Table 7.5 outlines four annexes of H.263 for higher compression efficiency or error resilience. In addition to the codec options, a different combination of multiplex levels and header types is used in each configuration to measure the relative performance of the various error control techniques.

Table 7.6 outlines the video encoding parameters of the six configurations for 3G-324M, all of which use QCIF as their video resolution. The bit-rates and frame lengths of the audio channels are specified for each configuration but the audio performance is not evaluated. Instead, dummy data is transmitted at 6.40–8.12 kbps. Level 3 of the multiplex layer, which is not used in 3G-324M, is used in configuration MPEG-4 (d) to illustrate the cost for excessive error resilience. In the configurations where the intra-refreshing of video is not used, the I-frames are encoded frequently enough, e.g., once

Table 7.6 3G-324M configurations.

Configuration	Video		Speech		Multiplex			Total Bit-rate (kbps)
	Bit-rate (kbps)	Frame rate (fps)	Bit-rate (kbps)	Frame length (ms)	Bit-rate (kbps)	Level	Header type	
H.263 (a)	49.37	9.53	7.60	20	7.04	2	Standard	64.01
H.263 (b)	51.04	9.05	6.40	30	6.56	2	Standard	64
MPEG-4 (a)	52.80	9.65	8	20	3.20	2	Extended	64
MPEG-4 (b)	50.04	15.63	8	10	6.31	2	Extended	64.35
MPEG-4 (c)	48.23	9.23	7.60	20	8.17	2	Standard	64
MPEG-4 (d)	39.50	6.53	8.12	10	17.28	3	Extended	64.9

per 2–3 seconds. A total bit-rate of 64 kbps is maintained by controlling the bit-rates for speech, video, and multiplex.

MPEG-4 (a) and (b) are similarly configured but they differ in the frame length of speech, and the rate control is set to generate a higher frame rate in MPEG-4 (b). For error resilience, both configurations resort to the extended header of the multiplex layer and frequent encoding of I-frames. In contrast, H.263 (a) and MPEG-4 (c) use the standard header, and support intra-refreshing with the videoFastUpdatePicture message. H.263 (b) has a relatively short list of features but is not at such a disadvantage in terms of video quality, as its speech bit-rate is the lowest. As the cost for multiplexing in level 3, the video bit-rate is reduced by about 20% in MPEG-4 (d) to less than 40 kbps. Likewise, a higher frame rate of video or shorter frame length of speech requires a higher multiplexing overhead.

For the video input, the *Overtime* sequence is sampled at 30 Hz. *Overtime* contains header-and-shoulder scenes with low motion activity typically found in video calls. Figure 7.11 shows the results of subjective tests taken for the six configurations, for six different transmission conditions [3GPP (2001)]. As the evaluation is not based on previously known references for quality, which did not exist for this kind of application at the time of testing, the MOS needs to be taken with a focus on the relative differences rather than on the absolute values.

In Fig. 7.11, the X axis represents the channel conditions labeled as F10-6, M10-6, F20-4, M20-4, F10-3, and M10-3. The M indicates the mobile-to-mobile channels including both the uplink and downlink of W-CDMA, and the F represents the fixed-to-mobile channels including only the downlink of W-CDMA. Therefore with the other parameters set to equal values, channel conditions whose labels start with F tend to represent friendlier channels for transmission. The M or F is followed by the interleaving depth, in ms, and further by the base-10 exponent of the bit error rate of the channel. For example, F20-4 represents a fixed-to-mobile link with 20 ms interleaving, and BER $= 10^{-4}$. In the tests, interleaving depths of 10 and 20 ms, and BER values of 10^{-6}, 10^{-4}, and 10^{-3} are considered.

In the most benign channel condition, F10-6, H.263 (a) outperforms all other configurations by noticeable margins, which is followed by MPEG-4 (c). Considering that H.263 (b) trails H.263 (a) by the largest margin, and H.263 (a) employs Annexes I

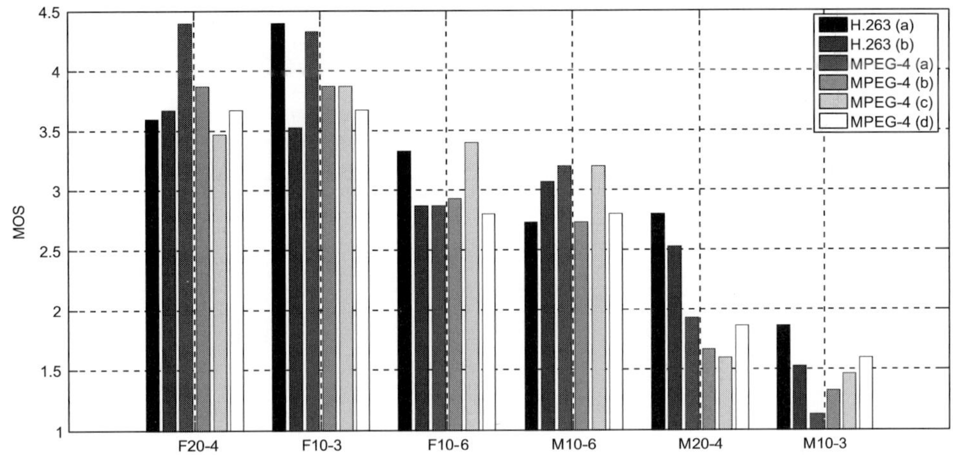

Fig. 7.11 Test results of 3G-324M (*Overtime*).

and J which H.263 (b) does not, advanced intra-coding (Annex I) and deblocking filter (Annex J) might have contributed to the outstanding performance of H.263 (a) significantly. In this condition rivaling the situations of error-free fixed networks, encoding methods increasing the error resilience contribute little to the overall performance, as the additional redundancy only reduces the bit-rate for the video. The MPEG-4 configurations perform similarly. In M10-6, MPEG-4 (a) exceeds H.263 (a) by a small margin but the results are similar to those of F10-6, with all configurations exceeding or approaching MOS = 3.5. In contrast, in the most hostile condition, M10-3, H.263 (a) exhibits the lowest quality, which may be understood since the encoding parameters of H.263 (a) are configured mainly for highest compression efficiency, with insufficient redundancy in the codec or the MUX-PDU bit-stream for error resilience. In this case, MPEG-4 (a) achieves the highest quality, as its combination of error resilient encoding and extended header in the multiplex layer matches the challenging situation properly. Considering that MPEG-4 (a) outperforms both MPEG-4 (b) and (c), it may be concluded that the higher frame rate or shorter header of MUX-PDU reduces video quality, and that intra-refreshing would not be of much help in poor channel conditions.

In the intermediate conditions, F20-4 and M20-4, where both compression efficiency and error resilience contribute to the performance, H.263 (b) shows higher performance than the other configurations. M10-3 is the condition closest to the actual operating environment of 3G-324M, and MPEG-4 (c) is the configuration most commonly used. Note that AMR 12.2 kbps with DTX turned on results in average bit-rates of 6–7 kbps, which matches the speech bit-rate of MPEG-4 (c).

7.3.2 Live Call Analysis

Figures 7.12 and 7.13 show the status of key parameters for voice service and 3G-324M, when they are observed in commercially operating W-CDMA networks in good channel conditions. The frequency spectrum used is in the 2.1 GHz band.

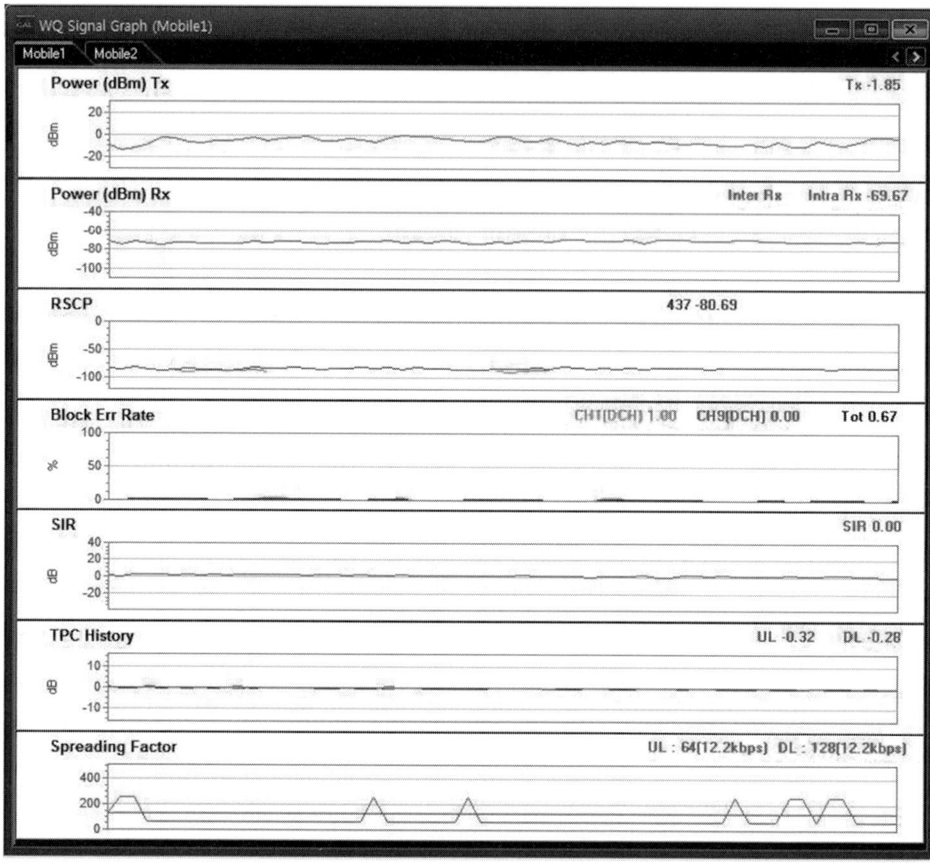

Fig. 7.12 W-CDMA call analysis (courtesy of Accuver Co., Ltd.).

Voice Service

The first and second plots of Fig. 7.12 show that the average power levels of the transmitted and received signals are -1.85 and -69.67 dBm, respectively. In W-CDMA, the maximum transmit power in voice service is assumed to be about 21 dBm. Therefore, it can be seen that the dynamic range of the power control in W-CDMA spans more than 80 dBm, and that the transmitted signal is attenuated in the channel by more than 60 dBm, assuming that the UE and the NodeB are transmitting at similar levels.

The third plot shows the variation of the downlink Received Signal Code Power (RSCP), which indicates the level of received power on the downlink physical channel reported to the NodeB for power control. It can be seen that the average value of RSCP is -80.69 dBm. In the fourth plot, it is shown that the BLER, the proportion of transport blocks that could not pass the CRC check, is limited to less than 1% for the downlink. The fifth plot shows that the downlink signal-to-interference ratio (SIR) is maintained at around 0 dB.

The transmit power control (TPC) history shows that the transmit power increases and decreases by 1 dB, under a stable operation of power control. In the last plot, it is seen

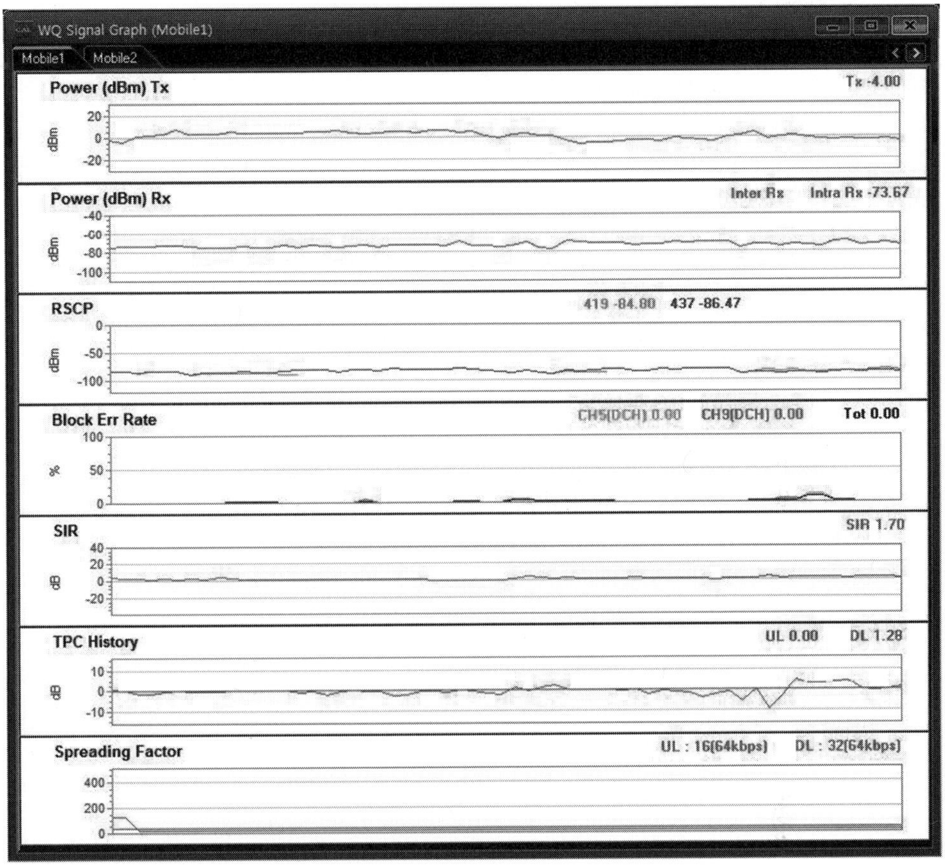

Fig. 7.13 3G-324M call analysis (courtesy of Accuver Co., Ltd.).

that a spreading factor of 64 is used for the uplink, and 128 is used for the downlink. Therefore, slot format 2 is used for the uplink, and slot format 11 is likely to be used for the downlink. It is seen that sometimes a larger spreading factor, 256, is used in the uplink, when there are no speech frames to transmit. On the other hand, the downlink spreading factor is kept at 128 throughout the measurement period.

3G-324M

Figure 7.13 shows the time-varying characteristics of the key parameters in a live 3G-324M call, using the same formats for the plots as in the voice call. For video compression, the MPEG-4 codec is configured for 48 kbps, with resynchronization and data partitioning used as the error-resilient encoding. In the adaptation layer and multiplex layer, AL2 and level 2 are used for both speech and video.

RVLC is rarely implemented or activated in 3G-324M, since the performance increase from applying RVLC does not always justify the required increase in complexity or delay. RVLC increases the size of the codebook table, and its codeword cannot be decoded as efficiently as that of the VLC. This observation might have led to the avoidance of error resilient encoding in later video codecs. In addition, the improved capability of the network that increases the bit-rate while reducing the error-rate makes error resilient encoding less useful.

As the voice and 3G-324M calls were measured in the same cell at similar network loading levels, several plots show similar characteristics in the two figures. However, noticeable differences can be found in the BLER and SIR values, which are lower than the corresponding values for the voice call. Since 3G-324M requires much higher QoS to ensure an acceptable video quality, the NodeB controls the transmit power to and from the UEs in the service so that the required QoS, typically a BLER of 0.1% or less, is maintained. In the fourth plot, the average BLER is 0%, in contrast to 0.67% for the voice call. Once the UE is in a stationary position, i.e., used in the hands-free mode, which is more common in the usage of 3G-324M than voice call, it is not difficult to maintain this level of remarkable link quality, due to the use of turbo code and fast power control.

For the voice call, power control is typically configured to maintain a BLER of 1–3%. In 3G-324M, as the speech is multiplexed with the video using the H.223 multiplex protocol, the additional power does not increase the speech quality, which is protected excessively. Lower values of transmit and receive power, and received signal code power can be understood since the UE in this 3G-324M call is under better channel quality or closer to the NodeB than the UE in the voice call shown in Fig. 7.12. It is shown that spreading factors of 16 and 32 are used in the uplink and downlink, respectively, for a bit-rate of 64 kbps in the two directions.

From the fourth plot, it can be seen that transport blocks are lost from time to time. According to Table 6.20, the size of each transport block is 640 bits in 3G-324M. As the receiver typically drops the transport blocks that fail the CRC check, loss of several consecutive transport blocks results in the loss of thousands of bits that have to be fed to the video decoder, which can be beyond the healing capability of error resilient encoding or the error concealment of the decoder. Therefore, intra-refreshing is the usual means for video error control in 3G-324M although it cannot prevent the quality loss or recover the quality immediately.

7.3.3 Voice Capacity

Table 7.7 shows the estimated capacity of voice service and 3G-324M in W-CDMA [Holma and Toskala (2010)]. It can be seen that the capacity of the downlink is smaller than that of the uplink, due to the limitation in the power headroom of the NodeB. As the number of UEs in the cell increases, the interference from UE transmissions becomes the major bottleneck of uplink capacity while the transmitter runs out of power in the downlink. As four cdma2000 channels can be packed into a single W-CDMA channel, the voice capacity of W-CDMA is clearly lower than that of cdma2000, considering the estimated capacity shown in Table 5.26 and ignoring the differences in the operation of codecs.

However, the larger channel bandwidth of W-CDMA enables a considerable number of higher bit-rate services such as 3G-324M. The burden on the network capacity from a

Table 7.7 Voice and video call capacity.

	AMR 12.2 kbps	3G-324M
Uplink	72–101	16–23
Downlink	70–94	7–14

video call is apparently less than that of ten voice calls, whose aggregate average bit-rate can be comparable to that of a video call. UEs at the cell edges or in poor channel conditions can drain the network capacity so that few other services can be supported. The fundamental differences in the required network capacity for voice and video services raise the issue of appropriate pricing, as a lower rate for a video call might invite so many users and increase the probability of a service request getting rejected, therefore ultimately reducing the overall quality of services. As the derivations of CDMA voice capacity, shown in Equations 5.10 and 5.17, do not assume any conditions that cannot be applied to W-CDMA, such as the effects of VBR speech coding, mathematical procedures to derive the upper bounds of voice capacity can also be used for W-CDMA.

7.4 References

3GPP. 2001. TR 26.912 V4.0.0 Codec for Circuit Switched Multimedia Telephony Service; Quantitative Performance Evaluation of H.324 Annex C over 3G. April.

Holma, H., and Toskala, A. 2010. *WCDMA for UMTS: HSPA Evolution and LTE*. 5th edn. Wiley.

ITU-T. 2006. H.324 Amendment 1: New Annex K Media Oriented Negotiation Acceleration Procedure and Associated Changes to Annex J. August.

8 Signal Processing in SC-FDMA/OFDMA Systems

8.1 Technical Background

Previous chapters have discussed the signal processing procedures for the second and third generations of digital circuit-switched mobile communications systems. In these systems, radio network controllers such as the BSC or the RNC assume full responsibility for assuring speech quality and maximizing network capacity. Circuit-switched mobile communications networks are similar to the PSTN in that the variance of the transmission delay is negligible since an end-to-end transmission path for speech is established before the conversation begins. Furthermore, the circuit assigned to an MS is not shared by other MSs during the call, regardless of voice activity or cell loading level. Therefore it can be said that circuit-switched networks provide stable quality at the cost of network capacity.

With the expansion of the Internet, the basic architecture of mobile communications networks has changed so that the entire transmission path can be paved with the Internet Protocol (IP) [Postel (1981a), Deering and Hinden (1995)]. Several layers of such protocols are present between the signal processing operations associated with the media compression and those for the radio transmission. Therefore, in packet-switched mobile communications networks, conventional techniques to control the tradeoff between quality and capacity, such as joint source-channel coding or unequal error detection and protection, cannot be easily applied. Such techniques require careful allocation of the bit-rate between encoded speech and the redundancy to meet the channel condition.

Moreover, the IP overhead generated when compressed media propagates through these protocols is significant, often larger than the encoded speech frame itself. Forty and sixty bytes of headers are added to a speech frame of a similar or smaller size while it is processed using the RTP/UDP/IPv4 or RTP/UDP/IPv6 protocols, in addition to one or two bytes for the RTP payload header. However, the overhead can be theoretically compressed to as small as one byte when the packets are transmitted over a wireless channel, exploiting the fact that most fields of the overhead, such as the IP address or port number, rarely change during a call, and need not be transmitted more than once or updated at all.

8.1.1 New Problem Description

In packet-switched networks, it is not necessary to reserve a dedicated end-to-end transmission path made up of several communication links during a call. Instead, each link can be independently used when necessary, and can be re-assigned when the transmission is over. A UE may be given an opportunity for later transmission if its priority is low, which may increase the transmission delay. Therefore channel assignment needs to be carefully coordinated by the network. To limit the sum of delays from each link to an acceptable level, the network nodes controlling the end-to-end transmission path need to be managed by a centralized controller. An efficient scheduling strategy is important in situations where the bandwidth is scarce or expensive, such as in mobile communications.

Once the media is packetized, i.e., compressed and complemented with the IP headers, it is delivered to the destination without decoding, as in the tandem-free operation of voice services or circuit-switched video telephony, with an important difference that the transmissions are not synchronized. To describe the transmission status of packet-switched networks, the *jitter* is defined as the behavior observed by the media receiver when the packets are received irregularly or not delivered at all. In packet-switched networks, network nodes, such as the routers, have to delay transmission or drop packets if they are fed more packets than they can handle at a given time. However, although packet-switched mobile communications networks are heavily dependent on the IP, their behavior should not be the same as that of the Internet where no quality of service, like a maximum delay or packet loss rate tolerated, is guaranteed. The mechanisms to deal with router congestion and control flow of the Internet can be borrowed, but have to be refined to handle more rapid fluctuations of link quality. Without a dominant network node given the authority and measures to control media quality and network capacity, the responsibility must be distributed over multiple network nodes, including the UEs.

Figure 8.1 compares the logical distance between speech encoding and decoding in mobile communications systems spanning three generations. The distance becomes longer at each generation, reaching its maximum at the Long Term Evolution (LTE). The longer the distance, the larger the savings in network capacity since a minimum bit-rate is used for transporting voice over more of the distance. The removal of transcoding to μ-law or A-law also reduces the delay and improves the speech quality. The network node that controls the bit-rate is shown for each generation with the signaling methods used. Figure 8.1 also illustrates that in LTE the bit-rate is controlled by both the network and the far-end UE.

The drastic changes in network architecture, transitioning from circuit-switched to packet-switched mobile communications, require more complicated signal processing procedures. However, the measures for control in conventional packet-switched networks are less accurate and responsive than their counterparts in circuit-switched networks. Considering the hefty technical price that must be paid, the gains need to be sufficient to justify the transition. However, combined with wider channel bandwidths and efficient scheduling strategies, packet-switched mobile communications networks provide a significant increase in both quality and capacity. Additional advantages include simplified network management by handling multiple packet-switched networks, such as mobile, fixed, or Wi-Fi, together and the corresponding reduction of cost. The *session* is a term commonly used to represent an instance of service or media call in packet-switched networks.

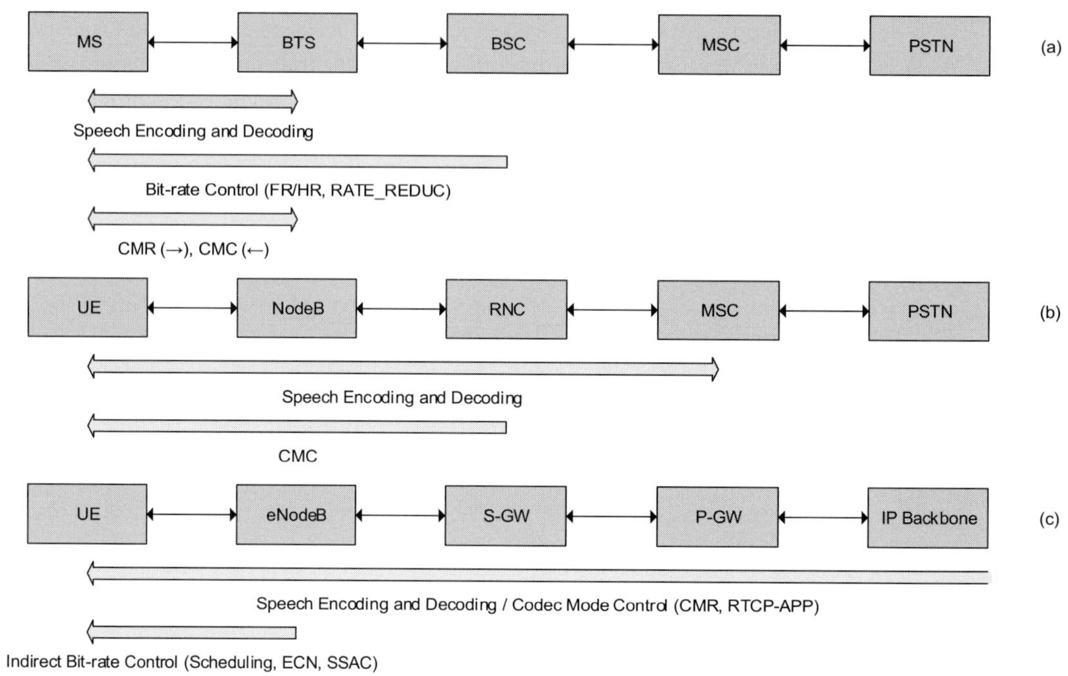

Fig. 8.1 Distance between speech encoding and decoding. (a) GSM/cdma2000. (b) W-CDMA. (c) LTE.

8.1.2 Packetization of Circuit-Switched Systems

As the first step to develop packet-switched mobile communications systems while recycling the existing infrastructure as much as possible, the radio signal processing procedures of the circuit-switched systems, GSM, cdma2000, and W-CDMA, were systematically modified to facilitate packet transmission. After these renovations, the second and third generation systems were renamed the General Packet Radio Service (GPRS) [Halonen *et al.* (2007)], cdma2000 1X Evolution-Data Only (EV-DO) [Vanghi *et al.* (2004)], and High-Speed Packet Access (HSPA) [Holma and Toskala (2006)]. These interim systems still maintain the system parameters inherited from their parent systems, such as their channel bandwidths, modulation techniques, and the error correction coding.

An important transmission technique introduced during the transition from circuit-switched to packet-switched systems is called the Hybrid Adaptive Repeat and Request (HARQ), in which the UE and the NodeB request the re-transmission of unsuccessfully received elements of each packet. With HARQ, information identical to the lost elements can be requested again so that the soft-combined, i.e., accumulated, energy increases the probability of successful decoding. It is also possible that information other than the lost elements can be requested so that the additional redundancy increases the likelihood of successful decoding. Since the number of re-transmissions increases as the transmit power decreases, and delayed packets also reduce media quality, a maximum number of re-transmissions is allowed based on the end-to-end delay budget.

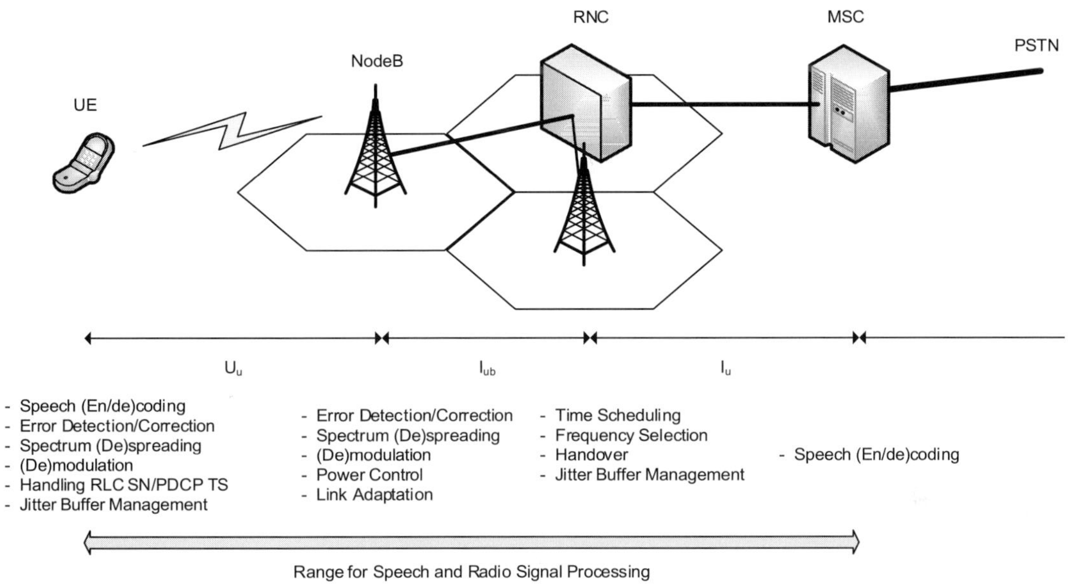

Fig. 8.2 Network architecture of CSoHS.

The conversion of circuit-switched networks is a lower-cost alternative to designing completely new packet-switched networks, but it still requires a significant infrastructure investment, if voice service with a quality similar to that of circuit-switched mobile or fixed networks is to be provided. An even lower-cost approach might be to apply the re-transmission techniques of packet-switched networks only to the wireless channel, leaving the other parts of the circuit-switched mobile communications networks intact.

Figure 8.2 shows the network architecture of the Circuit-Switched Voice Services over HSPA (CSoHS), where the buffer that manages the time-varying transmission delay, i.e., jitter, is located in the RNC for the uplink and in the UE for the downlink. Notice that the IP-based transport mechanisms are not employed anywhere in the transmission paths. In contrast to W-CDMA where speech frames are transmitted in the transparent mode of RLC, bypassing the intermediate radio protocols between speech codec and physical layer, in CSoHS, speech frames are typically transmitted in the unacknowledged mode in which the circuit-switched (CS) counter, sequence number (SN), and the transmission number (TN) are attached at the PDCP, RLC, and MAC, respectively, to handle the frames that are lost or delivered out-of-order. The five-bit CS counter is set to the first through fifth least significant bits (LSB) of the connection frame number (CFN) the RRC computes, at the time when the speech frame has been received. The CS counter, with the value of the Max CS delay which the RRC signals to the receiver, is used by the receiver to manage the jitter generated in the radio transmission.

The RNC of CSoHS has the same ability to control the AMR codec mode as the RNC of W-CDMA, using the codec mode control (CMC) and active codec-mode set (ACS) messages. In addition, the downlink of HSPA, High-Speed Downlink Packet Access (HSDPA), is a shared link whose radio resource should be scheduled by the NodeB but

Table 8.1 Voice capacity of W-CDMA.

Uplink	Downlink	
	(1-antenna)	(1-antenna with equalizer)
69	51	63

Table 8.2 Voice capacity of CSoHS uplink.

Basic	With DTX	With IC	With DTX and IC
93	121	118	165

Table 8.3 Voice capacity of CSoHS downlink.

Receiver	Basic	With equalizer
1-antenna	61	83
2-antenna	172	232

the uplink, High Speed Uplink Packet Access (HSUPA), is a dedicated link complemented with HARQ. Since the capacity of W-CDMA is limited by interference in the uplink, and by code and power in the downlink, the uplink capacity of the CSoHS is likely to be improved by suppressing transmit power as much as possible, e.g., using an equalizer or an interference canceller. The capacity of the downlink can also be improved by advanced scheduling algorithms that assign the full capacity of a 5 MHz bandwidth to the UE with the highest channel quality for shorter periods than the TTI values of W-CDMA while taking the delay constraints of the other UEs into account.

Therefore, the jitter of CSoHS in the uplink is mainly generated from re-transmission while that in the downlink is from both re-transmission and scheduling. Table 8.1 shows the estimated uplink and downlink voice capacity of W-CDMA Release 99 [Ozturk *et al.* (2010)], in which the downlink is the bottleneck. This capacity is estimated more conservatively than in Table 7.7 but is generally aligned with the results found in the literature and field trials. Since the voice capacity of CSoHS, outlined in Tables 8.2 and 8.3, is estimated using similar simulation environments, improvements of capacity from CSoHS can be easily identified.

With the maximum number of re-transmissions set to four, the uplink voice capacity is increased by 24 calls. Allowing more re-transmissions will undoubtedly increase the capacity but the quality is likely to suffer from the excessive delay. Applying discontinuous transmission (DTX) and interference cancelation (IC) increases the capacity further by 28 and 25 calls, respectively. Note that the DTX in CSoHS also refers to not transmitting an uplink pilot channel when voice activity or data transmission is absent, thereby reducing transmit power and the interference. Although the absence of an uplink pilot channel may reduce the channel estimation and interference cancellation capability of the receiver, it can be seen that DTX and IC do not interfere with each other, and additional gain can be achieved by employing both techniques together. In the downlink, CSoHS can boost the capacity by 10–20 calls, depending on the configuration of the receiver. Equipping the UE with an additional antenna has a dramatic influence on the capacity of the downlink, more than doubling the capacity from the single-antenna case. This optimistic estimate of the CSoHS capacity suggests that larger capacity will be achieved in packet-switched networks with multiple antennas and efficient scheduling algorithms. While CSoHS manages to provide significantly improved capacity while recycling the many network nodes of W-CDMA, it became apparent that longer-term solutions that allow the scheduling not only in the time domain but also in the frequency domain, and also allow variable channel bandwidth that is narrower or wider than the fixed 5 MHz of W-CDMA are necessary to meet the explosive requests for packet-switched media and data services.

8.2 Voice over Long Term Evolution

This chapter is confined to the signal processing of Long Term Evolution (LTE), a name which is more commonly used than the official name, the Evolved Universal Terrestrial Radio Access Network (E-UTRAN). Unlike HSPA, LTE does not have a parent network, and does not support any circuit-switched elements. The hierarchical network architecture of HSPA, where multiple NodeBs are managed by a controller at a higher layer, the RNC, is replaced by a flatter architecture consisting of just the UEs and the evolved NodeBs (eNodeBs). Figure 8.3 shows the network architecture of LTE. We focus on the media and radio signal processing operations of the new system, Voice over Long Term Evolution (VoLTE), taken between the UE and the eNodeB, and do not consider in detail the other transmission paths over which there is no further processing of the encoded media.

8.2.1 Network Architecture

The eNodeB is typically implemented as a Digital Unit (DU) connected to a set of Remote Radio Heads (RRH), which perform all RF-related operations. As the channel bandwidth increases and the signal processing procedures become more complex, this functional separation simplifies network installation and reduces management cost as the digital processing of signals to and from multiple RRHs can be located in one place. The cost can be further reduced by running the digital signal processing tasks on general-purpose processors rather than in hardware or on dedicated processors designed for specific tasks, and by using multi-core processors rather than multiple processors. The separation of RF and baseband signal processing appeared with the introduction of HSPA, becoming an established practice in the design and implementation of wireless network sites. Notice that the definition of the eNodeB includes neither a DU nor an RRH, nor the interface between them, allowing the separation to be at the discretion of the implementation. Therefore it is theoretically not impossible to locate the MAC and lower layers, and the RLC and upper layers at the RRH and the DU, respectively. As in

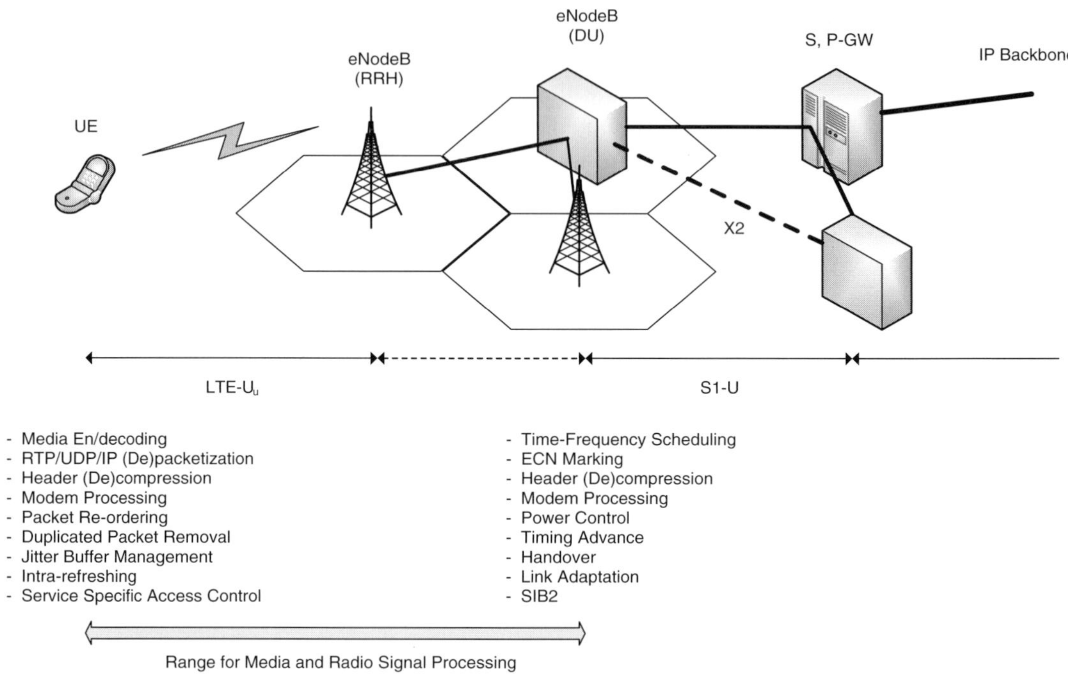

Fig. 8.3 Network architecture of LTE.

the intra-cell handover of circuit-switched networks, it is possible for a DU to hand a call from an RRH over to another RRH that it controls. However, the channel concept in LTE is quite different from that of TDMA or CDMA networks since the channel bandwidth over which the information is transmitted can be dynamically controlled even when the information bandwidth does not change.

In Fig. 8.3, a Serving Gateway (S-GW) is connected to the eNodeBs with the S1-U interface. It is further connected to a Packet Data Network Gateway (P-GW). These gateways do not have the responsibility of dynamically controlling radio resources or initiating handovers, as the BSC or RNC does in cdma2000 and W-CDMA, but they perform the important task of enabling or disabling services. An eNodeB is connected with neighboring eNodeBs via the X2 interface over which the packets are forwarded for handover. When handover occurs between two eNodeBs not connected with X2, the packets are indirectly forwarded via the S-GW. The P-GW is further connected to network nodes including the IP Multimedia Subsystem (IMS) [Poikselkä and Mayer (2009)], other S-GWs, P-GWs, the IP backbone networks, and the Internet. IMS is a key management system for various packet-switched networks, with tasks including the assignment of bandwidth resources over the wireless and wired channels, security assurance, and charging.

The new packet-switched core network, named the Evolved Packet Core (EPC) [Olsson *et al.* (2009)], consisting of the S-GW, P-GW and the Mobility Management Entity (MME), replaces the GPRS, the packet-switched core network for GSM and UMTS. Initially, the design of IMS began with GPRS in mind but as other wireless and wired networks adopted the management system, the IMS became independent of the radio access networks or media compression technologies. As shown in Fig. 8.4, the end-to-end transmission paths consist of wireless links between the UE and the eNodeB, and wired links, i.e., the core networks, which include many intermediate nodes, and the IP backbone networks that play the role of a central station for packet transport. While we focus on the wireless links and consider the wired sides of the networks as almost error-free, we do not completely exclude the possibility that the packets can be delayed or lost in the fixed networks, which may include the microwave links that connect isolated areas.

8.2.2 Functional Split

The signal processing procedures of LTE can be divided into two groups: (1) the media signal processing procedures which involve media (de)compression and IP (de)packetization, and (2) the radio signal processing procedures. Based on this classification, the UE can be logically divided into the media client responsible for the former procedures and the LTE modem for the latter ones. This division is similar to that in 3G-324M. Since the media signal processing procedures can be used for different packet-switched networks, albeit at different bit-rates or QoS parameters, and the media codecs can be replaced without much impact on the network architecture, it does not make sense to name the multimedia communications system after the packet-switched networks or the media codecs. The system shown in Fig. 8.4 is therefore called the Multimedia Telephony Service for IMS (MTSI) [3GPP (2017c)], whose name emphasizes that the IMS is an essential entity that can accompany any combinations of media processing and radio transmission techniques. The resource reservation procedures of IMS do not include many types of signal processing but they still govern the media and radio signal processing procedures over the entire transmission paths. A more commonly used name, however, is the Voice over LTE (VoLTE), although it also supports video. VoLTE is a subset of MTSI defined to facilitate the development and deployment of the new mobile multimedia communications system.

In the media client, one or more encoded speech frames are loaded onto a Real-time Transport Protocol (RTP) packet [Schulzrinne *et al.* (2003)], which is further loaded on top of a User Datagram Protocol (UDP) packet [Postel (1980)], and finally onto an IP packet. By way of contrast, an encoded video frame is, in general, too large to be transported over an RTP packet. Therefore, a video frame is segmented into smaller pieces that match the maximum transport capability of the network. The Real-Time Transport Control Protocol (RTCP) [Schulzrinne *et al.* (2003)] is used to exchange information on the status of the transmission that can be used to control media compression and transmission. For example, if the receiver informs the sender with the RTCP Receiver Report (RR) that a large proportion of the transmitted packets was not received, the sender may suspect the presence of congestion or unreliable links somewhere in the paths. RTCP can also be used to ask for specific actions of the far-end media client. The media adaptation, as illustrated in Fig. 8.4, includes a set of techniques to directly or indirectly control the media bit-rate or packet rate of the far-end client. The Session Initiation Protocol (SIP),

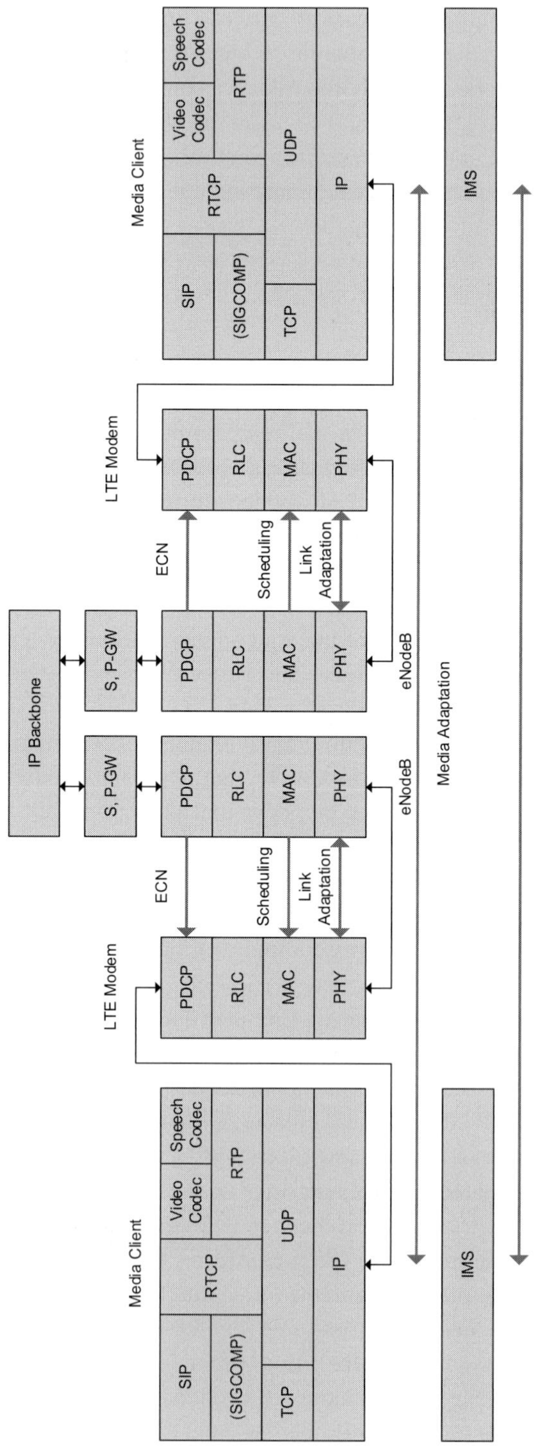

Fig. 8.4 End-to-end media compression and transmission paths of VoLTE.

used for the IMS signaling, has more relaxed delay requirements than the media but has to be received without error. Either RTP or Transport Control Protocol (TCP) [Postel (1981b)], which asks for re-transmission if packets are lost, can be used for SIP. The SIP messages typically contain a large amount of text and can be compressed with the Signaling Compression (SIGCOMP).

A key aspect of VoLTE, which has been absent in the previous circuit-switched networks, is the extensive use of IP. Techniques to reduce the overhead of IP were introduced to improve the spectral efficiency for radio transmission. In the case of speech, small packet-size, high frequency of transmission, and large number of UEs simultaneously using the services during busy hours make the bit-rate of the overhead from the headers of RTP/UDP/IP protocols significant. In the wireless channel where the cost for transmission is much higher than that of the wired channel, such a luxury cannot be tolerated. In GSM, 212 bits of overhead from radio signal processing are attached to 244 bits of speech in each frame, with the AMR speech codec operating at 12.2 kbps. The ratio of overhead can be increased if lower codec modes are used but, in general, the bit-rate is divided between the speech and the overhead in a balanced fashion. The spectral efficiency can be improved by reducing the bit-rate for the IP headers. In contrast, the overhead is not significant in video packets, as video frames are larger and less frequent than speech frames. A promising opportunity to increase the spectral efficiency can be located in the signal processing procedures of video conventionally not considered, such as in the reduction of quality loss from resizing by matching the resolution of the video encoded at the sender to that of the video displayed at the receiver using intelligent session negotiation procedures.

The LTE modem, as shown in Fig. 8.4, consists of four layers of radio protocols. Similar logical partitions can be seen in other packet-switched mobile communications systems but the detailed responsibilities of each layer may slightly differ. As the first step, the RTP/UDP/IP and UDP/IP headers of media and control packets are compressed by the Packet Data Convergence Protocol (PDCP) [3GPP (2017p)] and de-compressed by its counterpart across the wireless channel. At the Radio Link Control (RLC) [3GPP (2017o)], the header-compressed packets are concatenated or segmented to facilitate transmission over the wireless and wired channels, as a limitation of Maximum Transfer Unit (MTU) is typically enforced by the network. RLC may operate like TCP in that the radio protocol can ask its counterpart to retransmit lost data. However, such an operating mode, the acknowledged mode (AM), is rarely used in delay-constrained conversational services such as VoLTE. Instead the AM can be used for the SIP messages on UDP which should be delivered error-free but are more delay-tolerant. Note that the session set-up time is an important criterion of service quality and the exchange of SIP messages is not completely free from delay constraints.

Like the H.223 of 3G-324M, the Media Access Control (MAC) [3GPP (2017n)] multiplexes data streams from the RLC so that the radio bearers, i.e., the transmission opportunities provided by the lower layers, can be efficiently utilized. MAC pads the data if necessary, and also takes the responsibility of managing the HARQ procedures. When there are multiple streams of data competing to be transmitted as soon as possible, the MAC in the eNodeB determines the priority of each stream and schedules the transmission to and from each UE. In general, the speech, video, and RTCP packets of VoLTE receive higher priorities than the packets of non-real-time services. Therefore, the MAC in the eNodeB dynamically allocates available radio resource to UEs in the cell, using a variety of scheduling algorithms designed with various criteria, such as data throughput, fairness, or capacity for conversational services, in mind.

The link adaptation at the Physical Layer (PHY) [3GPP (2017j)] refers to the high-speed radio signal processing procedures that exploit the rapidly time-varying characteristics of the wireless channel between the UE and the eNodeB, in contrast to media adaptation which considers the slowly time-varying end-to-end transmission characteristics. The circuit-switched mobile communications networks also employ link adaptation techniques, such as power control, joint source-channel coding, or unequal error detection and protection. While the slow frequency hopping of GSM, which is performed in a pre-defined fashion, may not be considered as an adaptation technique that preemptively avoids the located channel impairments, it reduces the effects of frequency-selective fading and indirectly improves the performance of other radio signal processing procedures.

In addition to HARQ and scheduling, packet-switched networks provide other link adaptation opportunities, such as Adaptive Modulation and Coding (AMC), where a bit-rate matched pair of modulation and error correction code is dynamically selected from multiple pre-defined candidates, to best meet the instantaneous channel condition. HARQ exploits the fact that transmission over each link is not synchronized in packet-switched networks, and packets can stay longer at some network nodes than others as long as the total delay constraints are met. LTE provides another dimension for adaptation since unlike HSPA, which schedules only in the time domain, radio resources of LTE can also be scheduled in the frequency domain. With a channel bandwidth that can be as wide as 20 MHz, most frequency-selective fading can be avoided simply by scheduling transmission over frequencies not impacted by fading.

Although introduced with other quality-control techniques, media adaptation can be considered as a last resort when other media or radio signal processing procedures, especially those defined for a single wireless channel, fail to assure acceptable quality. In media adaptation, the media client in a UE asks the far-end client directly to adjust the compression or packetization methods but, unlike in the master–slave relationship between the MS and the BTS of GSM or in other circuit-switched networks, the far-end client may accept or reject the requests, making it difficult to estimate the results with confidence.

The Explicit Congestion Notification (ECN) [Ramakrishnan *et al.* (2001), Westerlund *et al.* (2012)] is a measure for the eNodeB to signal impending congestion to the UE, by marking on a two-bit space in the IP header of each packet. Like scheduling, ECN is a mono-directional technique usable only by the eNodeB, as illustrated in Fig. 8.4. Note that a UE cannot identify the network node or packet router that marked the received packets. For example, a UE may receive ECN-marked packets even when the downlink is of high quality or the cell is not loaded, caused by congestion somewhere in the paths or in the far-end uplink. Although each session is negotiated between two UEs via the IMS, the conditions can be re-negotiated or updated if the service quality negotiated is not met. Therefore session re-negotiation is another measure to deal with the variance of channel conditions or cell loading levels, and not to overcharge the UE for unfulfilled services.

Although the media client and the LTE modem are shown in Fig. 8.4 as two separate entities inside the UE, they can be implemented in a single processor. The majority of the complexity used in the media client will be for the video codecs, and the receiver, especially the turbo decoder, will consume most complexity of the modem. If the two entities are closely located, the media client may be able to control media processing more efficiently by taking the channel information, e.g., the status of data buffers in the LTE modem, into account. For example, a media client may be able to reduce the possibility of encoding more than can be transmitted by monitoring the level of transmit buffer, even though the information reflects only the status of the near-end uplink. If a buffer overflow is detected, the media client can reduce the encoding bit-rate immediately.

It takes more than a generation to construct complex mobile multimedia communications systems such as VoLTE, whose initial developments began with 3.5G radio access and core network technologies such as HSPA, new resource management system IMS, and 3G media compression algorithms. After LTE and EPC joined as new radio access and core network technologies, VoLTE started to be commercially deployed, albeit utilising media codecs built one or two generations earlier for circuit-switched networks. Enhanced Voice Services (EVS) and High Efficiency Video Coding (HEVC) are new speech and video coding standards that achieve higher compression efficiency and error resilience, using higher complexity, and meet the new situations of packet-switched networks.

8.3 Radio Signal Processing Procedures

In the LTE modem, packets to and from the media client pass through four layers of radio protocols. There are other important protocols not shown in Fig. 8.4, such as the Radio Resource Control (RRC) [3GPP (2017q)], which do not directly handle the media but play essential roles in the operation of LTE. RRC makes key decisions such as admission control and handover. The Non-Access Stratum (NAS), which interacts between the UE and the Mobility Management Entity (MME), is also one of such protocols. Figure 8.5 illustrates the responsibilities of each protocol layer, where the media and radio-related operations differ significantly from those in circuit-switched mobile communications networks. As in the W-CDMA modem, interfaces between two layers of radio protocols are named the *logical*, *transport*, and *physical channels*. Following the seven layer classification of the Open Systems Interconnection (OSI) [ISO/IEC (1994)] reference model, PHY is regarded as layer 1 (L1) while MAC, RLC, and PDCP are considered as layer 2 (L2). RRC is often regarded as layer 3 (L3) but NAS or other protocols above RLC do not exactly fit any layer in the classic model.

Figure 8.5 illustrates that the PHY is under the control of the RRC. The set of protocol layers from the media client to the PHY is referred to as the User Plane (UP) in that the encoded media and the end-to-end control signaling pass through these protocols. The set of protocol layers from the NAS to the PHY is referred to as the Control Plane (CP) in that the control signaling to and from the radio access and core networks pass through

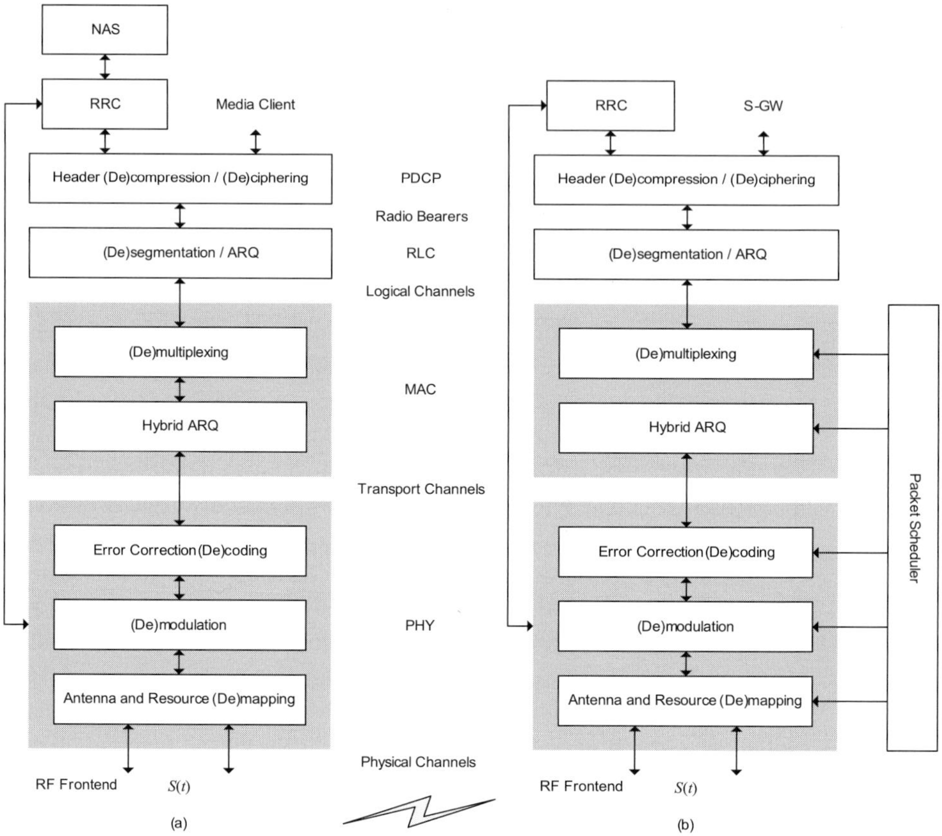

Fig. 8.5 LTE modem protocol architecture. (a) UE. (b) eNodeB.

those protocols. Here, we focus on the UP but discuss those aspects of CP to which the media and radio signal processing are related. As in the W-CDMA, we approach the radio signal processing of LTE in three steps: analysis of the PDCP, RLC, and MAC protocols commonly applicable to the uplink and downlink, and analysis of the radio signal processing operations in the PHY for each link. The radio signal processing operations of PHY are further separated into two stages.

8.3.1 Packet Data Convergence Protocol

At the Packet Data Convergence Protocol (PDCP) [3GPP (2017p)], packets from the media client have their accumulated IP headers compressed and ciphered. Note that the header compression, which is optional while the ciphering is mandated, is not applied to the control information even if it includes the headers. PDCP is also present in the W-CDMA modem but bypassed by the bit-stream in voice service or 3G-324M as the transmission paths are not paved with the IP. As in the error resilient encoding of video, the PDCP has to compress the IP headers without losing the error resilience significantly, i.e., maintaining their *robustness* to some extent. Adding a controlled amount of redundancy is a proven measure against channel impairments but needs to be performed with the required level of error resilience in mind. The procedures used to reduce the bit-rate of IP headers are taken in the Robust Header Compression (ROHC), an internal module of PDCP that consists of three operating modes: Uni-directional mode (U-mode), Bi-directional Optimist mode (O-mode), and Bi-directional Reliable mode (R-mode).

ROHC consists of a set of Internet technologies to reduce the overhead of protocol headers that were standardized by the Internet Engineering Task Force (IETF), the organization managing the Internet standards, but ROHC was designed with mobile communications networks in mind and has rarely been adopted in fixed networks where the bandwidth is abundant, or lower-tier wireless networks such as Wi-Fi where the quality of service needs not be guaranteed. By selecting an appropriate operating mode that best meets the situation, ROHC can control the tradeoff between spectral efficiency and error resilience. The ROHC profiles and parameters of each session are configured by the RRC.

ROHC reduces the bit-rate of media packets based on observations of the fields in the protocol headers. Figure 8.6 shows the headers of the RTP/UDP/IP protocols. Analyzing the headers of the protocols in the bit-stream identifies each field's expected update frequency and locates the opportunities for bit-rate reduction where the frequency of updating the value is lower [Rein *et al.* (2003), Taylor *et al.* (2005), Fitzek *et al.* (2005)]. The fields can be grouped into five classes [Bormann *et al.* (2001)], based on the expected frequency of the updates during a call, or a *session*, a term used in packet-switched networks. The values of INFERRED fields, such as the packet length, can be inferred from other fields or directly counted, and need not be transmitted. The

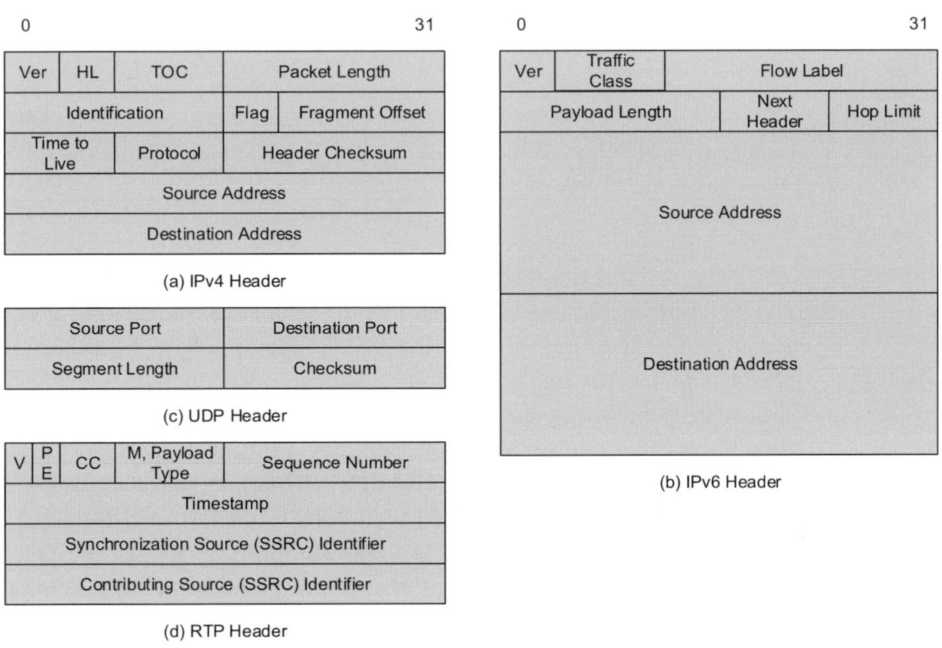

Fig. 8.6 RTP, UDP, IPv4, and IPv6 header fields.

Table 8.4 Class of RTP/UDP/IPv4 header fields.

	Field	Size (bits)	Class
IPv4	Version (Ver)	4	Static
	Header length (HL)	4	STATIC-KNOWN
	Type of service (TOC)	8	CHANGING
	Packet length	16	INFERRED
	Identification	16	CHANGING
	Reserved flag	1	STATIC-KNOWN
	Don't fragment flag	1	STATIC
	More fragments flag	1	STATIC-KNOWN
	Fragment offset	13	STATIC-KNOWN
	Time to live	8	CHANGING
	Protocol	8	STATIC
	Header checksum	16	INFERRED
	Source address	32	STATIC-DEF
	Destination address	32	STATIC-DEF
UDP	Source port	16	STATIC-DEF
	Destination port	16	STATIC-DEF
	Length	16	INFERRED
	Checksum	16	CHANGING
RTP	Version (V)	2	STATIC-KNOWN
	Padding (P)	1	STATIC
	Extension (E)	1	STATIC
	CSRC counter (CC)	4	CHANGING
	Marker (M)	1	CHANGING
	Payload type	7	CHANGING
	Sequence number	16	CHANGING
	Timestamp (TS)	32	CHANGING
	SSRC	32	STATIC-DEF
	CSRC	0–480	CHANGING

STATIC fields, such as the version of the protocol, are expected to be constant throughout a session. Therefore, values of the STATIC fields need to be transmitted only once if the channel is error-free or transmitted until successfully received. The STATIC-DEF fields, such as the source and destination addresses, are the fundamental parameters that define the operation of each session, and they are handled as STATIC, as they do not change during a session. The STATIC-KNOWN fields typically have well-known values and do not need to be transmitted at all. Finally, the CHANGING fields are expected to vary during a session in some fashion. The bit-rate can also be reduced with the CHANGING fields if the pattern of change can be exploited. For example, if the step of a field by which the value is updated for each packet is known, only an initial value and the step need to be transmitted. Table 8.4 outlines the fields of the RTP/UDP/IP protocols and their classes.

The sequence number field in the RTP header is incremented by one for each new packet but can still be inferred if the packets are successfully received. Therefore, this number needs to be transmitted only when some packets are lost and the receiver loses the current value, or when the sequence number reaches its maximum value, to re-initialize the packet counter at the receiver. The Timestamp field, which can be used to synchronize the speech and video frames or control the playback time, is set to the instant when the first octet of media data in the payload is generated. The Timestamp is incremented by a unit, TS_{stride} ms, and by $TS_{\text{stride}} \times n$ ms if n speech frames are transmitted per packet. A period of silence also increases the value of Timestamp.

Therefore Timestamp (TS) can be modeled as a linear function of a scaled version of the Timestamp, $TS_{\text{scaled}} \times n$, and an offset, TS_{offset},

$$TS = TS_{\text{scaled}} \times n \times TS_{\text{stride}} + TS_{\text{offset}}. \tag{8.1}$$

Note that in the case of video codec, multiple video packets may share a Timestamp value if the packets constitute a large video frame. From the above relationship, it is clear that once TS_{offset} is known to both sides, only n needs to be transmitted. Likewise, further savings of bit-rate can be obtained if these principles are applied to the UDP header where the source and destination ports are static, and the segment length can be inferred.

Robust Header Compression

In ROHC, the entity located at the sender of packets that compresses the headers is called the *compressor*, and its counterpart located at the receiver that reconstructs the headers is called the *decompressor*. It is essential that the compressor and the decompressor share a common set of key parameters, *context*, for a synchronized operation. As header compression proceeds, the need for a change of context naturally occurs, of which the decompressor needs to be informed if the compressor and the decompressor are to operate using the same context. During transmission, loss of packets can make the context of the decompressor obsolete, triggering error propagation in the reconstruction of the headers. In ROHC, three states are defined for the compressor: Initialization and Refresh (IR), First-Order (FO), and Second-Order (SO). Figure 8.7 illustrates the state transition diagram showing the three states of the compressor.

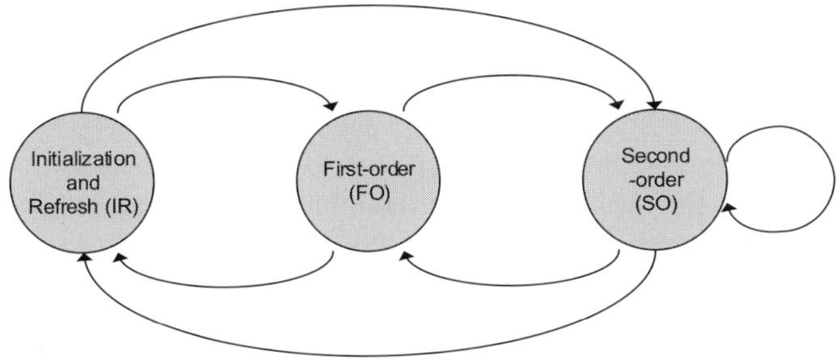

Fig. 8.7 Compressor state transition diagram.

In the IR state, the compressor sends all headers in uncompressed formats, with additional information to initialize the context at the decompressor. Therefore the packets transmitted in this state are larger than their original sizes. In the FO state, the headers are sent in partially, but not completely, compressed formats. The compressor changes its state from IR to FO, when it has confirmed that the decompressor has received the necessary context, and from SO to FO, when new headers no longer follow the observed patterns or cannot be efficiently compressed with the context the decompressor currently maintains. In the FO state, negative feedback from the decompressor that the context is no longer valid or that an update was not received properly can also trigger transitions to the lower state. In the SO state, the efficiency of compressing headers reaches the maximum, able to predict the headers completely only from the sequence number of the RTP header. The compressor enters the SO state when it becomes apparent that the decompressor successfully identified the relationships between the fields in the headers and the sequence number. When the compression efficiency reaches its maximum, it is sufficient to transmit only a one-byte ROHC header, as shown in Fig. 8.12, instead of the 40 bytes consisting of the RTP (12), UDP (8), and IPv4 (20) headers. Such a compression ratio is no less significant than those of typical speech codecs, contributing to the capacity of packet-switched mobile communications networks. In the case of IPv6, a two-byte UDP checksum should accompany the ROHC header, making three bytes as the minimum header size achievable.

An interesting analogy can be found between header compression and video compression. In 3G-324M, encoding a series of P-frames after an I-frame increases temporal quality without reducing spatial quality, since a P-frame is typically much smaller than an I-frame. However, such a saving comes with a more fragile data structure as the loss of video data belonging to a P-frame can initiate noticeable error propagation to the following P-frames, even when there is no further error, until a clean I-frame stops it. In ROHC, the IR state of the compressor is conceptually similar to the state of video encoder generating I-frames, which reduces spectral efficiency but increases error resilience in return. On the other hand, the SO state can be considered as similar to that of encoding P-frames, which increases spectral efficiency but reduces error resilience. The FO state can be considered as an intermediate state between the two extreme cases. As the loss of video data over W-CDMA initiates the encoding of one or more I-frames, through issuing the videoFastUpdatePicture messages, loss of packets over LTE initiates the transmission of uncompressed IP headers. In Fig. 8.7, each branch represents the transition between two states. Although only the SO state is drawn to include a branch for self-transition, the other two states can also stay in the same state for consecutive packets if a transition is unnecessary. The compressor can switch itself from one state to another in one step to handle an abrupt loss of synchronization.

Like the compressor, the decompressor at the other side of channel has three states: No Context (NC), Static Context (SC), and Full Context (FC), as shown in Fig. 8.8. Each state is matched to a counterpart state of the compressor. Operation of the decompressor starts with the NC state, where it has yet to successfully decompress the headers of at least one packet. In the SC state, it is assumed that only the static information, which does not change during a session, such as the IP address or port number, is valid. It is

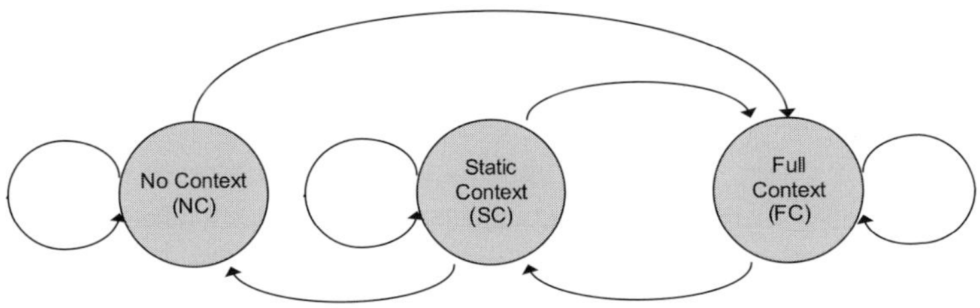

Fig. 8.8 Decompressor state transition diagram.

not possible for the decompressor to switch its state directly from the NC state to the SC state. It is only allowed to switch from the NC state to the FC state, where both sides have fully synchronized the contexts. From the FC state, once the decompressor repeatedly fails to reconstruct the headers, the state is first shifted down to the SC state, and further to the NC state if the failure continues. In Fig. 8.8, a branch for self-transition is available for each state but unlike that of the compressor, the structure of the state transition diagram is not symmetric as there is no direct path from NC to SC, or from FC to NC. For example, the state can be switched from FC to SC or from SC to NC if the proportion of CRC failures exceeds certain thresholds. As soon as the headers are better decompressed, the state can be switched from either NC or SC to FC. The operation of the compressor and the decompressor, each of which has three internal states, can be classified into three operating modes. In each operating mode, the compressor and the decompressor interact in a unique fashion.

In the U-mode, a feedback channel is not required. Instead the compressor periodically refreshes the context at the decompressor by sending up-to-date information. The compressor switches its state based only on the periodic time-out and the irregularities in the header field patterns it generates. For example, the 16-bit sequence number in the RTP header can be, depending on the initial values, re-set to zero within a small number of packets. Once some packets are lost and the context at the decompressor becomes obsolete, error propagation can occur, which would continue until the periodic context refreshing stops the propagation. In this mode, the efficiency of header compression is naturally low, but the mode can handle situations where a feedback channel from the decompressor is not available. Conceptually, this mode can be considered as similar to the video encoding strategy of 3G-324M, in which an I-frame followed by a fixed number of P-frames is periodically encoded but the videoFastUpdatePicture message is neither sent nor received. Even in situations where a feedback channel is present, the ROHC must start from the U-mode and switch to other bi-directional modes, such as the O- or R-mode after some warming up period.

After entering the O-mode, the compressor does not refresh the context at the decompressor unless explicitly requested. This mode provides the highest header compression efficiency as long as the error rate is limited to an acceptable level. Conceptually, this mode is similar to the video encoding strategy, in which all frames, after an I-frame, are encoded as P-frames until an error occurs and a videoFastUpdatePicture message is received. As illustrated in Fig. 9.24, in the O-mode, a series of packets can be lost if the context at the decompressor is not recovered soon. Therefore the performance of O-mode depends upon the availability and reliability of the feedback channel. The duration of error propagation can be reduced if a sufficient bit-rate is reserved for the feedback channel. In the R-mode, all context updates at the decompressor are acknowledged to the compressor, which requires a higher-capacity feedback channel than the O-mode. The compressor considers the synchronization of context to be more important than the efficiency of header compression. In VoLTE, the R-mode is generally regarded as consuming excessive bit-rates for the quality improvement it provides and the O-mode is the de-facto standard operating mode of ROHC.

Figure 8.9 illustrates the full-duplex operation of ROHC at the PDCP. In the reverse link direction at the UE, packets with uncompressed IP headers are sent from the media client. After header compression, the packets are ciphered and handed down to the MAC. In the forward link direction at the UE, packets with compressed IP headers are received from the RLC. After deciphering and header decompression, the packets, now with full IP headers, are sent to the media client. Note that in Fig. 8.9, the ROHC maintains a set of contexts for each direction. The compressor attaches the context information requested by the decompressor to the compressed headers such that the total bit-rate is within the granted QoS level of the session.

ROHC Packet Type

As multiple packet streams consisting of speech, video, or control packets may flow simultaneously between the compressor and the decompressor, ROHCs at both sides

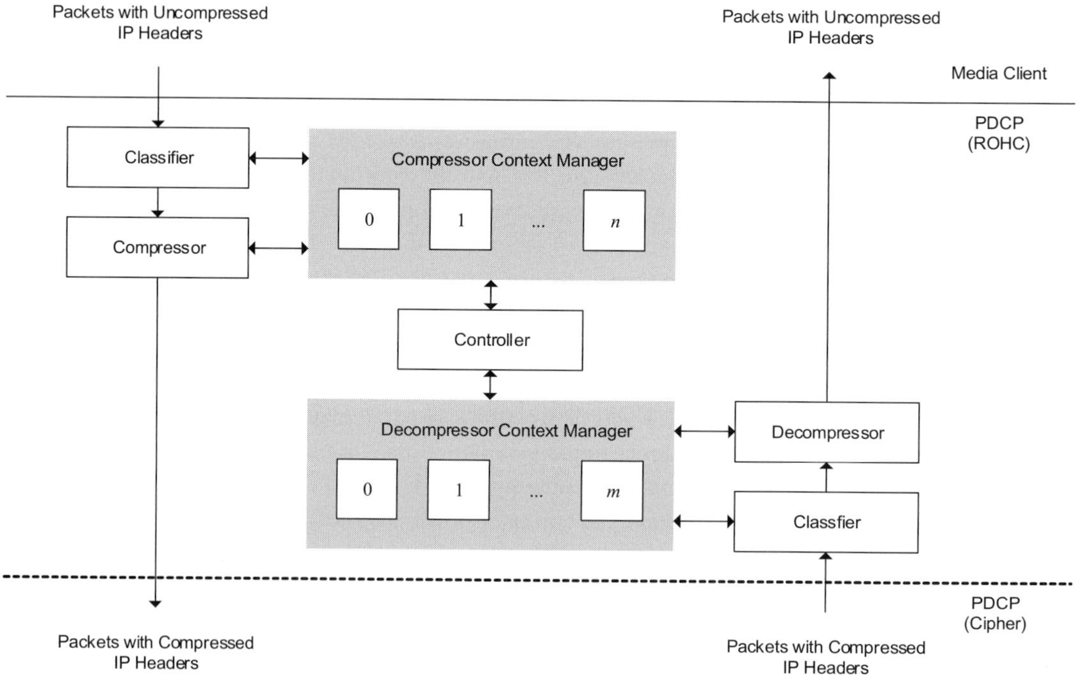

Fig. 8.9 Full-duplex ROHC operation.

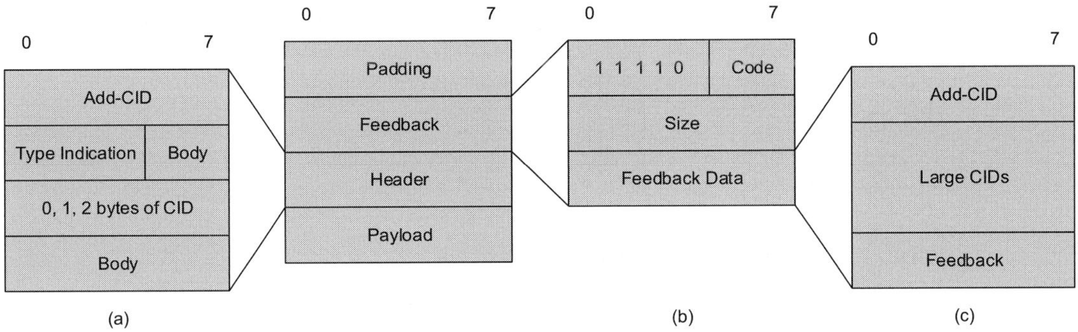

Fig. 8.10 ROHC packet formats. (a) Header. (b) Feedback. (c) Feedback data.

of the channel have to handle the IP headers of each stream separately. To identify the context of each stream, a field, the Context Identifier (CID), is included in each ROHC packet that has a format made up of four fields: padding, feedback, header, and payload, as shown in Fig. 8.10. When a radio protocol logically located below the ROHC, such as the RLC, requires the size of the data input, the service data unit (SDU), to be of lengths longer than the header-compressed packets, the packets are padded with an integer number of a pre-determined byte, 1110000.

The capability of an ROHC is classified into several profiles based on the headers it can compress. For example, profile 1 compresses the RTP/UDP/IP headers while profile 2, used to handle the RTCP packets, compresses the UDP/IP headers. Profile 3 compresses the Encapsulating Security Payload (ESP) and the IP headers, and profile 0 covers the remaining packet types that cannot be compressed by the other profiles. The ROHC header can have formats specific to the profile used. The header format for profile 1 is shown in Fig. 8.10(a). The header starts with an Add-CID byte and optionally another byte whose bits are divided for type indication (5 bits) and body (3 bits), followed by zero, one, or two bytes of CID. Note that the CID here refers to a context different from that in the feedback. The CID is followed by a header body of variable-length.

There are formats for packets to initialize and refresh (IR) the context. The IR packets can be classified into two types: Type IR, shown in Fig. 8.11(a), fully initializes the context at the decompressor, regardless of the previous state while Type IR-DYN, as shown in Fig. 8.11(b), refreshes only the dynamic part of the context, such as the timestamp and sequence number of RTP. The bit-rate is reduced from ROHC as the proportion of Types IR and IR-DYN packets are very small, typically less than 1–2%.

Figure 8.12 shows the ROHC header in what is theoretically the most compressed form of the O-mode: a one-byte UO-0 packet consisting of a four-bit SN and a three-bit CRC. The sequence number (SN) is the sequence number of RTP header encoded in the window-based least significant bit (LSB) encoding. In this state, the compressor and the decompressor have reached a complete synchronization where each attribute in the IP headers can be restored from the SN only. In practice, the size of compressed RTP/UDP/IP headers is three to four bytes under normal ROHC operations. With IPv6,

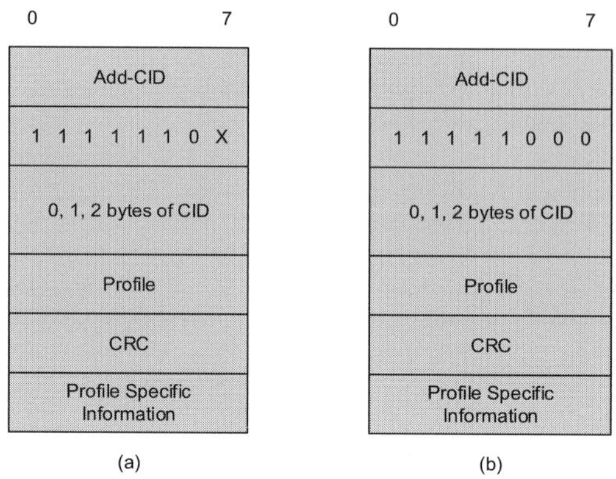

Fig. 8.11 IR packet formats. (a) Type IR. (b) Type IR-DYN.

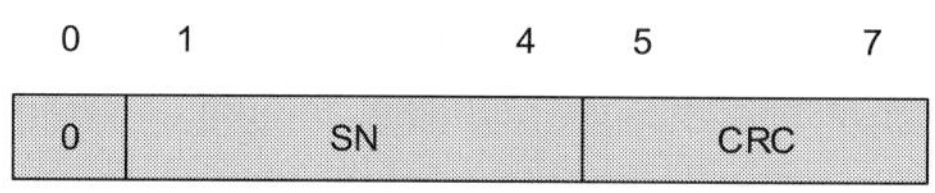

Fig. 8.12 Fully compressed ROHC header.

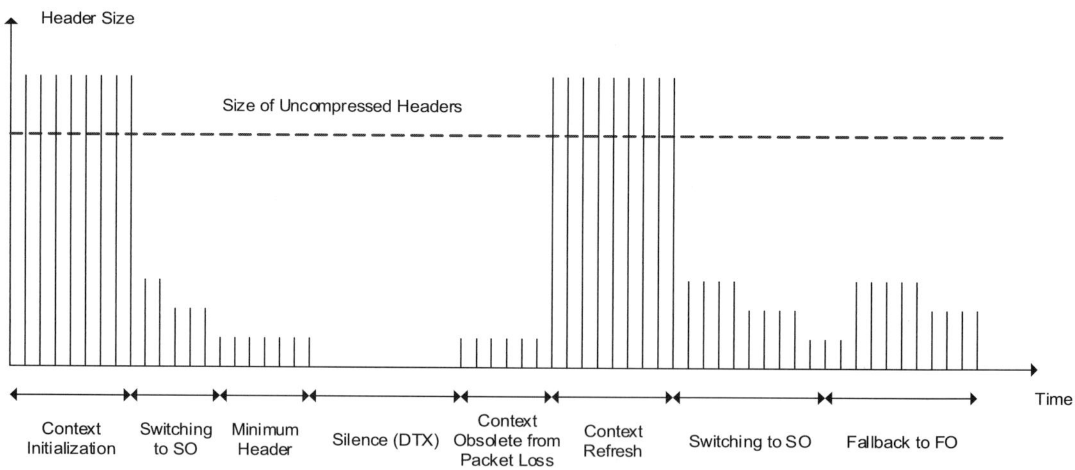

Fig. 8.13 ROHC header-size variation.

it is a common practice to complement a UO-0 packet with a two-byte checksum of the UDP.

Figure 8.13 illustrates the variation of bit-rate for IP headers when the ROHC is employed, assuming the same payload size for each packet. The variation starts from the initialization of context, followed by an abrupt reduction of the overhead as the context is established, and a re-initialization requested by the decompressor after a loss of packets. After the compressor receives an acknowledgment of successful context initialization from the decompressor, its state is switched from IR to SO, where the header size

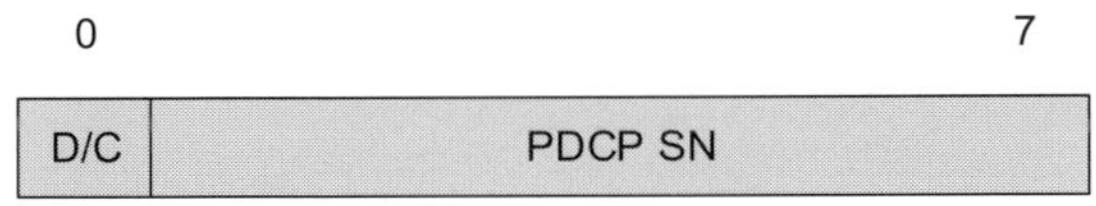

Fig. 8.14 PDCP header with 7-bit SN.

approaches the minimum or even zero during silence. After the packet loss completely obsoletes the context at the decompressor, the compressor, whose state now falls back to IR, refreshes the damaged context with a series of IR packets. With a confirmation of successful context resynchronization from the decompressor, the compressor returns its state to SO. Then the reduction of IP overhead resumes but the state of the compressor falls back to FO after the decompressor reports that it has failed to decode a proportion of received packets, which is more than acceptable.

It can be seen that when the ROHC is used, the overhead from IP headers can be temporarily higher than that from the uncompressed headers, for example, when Type IR packets containing the information for full context and additional control information are transmitted at the beginning of the session. With a set of inconsistently defined data compression techniques to represent the fields of the IP headers, ROHC shows the challenges of extending the use of protocols designed for the Internet to mobile communications networks.

After ROHC operation, a one-byte PDCP header is attached to the header compressed packets before they are handed down to the RLC. The header, as shown in Fig. 8.14, starts with a one-bit field that indicates that the payload is Data (D) or Control (C), followed by a seven-bit Sequence Number (SN) used to deliver the PDUs in a proper order. The SN is used in the packet re-ordering when the packets are received out of order, e.g., when resetting the RLC in the handover or re-establishing an RRC connection. If a longer SN, 12 bits, is used, up to 2048 packets can be monitored for re-ordering. Such a long SN is likely to be used for non-real-time applications.

During handover, the PDCP SNs of the uplink and downlink are passed from the source eNodeB to the target eNodeB. The minimum amount of overhead attached to a PDCP SDU, when it leaves the PDCP toward the RLC, is therefore two or four bytes: one-byte ROHC header, two-byte UDP checksum when IPv6 is used, and one-byte PDCP header. The bit-rate saving from the ROHC would be largest in the handling of frequently transmitted small packets such as speech. The headers attached in radio protocols logically lower in the architecture than PDCP, i.e., the RLC and MAC, also play essential roles for packets to safely traverse the wireless channel. PDCP packets are ciphered with the Advanced Encryption Standard (AES) algorithms.

8.3.2 Radio Link Control

The header-compressed and ciphered packets in the PDCP are passed to the Radio Link Control (RLC) [3GPP (2017o)], which has three operational modes: Transparent Mode (TM), Unacknowledged Mode (UM), and Acknowledged Mode (AM), as in the RLC of W-CDMA. In the operational modes other than TM where no header is attached to the

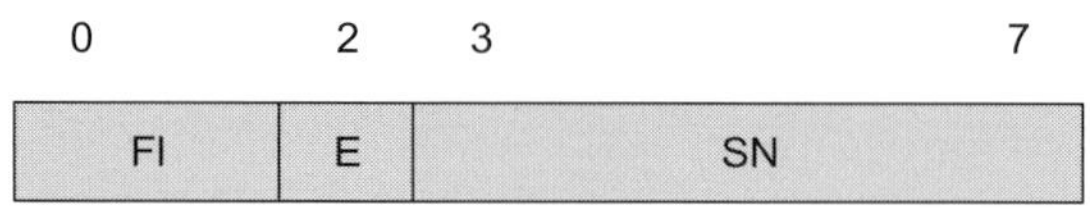

Fig. 8.15 RLC header with 5-bit SN.

payload, RLC may re-order packets delivered out of order or detect duplicated packets, due to the HARQ or handover. In VoLTE, the UM is typically used for transmission of RTP and RTCP packets, leaving the major responsibility for reliable transmission to the MAC and PHY. At the RLC, SDUs can be segmented or concatenated, depending on the decision of the scheduler to match the available radio resource.

As shown in Fig. 8.15, an RLC PDU starts with a byte whose bits are allocated as: a two-bit Framing Information (FI), a one-bit Extension (E), and a five-bit Serial Number (SN). The FI bits indicate whether the first byte of data corresponds to the first byte of an RLC SDU, and whether the last byte of data corresponds to the last byte of an RLC SDU. For example, FI = 00 represents that the first byte of data corresponds to the first byte of an RLC SDU, and the last byte of data corresponds to the last byte of an RLC SDU. Likewise, FI = 11 represents that the first byte of data does not correspond to the first byte of an RLC SDU, and the last byte of data does not correspond to the last byte of an RLC SDU either. Here FI = 01 and 10 represent the other two possible situations.

The E bit indicates whether data or a set of control field and length information follows the RLC header. The SN bits are used to facilitate the in-sequence delivery of packets. With the five-bit SN, up to 32 PDUs can be ordered at the RLC but it is very unlikely that two different PDUs with the same SN will arrive within such a small time window. Although lost PDUs can be identified by inspecting the SN in the UM, no further actions are made by the RLC since the UM is used mainly for delay-constrained services such as VoLTE. The interfaces between the RLC and the MAC are called logical channels. The logical channel over which the data of the uplink and downlink is transported is the Dedicated Traffic Channel (DTCH).

8.3.3 Medium Access Control

Medium Access Control (MAC) [3GPP (2017n)] is assigned the responsibility for the reliable transmission of data. MAC multiplexes the SDUs received from the RLC by prioritizing the logical channels, and de-multiplexes the received PDUs. MAC also controls the HARQ procedures at the PHY to meet the required QoS targets, setting performance objectives such as the maximum packet loss ratio or delay tolerated. Unlike the PDCP or RLC, which performs similar tasks when the protocol is located at the UE or the eNodeB, the MAC at eNodeB has far more responsibilities than those of its counterpart at the UE.

In LTE, each transmission, over the uplink or downlink, requires a permission from the MAC at the eNodeB. The MAC at the UE periodically reports the status of the uplink buffer to the MAC at the eNodeB, which also controls the downlink scheduling.

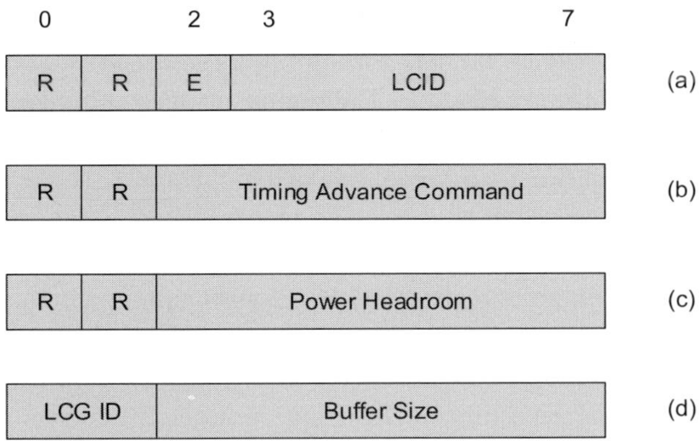

Fig. 8.16 MAC header and control elements. (a) Subheader. (b) Timing advance command. (c) Power headroom. (d) (Short) buffer status.

The MAC has the additional responsibilities of controlling the Random Access, Discontinuous Transmission (DTX), Discontinuous Reception (DRX), and Semi-Persistent Scheduling (SPS) procedures.

As in the other radio protocols of LTE, a MAC PDU starts with a header, which consists of one or more MAC PDU subheaders. The length of each subheader can be one, two, or three bytes. Each subheader type is used in front of a MAC SDU, a MAC control element, or padding bits, as a MAC header can be followed by zero or more MAC SDUs, zero or more MAC control elements, or padding bits. Figure 8.16 shows the formats of the MAC subheader and the MAC control elements. The minimum amount of overhead that can be attached to a MAC SDU, when it leaves the MAC toward the PHY, is one byte. However, it is a common operating practice of the MAC at the eNodeB to assign one or two additional bytes to provide the room for control information that the UE may want to transmit, such as a timing advance command or a power headroom report.

In Fig. 8.16(a), the one-byte subheader consists of a two-bit Reserved (R), a one-bit Extension (E), and a five-bit Logical Channel ID (LCID) fields. Both R bits are set to 0. If set to 0, the E bit indicates that more fields are present in the MAC header. The LCID bits represent the identity of the logical channel, type of control element, or the padding. When a MAC SDU is to be sent to the uplink or downlink, for example, a length field of seven or fifteen bits following one Format (F) bit, which represents the size of the SDU in bytes, is attached to the LCID. During typical VoLTE sessions, a speech packet is transmitted each 20 ms when voice activity is present, exploiting the 1 ms Transmission Time Interval (TTI) of LTE. However, there can be situations when multiplexing multiple MAC SDUs into a MAC PDU is preferred.

The uplink timing advance originated in TDMA networks such as IS-54 and GSM, while cdma2000 and W-CDMA, which do not require the uplink signals from the MSs or UEs to be synchronized, do not need this technique. The radio signal processing procedures of LTE, however, do resort to the timing advance. As in GSM, where the BTS controls the timing advance in 64 increments, the eNodeB controls the transmission timing of each UE with a command of six bits as shown in Fig. 8.16(b), albeit with a much finer time resolution. The LCID field is set to 11101 in the MAC subheader if the control element is used for timing advance. The timing advance command, the six bits after the two R bits in the MAC control element, indicates a requested change of uplink timing relative to the current timing, as a multiple of $16T_s$ where T_s is one of the fundamental time units of LTE, defined to be $\frac{1}{30{,}720{,}000}$ seconds. When faster adjustments are necessary to establish synchronization, for example, when the eNodeB responds to the random access requests from an unregistered UE, a longer timing command of eleven bits can be used to signal larger step sizes.

The Power Headroom Report (PHR), which has a similar structure as the timing advance command, indicates the estimated difference between the current and the maximum transmit powers of the UE, in steps of 1 dB. For example, the lowest level represents that the measured quantity is between -23 and -22 dB while the next level represents that it is between -22 and -21 dB. The highest level, 40 dB, means that the transmit power can be increased by more than 40 dB from the current value. PHR can be transmitted periodically or in an event-driven fashion, as configured by the RRC. LCID is set to 11010 for PHR.

The Buffer Status Report (BSR) represents the total amount of data left in the transmission buffer, in bytes, across all logical channels of a logical channel group, measured after all MAC PDUs for the current TTI have been built. The report includes the information on all data available for transmission in the RLC and PDCP, except for the RLC and MAC headers. The two-bit Logical Channel Group ID (LCG ID) identifies the group of logical channels whose buffer status is being reported. The control element shown in Fig. 8.16(d) corresponds to the short or truncated BSR control element but a three-byte BSR is also available for situations when detailed information on the uplink buffer status is required. LCID is set to 11101 and 11100 for short and truncated BSR, respectively, while 11110 is used for long BSR.

When the data arrives in an empty transmission buffer, or a UE is handed over to a new eNodeB, a BSR needs to be sent. In this case, a Scheduling Request (SR) can also be transmitted using a one-bit flag on the Physical Uplink Control Channel (PUCCH), or with random access procedures when the uplink physical channel is not yet established. PUCCH includes a set of formats in which several types of control information can be efficiently packed. The interfaces between the MAC and the PHY are called transport channels. The transport channel over which the uplink data from each UE is transported is the Uplink Shared Channel (UL-SCH). The counterpart of the UL-SCH in the downlink direction is the Downlink Shared Channel (DL-SCH).

Figure 8.17 illustrates the radio signal processing procedures of LTE in the radio protocols, where two packets and another packet, from bearers 1 and 2, respectively, are header-compressed and ciphered at the PDCP, segmented or concatenated at the RLC, multiplexed at the MAC, and finally sent to the PHY, where the size of each MAC PDU needs to match the transmission capability of the PHY, i.e., the Transport Block Size (TBS). There can be only a finite number of such sizes, each of which is determined based on the modulation scheme, error correction coding, and the channel bandwidth. In the TDMA and CDMA mobile communications systems discussed so far, the channel

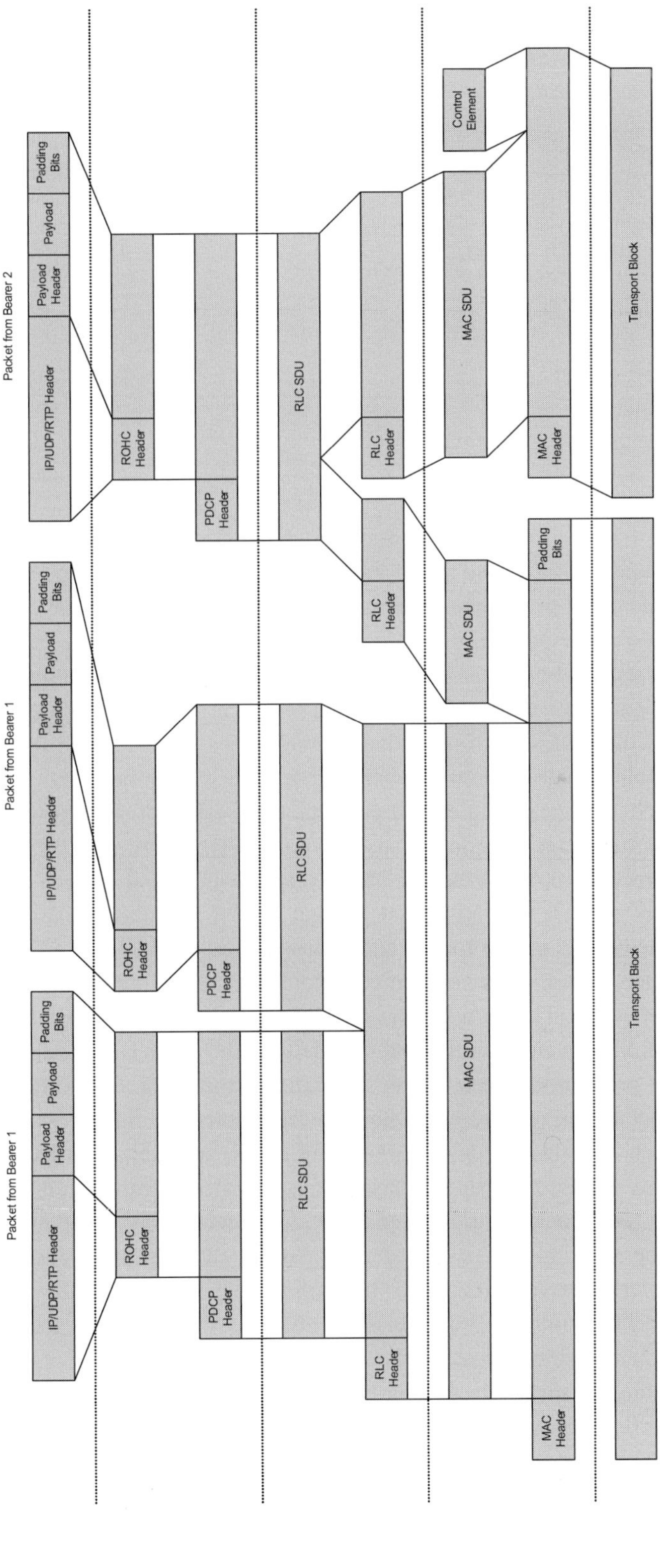

Fig. 8.17 Step 1: L2 header processing and data concatenation/separation in LTE baseband modem.

Table 8.5 Minimum overhead from packetization and L2 header processing (IPv4).

	AMR		AMR-WB	
	12.2 kbps	SID	12.65 kbps	SID
Encoded speech	244	39	253	40
Payload header	16	16	16	16
RTP padding	4	1	3	0
Minimum ROHC header	24	48	24	48
PDCP header	8	8	8	8
RLC header	8	8	8	8
MAC header	8	8	8	8
Other overhead	16	16	16	16
Total (bits)	328	144	336	144
LTE transport block size (bits)	392	208	392	208

bandwidth is fixed and therefore not considered as a controllable parameter. Similar, but not identical, procedures are used in the uplink and downlink.

A separate Quality of Service (QoS) control of each packet stream can be tricky when the packets are concatenated or segmented in the intermediate protocol layers. At the PDCP, the ROHC profile is matched to the required QoS of each packet stream. At the RLC, the SDUs from the same bearer can be concatenated into a single PDU. When multiple speech packets are bundled to reduce the consumption of the control channel for scheduling, such a concatenation can be efficient. For speech, packets belonging to the same bearer include the speech frame, SID, and the RTCP packet. At the RLC, the transmission mode is matched to the QoS requirement of each SDU.

It is possible that RLC SDUs larger than the available transmission capacity can be segmented, as shown for the packet from bearer 2. At the MAC, RLC PDUs with different QoS requirements can be concatenated, depending on the scheduling policy or the buffer status. The MAC at the eNodeB can exercise a complete control over the priority of packets in its buffer but the MAC at the UE has to multiplex packets based on a set of pre-defined rules, as the scheduling grant from the eNodeB does not specify how to allocate the resource over bearers. In circuit-switched networks, when the MS needs to transmit other types of data during a call, such data can be transmitted when voice activity is absent. In an urgent situation, the speech frames can be reduced or dropped to make room for the control signal, i.e., by frame stealing. The multiplexing rules defined for the MAC at the UE are generalizations of techniques designed to meet the needs of diverse service mixes over the uplink, e.g., sacrificing the quality of lower-priority traffic in a controlled fashion if the transmission of higher-priority traffic is urgent but the transmission opportunities are limited.

Table 8.5 shows the header-processing procedures where a speech frame or an SID of AMR and a new speech codec, AMR-WB, is systematically wrapped with a minimum amount of header or control information at each protocol layer to be sent over the wireless channel. The octet-aligned format, in which an RTP payload header of two bytes is inserted between a speech frame and an RTP header, of AMR and AMR-WB is assumed in the packetization. The use of IPv6 would incur a couple of bytes more in the ROHC header when the headers are fully compressed. Otherwise, additional overhead from the longer IP header would be significant.

Note that the ROHC header sizes of SIDs are in general larger than those of speech frames. Since an SID is transmitted following a period of silence, after either a last speech frame or another SID is sent, the timestamp in the RTP header is likely to have changed by a value that may not be predictable by the decompressor, which can break the synchronization of contexts between two sides. Therefore, the compressor has to include the difference from the new timestamp to the old timestamp of the last successfully transmitted packet in the ROHC header, making it longer than the minimum size. The values of other fields may have been changed too. The difference between the LTE transport block size and the total payload bits can be used to hold additional control information such as BSR or PHR. Note that typically longer, 2-byte headers are attached in the PDCP and RLC, and typical sizes of BSR and PHR are 4 and 2 bytes, respectively. Then the transport block containing a speech frame is transmitted from the PHY, using two resource blocks spanning a bandwidth of 360 kHz. For video or RTCP packets, which do not have fixed sizes, it can be difficult to predict the packet patterns as precisely as speech, shown in Table 8.5.

8.3.4 Physical Layer

Figure 8.18 illustrates the time-frequency structure of LTE where a chunk of bandwidth up to 20 MHz is divided temporally and spectrally to construct a set of radio resources. It is possible to simultaneously utilize additional chunks of bandwidth, through the Carrier Aggregation (CA), but media packets in VoLTE sessions are likely to be accommodated in the lower bandwidth to exploit the favorable propagation and coverage at the expensive spectrum. In the spectral domain, each frequency band consists of a number of subcarriers, each of which spans 15 kHz. In comparison with the carriers of GSM whose width is 200 kHz, the subcarriers of LTE are narrower but support multiple modulation schemes. The width of 15 kHz was chosen as the narrowest that would avoid interference from neighboring subcarriers with the filtering technologies available at the time. With a maximum 20 MHz of the channel bandwidth, there can be more than 1300 subcarriers.

In the temporal domain, a 0.5 ms slot includes six or seven Orthogonal Frequency Division Multiplexing (OFDM) symbols, and a subframe consists of two consecutive slots on the same subcarrier. Such a refined temporal and spectral resolution enables very diverse strategies for allocating radio resources, depending on the link quality and service policy. Just as a portion of time slots in the TDMA frames of GSM has to be spent for control signaling, a portion of radio resources in LTE has to be used for management purposes. In general, the finer the time-frequency resolution of radio resources, the higher the proportion of radio resources required for network management. This means that a design compromise has to be made on the minimum amount of radio resources an eNodeB can assign to a UE, which LTE defines to be a pair of resource blocks consisting

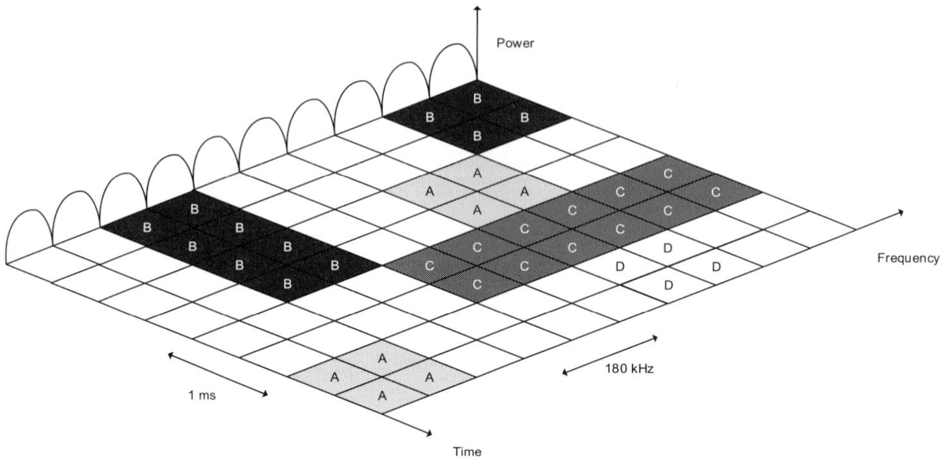

Fig. 8.18 Frequency-, time-, and power-domain representation of physical layer.

of twelve subcarriers for a duration of one subframe. Scheduling radio resources by a unit smaller than this is considered to be spectrally inefficient, and thus used only in limited cases.

The rectangular structure of the radio resources of LTE in the spatiotemporal domain applies to both the uplink and downlink, except that in the uplink, resource blocks transmitted at the same time are not separated in the frequency domain unless the CA is used. In this regard, the uplink of LTE uses a class of multiple access techniques called the Single-Carrier (SC) FDMA. The key reason to differentiate the uplink is to limit the variation of transmit power to the linear-operating range of the power amplifier at the UE, which is likely to be less efficient than its counterpart at the eNodeB. In Fig. 8.18, the resource blocks are dynamically assigned to four UEs: A, B, C, and D. Simultaneous scheduling of spectrally separated blocks, as done for UE B, is allowed only in the downlink. Therefore the spectral efficiency from scheduling is reduced in the uplink.

Figure 8.19 shows the structure of resource blocks spanning twelve subcarriers and two slots, i.e., a Resource Block (RB) pair. The smallest unit, consisting of a 0.5 ms resource block, is called the resource element. RB is the minimum scheduling size for the uplink and downlink. Modulation and error correction coding can be applied separately to each RB. A 20 MHz bandwidth is divided into 110 resource blocks in the frequency domain.

From the point of view of the receiver, the signals of the uplink and downlink would be seen in the frequency-power domain as illustrated in Fig. 8.20, where the allocation of RBs to five UEs, A–E, is illustrated. Although the power is drawn at similar levels, it is not necessary to align the received power level as strictly as in CDMA since the signals to and from each UE are separated in the temporal and spectral domains. Such an alignment may be necessary in the layered transmission modes, such as the Multiple-Input and Multiple-Output (MIMO), to help the receiver differentiate the overlaid layers, but generally opportunistic transmission techniques like MIMO are not considered for delay-constrained, conversational services with a guaranteed QoS.

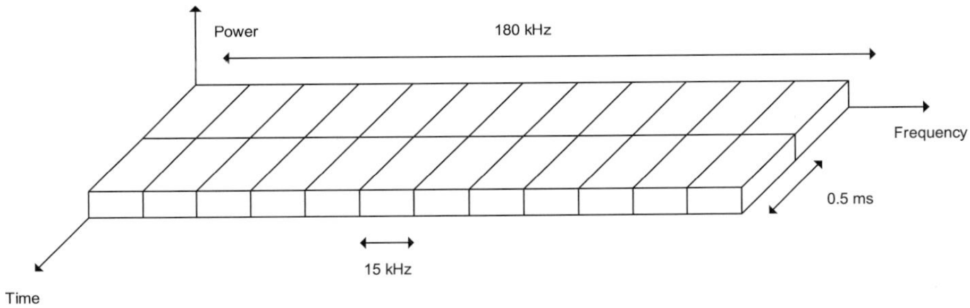

Fig. 8.19 Resource-block pair.

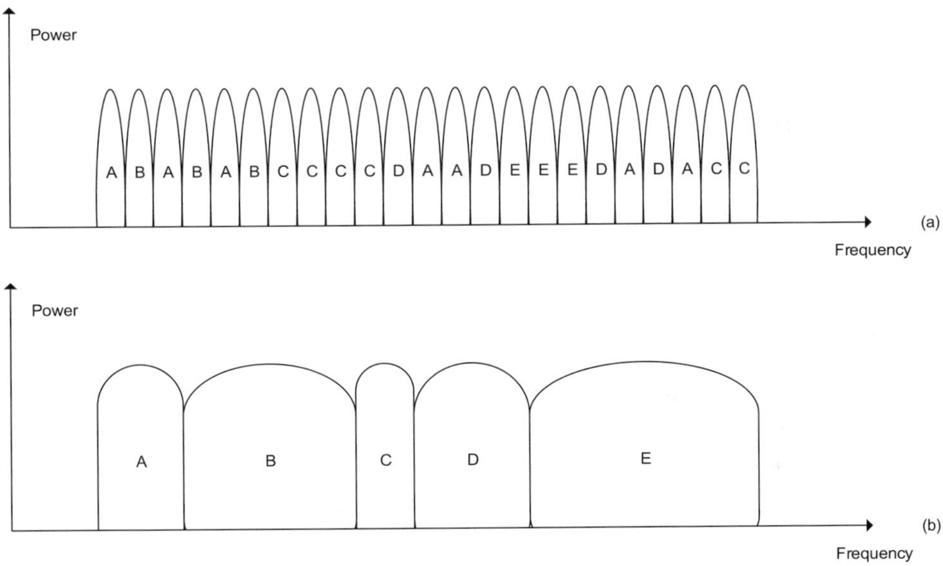

Fig. 8.20 Frequency- and power-domain representation of physical layer. (a) Uplink. (b) Downlink.

As in GSM, where hierarchical time-frame structures, ranging from a 120 ms 26-multiframe to a 0.577 ms time slot, are defined, LTE provides a set of time units to facilitate the management of radio resources. The largest unit is the 10 ms frame, which is divided into ten 1 ms subframes. As illustrated in Fig. 8.21, each slot, one half of a subframe, is made up of either six or seven symbols, depending on the length of the Cyclic Prefix (CP). CP is commonly used in OFDM to reduce the influence of interference from the delay spread. While a longer CP can be useful for larger cells, in typical operations, the eNodeB employs a normal CP as commercial networks are planned so that the maximum distance between the eNodeB and the UE is managed. Each time unit of LTE is an integer multiple of a basic time unit, $T_s = \frac{1}{15{,}000 \times 2{,}048} = \frac{1}{30{,}720{,}000}$. T_s was selected as the sampling time, assuming that a size 2048 fast fourier transform (FFT) is employed.

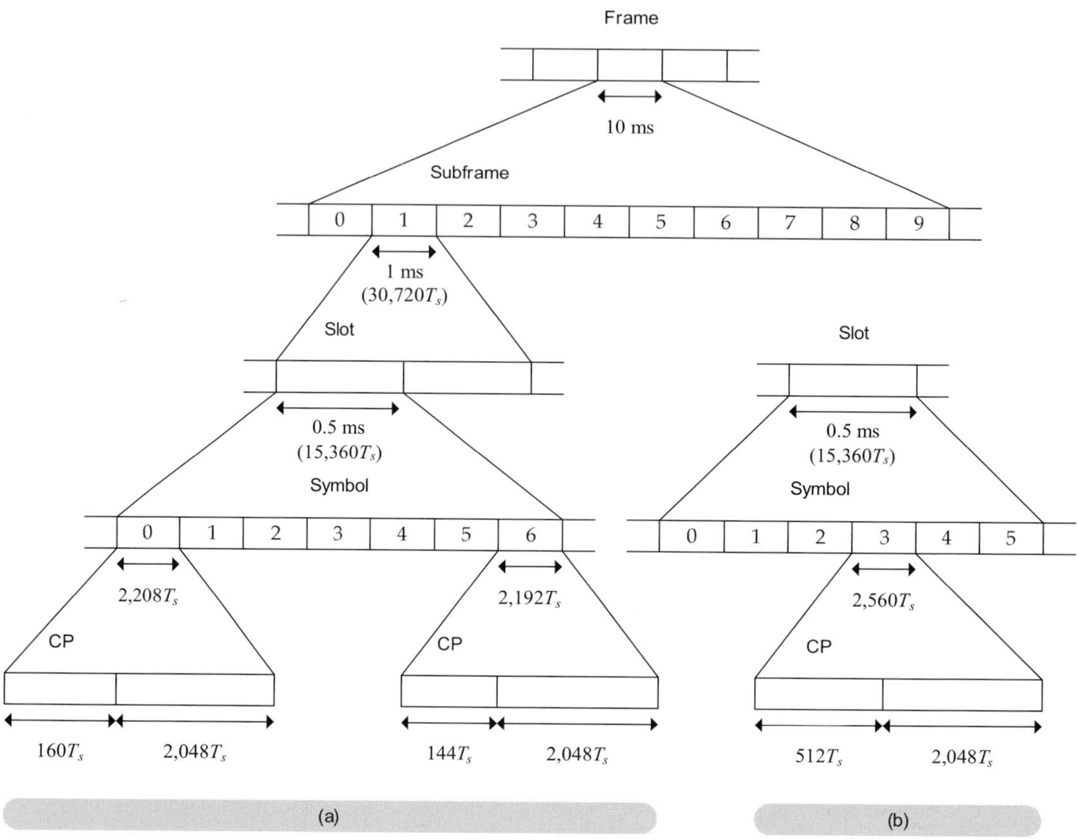

Fig. 8.21 Hierarchical frame structure: Type 1 (FDD). (a) 7-symbol CP. (b) 6-symbol CP.

The lengths of the CPs with seven and six symbols, i.e., the normal cyclic prefix and the extended cyclic prefix, are $144T_S$ and $512T_S$, respectively. With a normal cyclic prefix, a longer CP of length $160T_S$ is used in symbol 0 to fill the 0.5 ms slot completely. The duration of each symbol is $2,048T_S$ for both prefix types. Although the extended cyclic prefix is defined for larger cells, the normal cyclic prefix is used in most cases, as even in the larger cells, the link quality can degrade more from sources other than the delay spread, such as the noise or interference. Therefore, the lengths of a slot, a subframe, and a frame are $15,360T_S$, $30,720T_S$, and $307,200T_S$, respectively. Note that in the time division duplex (TDD) mode of LTE, Type 2, a slightly different frame structure with additional fields, is used.

Table 8.6 summarizes the key parameters used by LTE, where the channel bandwidth is scalable from 1.4 to 20 MHz in steps of 180 kHz. Among the four modulation schemes LTE provides, BPSK is used only for the control channels. A rate $\frac{1}{3}$ turbo code, which shares a similar decoder structure with that used in W-CDMA, is used for error control coding. Although theoretically more-efficient codes were available when LTE was designed, minimizing the need for new hardware to implement the iterative decoders for such codes was one of the key design objectives. This could change if significantly

Table 8.6 System parameters of LTE.

Multiple access	SC-FDMA (uplink), OFDMA (downlink)
Modulation	BPSK (control data), QPSK, 16QAM, 64QAM
Channel bandwidth (MHz)	1.4–20
Subcarrier width (kHz)	15

Table 8.7 Scalable channel bandwidth.

Channel bandwidth (MHz)	Number of resource blocks
1.4	6
3	15
5	25
10	50
15	75
20	100

improved performance of new channel codes would justify the additional complexity. As the radio signal processing procedures of the modem become more complex, the return from enhancing only one step of the chain of procedures becomes smaller. The introduction of HARQ to HSPA, where re-transmission and energy accumulation share the responsibility of channel coding to ensure successful transmission, further reduced the necessity of replacing the error correction code proven in W-CDMA.

Although the physical layer of LTE enables flexible channel bandwidth to reduce complexity and simplify implementation, only six bandwidths are defined in the RF specifications of LTE, as outlined in Table 8.7. The two lowest bandwidths, 1.4 and 3 MHz, are defined for easy migration from GSM and cdma2000. Notice that the higher bandwidths are integral multiples of W-CDMA channel bandwidth.

Uplink Signal Processing Procedures

Figure 8.22 illustrates a portion of the radio signal processing associated with the uplink [3GPP (2017k), 3GPP (2017l), 3GPP (2017m)]. The physical channel over which the uplink data from each UE is transported is the Physical Uplink Shared Channel (PUSCH). Each UE cannot have more than a single PUSCH. The counterpart of the PUSCH in the downlink is the Physical Downlink Shared Channel (PDSCH), which can also be used for paging information.

Suppose that some MAC PDUs, shown as the transport blocks in Fig. 8.17, are sent to the PHY. As the first step, a 24-bit CRC is computed from each transport block. If the CRC-appended transport block, called a code block, includes more than 6144 bits, the code block is segmented into two blocks. Then two 24-bit CRCs are computed and attached to the segmented code blocks. LTE provides two rate $\frac{1}{3}$ channel coding schemes: tail-biting convolutional coding and turbo coding. In the tail-biting convolutional coding with constraint length K, the initial values of the shift registers in the convolutional encoder are set to the last $K - 1$ information bits so that the initial and

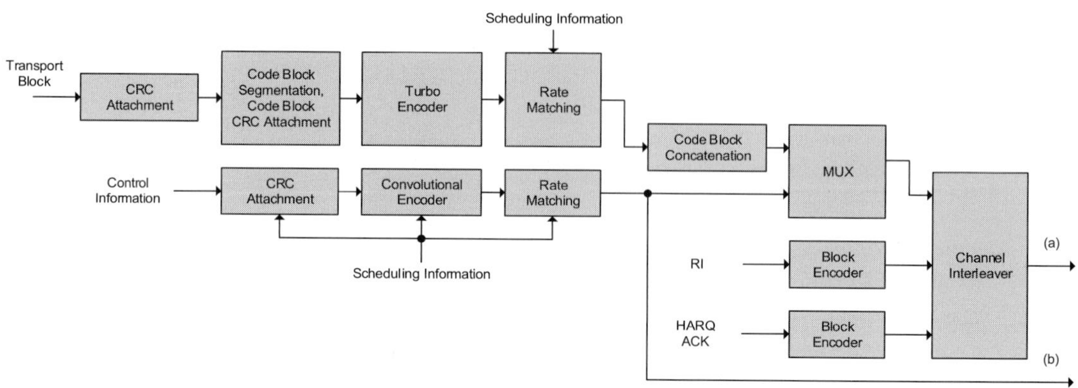

Fig. 8.22 Step 2: radio signal processing procedures of LTE uplink (1).

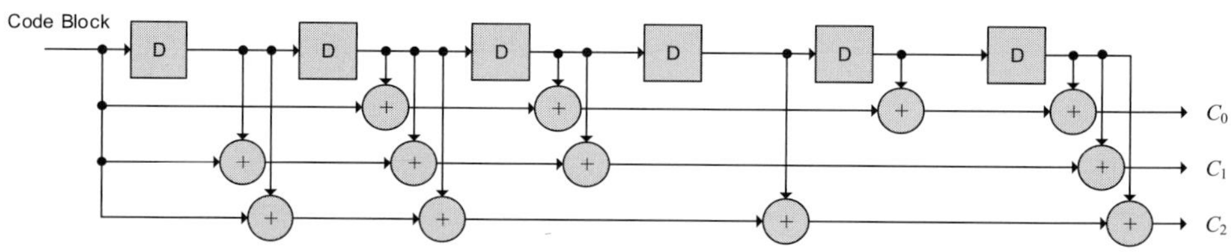

Fig. 8.23 Rate $\frac{1}{3}$ convolutional encoding.

final states of the shift register are identical. To avoid the loss of spectral efficiency, zero padding, which is commonly employed in convolutional coding, is not applied but nevertheless the error correction performance is degraded only slightly. This coding scheme is also used for the Broadcasting Channel (BCH) or control signaling. Figure 8.23 shows the $K = 7$, rate $\frac{1}{3}$ tail-biting convolutional encoder, whose constraint length is shorter than those, $K = 9$, of the rates $\frac{1}{2}$ and $\frac{1}{3}$ convolutional codes used in W-CDMA.

Turbo coding is applied to the data, e.g., the code blocks from the uplink shared channel. Figure 8.24 shows the shift register circuitry of the rate $\frac{1}{3}$ turbo encoder, consisting of two constituent encoders that are identical to the rate $\frac{1}{2}$ parallel concatenated convolutional code (PCCC) of W-CDMA, and a new interleaver. The initial values of the shift registers of the PCCC encoder are set to zeros and the tail bits are zero-padded after the code block data is encoded. The interleaver, based on Quadrature Permutation Polynomial (QPP) principles, maps the input to the output so that the following relationship is met:

$$c(i) = f_1 i + f_2 i^2 \bmod K, \tag{8.2}$$

where $c(i)$ is the index for the input bit which is mapped to the ith output bit. f_1 and f_2 are the parameters determined by K, the size of the code block. The QPP interleavers are known to have the *maximum contention-free* property, which enables parallelized decoding and simplified implementation. In the downlink, the same turbo code is used.

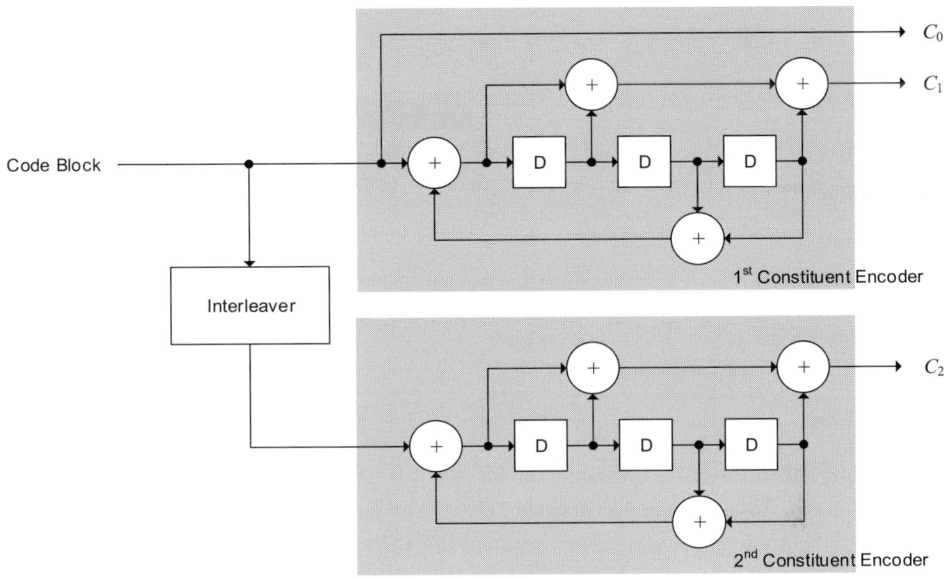

Fig. 8.24 Rate $\frac{1}{3}$ turbo encoding.

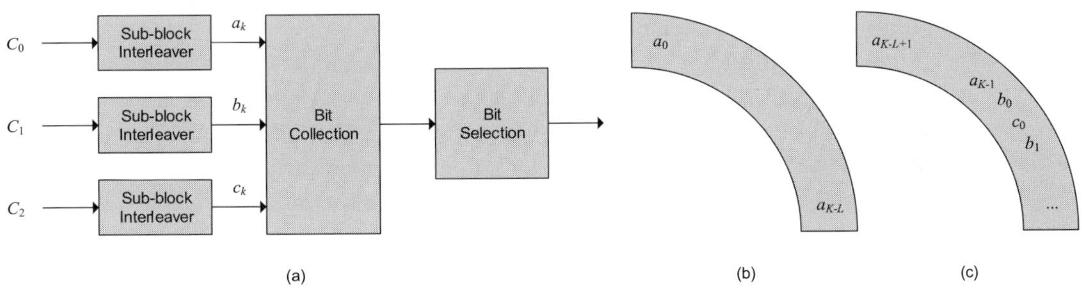

Fig. 8.25 Rate matching of turbo-coded transport channels.

The three bit-streams from the turbo encoder, a systematic bit-stream and two parity bit-streams, are separately interleaved and assembled in a circular buffer. The three bit-streams are concatenated in the order of the systematic bits, and a series of two parity-bit pairs generated from the two parity bit-streams. From the circular buffer, any number of consecutive bits that matches the transmission capability of the resource blocks available at the time can be selected. Figures 8.25(b) and 8.25(c) illustrate two sets of bits stored in the circular buffer. If the end of the parity bit-streams is reached during bit-selection, the bit-selection continues from the beginning of the systematic bits.

For each code block, four non-overlapping sets are defined, called the Redundancy Versions (RV). Each of these starts at a different location in the circular buffer. The MACs at the UE and the eNodeB exchange the information about RV so that the same RV reported as having been incorrectly received or a different RV containing the parity bits not transmitted, depending on the HARQ strategy, is sent to complete the

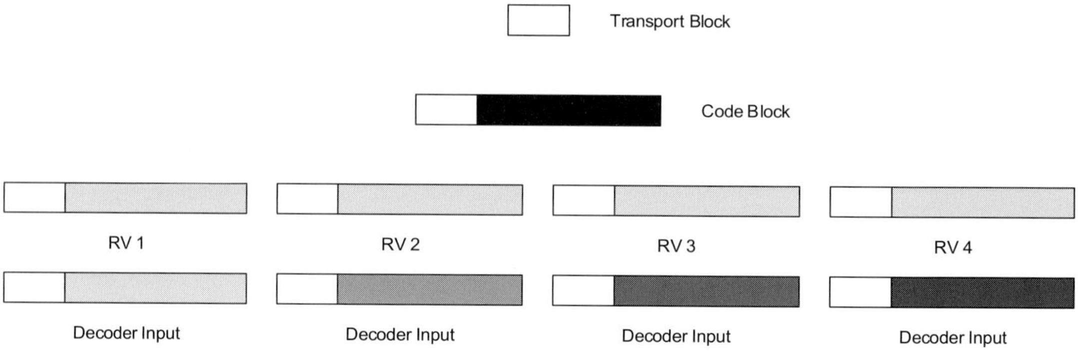

Fig. 8.26 Chase combining HARQ.

transmission of each code block. RV = 0 represents the systematic bits initially transmitted. The strategies to operate HARQ can be classified into two groups. *Chase combining* refers to a set of strategies where the same bits are repeatedly transmitted if a previous transmission was unsuccessful. The channel decoder accumulates the energy of each transmission before attempting the next decoding. If the number of transmission failures exceeds a threshold, the code block is considered to be lost and removed from the buffer for the uplink. Unless the RLC is configured in the acknowledged mode, further re-transmission is not attempted.

Figure 8.26 illustrates the operation of chase combining. Although the code block is drawn as consisting of a systematic element and a parity element, non-systematic channel codes can also be used for chase combining. In the figure, the parity element is illustrated as a darker and larger block following a systematic element. The parity element in the channel decoder becomes darker as the energy is accumulated with each re-transmission. Although the color of systematic bits remains white, the energy is also accumulated in the systematic bits.

Another strategy for HARQ is *incremental redundancy*, with which a different RV is transmitted if a previous attempt failed. As additional parity bits are available after each transmission, the channel decoder can exploit the newly available information to increase its probability of successful decoding. Therefore in a strict sense, this is not exactly a strategy for re-transmission. Figure 8.27 illustrates the operation of incremental redundancy. Unlike chase combining, each RV adds completely new information to the decoder. The number of bits fed to the channel decoder is identical at each re-transmission since the padding bits fill the unused space. Unlike the all-zero bits typically used to flush, i.e., set the values of shift registers inside the convolutional encoders to zeros, the padding bits for the soft-decision channel decoders used in HARQ are real numbers.

Figure 8.28 illustrates the timing relationships of LTE for the uplink and downlink, assuming a single HARQ process. Up to eight HARQ processes can be run simultaneously in both links. Although Figures 8.28(a) and 8.28(b) look similar, they operate under quite different principles. The HARQ of the uplink is *synchronous*, as the timing for re-transmission is fixed to eight subframes after the initial transmission or

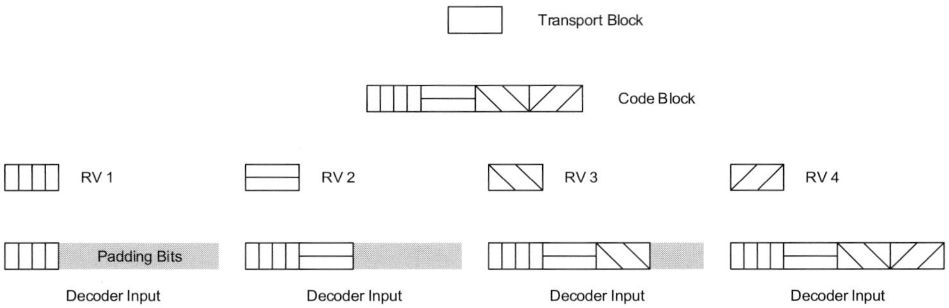

Fig. 8.27 Incremental redundancy HARQ.

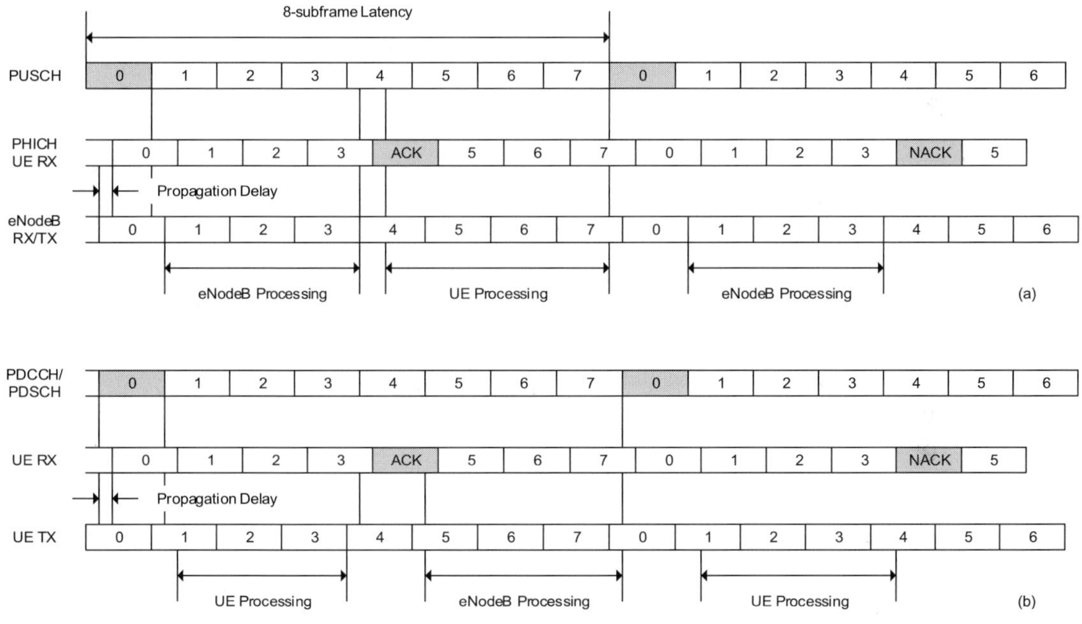

Fig. 8.28 HARQ timing relationship. (a) Uplink. (b) Downlink.

the previous re-transmission, in the case of FDD. As the eNodeB has to schedule the resource for each re-transmission from the UE, the synchronous scheduling saves control channel capacity as the eNodeB does not have to inform the UE of the next opportunity for re-transmission. Moreover, since the same set of resource blocks as used for the initial transmission are assigned for re-transmission, the only information to be transmitted, via the Physical HARQ Indicator Channel (PHICH), is the positive or negative acknowledgment. As seen in Fig. 8.28(a), the processing times allowed for the eNodeB and the UE are tightly limited to enable the synchronous HARQ. To meet such a tight timing relationship, re-transmitted speech packets receive a higher priority in uplink scheduling than new ones.

In contrast, the HARQ of the downlink is *asynchronous* in the sense that each re-transmission has to be explicitly scheduled by the eNodeB as new data in order to maximize the scheduling efficiency and reduce the probability of collision with other downlink data or control signaling. Therefore, although it has been drawn similar to the relationship of the uplink, the timing relationship illustrated in Fig. 8.28(b) corresponds to the minimum HARQ delay realizable in the downlink. In both cases, a processing delay of about 3 ms is assumed. This time has to be spent in the computation of the CRC and in the preparation of old or new data for transmission depending on the result of the CRC check. Although the propagation delay illustrated is shorter than the duration of a subframe, 1 ms, it may take longer, depending on the deployment scenarios with large cells.

In the next step, code block concatenation, the rate-matched outputs for the code blocks are concatenated. Notice that the architecture and operation of the channel decoder does not depend on the HARQ strategy. In the incremental-redundancy HARQ, the unoccupied space is filled with padding bits, zeros, which do not bias the soft-decision channel decoding process. Therefore, regardless of the number of re-transmissions, the channel decoder in HARQ operates in exactly the same fashion. Typically incremental redundancy requires lower bit-rates than chase combining but it can be more vulnerable when the feedback channel carrying the RV information is corrupted in the air.

The control signaling for the uplink, such as the Channel Quality Indicator (CQI) or the Precoding Matrix Indicator (PMI), is rate $\frac{1}{3}$ tail-biting convolutional-coded and multiplexed with the concatenated code blocks. The rate-matched control information and the concatenated code blocks are multiplexed so that the information on HARQ Acknowledgment (ACK) is present in both slots and also mapped to the resource blocks around the Demodulation Reference Signal (DMRS), which helps the receiver at the eNodeB synchronize with the uplink transmission. In other words, the uplink DMRS plays a similar role as the uplink pilot channel of W-CDMA. It ensures that the data and control information are mapped to different modulation symbols. Then the output from the multiplexer is entered into the channel interleaver with other control information, including the HARQ ACK and Rank Indication (RI).

The block coding schemes used for the RI, ACK and CQI, which include repetition coding, simplex coding, and Reed–Muller coding, depend on the length of the control information. The key objective of channel interleaving is to facilitate the resource allocation procedures, as pre-processing before the modulation mapping stages. In the channel interleaver, the modulation symbols are mapped onto the transmit waveform so that the HARQ ACK and RI are present on both slots in the subframe. The HARQ ACK is mapped to the resource blocks around the uplink DMRS while the RI is mapped to the resource blocks around those used by the HARQ ACK. The output of the channel interleaver is read out column-by-column from a matrix, in which a few columns are reserved to include ACK and RI, as outlined in Table 8.8 for the configurations of the normal and extended cyclic prefixes. When there remain no concatenated code blocks with which to be multiplexed, after applying the measures for error detection and correction, the control information is fed to branch (b) for further processing of PUCCH. PUSCH and PUCCH are not simultaneously transmitted as only a single carrier is allowed in the uplink of SC-FDMA when the CA is not used. They can, however, be transmitted consecutively in the time domain.

Table 8.8 Position for control information.

Cyclic prefix	Column for RI	Column for ACK
Normal	1, 4, 7, 10	2, 3, 8, 9
Extended	0, 3, 5, 8	1, 2, 6, 7

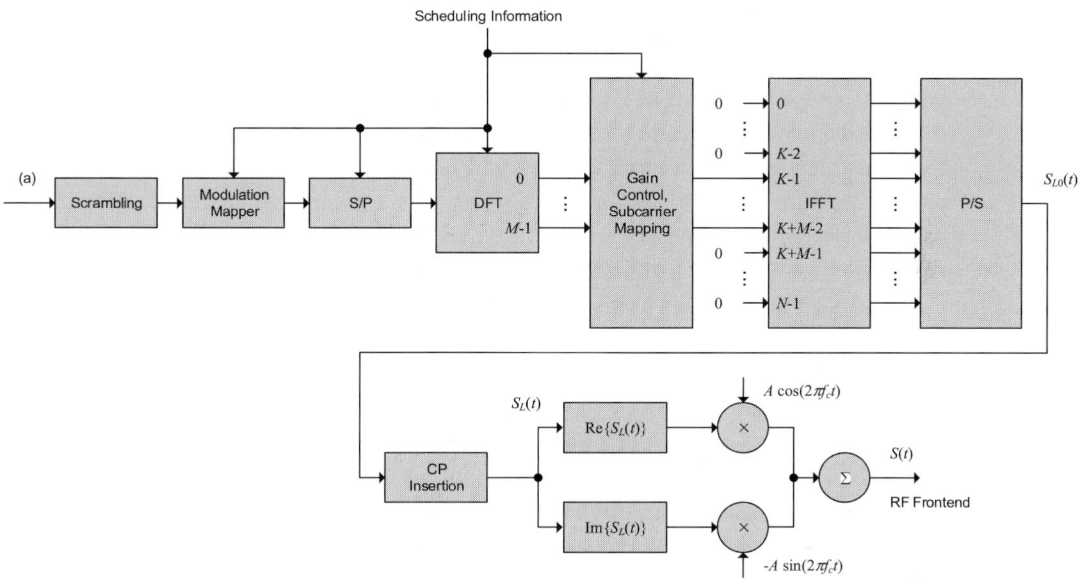

Fig. 8.29 Step 3: radio signal processing procedures of LTE uplink (2).

Figure 8.29 shows the remaining radio signal processing procedures for the uplink. Each codeword entered into branch (a) is scrambled, i.e., exclusive-ORed with a UE-specific scrambling sequence, before modulation. The scrambling sequence is the length-31 Gold sequence also used in the uplink of W-CDMA. One objective of scrambling is the randomization of interference to increase the error-correcting performance of turbo coding. The Walsh encoding in the reverse link of IS-95 plays a similar role of randomizing transmissions but the uplink of LTE is scrambled at the bit-level and the bit-rate is not changed. At the beginning of each subframe, the generator of the scrambling sequence is initialized with parameters related to the identity of the cell and the network, and the number of codewords from the turbo encoder. Depending on the modulation scheme, two, four, and six bits are mapped to a complex-value modulation symbol $I + jQ$, corresponding to QPSK, 16QAM, and 64QAM, respectively.

While GSM supports multiple modulation schemes, such as GMSK, 8-PSK, or AQPSK, the modulation is rarely changed once a call is established. Instead, the mechanisms of GSM to adapt to the time-varying channel quality depend on power control, slow frequency hopping, and joint source-channel coding if AMR is employed. In LTE, where the compressed media can be transported over multiple wireless links, the eNodeB cannot influence the source encoding at the UE directly, other than by resorting to indirect measures such as packet dropping or ECN. However, LTE can utilize methods for adaptation that were not exploited in previous circuit-switched networks, such as adjusting the modulation and bandwidth.

As a method to increase the throughput without spending additional bandwidths, when the link quality and location of UEs, and the cell loading level are favorable, it is possible to separate the complex-valued modulation symbols into multiple groups, i.e., layers, and apply spatial multiplexing with multiple antenna ports. It is also possible to assign resource blocks belonging to the same time-frequency grid to UEs, and compensate for the effects of channels with feedback information. These types of transmission techniques are called Multiple-Input and Multiple-Output (MIMO). However, for typical low bit-rate conversational services like VoLTE, a single layer and a single antenna port are generally assumed, setting RI to 1. Among the nine transmission modes of PUSCH, the single antenna-port mode corresponds to the transmission mode 1 [3GPP (2017m)]. The scheduling grant received from the eNodeB includes the information necessary to map the symbols to resource blocks. To signal a set of contiguously allocated resource blocks, the UE receives a Resource Indication Value (RIV) which refers to a starting resource block and a length in terms of contiguously allocated resource blocks. Other key information includes the Modulation and Coding Scheme (MCS), and the RV.

Suppose there are M_{symb} symbols to be mapped to the resource blocks. Let the scheduled bandwidth and size of the resource blocks, as a number of subcarriers, be M_{sc}^{PUSCH} and N_{sc}^{RB}, respectively. Then $M_{sc}^{PUSCH} = M_{RB}^{PUSCH} N_{sc}^{RB}$ where M_{RB}^{PUSCH} represents the bandwidth as a multiple of the number of resource blocks. Then the M_{symb} symbols are divided into $N_{symb}^{UL} = M_{symb}/M_{sc}^{PUSCH}$ sets, where each set corresponds to an SC-FDMA symbol. The number of SC-FDMA symbols, except for the cyclic prefix, equals the number of symbols transported in a subframe. The signal processing operations from this step shape the spectrum of the symbols onto a set of consecutive subcarriers, which are placed onto the scheduled location of the uplink bandwidth. Then a length M_{sc}^{PUSCH} Discrete Fourier Transform (DFT) is applied to each set of M_{sc}^{PUSCH} symbols, according to

$$Y^l(k) = \frac{1}{\sqrt{M_{sc}^{PUSCH}}} \sum_{i=0}^{M_{sc}^{PUSCH}-1} x(i) e^{-j\frac{2\pi ik}{M_{sc}^{PUSCH}}}, \tag{8.3}$$

$$k = 0, \ldots, M_{sc}^{PUSCH} - 1, \tag{8.4}$$

$$l = 0, \ldots, N_{symb}^{UL} - 1 \tag{8.5}$$

where k and l are the indexes for the subcarriers and the SC-FDMA symbols respectively.

The primary objective of applying the DFT is to spread the symbol information over the uplink bandwidth that has been allocated. A significant amount of computational savings can be achieved if the length M of the DFT, as shown in Fig. 8.29 as $M = M_{sc}^{PUSCH}$, can be expressed as a power of two. However, using $M = 2^n$, where n is a non-negative integer, as the only allowable DFT sizes makes the allocation of resource blocks very inefficient. For example, only ten sizes are available between 2 and 1024, with exponentially increasing jumps between two sizes. Such an inefficiency could be observed in the supported resolutions of H.263 in 3G-324M, shown in Table 6.14, where both width and height double at each step. On the other hand, enforcing no constraints on the DFT size increases the implementation complexity prohibitively. As a compromise, reasonably efficient algorithms for computing the DFT can be found if the bandwidth of the PUSCH resource blocks satisfies the equality,

$$M_{sc}^{PUSCH} = 2^{\alpha_2} 3^{\alpha_3} 5^{\alpha_5} N_{sc}^{RB} \tag{8.6}$$

for non-negative integer values of α_2, α_3, and α_5. The symbols output from the DFT are multiplied by an amplitude-scaling factor, β_{PUSCH}, to correct the transmit power of PUSCH in relation to the control channels. A length-N Inverse Fast Fourier Transform (IFFT) is then applied to the amplitude-adjusted symbols. Unlike M, which is determined by scheduling, $N = N_{RB}^{UL} N_{sc}^{RB}$ is the number of all subcarriers in the uplink.

The consecutive input ports of the IFFT into which the output of the DFT are entered determine the relative locations of the consecutive, i.e., localized, signal bandwidth over the uplink bandwidth. The unfed input ports of the IFFT are set to zero. If separate sets of input ports are selected, it might be possible to distribute the signal bandwidth and obtain some diversity effects against frequency-selective fading. However, such configurations will reduce the efficiency of the power amplifier and increase the overhead for the scheduling grant. Therefore, such transmission techniques are not allowed in the uplink of LTE, but with the introduction of carrier aggregation it became possible to transmit data on separate sets of spectra belonging to different carriers. In Fig. 8.29, the first output of the DFT, $Y(0)$, is fed to the Kth input port of IFFT.

Figure 8.30, in which the normal cyclic prefix is assumed, illustrates the time-frequency structure of the SC-FDMA symbols before the IFFT. The same structure can be applied to the downlink if N_{RB}^{UL} and N_{symb}^{UL} are replaced by N_{RB}^{DL} and N_{symb}^{DL}, respectively. The DC subcarrier at the center of the bandwidth is not used to allow for the direct-conversion receiver implementations.

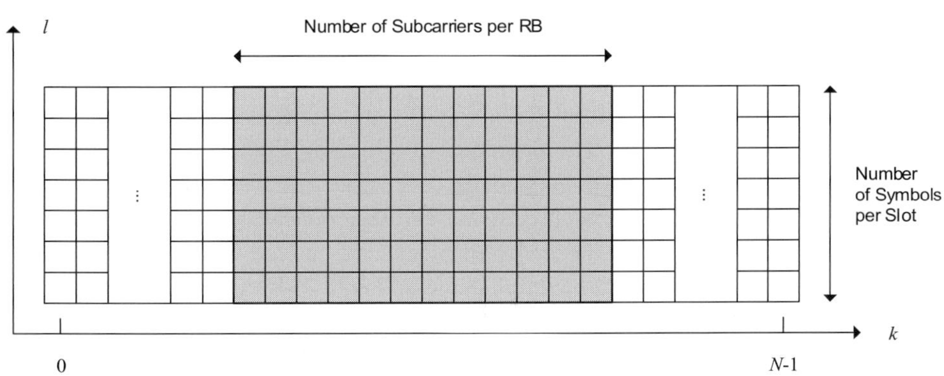

Fig. 8.30 Uplink resource grid.

Let the complex value $a_{k,l}$ be the contents of a resource element (k, l), which is zero for $k \leqq K - 2$ or $k \geqq K + M - 1$. Then each row of the resource elements are entered into the IFFT, where the elements are modulated with a time-continuous exponential function whose ordinary frequency is proportional to k. For each l, whose duration is $0 \leq t \leq NT_s$, the output from the IFFT is summed to generate the baseband signal,

$$S_{L0}(t) = \sum_{k=-\lfloor N_{RB}^{UL} N_{sc}^{RB}/2 \rfloor}^{\lceil N_{RB}^{UL} N_{sc}^{RB}/2 \rceil - 1} a_{k^{(-)},l} e^{j2\pi(k+1/2)\Delta f t}, \tag{8.7}$$

where $a_{k^{(-)},l} = k + \lfloor N_{RB}^{UL} N_{sc}^{RB}/2 \rfloor$, $N = 2048$, and $\Delta f = 15$ kHz. (The operation $\lfloor \cdot \rfloor$ truncates the quantity between the brackets to a nearest integer value.) After the rows corresponding to each slot are processed, the CP is copied to the front position of the time-frequency structure, the uplink resource grid.

Inserting a CP of length $N_{CP,l}T_s$ at the beginning of the SC-FDMA symbols, by copying the waveform of the same length from the last part of $S_{L0}(t)$, mathematically corresponds to shifting $S_{L0}(t)$ in the positive direction in the time domain by $N_{CP,l}T_s$. In Fig. 8.29, the baseband signal $S_L(t)$ is

$$S_L(t) = \sum_{k=-\lfloor N_{RB}^{UL} N_{sc}^{RB}/2 \rfloor}^{\lceil N_{RB}^{UL} N_{sc}^{RB}/2 \rceil - 1} a_{k^{(-)},l} e^{j2\pi(k+1/2)\Delta f(t - N_{CP,l}T_s)} \tag{8.8}$$

for $0 \leq t \leq (N_{CP,l} + N)T_s$.

As the CP is inserted as a type of guard time against the delay spread, where the tails of the previous symbols corrupt the current symbols, the CP is not considered in the decoding process. Therefore the above definition of $S_L(t)$ is required to be valid only for $N_{CP,l}T_s \leq t \leq (N_{CP,l} + N)T_s$. In fact, it does not matter even if signals unrelated to the symbols are used as the CP. Finally, the real and imaginary components of $S_L(t)$ are separated, modulated to RF, summed, and filtered before transmission over the wireless channel. The filtering applied prior to transmission is defined by the requirements in [3GPP (2017h)].

Downlink Signal Processing Procedures

Figure 8.31 illustrates a portion of the radio signal processing procedures of the downlink [3GPP (2017k), 3GPP (2017l), 3GPP (2017m)]. The physical channel over which the downlink data from the eNodeB is transported is the Physical Downlink Shared Channel (PDSCH). The procedures outlined in Fig. 8.31(b) are applicable also to the paging and multicasting channels.

After the L2 header processing and data concatenation/separation procedures illustrated in Fig. 8.17, the transport blocks are entered into the physical layer. Using the same shift register circuitry as in the uplink, a 24-bit CRC is computed from each transport block. Again as in the uplink, if the CRC-appended transport block, the code block, is larger than 6144 bits, the block is segmented. Two 24-bit CRCs are then computed and attached to the segmented code blocks. The rate $\frac{1}{3}$ turbo coding, using the same encoder as in the uplink, is applied. The output symbols are rate-matched as in the

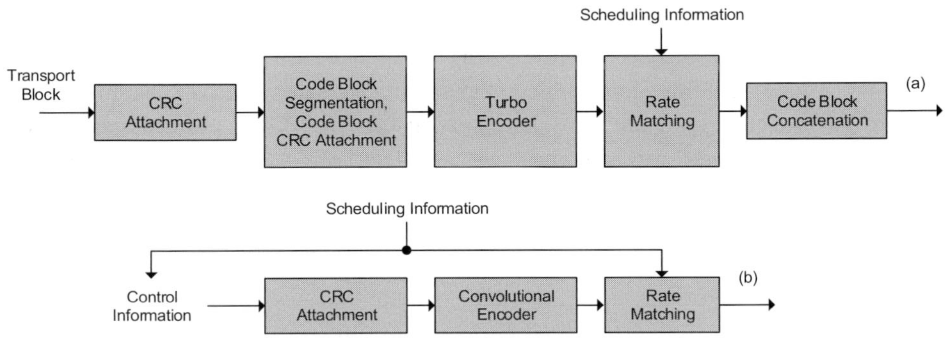

Fig. 8.31 Step 2: radio signal processing procedures of LTE downlink (1).

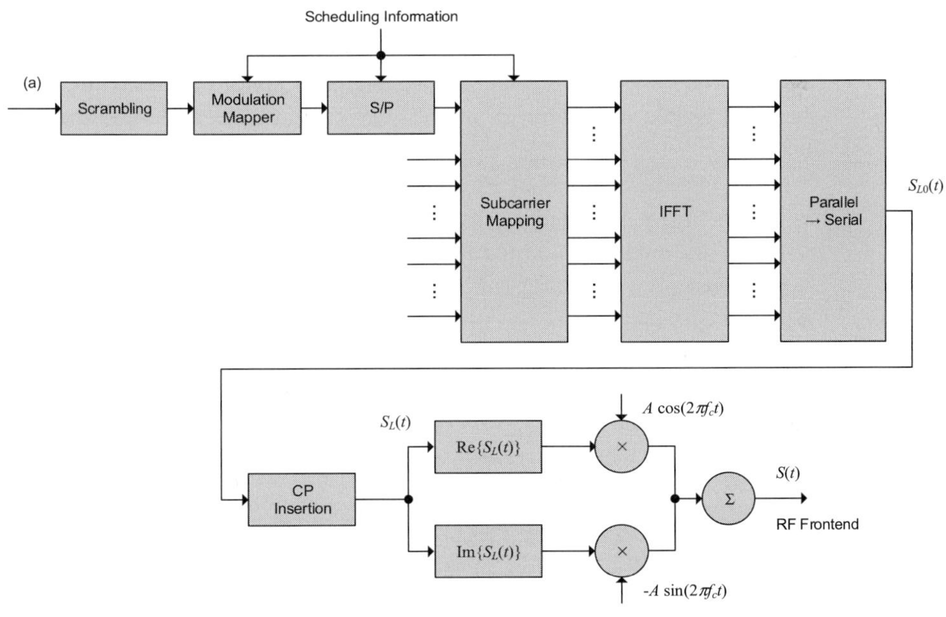

Fig. 8.32 Step 3: radio signal processing procedures of LTE downlink (2).

uplink, to meet the scheduled capacity. If necessary, the code blocks are concatenated to generate the codewords. As shown in Fig. 8.31(b), the Downlink Control Information (DCI), which includes the scheduling information for the uplink and downlink, the request for an aperiodic CQI report, or the uplink power control command, undergoes channel coding procedures simpler than those for the transport block.

Figure 8.32 shows the remaining radio signal processing procedures for the downlink. As in the uplink, we assume a single layer and a single antenna port for typical low bit-rate conversational services. Among the nine transmission modes of the PDSCH, the single antenna-port mode corresponds to the transmission mode 1 [3GPP (2017m)]. The procedures, based on the conventional OFDMA, are similar to those for the uplink except that the DFT stage of SC-FDMA is now absent. Before modulation, each codeword entered into branch (a) is scrambled with a UE-specific sequence, whose generation circuitry is initialized with the same parameters as those used in the uplink. After the scrambling, depending on the modulation scheme, two, four, and six bits are mapped to a complex-value modulation symbol $I + jQ$, which corresponds to QPSK, 16QAM, and 64QAM, respectively.

Time-frequency structures, similar to that shown in Fig. 8.30 for the uplink, are used for the downlink, where the modulation symbols from the transport channels and the control channels are fed into the resource grid. However, unlike the case in the uplink, the subcarriers assigned to a UE need not be consecutive in the frequency domain at an instant of time. The eNodeB determines the locations on the resource grid to which the modulation symbols are mapped. It considers diverse factors such as the availability of radio resources, the information on link quality to and from the UE, or the scheduling information of neighboring eNodeBs if the information is exchanged via the X2 interface. As higher-order modulation schemes, such as 64QAM, require higher link quality, they can be used only in relatively small cell areas. In practice, some of them are usable only in areas where LOS paths between the UE and the eNodeB are available.

Let $a_{k,l}$ be the contents of a resource element (k, l), assuming a resource grid similar to that in Fig. 8.30. Then the resource elements are entered into the IFFT, where the elements are modulated with a time-continuous exponential function whose frequency is proportional to k. For each l, whose duration is $0 \leq t \leq NT_s$, the outputs from the IFFT are summed to generate the baseband signal,

$$S_L(t) = \sum_{k=-\lfloor N_{RB}^{DL} N_{sc}^{RB}/2 \rfloor}^{-1} a_{k^{(-)},l} e^{j2\pi k\Delta f t} + \sum_{k=1}^{\lceil N_{RB}^{DL} N_{sc}^{RB}/2 \rceil} a_{k^{(+)},l} e^{j2\pi k\Delta f t}, \tag{8.9}$$

where $k^{(-)} = k + \lfloor N_{RB}^{DL} N_{sc}^{RB}/2 \rfloor$, and $k^{(+)} = k + \lfloor N_{RB}^{DL} N_{sc}^{RB}/2 \rfloor - 1$. $N = N_{RB}^{DL} N_{sc}^{RB}$ is the number of subcarriers in the downlink. The center subcarrier in the middle of the downlink bandwidth, i.e., the DC subcarrier, is not used for transmission, as the subcarrier might be influenced by high-level interference caused by inaccuracies of the oscillators in the RF circuitry.

Following similar derivations as in the uplink, after the insertion of the CP, the baseband signal $S_L(t)$ is

$$S_L(t) = \sum_{k=-\lfloor N_{RB}^{DL} N_{sc}^{RB}/2 \rfloor}^{-1} a_{k^{(-)},l} e^{j2\pi k\Delta f(t-N_{CP,l}T_s)} + \sum_{k=1}^{\lceil N_{RB}^{DL} N_{sc}^{RB}/2 \rceil} a_{k^{(+)},l} e^{j2\pi k\Delta f(t-N_{CP,l}T_s)} \tag{8.10}$$

for $0 \leq t \leq (N_{CP,l} + N)T_s$. As using different CP types in a cell will prevent the use of computationally-efficient IFFT, only a single CP type is used in the uplink or in the downlink. Finally, the real and imaginary components of $S_L(t)$ are separated, modulated to RF, summed, and filtered before transmission over the wireless channel. The filtering applied prior to transmission is defined by the requirements specified in [3GPP (2017i)].

8.3.5 Link Management

Since the subcarriers of LTE can be switched at any time by scheduling, the slow frequency hopping concept of GSM can be borrowed, but further measures are required as now both the bandwidth and the modulation can be adjusted at each transmission. Since the service quality and network capacity of LTE depends, to a large extent, on efficient scheduling over the time-frequency domain, the eNodeB needs to be supplied with as much up-to-date information on the channel condition and the loading level as possible, but has to spend minimum bit-rates for the measurement and signaling of that information. The information to enable the management of links should be transmitted in both the uplink and downlink.

Uplink Control Information

We assume a single layer and a single antenna port in the configuration shown in Fig. 8.22; the control information related to spatial multiplexing, such as the precoding matrix indicator and the rank indication, is not considered. Then the control information transmitted in the uplink includes the CQI, HARQ (N)ACK, DMRS, SR, and SRS.

The Channel Quality Indicator (CQI), which is inherited from HSPA, is the basic feedback information about the reception status from the UE. Using the CQI, the UE informs the eNodeB of the key parameters for link adaptation that the UE expects to be valid at the next scheduling opportunity. CQI is a four-bit parameter, with values ranging from 0 to 15, which specifies a combination of the modulation scheme, transport block size, and the number of bits per resource element. Based on the capability of the receiver, transmission mode, number of antennas, and the interference level, the UE derives the highest index for each subframe from Table 8.9 that satisfies the following condition: *a single PDSCH transport block with a combination of modulation scheme and transport block size corresponding to the CQI index and occupying a group of downlink physical resource blocks termed the CSI reference resource, should be received with a transport block error probability not exceeding 0.1.* How to compute the CQI is left to the discretion of the implementation, to enable further innovations in the channel estimation techniques, as in the codec mode adaptation of GSM.

The coding rate in the third column of Table 8.9 refers to the number of bits applied to the rate $\frac{1}{3}$ turbo encoder divided by the number of bits after rate matching, which is multiplied by 1024. A lower CQI index implies a lower link quality. When even CQI 1 does not meet the situation, the UE transmits CQI 0, to which the eNodeB would not assign any radio resource. If the channel condition improves within a few speech frames, then the accumulated frames within delay constraints may be transmitted in a bundled form, but some of the old frames may have to be thrown away if the CQI is not high enough or if the eNodeB grants less resource than necessary to transmit all accumulated frames.

Note that the coding rates of CQIs 7 and 8, which use 16QAM, are, in fact, lower than that of CQI 6 using QPSK. Likewise, CQIs 10 and 11, which use 64QAM, achieve lower coding rates than CQI 9 using 16QAM. The partitioning of CQI space with overlapping

Table 8.9 Modulation and coding scheme for CQI.

CQI index	Modulation	Coding rate (×1024)	Bits per resource element
0		Out of range	
1	QPSK	78	0.1523
2	QPSK	120	0.2344
3	QPSK	193	0.3770
4	QPSK	308	0.6016
5	QPSK	449	0.8770
6	QPSK	602	1.1758
7	16QAM	378	1.4766
8	16QAM	490	1.9141
9	16QAM	616	2.4063
10	64QAM	466	2.7305
11	64QAM	567	3.3223
12	64QAM	666	3.9023
13	64QAM	772	4.5234
14	64QAM	873	5.1152
15	64QAM	948	5.5547

is based on the observations that under certain channel conditions, switching to a higher-order modulation does not guarantee a higher data-rate as the higher-order modulation is likely to require a higher SNR. Therefore, in Table 8.9, CQIs 5–8 may constitute a gray region where either QPSK or 16QAM can be used with little difference in performance.

Depending on the direction of any changes in the estimated channel condition, it can be a conservative but efficient operating strategy to maintain the same modulation scheme if the expected outcome of switching to a lower- or higher-order modulation is not clearly indicated. In the codec mode adaptation of AMR in GSM, a conceptually similar strategy was implemented by inserting a hysteresis band between two bit-rate allocations, between encoded speech and channel code. One key difference from GSM is that there is no requirement for the neighbor-only transition as the CQI can be set to any value that matches the situation regardless of the current or previous CQI values, to track rapidly time-varying channel conditions and traffic levels faster than the AMR in GSM does, i.e., once per 20 ms.

As the 26-bit training sequence of GSM is located in the center of each 148-bit slot, the demodulation reference signal (DMRS) occupies the 4th SC-FDMA symbol in each slot of the PUSCH, as illustrated in Fig. 8.33, where the normal CP is assumed. The 4th SC-FDMA symbol is also used with the extended CP. DMRS covers a series of consecutive resource blocks in the frequency domain, so that the length of each DMRS is equal to an integer multiple of 12 resource blocks. However, in contrast to the training sequence of GSM whose length is fixed, it is necessary to generate DMRS sequences of variable lengths, because the uplink bandwidth of LTE varies depending on the scheduling decisions.

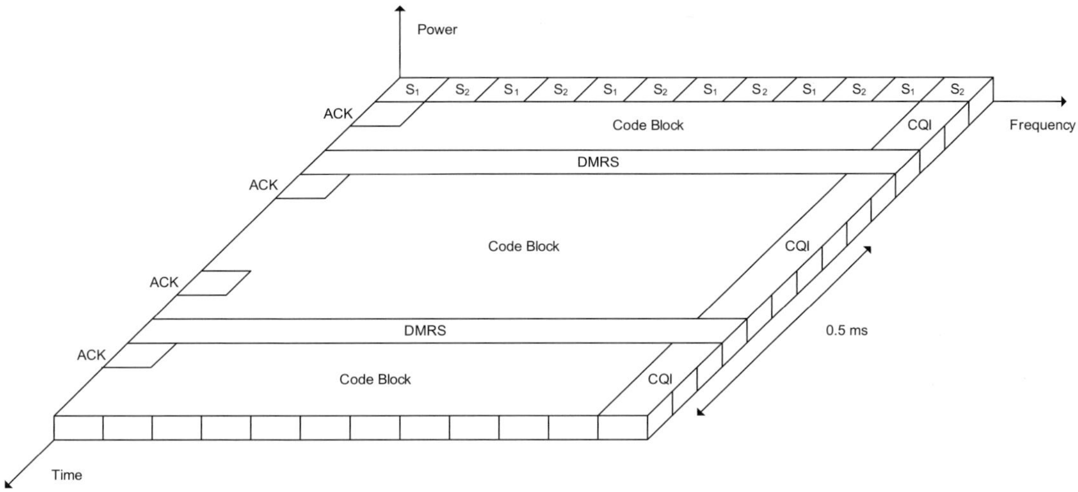

Fig. 8.33 PUSCH resource allocation.

For DMRS sequences whose length is equal to or larger than 36, cyclic extensions of the Zadoff–Chu sequences of length M_{zc} in the frequency domain are used, which are defined as

$$X_k^{zc} = e^{-j\pi u \frac{k(k+1)}{M_{zc}}} \tag{8.11}$$

for $0 \leq k \leq M_{zc}$. The Zadoff–Chu sequence, which is also called the Constant Amplitude Zero Autocorrelation (CAZAC) sequence, is known to have the favorable property of constant power in both the frequency and time domains, as well as the property of orthogonality. The number of Zadoff–Chu sequences is equal to the number of integers which are relatively prime to the length. For DMRS sequences of length 12 and 24, sequences found computationally to have similar properties as the Zadoff–Chu sequences are used instead, as there are not enough Zadoff–Chu sequences for such short lengths. Another method to increase the number of sequences is phase rotation, which is mathematically equivalent to cyclic extension in the time domain.

A key advantage of LTE that the previous generations of mobile communications systems did not have is the gain from rapidly shifting the channel bandwidth when frequency-selective fading is observed near the current bandwidth. Although the intra-cell handover in IS-54 or GSM can provide similar bandwidth-relocating mechanisms to avoid fading, such switchings have a large latency since the synchronization to the target frequency needs to be established before shifting the channel frequency. This limits the capability of handling rapid variations in the link quality. Moreover, such a handover is controlled by the BSC that manages the available time-frequency slots of multiple cells, requiring a substantial signaling delay in the communications with the MS and the BTS. Notice that in the slow frequency hopping of GSM, the channel frequency is hopped in a pre-defined fashion and may not avoid the frequencies that are under a wide or deep fading. Although LTE can enjoy a higher degree of freedom in the use of frequency spectrum, measurement of the channel quality to and from each UE, whose width can vary for each transmission, over a wide spectrum poses a serious challenge.

In LTE, the eNodeB can detect whether the uplink bandwidth assigned to a UE is being impaired by monitoring elevated packet-loss rate or low correlation with the DMRS. However, it is not feasible to estimate the situation in other parts of the frequency spectrum, as the transmission of DMRS is confined to the scheduled bandwidth. Therefore the UE needs to transmit certain reference signals outside the uplink bandwidth currently being scheduled for the eNodeB to gather information on the link quality of the other uplink bandwidths and take that information into account in the uplink scheduling. The Sounding Reference Signal (SRS), which uses the same Zadoff–Chu sequences as the DMRS, is transmitted for this objective, particularly when the uplink bandwidth is narrow or transmission on the uplink is not frequently attempted.

SRS complements the operation of DMRS since the SRS can be transmitted even when there is no transmission on the PUSCH. At the last symbol of SC-FDMA, the SRS is transmitted on every second subcarrier over a bandwidth which is a multiple of four resource blocks. Therefore the length of each SRS sequence is an integer multiple of 24. In Fig. 8.33 where S_1 and S_2 represent the SRS sequences from two UEs, it is shown that a half of the SRS symbols at the end of each SC-FDMA symbol can be transmitted by another UE located in the same cell. As the width and frequency of the SRS contribute to the accuracy of channel estimation, a tradeoff should be made in the assignment of radio resources over the two dimensions to meet the management needs of the eNodeB while minimizing the transmit power.

SRS can be transmitted periodically, every 2 ms to 160 ms, or aperiodically upon request from the eNodeB. In practise it is unnecessary to cover the entire uplink bandwidth with an SRS from each UE, which will be prohibitive as the bandwidth can be as wide as 20 MHz, and multiple bandwidths can be simultaneously used when the carrier aggregation is used. In addition, with the limited transmit power of the UE, stretching the coverage of the SRS as widely as possible will reduce the received energy density, which can render channel estimation difficult when the channel condition is unfavorable.

When it is not possible to transmit the SRS over a wide bandwidth, an estimate of the channel condition can still be made by transmitting SRS over a series of smaller sub-bandwidths and interpolating the estimated condition of each sub-bandwidth. The channel estimation using this approach can be practical as long as the statistical nature of the channel remains stationary during the measurement and interpolation process. Figure 8.34 illustrates two strategies for SRS transmission. Combined with the scheduling request, uplink control information enables the eNodeB to identify the link quality and buffer status of each UE, which can be taken into account in the management of radio resource. The two strategies for deploying SRS can be simultaneously used.

Downlink Control Information

In the downlink, the Cell-specific Reference Signal (CRS) is transmitted for the UEs to demodulate the signals the eNodeB transmits coherently. Each cell is assigned a set of downlink reference signals, whose number is related to the number of antenna ports. Each reference signal consists of reference symbols periodically located over the

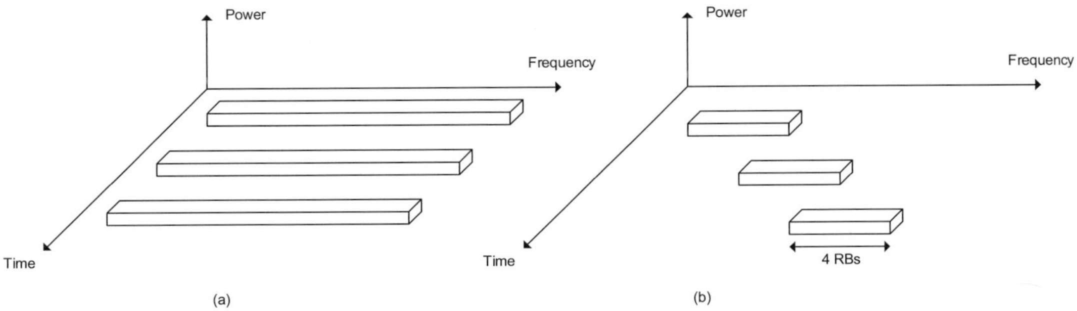

Fig. 8.34 SRS deployment patterns. (a) Periodic. (b) Interpolated.

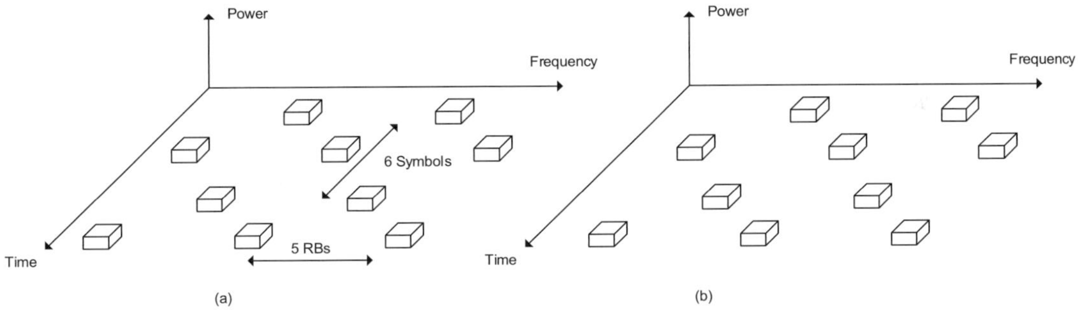

Fig. 8.35 CRS deployment patterns. (a) Frequency shift of n. (b) Frequency shift of $n + 1$.

downlink bandwidth. If a subcarrier contains the reference symbols, then five neighboring subcarriers on either side of the subcarrier cannot have reference symbols at the same temporal location.

The value and location of the reference signals are chosen to differentiate each cell. As shown in Fig. 8.35, the CRS can be shifted by up to six resource blocks. In the cell search procedures, each frequency shift corresponds to a group of 84 cell identities. The period of each reference signal is 10 ms in the time domain, but in the frequency domain the reference signal is defined assuming the maximum bandwidth of LTE, 20 MHz, which corresponds to 110 resource blocks. As there can be four reference symbols in each resource block, the length of the reference signal is $4 \times 110 \times 10 \times 2 = 8880$ symbols. The eNodeBs with downlink bandwidth less than 20 MHz transmit only a portion of each sequence over the bandwidth. For some layered transmission modes, UE-specific reference signals can also be used.

Timing Advance

When the layered transmission modes are not used, orthogonality is achieved mainly through the separation of signals in the temporal and spectral domains. As in D-AMPS or GSM, timing advance is essential in the uplink of LTE to align the signals from the UEs. The link control information such as the DMRS or SRS enables the eNodeB to estimate the propagation delay from each UE. The timing advance command MAC control element, as shown in Fig. 8.16, can be used so that any timing misalignments among the received signals can be resolved with cyclic prefixes.

As shown in Fig. 2.36 for GSM, the eNodeB advances the transmission of the UE by twice the propagation delay to synchronize the frames sent by the UEs. The timing advance command can be issued only when it is necessary, at a typical rate of once every few seconds. A single timing advance command is used even when multiple uplink bandwidths are used in the carrier aggregation, and the bandwidths are separated enough for different propagation characteristics, as the complexity of managing multiple control signaling outweighs the gain from separate timing control commands.

8.3.6 Operational Strategy

In LTE, a larger degree of freedom is available in the transmission of packets, compared with the tight speech and radio signal processing procedures of circuit-switched networks. With a hierarchical protocol architecture consisting of many layers, LTE provides more dimensions in which the parameters can be adapted to meet time-varying channel conditions and cell-loading levels. There can be many strategies to transmit packets, but it is the eNodeB that decides the timing and condition of any transmission to and from each UE under its control.

Packet Scheduling

The scheduling refers to the process of allocating finite radio resource to the UEs for the transmission and reception of packets over the uplink and downlink. As the radio resource is limited, the sum of capacity requirements for the UEs may exceed the amount available in some cases. More UEs can be sequentially served if the time at which each UE needs the resource is limited and distributed, and the once-assigned resource can be retrieved and re-assigned promptly to other UEs in need. Such practises are the basic operating principles used in packet-switched networks. Typical scheduling strategies include round-robin, Max-CQI, proportional-air, and delay-constrained scheduling.

Round-robin scheduling is a basic strategy in which each UE is assigned an equal amount of resource periodically. Depending on the link quality, the assigned resource may result in successful transmission or it may not. Although this simple strategy may look to be the fairest, the UEs with poor link quality can be at a serious disadvantage and are not likely to be treated properly. In situations where the channel condition changes rapidly and the transmission cost is high, as in LTE, this primitive strategy for scheduling is considered insufficient in terms of both media quality and network capacity.

In *Max-CQI* scheduling, the eNodeB selects a few UEs with the highest link quality at each scheduling opportunity, disregarding those with poor link quality. With this strategy, the gap in service quality can be further widened between the fortunate and the unfortunate groups of UEs. While the service quality of a subset of UEs with good link quality can be guaranteed, it may not be feasible to provide even a minimum level of service quality to the poor UEs. Therefore Max-CQI is considered effective in maximizing the data throughput at the sacrifice of the fairness when the delay constraints are relaxed.

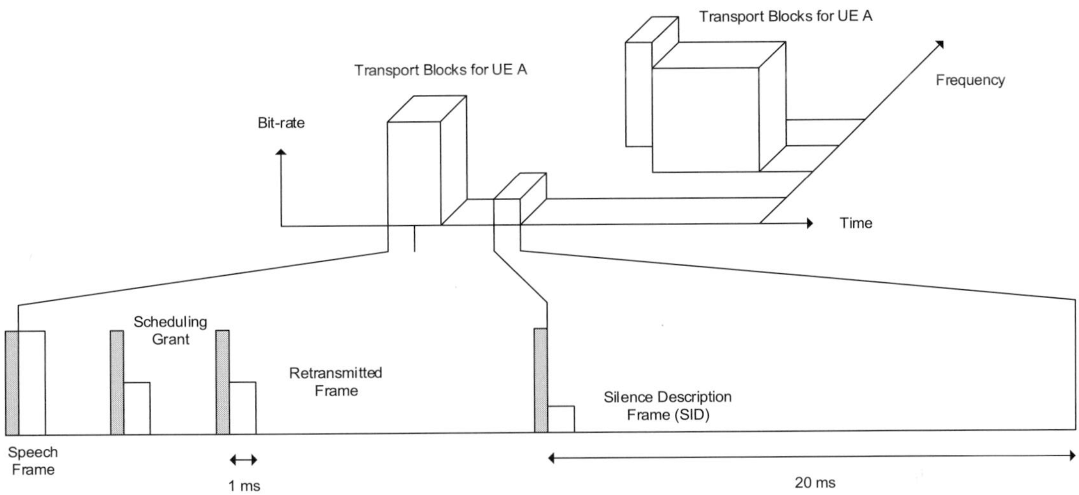

Fig. 8.36 Resource allocation in dynamic scheduling.

Proportional-fair and *delay-constraint* schedulings are strategies whose allocation philosophy lies between those of round-robin and Max-CQI, in that both try to provide a minimum level of service to most UEs in the cell. As the name suggests, delay-constraint scheduling pays attention to certain UEs whose packets should be delivered within a time limit, such as those engaged in VoLTE sessions. In practice, schedulers of the eNodeB operate based on a mix of several strategies, and may change the strategy to match the situation. Detailed analysis of these scheduling algorithms can be found in [Chakraborty *et al.* (2007)].

In LTE, there are two major classes of scheduling methods: Dynamic Scheduling (DS) and Semi-Persistent Scheduling (SPS). DS is the basic scheduling strategy of LTE, where an appropriate amount of radio resource is assigned to each UE explicitly by the eNodeB for the uplink and downlink. For a wide variety of services, this strategy provides an acceptable level of service quality and network capacity. Figure 8.36 illustrates the principles of DS for a VoLTE session, in which the eNodeB assigns resource for a speech frame, an RV transmitted after a HARQ failure, or an SID, via separate scheduling grants. If the switching of subcarriers is necessary, more overhead is taken to signal the change than when the same subcarriers as used for a previous transmission are used.

Note that in the figure, a packet including an entire speech frame or an SID is transmitted in a 1 ms subframe. In the compressed DPDCH of W-CDMA+, the time periods of speech compression and wireless transmission began to differ. The relationship between the two periods almost disappeared in the LTE as delay becomes a key parameter as well as bandwidth and modulation to control for efficient adaptation to link quality and orderly allocation of radio resource.

In the operation of packet-switched networks, HARQ with a controlled number of re-transmissions is, in fact, preferred to completing the transmission with one attempt, as the average transmit power and interference level can be reduced by HARQ. However, if the proportion of voice traffic increases, the required capacity of the Physical

Downlink Control Channel (PDCCH) for the scheduling grants becomes excessive, thereby upper-bounding the maximum voice capacity. Semi-persistent scheduling (SPS) is a novel strategy introduced in LTE to reduce the consumption of PDCCH especially in the transmission of speech packets. SPS exploits the fixed bit-rate (FBR) speech coding principles that have been used from FR to AMR. The principles were initially designed to fill the periodically reserved time slots of GSM with speech frames. However, as SPS does not depend on any characteristics of the bit-stream structures, the use of SPS is not confined to these speech codecs but can be applied to any services that require periodic assignment of radio resource in similar amounts.

If the SPS is enabled by eNodeB using the RRC signaling after receiving a scheduling request, the eNodeB assigns a periodic series of identically-sized resource blocks, on the same set of subcarriers, with the same modulation and coding scheme, using a single scheduling grant. Other UEs are not allowed to transmit or receive on those assigned resource blocks until the assignment is released. The parameters of SPS are computed based on information broadcast by the eNodeB. The System Frame Number (SFN) is a ten-bit timing information contained in the Master Information Block (MIB), which is broadcast periodically for the UE to acquire basic system information. Another SPS parameter is the Subframe_Offset, whose value is set according to the configuration of the TDD mode. When the LTE is operating in the FDD mode, the Subframe_Offset is set to zero.

In SPS, the scheduling grant includes the semiPersistSchedIntervalUL parameter, which represents a period of automatic scheduling for which neither a request nor a grant is necessary. In the TDD mode, separate intervals can be used for the uplink and downlink. After receiving an uplink scheduling grant that initializes the SPS, the UE should consider that the following, Nth, grant occurs in the subframe for which the following relationship holds,

$$\begin{aligned}(10 \times \text{SFN} + \text{subframe}) = [(10 \times \text{SFN}_{\text{start time}}) &+ \text{subframe}_{\text{start time}} \\ &+ N \times \text{semiPersistSchedIntervalUL} + \text{Subframe}_{\text{Offset}} \\ &\times (N \bmod 2)] \bmod 10240, \end{aligned} \tag{8.12}$$

where $\text{SFN}_{\text{start time}}$ and $\text{subframe}_{\text{start time}}$ are the SFN and the subframe at the time the configured uplink grant was (re-)initialized.

If the UE fails to transmit implicitReleaseAfter times, which can be set to 2, 3, 4, or 8, the SPS is implicitly released, allowing the pre-allocated radio resource to be used for other purposes. For the downlink, similar procedures including the semiPersistSchedIntervalDL parameter are used:

$$\begin{aligned}(10 \times \text{SFN} + \text{subframe}) = [(10 \times \text{SFN}_{\text{start time}}) &+ \text{subframe}_{\text{start time}} \\ &+ N \times \text{semiPersistSchedIntervalDL} + \text{Subframe}_{\text{Offset}} \\ &\times (N \bmod 2)] \bmod 10240. \end{aligned} \tag{8.13}$$

Figure 8.37 illustrates the operation of SPS, in which the SPS is initiated upon the reception of a scheduling request to send a MAC PDU containing at least one speech frame. The SPS operation is interrupted and terminated by an irregular event, such as silence or HARQ. For MAC PDUs smaller than those capable of containing a speech

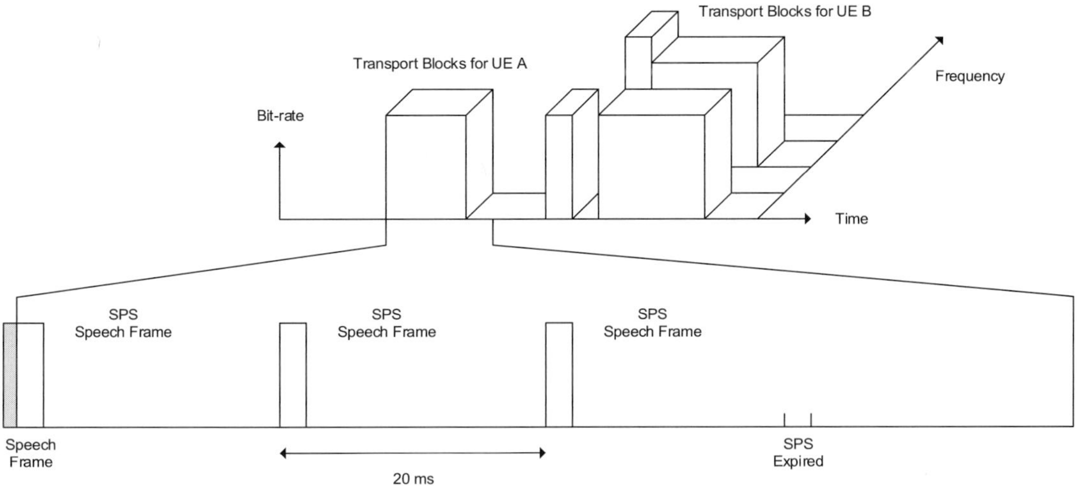

Fig. 8.37 Resource allocation in semi-persistent scheduling.

frame, such as the one containing only an SID or an RV, SPS is unlikely to be triggered. Note that SPS will not be activated for MAC PDUs not including speech frames even if their sizes are similar to those including speech frames, as the eNodeB is implicitly aware of the type of PDUs from the QoS information IMS provides. In Fig. 8.37, it is shown that after the first speech frame is dynamically scheduled, the two following frames are served with SPS. The dynamic scheduling is resumed when the initial transmission of the third speech frame fails. SPS can be used in either the uplink or downlink, or both. This operation of SPS is conceptually similar to that of the cruise control, which manages the throttle of a vehicle to maintain a steady speed, and releases the control if the fixed speed is no longer necessary or supportable, e.g., when there are curves that require slowing down or the brakes are applied.

Considering that a typical burst of speech, i.e., an onset, lasts for a few hundred milliseconds, SPS is likely to save the capacity of the PDCCH for the tens of scheduling grants required to transmit a fragment of speech waveform. Since the scheduling is neither requested nor granted, transmit power can also be saved, which also reduces the interference. With dynamic scheduling, the UE has to be powered on for 100% of a 20 ms time interval but SPS enables turning off the transmitter and receiver circuitry of the UE when voice activity is absent. However, in return, as the scheduler has fewer degrees of freedom in resource allocation when SPS is activated for some UEs, the overall throughput of each cell, the amount of total data for conversational services and other non-real-time data services, decreases. Moreover, it is often observed that the SPS exhibits limitations when the link quality deteriorates rapidly, as the proper operation of the SPS depends on the statistical nature of the currently assigned subcarriers that should remain stationary and reliable while transporting all the speech frames. If relocation to another set of subcarriers or change of MCS is required, the SPS has to be released and the DS or other algorithms take over the scheduling operation.

Therefore for the scheduler to operate in SPS as long as possible, the MCS needs to be set rather conservatively, i.e., to as low-order modulation and high coding rate as possible, which will reduce the network capacity correspondingly. Similarly, to avoid frequency-domain relocation as much as possible, resource blocks with the highest link-quality need to be assigned, which also limits the scheduling efficiency. Therefore the operation of SPS can prevent DS from maximizing the frequency-domain scheduling gain for non-real-time data traffic, by fragmenting the spectrum, especially in the uplink where radio resources are scheduled contiguously in the frequency domain. Furthermore, variations in the ROHC overhead may interfere with the capacity assigned for SPS. Even without packet loss, ROHC may enter the initialization and refresh (IR) status if some fields reach the end of their representable ranges or the synchronization becomes unstable, making packets larger than the SPS granted. As a result, SPS has been found to be less useful for UEs in motion than DS.

Applying SPS to the VBR speech codecs such as EVRC will limit the scheduling efficiency as the resource blocks assigned for a Rate 1 frame will not be fully used by the following frames of lower rates. For the case of video, although the video encoder is rate-controlled to compress video at near-constant bit-rates, and therefore does not have to handle the effects of silence, SPS is unlikely to save the capacity for the control channel, as the fluctuation of bit-rate around a target bit-rate is much more than that of the speech codecs, because of the variations in motion activity or frame type. Another disadvantage of SPS is that if speech frames arrive in the buffer for the uplink just after a buffer status report (BSR) was sent to the eNodeB, they have to wait there for almost 20 ms while DS will enable immediate transmission of the frames at the next opportunity.

Link Adaptation

Figure 8.38 illustrates a hypothetical procedure of downlink link adaptation in which the transport block size and the number of assigned resource blocks are dynamically controlled to match the rapid variation in channel quality while maintaining a fixed source encoding bit-rate or data throughput. In Fig. 8.38(b), fluctuation of the CQI value shows that the link quality gradually deteriorates but slowly rebounds after reaching the lowest level. Lower CQI values make the scheduler at the eNodeB reduce the transport block size but assign more resource blocks to compensate for the reduction in transport capability, as a lower-order modulation and a higher coding rate than usual are used. As the CQI increases, so does the TBS while fewer RBs are required. In Fig. 8.38, fluctuation of the MCS is not illustrated but is likely to be directly affected by that of the CQI.

In practise, neither speech nor video in VoLTE generates a constant source encoding bit-rate. Even in the fixed bit-rate operation of the speech encoder, DTX interrupts the generation of a constant bit-rate whenever voice activity is absent. For a short-term period during which video content or its motion activity does not vary significantly, the output bit-rate of the video encoder can approximate constancy. As illustrated in Figures 8.38(c) and 8.38(d), changes in the transport block size and the number of assigned resource blocks are generally not completely synchronized. If the combined actions of controlling the TBS, number of assigned RBs, and the transmit power cannot sustain the negotiated source bit-rate, either media adaptation or session re-negotiation may

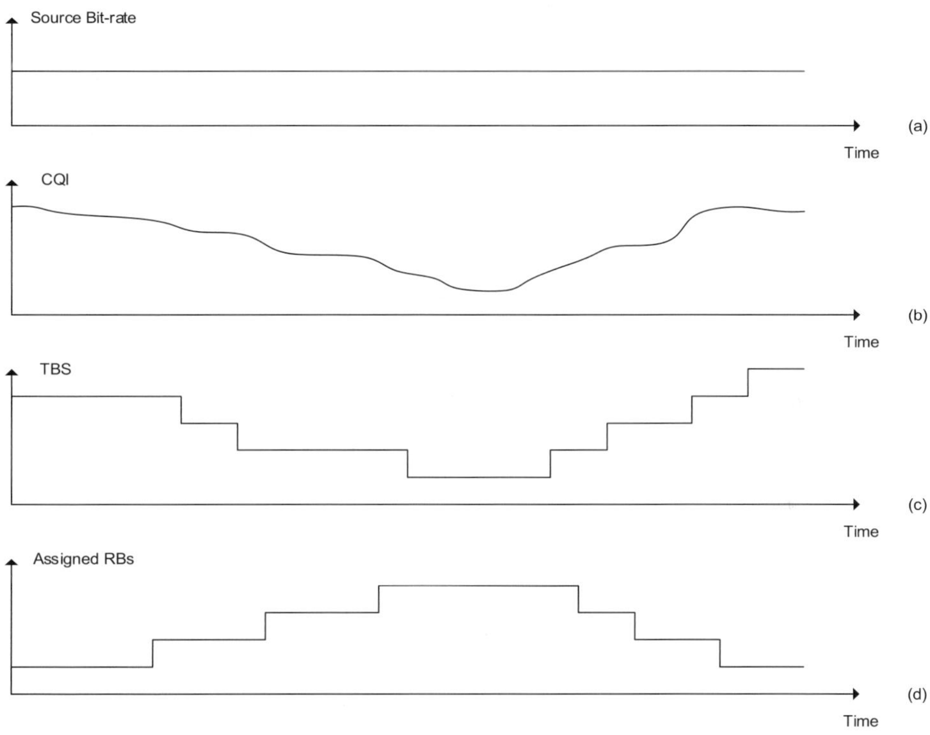

Fig. 8.38 Link adaptation for constant source bit-rate.

be activated to reduce the source bit-rate or discontinue the service, depending on the service policy.

Explicit Congestion Notification

Explicit Congestion Notification (ECN) can be marked on the header of an IP packet by the eNodeB but it is not specified which protocol of the eNodeB should be responsible for the action. If two UEs agree to use the ECN during a session for all RTP packets except the RTCP packets, two bits in the TOC (IPv4) or Traffic Class (IPv6) field of the IP header, as shown in Fig. 8.6, should be marked as 01 (ECT(1)) or 10 (ECT(0)), to signal that the UEs sending or receiving the packets are capable of handling ECN. Otherwise, the two bits should be marked as 00.

When the eNodeB realizes that its loading level is approaching the maximum capacity or that reliable transmission of packets can no longer be guaranteed, it starts taking elementary measures such as assigning the UEs at cell edge to frequency subbands with low measured inter-cell interference or handing them over to neighboring cells. If the measures have turned out insufficient, the eNodeB can start marking each packet sent in the congested direction, as 11, representing the ECN Congestion-Experienced (ECN-CE). Then the UE considers the marked packets as lost, even when they are received without error and played out, and takes measures such as asking the far-end UE to reduce the bit or packet rate. Figure 8.39 illustrates the ECN operation for the uplink, in which

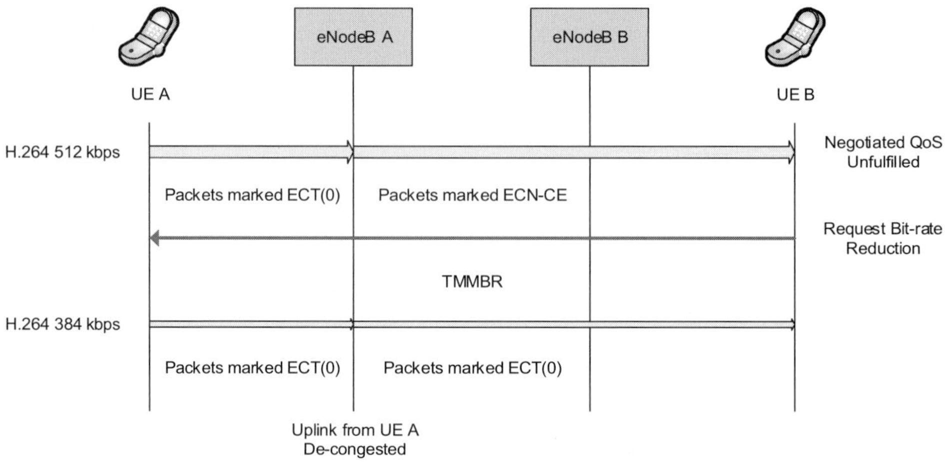

Fig. 8.39 Uplink ECN operation.

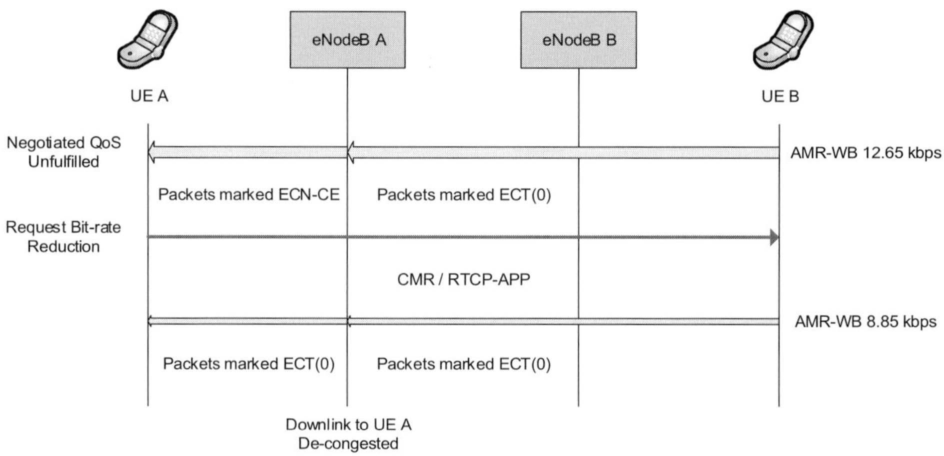

Fig. 8.40 Downlink ECN operation.

eNodeB A located in the uplink of UE A marks ECN on the video packets encoded by H.264 at 512 kbps and transmitted from UE A to UE B. UE B, upon detecting the ECN-CE marking, asks UE A to reduce the bit-rate with the Temporary Maximum Media Bit-rate Request (TMMBR), a type of RTCP messages, to 384 kbps. Note that the network is not directly informed of this reaction from UE B.

Once an eNodeB or a packet router marks ECN on the packets, other eNodeBs or routers should not overwrite the markings. As ECN can be marked by any network node in the end-to-end transmission path, the UE cannot identify which node marked 11, or the locations of congestion or link failure. A straightforward strategy to implement the ECN capability at the eNodeB would be to let the PDCP do the marking, with the scheduling information of the MAC. Figure 8.40 illustrates the ECN operation for the downlink, in which eNodeB A, located in the downlink of UE A, marks ECN on the speech packets encoded by AMR-WB at 12.65 kbps and transmitted from UE B to UE A. UE A, upon detecting the ECN-CE marking, asks UE B to reduce the bit-rate, with the Codec Mode Request (CMR) field in the RTP payload header, to 8.85 kbps.

Bundled Transmission

In the downlink, the MAC may bundle several PDUs and send the chunk in a single transmission. While the chunk of PDUs should wait until an appropriate transmission opportunity is located, it needs to be transmitted successfully, e.g., using high transmit power, if the loss of several speech frames and noticeable quality degradation are to be avoided. Figure 8.41 illustrates the action of MAC that bundles an 80 ms amount of PDUs for each downlink transmission. Notice that four or less speech frames can be bundled in this strategy for the delay constraints.

As a result, this bundling technique is likely to be used for UEs with high SNRs, with which higher-order modulations can be applied. Since the packet bundling increases the time during which the PDUs have to wait in the buffer for the downlink, it is important not to exceed the total delay budget. SPS or packet bundling is designed for situations where an eNodeB has to serve many UEs requesting resources frequently, or where the proportion of UEs engaging in VoLTE is very high. It is essential to save PDCCH as much as possible in such conditions.

In the early stage of VoLTE deployment or in areas where the coverage of the LTE network is limited, situations are often encountered in which the link quality of the uplink is unsatisfactory but the transmit power of the UE is already at maximum and cannot be increased further. Transmit time interval (TTI) bundling is a transmission strategy that can overcome this kind of difficulty, in which a single transport block is repeatedly transmitted for up to four consecutive TTIs. The probability of successful decoding is increased from the accumulated energy at the turbo decoder. A key difference from the chase combining HARQ is that in the TTI bundling, an ACK or a NACK is transmitted only after the eNodeB receives the last bundled transport block. The TTI bundling has the effects of reducing the Carrier-to-Interference Ratio (CIR) of the uplink required for proper operation, therefore effectively extending the coverage by 2–3 dB. In the reverse

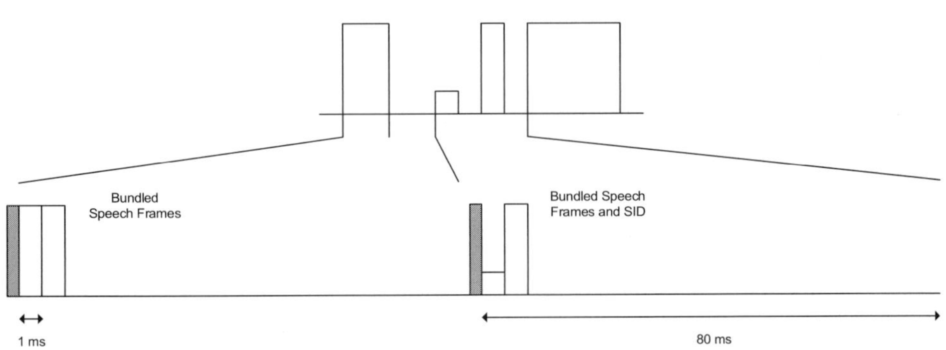

Fig. 8.41 Resource allocation in packet bundling.

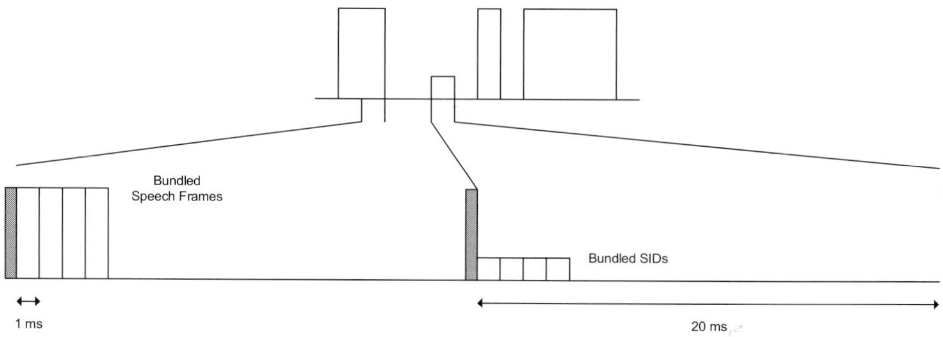

Fig. 8.42 Resource allocation in TTI bundling.

link of IS-95 and cdma2000, the symbols are repeated in such a way that the number of repetitions is inversely proportional to the rate. However, in that case the energy per information bit is maintained at the same level. Figure 8.42 illustrates the operation of TTI bundling, where a transport block is transmitted four times. The price paid for extended coverage is the higher power consumption from multiple transmissions. Therefore this bundling technique is likely to be used for UEs with low SNRs, typically at the cell edges.

Access Barring

eNodeB is endowed with only indirect measures to influence the source encoding of UE, such as packet dropping or ECN, but it is possible for eNodeB to influence the session negotiation of UE with indirect measures. When a large number of service requests is expected, e.g., around stadiums where important sport or social events are in progress, the eNodeBs in charge of the areas may periodically broadcast access barring information via the System Information Block 2 (SIB2) of the RRC messages, whose format is shown in Fig. 8.43.

The access barring can be considered as an indirect measure for resource management, in contrast to the more direct measure of scheduling, since the technique suppresses the traffic only statistically. In other words, the eNodeB cannot explicitly select the UEs whose service requests will be denied. SIB2 includes the information for the Service Specific Access Control (SSAC) barring, *ssac-BarringForMMTEL-Video-r9* and *ssac-BarringForMMTEL-Voice-r9*, for originating video and voice calls, respectively. The information elements are configured with two parameters, *ac-BarringFactor* and *ac-BarringTime*.

When the SSAC barring is enforced, upon a request for a video or voice call, the UE draws a random number, *rand*, which is uniformly distributed over [0, 1). Access for the requested media service is allowed if *rand* is smaller than the broadcast *ac-BarringFactor*. Otherwise, the access is delayed for $(0.7 + 0.6 \times rand) \times ac-BarringTime$ seconds, before the UE can request the service again. Figure 8.44 outlines the flow of this barring algorithm. By controlling the value of *ac-BarringFactor*, the

```
SystemInformationBlockType2 ::= SEQUENCE {
 ac-BarringInfo  SEQUENCE {
  ac-BarringForEmergency BOOLEAN,
  ac-BarringForMO-Signalling AC-BarringConfig OPTIONAL, -- Need OP
  ac-BarringForMO-Data AC-BarringConfig OPTIONAL -- Need OP
 }              OPTIONAL, -- Need OP
 radioResourceConfigCommon RadioResourceConfigCommonSIB,
 ue-TimersAndConstants UE-TimersAndConstants,
 freqInfo  SEQUENCE {
 ul-CarrierFreq ARFCN-ValueEUTRA  OPTIONAL, -- Need OP
 ul-Bandwidth ENUMERATED {n6, n15, n25, n50, n75, n100}
OPTIONAL, -- Need OP
  additionalSpectrumEmission  AdditionalSpectrumEmission
 },
 mbsfn-SubframeConfigList  MBSFN-SubframeConfigList  OPTIONAL,  -- Need OR
 timeAlignmentTimerCommon  TimeAlignmentTimer,
 ...,
 lateNonCriticalExtension OCTET STRING (CONTAINING SystemInformationBlockType2-v8h0-IEs)     OPTIONAL, -- Need OP
 [[ ssac-BarringForMMTEL-Voice-r9 AC-BarringConfig   OPTIONAL, -- Need OP
  ssac-BarringForMMTEL-Video-r9 AC-BarringConfig   OPTIONAL -- Need OP ]],
 [[ ac-BarringForCSFB-r10   AC-BarringConfig  OPTIONAL -- Need OP ]]
}
...
AC-BarringConfig ::=  SEQUENCE {
 ac-BarringFactor ENUMERATED {
    p00, p05, p10, p15, p20, p25, p30, p40,
    p50, p60, p70, p75, p80, p85, p90, p95},
 ac-BarringTime ENUMERATED {s4, s8, s16, s32, s64, s128, s256, s512},
 ac-BarringForSpecialAC BIT STRING (SIZE(5))
}
...
```

Fig. 8.43 SIB2 format.

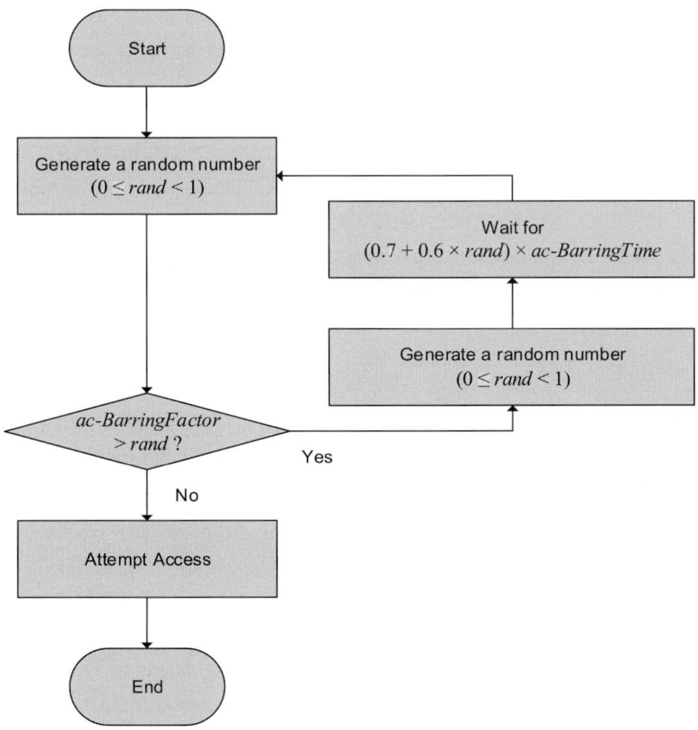

Fig. 8.44 Access barring algorithm.

eNodeB can indirectly control the probability and duration that service requests are temporarily denied for.

The *ac-BarringFactor* for video is likely to be set lower than that for speech, as video services consume much higher bandwidth and need to be suspended earlier than voice services when the network becomes loaded. Although the principles of SSAC can be used for a wide variety of services, a typical usage of the technique is to limit demanding services which require low delays or high bit-rates, such as voice or video telephony, when the eNodeB senses the activity of a large number of UEs in its cell. If a session is negotiated, the SSAC barring will influence the IMS such that speech or video is temporarily excluded in the selection of an SDP offer or in the construction of an SDP answer. If the level of cell loading is intensified further, the network may allow only the most essential services such as emergency or voice services at low bit-rates.

By broadcasting the SIB2 with very low *ac-BarringFactor* values, some services can be effectively prohibited from the cell. The SSAC barring can play a conceptually similar role to that taken by the Codec_Mode_Command message of W-CDMA or the RATE_REDUC message of cdma2000, as an emergency measure to handle highly concentrated service requests when temporarily sacrificing the quality to some extent can be tolerated for larger capacity. eNodeB may consider SSAC barring when the UEs do not seem to react to the received ECN-CE. However, neither ECN nor SSAC barring provides mechanisms to control the bit-rate as finely and in such a timely manner as those of W-CDMA or cdma2000. The indirect mechanisms designed for VoLTE provide actions that are cruder than those of their circuit-switched counterparts. For example, allowing services at lower bit-rates than usual is not possible. This inflexibility originated from the design requirements set for SSAC barring, which were based on the media configurations of W-CDMA where the video bit-rate of 3G-324M is fixed at 48–49 kbps, and the speech bit-rate is equal to or less than 12.2 kbps. In VoLTE, the speech and video bit-rates can be higher than the 3G values. Therefore the SSAC barring can play only a limited role in the management of radio resources.

Handover

Figure 8.45 illustrates the handover procedures with which a UE connected to the source eNodeB, eNodeB A, moves toward a neighboring, target eNodeB, eNodeB B. Throughout the handover procedures, the UE establishes new connections with eNodeB B, closes old connections with eNodeB A, and starts sending and receiving packets using the new connections.

Handover in LTE differs from either the hard handover of TDMA or the soft handover of CDMA. In a TDMA handover, the MS releases the channels assigned by the source BTS and then establishes new channels with the target BTS, typically losing a few speech frames during the transition that might be audible. This strategy is called *break before connect*. In the handover of CDMA, the MS establishes new channels with the target BTS while maintaining the current channels with the source BTS. This prevents the loss of speech frames but the network capacity is compromised. This strategy is called *connect before break*. Note that in CDMA, connections with multiple BTSs can

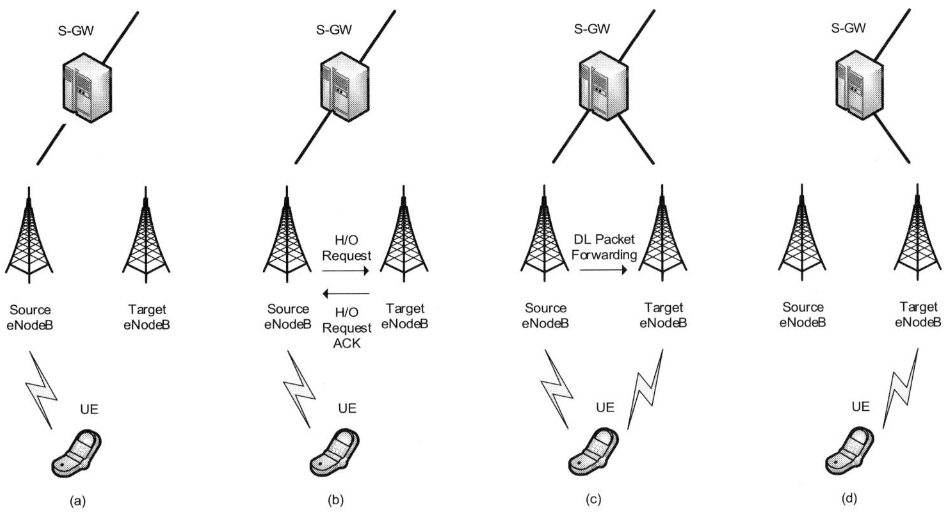

Fig. 8.45 Handover procedures over X2 interface.

be established and maintained before a target BTS is selected, consuming the capacity further. In the handover of LTE, the UE releases the channels assigned by the source eNodeB and establishes new channels with the target eNodeB. However, the speech frames en route to the UE when it changes the eNodeB are forwarded to the target eNodeB, therefore effectively combining the advantages of both handover strategies. This handover technique exploits the asynchronous nature of packet networks in which the packets can be buffered in the middle of transmission and wait until the next path is cleared.

In Fig. 8.45, the source eNodeB configures the conditions at which the UE measures the link quality of the neighboring eNodeBs and the formats of reports, with the measurement control message. The UE may report periodically or upon the occurrence of specific events, with a measurement report message. Based on the received reports, the source eNodeB evaluates the necessity of a handover. If the handover of a UE is decided upon, the source eNodeB determines a target eNodeB from several candidates, and the frequency band if multiple bands are used by the target eNodeB. When the measurement report contains the information on the link quality of several neighboring eNodeBs, the source eNodeB is likely to choose the one with the highest quality. Then the source eNodeB sends a handover request message, which contains the information on QoS that the UE is currently using to the target eNodeB. The target eNodeB then decides whether or not to accept the request, depending on the available radio resource and the level of QoS requested by the source eNodeB. If the request is accepted, the target eNodeB returns a handover request acknowledge message to the source eNodeB. Upon receiving the acknowledge message, the source eNodeB sends the handover command message to the UE and starts forwarding packets heading for the UE to the target eNodeB. A list of packets already delivered to UE via the downlink is sent to the target eNodeB so that the eNodeB can remove such packets from its downlink buffer.

It is possible that some packets successfully received by the UE but not included in the list are forwarded to the target eNodeB. In that case, it is the responsibility of the UE to re-order the received packets and remove any duplicated ones. Then the UE completes synchronization with the target eNodeB and asks for scheduling using a scheduling request message. The target eNodeB sends an uplink scheduling grant and a timing advance command to the UE, which, in return, sends a handover confirm message to the target eNodeB. Then the target eNodeB sends a path switch message to the MME, reporting the switching of eNodeB serving the UE. The MME sends a user plane update request message to S-GW, which switches the path of the downlink transmission to the target eNodeB. The target eNodeB sends a release resource message to the source eNodeB so that the radio resource assigned to the UE is released, completing the handover procedures. Note that if direct paths between the source and target eNodeBs are not available, the packets need to be forwarded to S-GW and redirected there. This fail-safe procedure enables a so-called lossless handover in which theoretically no packet is lost during the transition of eNodeB. However, even if all packets are delivered to the respective destinations, it still cannot be guaranteed to deliver the packets in the proper order. Therefore even if the UE typically re-orders the received packets, packets arriving too early or too late have to be removed from the buffer, if they are unlikely to meet their expected play-out time.

Note that during handover, information on the status of ECN marking at the source eNodeB is not transferred to the target eNodeB. Suppose, for example, that handover is initiated while the source eNodeB is marking ECN-CE. Therefore the packets are in the ECN(0) or ECN(1) state until arriving at the source eNodeB via the uplink or downlink. After handover, the forwarded packets may still be marked with ECN-CE but the target eNodeB may not overwrite the ECN(0) or ECN(1) bits of new packets received from the UE or S-GW, if it does not find the need in the uplink or downlink. As the UE is likely to consider the disappearance of ECN-CE after handover as a sign of channel improvement, it may ask the far-end UE to increase the bit-rate, after observing the reception status for a period long enough for the decision.

It is also possible that the target eNodeB cannot handle the ECN. The ROHC context at the source eNodeB is not transferred to the target eNodeB during handover either. The RTP/UDP/IP headers of packets forwarded via the X2 interface or S-GW are not compressed but the expanded headers do not incur serious capacity issues as the number of forwarded packets is usually small. Although ECN was introduced to LTE after the handover procedures were designed, the necessity to include the status information of ECN in the handover messages was not recognized, as the ECN is considered in the media adaptation in which it takes much longer than handover until the expected changes are realized.

MIMO

The Multiple-Input and Multiple-Output (MIMO) can be considered as an opportunistic transmission strategy for carrying more data than usually possible if the channel conditions are favorable, as illustrated in Fig. 8.46. With MIMO, which can be considered as a multidimensional extension of VAMOS in the GSM, multiple layers of resource

Table 8.10 LTE UE category and downlink peak data rate.

UE category	4	6	9	12	16
CA configuration	10 × 2 (20 × 1)	20 × 2	20 × 3	20 × 3	20 × 4
MIMO antenna configuration	2 × 2	2 × 2	2 × 2	2 × 2 (4 × 4)	2 × 2 (4 × 4)
Peak data rate (Mbps)	150	300	450	600	980
Highest-order modulation	64QAM	64QAM	64QAM	256QAM	256QAM

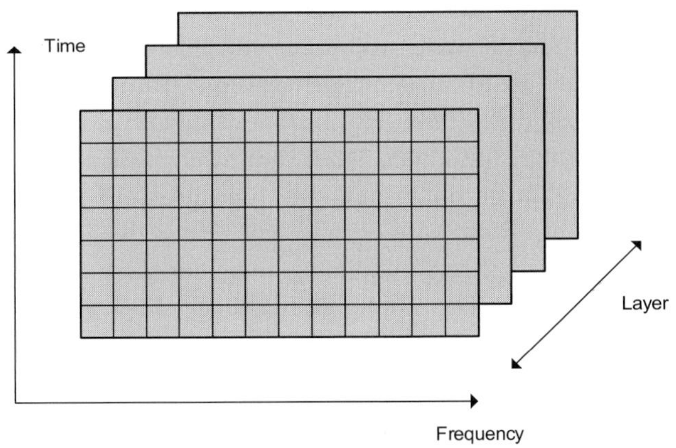

Fig. 8.46 Overlapped resource blocks in MIMO.

blocks belonging to the same time-frequency grid are assigned to multiple UEs or to a single UE. The orthogonality is provided in the spatial domain, which is complemented with orthogonal patterns adaptively applied to the resource blocks. MIMO is basically a transmission technique more suitable to the downlink than the uplink, as in the downlink, the eNodeB can take the link quality reported by UEs into account for scheduling.

MIMO is able to theoretically double or even quadruple the peak data rate with additional antennas at the receiver and the transmitter. Table 8.10 outlines the categories of UE and the corresponding configurations of antenna and carrier aggregation (CA) that maximize the rate in the downlink. However, MIMO is rarely employed for real-time conversational services or live multimedia streaming services since a constant level of QoS cannot easily be guaranteed with this approach. Note that at Categories 12 and 16, 4 × 4 MIMO is supported for only one and two carriers respectively. In addition to the complexity, installing four or more antennas in the UE is challenging, and the highest-order modulation is rarely chosen in typical operations. The bit-rates experienced in commercially operational networks are typically less than 20–30% of the maximum rate.

8.4 Media Signal Processing Procedures

This set of signal processing procedures takes place only at a UE for the uplink. The intermediate network nodes between two UEs, including eNodeB, S-GW, and P-GW, do not modify the bit-stream generated in this step. The media client of the VoLTE UE, as shown in Fig. 8.4, consists of a speech codec and a video codec, logically located on top of several layers of IP protocols. While AMR still remains the mandated speech codec of VoLTE, Adaptive Multi-Rate Wideband (AMR-WB) was a de-facto speech codec for the services in the initial deployments. As the name of the codec implies, the speech spectrum of AMR-WB is twice that of the narrowband speech codecs, which is enabled by sampling the speech waveform at 16000 Hz. In practice, only information included in the bandwidth of 50–7000 Hz can be encoded.

Figure 8.47 shows two speech waveforms sampled at 8000 and 16000 Hz. Figure 8.47(a) shows the case when the spectrum of the speech signal is confined completely within the narrowband, i.e., up to 3500 Hz. In this case, the waveform constructed by connecting samples digitized at 16000 Hz is almost identical to the waveform of 8000 Hz. As a result, there is no benefit from using a higher sampling rate, which necessitates increased processing power and higher bit-rate. Figure 8.47(b) shows the case when a significant amount of spectrum extends to the higher band. Now the two waveforms do not overlap completely. The more high frequency elements the speech signal contains, the more different the two waveforms become. If the bit-rate budget allows, incorporating the spectrum in the higher band into the speech compression is likely to improve quality, especially in the personal and emotional aspects.

Wideband speech coding is intended to include the frequency components of a speech signal up to 7000 Hz while suppressing the bit-rate as much as possible. Figure 8.48 shows wideband audio waveforms and their spectrograms. In comparison with the

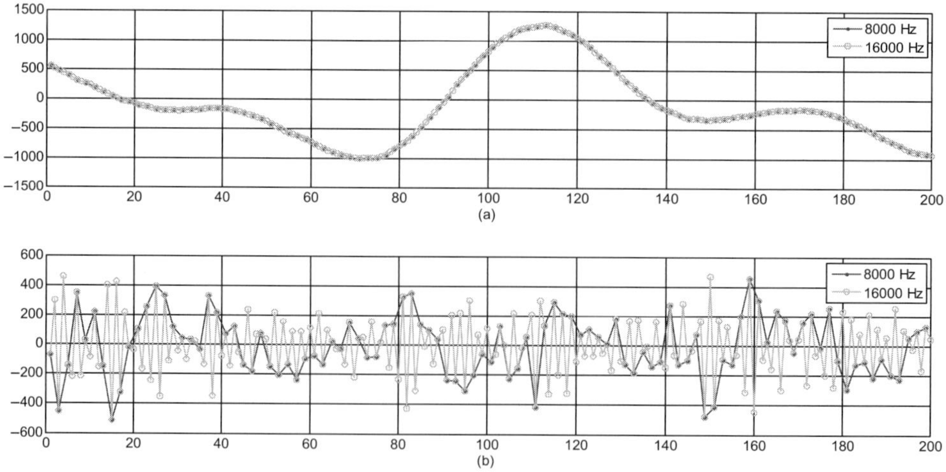

Fig. 8.47 Narrowband and wideband speech sampling. (a) Information contained in narrowband. (b) Information present outside narrowband.

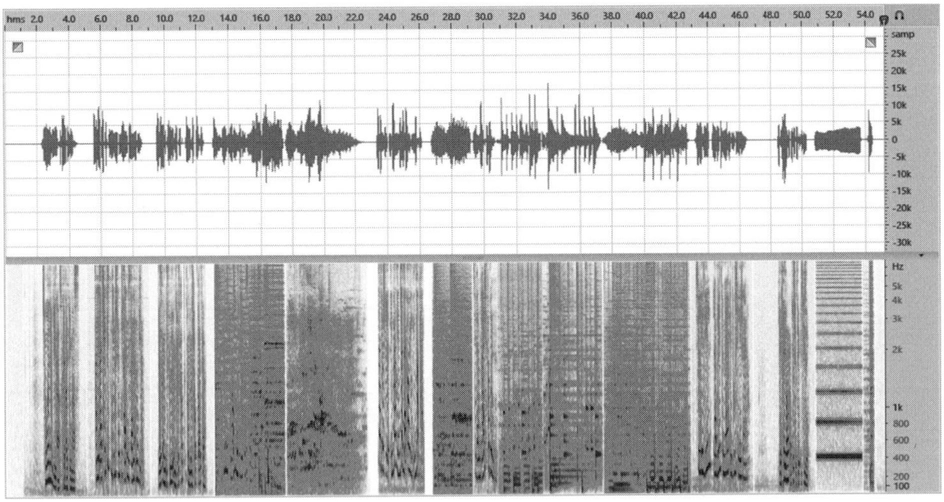

Fig. 8.48 Spectrogram of wideband speech waveforms.

spectrograms of the narrowband audio waveforms shown earlier in Fig. 2.1, which was generated from the same input signal, a greater amount of higher frequency content can be seen.

If audio information with wider bandwidth could be encoded with comparable bit-rates as for the narrowband version, then it would open a new path for codec evolution, assisted by the independence of LTE from circuit-switched core networks and PCM formats. The necessity of suppressing the bit-rate is evident as even the capacity of LTE will be ultimately challenged by traffic expansion. Although service providers operate LTE networks over multiple frequency bands, typically a single band, i.e., the primary band, has the largest coverage and receives the highest level of network maintenance as this band carries the most sensitive traffic such as the emergency, voice, and high-priority data.

Unfortunately, only a selected set of frequency bands can serve as the primary band, whose requirements include very good propagation characteristics for coverage, and availability in a large number of geographic areas or countries to enable global roaming. Moreover, the network capacity of the primary band cannot be increased simply by installing more base stations and splitting cells, because this would increase the frequency of handover and reduce the media quality. As the precious frequency bands are already crowded, it is more practical to increase the capacity of the existing primary band by reducing its traffic. Higher complexity is a relatively small price considering the benefits of higher audio bandwidth in quality and capacity. There is no reason to stop at wideband, if further improvements can be achieved at even higher sampling rates. Figure 8.49 illustrates how the voice bandwidth can be partitioned.

In VoLTE, progress was also made in the compression of video. The H.263 video codec was mandated while the MPEG-4 and H.264 [ITU-T (2010)] video codecs were optionally supported when the design of VoLTE began, assuming that HSPA would be

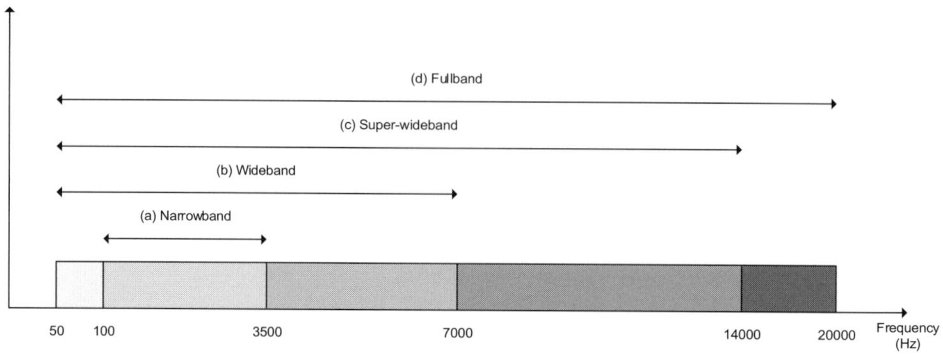

Fig. 8.49 Voice bandwidth partitioning. (a) Narrowband (NB). (b) Wideband (WB). (c) Super-wideband (SWB). (d) Fullband (FB).

the radio access network. This configuration was inherited from 3G-324M. However, unlike speech codecs, which are typically implemented in software, continuing the support of video codecs with low compression efficiency to the next generation of mobile multimedia communications systems to ease the interworking and compatibility pressure was considered to be worth neither the effort nor the cost. Therefore the support of the two obsolete video codecs used in 3G-324M was terminated in VoLTE, keeping only H.264 as the minimum video codec. Giving up compatibility is not a common decision in the design of mobile communications systems but it was still acceptable as the video codec is not an essential part of the radio access or core networks. A new video codec, High Efficiency Video Coding (HEVC), was introduced but video compression techniques were already mature, and gains from the new codec were expected mainly at bit-rates and resolutions much higher than those used for H.264. As illustrated in Fig. 8.4, the compression and transmission of media are managed by controllers that use the IMS and RTCP for session negotiation and media adaptation. The two procedures are similar in concept to the call setup and intra-refreshing of 3G-324M using the H.245 protocol, but they also include aspects that satisfy the new needs of packet-switched networks.

8.4.1 Adaptive Multi-Rate Wideband Speech Codec

Like AMR, AMR-WB is a multi-rate speech codec in which nine bit-rates, ranging from 6.6 to 23.85 kbps, are defined [3GPP (2017d)]. AMR-WB was designed for GSM and W-CDMA but unlike AMR, no codec mode is borrowed from other speech codecs. AMR-WB, also based on the ACELP principles, has a similar architecture as AMR, as shown in Fig. 8.50. Noticeable differences from AMR are the down-sampling and high-pass filtering (HPF) operations in the input stage, and the computing of high-band gain, g_{HB}, which is used only in the 23.85 kbps.

Even though the speech is sampled at 16 kHz, the AMR-WB speech encoder performs the analysis of linear prediction, long term prediction, and fixed codebook parameters at 12.8 kHz to limit the computational complexity. This means that the input signal has to

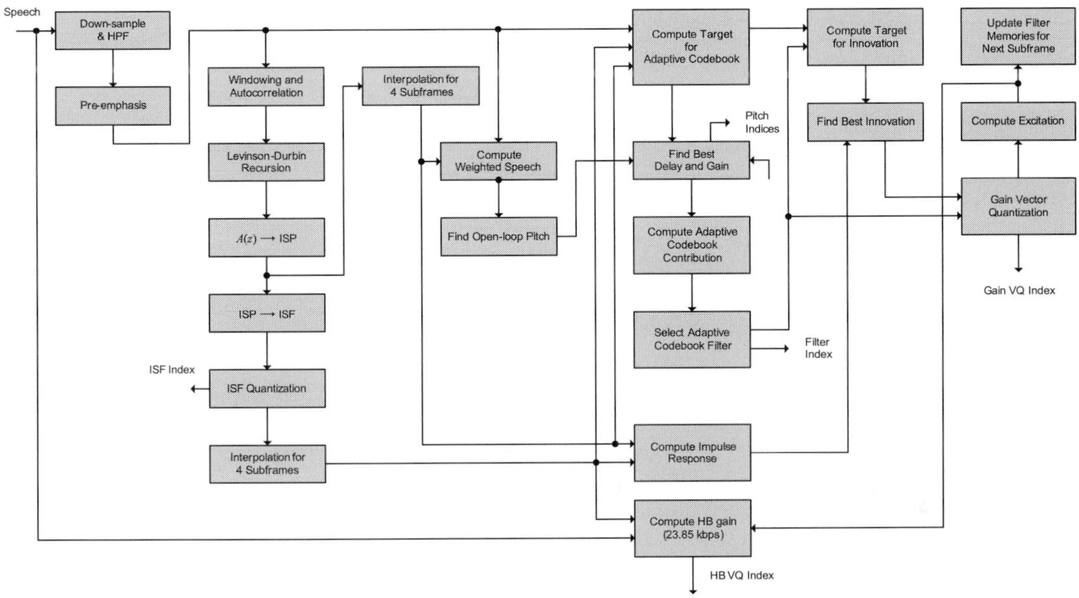

Fig. 8.50 Block diagram of AMR-WB speech encoder.

be decimated from 16 to 12.8 kHz. The decimation is done in three steps: the input signal is up-sampled by a factor of four, then filtered with a FIR low-pass filter that has a cut-off frequency at 6.4 kHz. Finally, the signal is down-sampled by a factor of five, achieving the 12.8 kHz sampling rate. The decimation to 12.8 kHz limits the audio bandwidth to 6.4 kHz, but the higher frequency band, 6.4–7.0 kHz, contains important spectral components that contribute to the wideband quality. At the decoder, an excitation is generated to model the high band, using the information contained in the low band. In the 23.85 kbps, the gain in the high band is computed and transmitted to complement the modeling, which is obtained using the relation,

$$g_{HB} = \frac{\sum_{i=0}^{63} (s_{HB}(i))^2}{\sum_{i=0}^{63} (s_{HB2}(i))^2}, \tag{8.14}$$

where $s_{HB}(i)$ is the band-pass filtered input speech and $s_{HB2}(i)$ is the high-band speech synthesis.

AMR-WB has a similar bit allocation as AMR, which is outlined in Table 8.11, with slightly different terminology. For an SID, the AMR-WB speech encoder generates 35 comfort noise bits, which include five indexes of the Immittance Spectral Frequency (ISF) subvectors, a six-bit index representing the logarithmic frame energy, and one bit for a dithering flag [3GPP (2017f)]. Then a one-bit SID type indicator and a four-bit mode indication are added to the 35 noise bits to construct an SID frame. As in AMR, the SID falls into two classes, SID_UPDATE, an ordinary SID containing the comfort noise parameters, and SID_FIRST, a marker for the end of a talkspurt in which all 35 bits are set to zero [3GPP (2017g)].

Table 8.11 Bit allocation of AMR-WB speech codec.

Mode	VAD flag, ISP sub-vector index	Adaptive codebook index	LTP-filtering flag	Codebook index	Codebook gains	High-band energy	Pulse selector	VQ gain	Total
23.85	47	30	4	352	28	16			477
23.05	47	30	4	352	28				461
19.85	47	30	4	272			16	28	397
18.25	47	30	4	224			32	28	365
15.85	47	30	4	208				28	317
14.25	47	30	4	176				28	285
12.65	47	30	4	144				28	253
8.85	47	26		80				24	177
6.60	37	23		48				24	132

Error Concealment

AMR and AMR-WB can operate with three generations of mobile communications systems, GSM, W-CDMA, and LTE. However, unsuccessful transmission results in quite different effects on the received speech frames even when the loss rates are similar. Figures 2.21(b), 2.21(c), and 2.21(d) illustrate the inputs to the speech decoder when speech frames are transmitted over a bit/frame-error channel, an erasure channel, and a packet-lossy channel with jitter, respectively. GSM and W-CDMA can be configured to generate error patterns for either of the first two channel types, depending on the behavior of the receiver to handle erroneous frames. The bit-error channel fits the typical deployments of GSM where unequal error detection and protection are employed, and the erasure channel matches the situations of W-CDMA where the radio link control (RLC) in the unacknowledged mode (UM) drops erroneous data. Finally, the packet-lossy channel matches the LTE situations where packets can be lost, received in a different order, and even duplicated during transmission. In addition, the transmission time can vary.

With the error concealment of the speech decoders in VoLTE to improve subjective quality, lost speech frames are substituted with either a repetition or an extrapolation of previously decoded speech frames. These frames are substituted so that the output level is gradually decreased, to avoid the risk of loudly playing out an unexpected sound. Likewise, a lost SID can be approximated by using the information from previously decoded SID frames. If many subsequent SIDs are lost, the output level of the comfort noise is gradually decreased [3GPP (2017e)].

8.4.2 H.264 Video Codec

The two video codecs with which 3G-324M is equipped, H.263 and MPEG-4, are based on the early compression algorithms designed in the 1990s when the available computational complexity was very limited. The key features of MPEG-4, error resilience tools and object oriented coding, did not exhibit a noticeable performance advantage in the mobile applications of W-CDMA. In the development of the successor

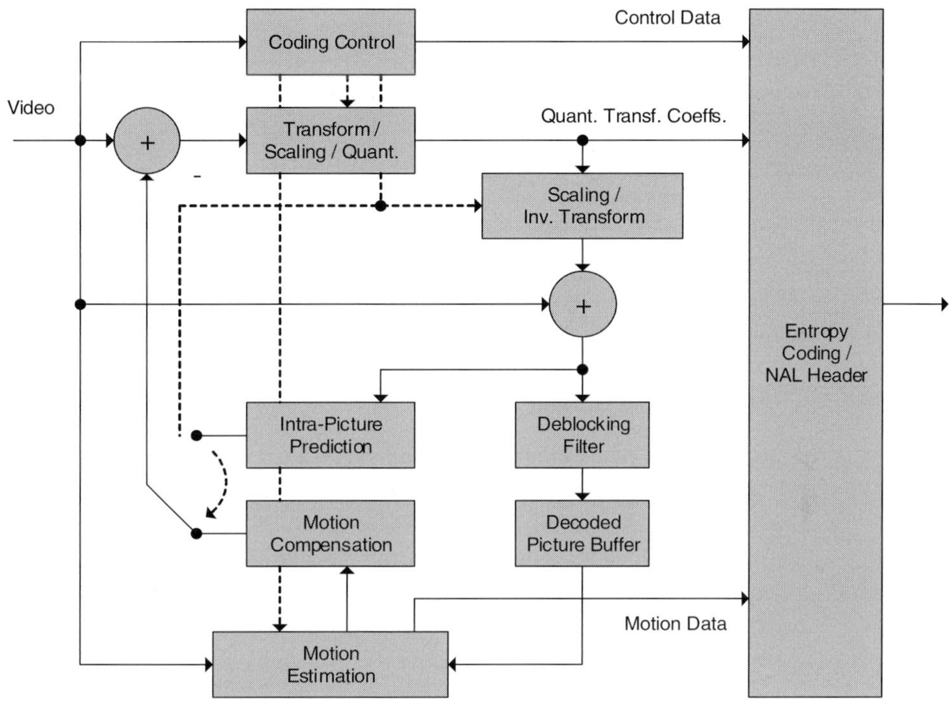

Fig. 8.51 Block diagram of H.264 video encoder.

to the two codecs, H.264, the design focus shifted back to the compression efficiency [ITU-T (2010)].

Figure 8.51 shows the encoder structure of H.264. It was initially added as the third video codec to 3G-324M but was not implemented as such since the fixed, low bit-rate of 3G-324M made the introduction of a new video codec not worth the effort or the cost. Increasing the video bit-rate of 3G-324M is quite challenging since the total bit-rate cannot be increased in an elegant fashion but can only be doubled or quadrupled, by applying smaller spreading factors in the radio access network and aggregating 64 kbps channels in the core network. With the limited capacity of W-CDMA, such an abrupt increase of bit-rate can leave insufficient capacity for other services. As a total bit-rate of 64 kbps is widely available in the PSTN, it is not economical not to use this well-established bit-rate in the wireless channel.

When the design objectives of H.264 were set, it was assumed that the error resilience would be taken care of mainly by the network, so that the loss of video data would always be below an acceptable level. Thus, features such as the error resilience tools of MPEG-4, except for data partitioning, were excluded. To facilitate the transmission of video, H.264 provides an interface, the Network Adaptation Layer (NAL), which simplifies the interaction with the network. The RTP payload header of H.264, shown in Fig. 8.53, is an example of the NAL.

In H.264, each step of the video encoding process is refined, albeit using increased computation. To achieve higher compression efficiency, H.264 allows motion estimation and compensation on smaller block sizes, down to 4×4, than allowed in the two previous video codecs. An important feature of H.264 that contributes to its higher quality is the in-loop adaptive deblocking filter, which smooths out blocking artifacts around block edges in the reconstructed picture. In addition to conventional variable-length coding (VLC), Context-based Adaptive Binary Arithmetic Coding (CABAC) is used to compress the syntax elements efficiently by selecting probability models for each element. Compared with the 48 kbps video at QCIF, which was supported by 3G-324M, VoLTE requires at least level 1.2 of H.264, which supports bit-rates up to 384 kbps. In commercial VoLTE deployments, 300–500 kbps at VGA, and 20–30 fps is commonly used for H.264, as shown in Fig. 8.59.

8.4.3 RTP/UDP/IP Packetization

Figure 8.52 illustrates the packetization procedures of VoLTE in which the RTP, UDP, and IP headers are sequentially attached to a payload, a frame from the AMR-WB speech encoder. A 253-bit frame of AMR-WB 12.65 kbps is zero-padded to make the size of payload octet-aligned, i.e., an integer number of bytes. To assist the handling of packets at the receiver, each payload is preceded by a payload header defined exclusively for the medium. For AMR or AMR-WB operating in the octet-aligned format, a 16-bit payload header, which contains the information on the codec mode and control information, is attached before the zero-padded payload is loaded onto an RTP packet [Sjoberg *et al.* (2007)].

Then, additional protocol headers are attached to the RTP packet, which is now ready to be transported to any location linked to the Internet. The combined overhead from the headers is larger than the encoded speech frame, which necessitates reducing the overhead especially when the packets are transmitted over a wireless channel. Note that if the payload is larger and less frequent, as is the case with video frames, the relative overhead from the protocol headers is reduced. In the case of speech, if multiple frames are

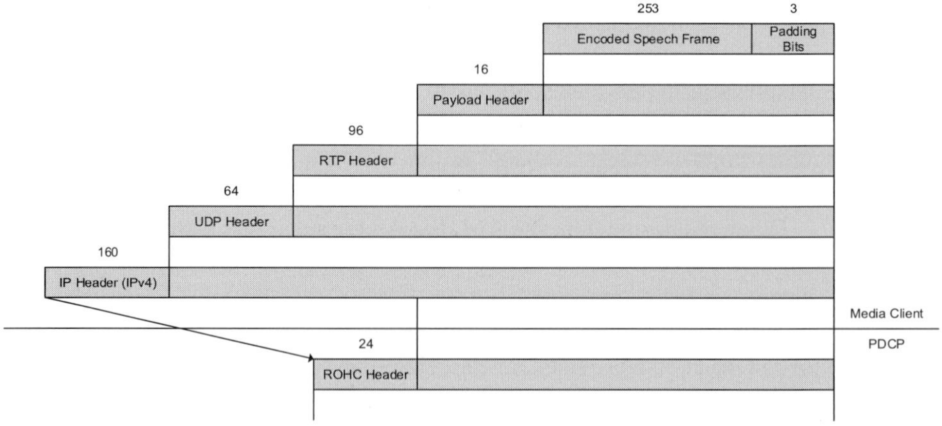

Fig. 8.52 RTP/UDP/IPv4 packetization and header-compression procedures for AMR-WB 12.65 kbps.

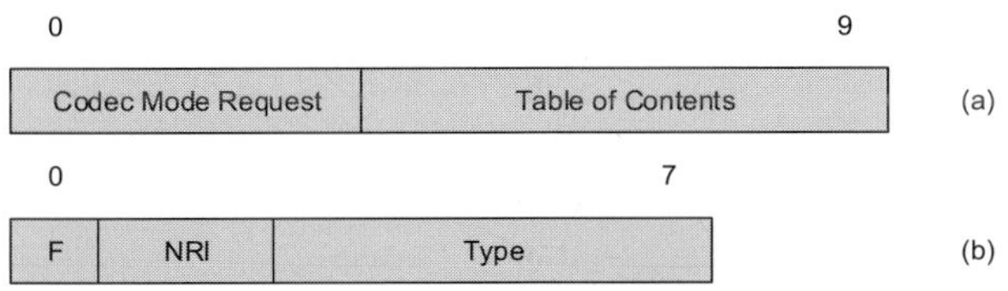

Fig. 8.53 RTP payload header. (a) AMR-WB (bandwidth-efficient). (b) H.264.

loaded onto an RTP packet, the overhead decreases but so does the quality, as the average end-to-end delay increases. In the ROHC of the LTE modem, 320-bit RTP/UDP/IPv4 headers can be compressed to as few as 24 bits under favorable channel conditions, by exploiting the statistical characteristics of the fields of the headers.

Figure 8.53(a) shows the payload header of AMR-WB in its bandwidth-efficient format. The first four bits of the header constitute the codec mode request (CMR). Unlike the two-bit CMR message of GSM or the two-bit CMC message of W-CDMA, which can specify a codec mode from an ACS including up to four modes, the four-bit CMR in the payload header enables the change of codec mode to any mode. The two-bit CMR or CMC in GSM and W-CDMA designates the requested codec mode as a relative order of the modes included in the ACS while the four-bit CMR in the payload header designates the absolute codec mode directly.

In GSM, position of the in-band signaling was secured by reducing the number of bits for encoded speech and channel code, and re-organizing the bit allocation. The practice of using only two bits was inherited by W-CDMA to enable direct interworking or handover with GSM. The RTP payload header of AMR and AMR-WB was initially defined for the Internet where flexibility was considered to be more important than spectral efficiency. Therefore, the two-bit limitation is not enforced in the payload header. Note that CMR is not the only method that can be used to signal the requested codec mode to the far-end speech encoder. CMR is a type of *in-band signaling* method where control information is embedded in the data bit-stream. In contrast, *out-of-band signaling* refers to transmission strategies where control information is sent in bit-streams separate from data. The spectral efficiency can be increased with in-band signaling but its flexibility is compromised. In contrast, new features or capability can be introduced relatively easily to out-of-band signaling at the cost of higher bit-rates. For example, the out-of-band signaling will require a whole new set of IP headers for the control information.

The six bits following the CMR constitute the Table of Contents (ToC), which indicates whether this frame is followed by another in the payload, whether the speech is encoded in AMR or AMR-WB, or whether a speech frame or an SID is carried. When multiple speech frames are loaded onto an RTP packet, each frame needs to be accompanied by a ToC but only a CMR can be transported in an RTP packet. Whether the octet-aligned or bandwidth-efficient packetization format is used, the size of the generated payload should be an integer number of bytes. In the bandwidth-efficient format, a minimum number of padding bits are added to the speech frame so that the combined length of the payload header, speech frame, and the padding bits is an integer number of

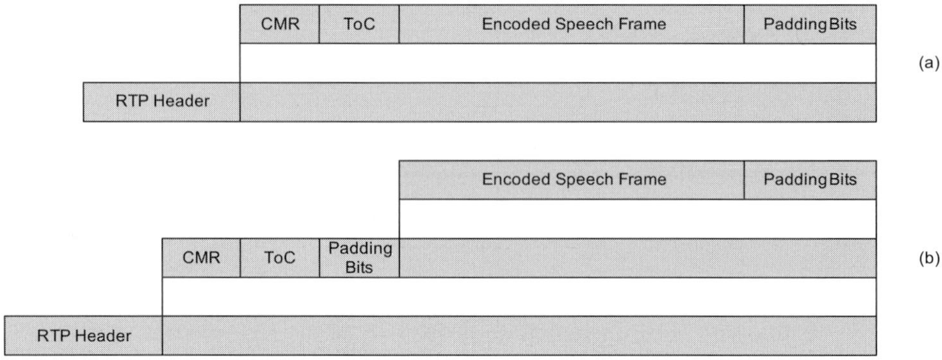

Fig. 8.54 RTP packet formats of AMR and AMR-WB. (a) Bandwidth-efficient. (b) Octet-aligned.

bytes. In the octet-aligned format, the padding bits are added separately to the payload header and the speech frame. Figure 8.54 illustrates how RTP packets are built when the bandwidth-efficient and octet-aligned formats are used.

It is possible to save the bit-rate by using the bandwidth-efficient format but the gain is negligible, typically 1–2 kbps, at best. On the other hand, the header processing can be simplified somewhat if the octet-aligned format is used. If a single bit-rate of AMR or AMR-WB is used, which is a common deployment scenario, information in the payload header is redundant as the speech decoder can reconstruct the frame without inspecting the header and does not have to ask for another codec mode. Therefore, in the payload format design of EVS, a new speech codec succeeding AMR-WB, this observation was taken into account so that a speech frame can be loaded onto an RTP packet without a payload header to save both the bit-rate and the processing overhead.

In contrast to speech frames which are independently encoded for each 20 ms period, the length or structure of encoded video frames depends on the video content and the motion activity. Thus, the strategy used to partition video frames can influence the error resilience and spectral efficiency. A series of encoded macroblocks, which may not be contiguously located in a frame, is referred to a segment and each individual segment is loaded onto an RTP packet, to limit the visual impact of packet loss. Figure 8.53(b) shows the payload header of H.264 [Wang *et al.* (2011)].

However, partitioning the video into segments that are too small increases the relative proportion of IP overhead and decreases the spectral efficiency. On the other hand, segments that are too large will reduce the error resilience since the loss of a single packet can result in significant quality degradation or long error propagation. The structural dependence between encoded video frames is another characteristic that video does not share with speech. With a fixed budget for the bit-rate, encoding too many independent frames will increase error resilience but reduce video quality, especially in the temporal domain. Therefore the granularity of packet sizes or the order of frame types can be used to control the tradeoff between error resilience and spectral efficiency, which can be used to match the characteristics of the network. For example, in each mobile communications system, it is recommended not to exceed the maximum transfer unit (MTU), the largest block size allowed for each transmission.

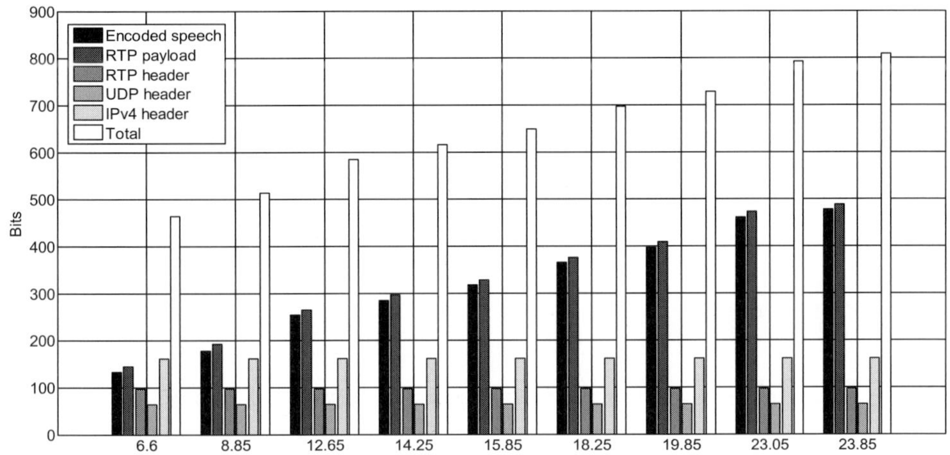

Fig. 8.55 Bits for encoded speech (AMR-WB), RTP, UDP, and IPv4 headers.

Figure 8.55 illustrates the process of bit-rate expansion as the three protocol headers are attached to speech frames in the bandwidth-efficient format, showing that the total bit-rates are several times larger than speech bit-rates. It can be seen that the relative overhead per packet gets smaller as the speech bit-rate increases.

8.4.4 Jitter Buffer Management

Jitter Buffer Management (JBM) is an integrated control mechanism that combines the buffer for received media frames, time-domain media scaling, i.e., compressing or stretching of a period covered by each media frame, algorithms that process decoded media signals to meet the target play-out schedule, channel estimation algorithms that monitor the transmission and reception status of packets and track the change of situations, and the media adaptation algorithms, if media compression or packetization of the far-end encoder can be remotely controlled. In typical implementations of voice services, JBM is installed around the speech decoder, as shown in Fig. 8.56. The channel estimation block gathers information about channel conditions, such as packet loss rate, by inspecting the SN field of the RTP header or comparing the RTCP SR with the list of received packets, to assist the decision of the control block. If available, channel estimation may also use the information from the LTE modem, for example, the buffer status at the MAC. If the level of the frame buffer, the number of stored speech frames, becomes too low while the loss of packets is detected, the channel estimation block may suspect the presence of congestion somewhere in the transmission path and inform the control block, which triggers the time-scaling block so that the decoded frames are played back for longer than encoded.

The time-scaling operations can be implemented in two approaches, frame-based and signal-based scaling. In frame-based scaling, time scaling is performed on the encoded speech frames, with a granularity of one speech frame per 20 ms duration. Techniques

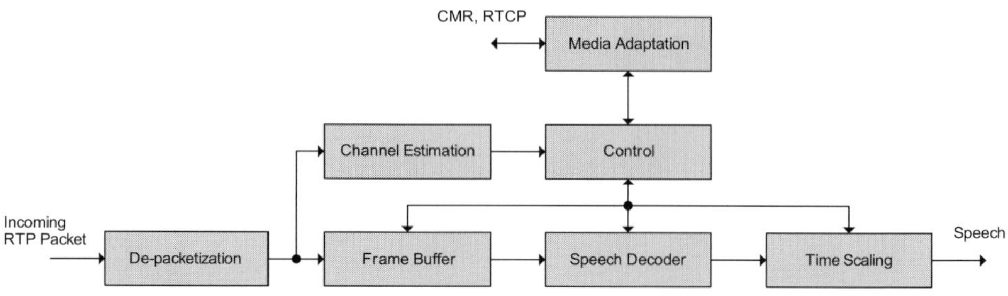

Fig. 8.56 Block diagram of JBM.

of frame-based scaling include inserting concealed frames, dropping overly-delayed frames, and inserting or deleting comfort noise frames. Inserting or deleting speech frames in the frame buffer results in higher distortion than in the processing of PCM samples but the media can be scaled faster. Inserting or deleting NO_DATA frames during DTX also enables faster scaling while minimizing the distortion. In signal-based scaling, the speech is time-scaled on the PCM samples. To meet the target play-out delay, the output signal of the speech decoder can be time-warped, i.e., shrunk if more samples than necessary are available and stretched otherwise.

8.5 Resource Reservation Procedures

8.5.1 IP Multimedia Subsystem

Figure 8.57 shows how the basic elements of the IP Multimedia Subsystem (IMS), Proxy Call Session Control Function (P-CSCF), Session Call Session Control Function (S-CSCF), Interrogating Call Session Control Function (I-CSCF), and the Policy and Charging Rules Function (PCRF), are interconnected between the radio access (eNodeB) and core (EPC) networks of two service providers. For each UE, P-CSCF is the gate-keeper or the first contact point to the IMS, as all signaling messages to and from the UE will pass this node. The major responsibilities of P-CSCF include the compression and ciphering of the Session Initiation Protocol (SIP) messages, interaction with PCRF, and the detection of emergency situations.

S-CSCF is in charge of registering the identity of the UE to the IMS, routing the SIP messages to other IMS nodes and maintaining the session information. I-CSCF assigns an S-CSCF when a UE registers with the IMS, based on the information from the Home Subscriber Server (HSS) and the Subscription Locator Function (SLF). I-CSCF also routes messages to the assigned S-CSCF. PCRF inspects the SIP messages exchanged between the UEs, enforcing the policies of the service provider. For example, a service provider may not allow speech to be encoded by certain codecs or may prohibit video at bit-rates higher than a limit. The Application Server (AS), which is not an intrinsic part of the IMS, provides services that facilitate the operation of the IMS and improve the quality of services. AS may provide information on the presence or location of the UE to the S-CSCF. The Packet Data Network Gateway (P-GW) includes the Policy

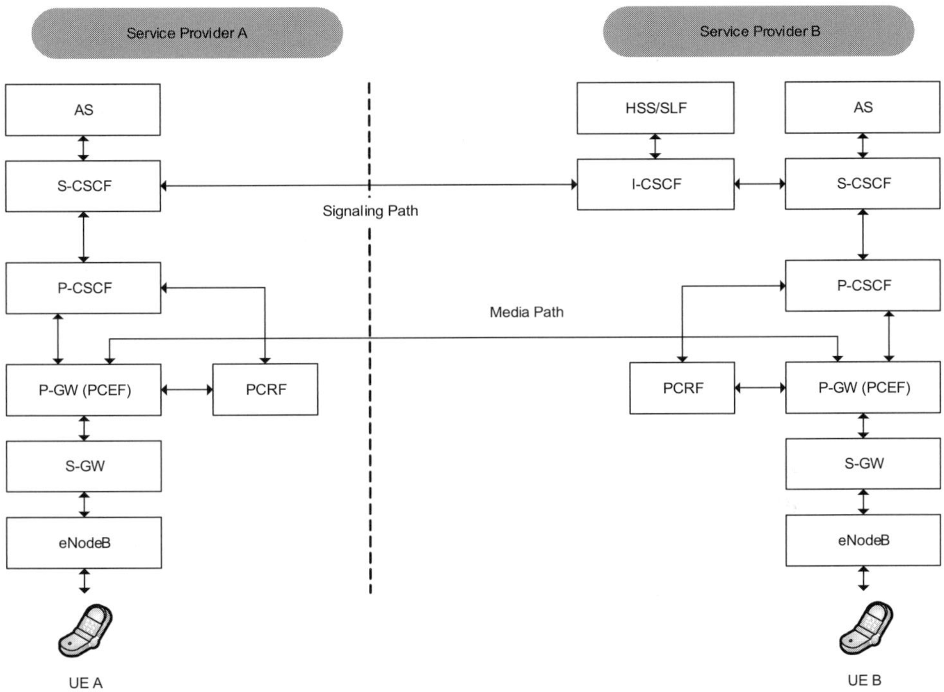

Fig. 8.57 IMS/EPC architecture.

and Charging Enforcement Functionality (PCEF). The history of IMS development and detailed descriptions of each node can be found in [Poikselkä and Mayer (2009)].

As illustrated in Fig. 8.57, the SIP messages are routed to the far-end UE over different paths from the media packets, as the IMS nodes are typically installed at a secure, centralized facility. The media packets are transmitted over the shortest paths possible. A small amount of radio resource for the SIP messages is reserved as the signal bearer when a UE is attached to the IMS. The radio resource for media packets, the dedicated bearer, becomes available after the session negotiation and the resource reservation procedures are completed.

8.5.2 SDP Offer

A Session Description Protocol (SDP) is used to describe the configuration of the medium [Handley *et al.* (2006)]. A set of configurations is offered by one UE and answered by another. We assume a typical session negotiation situation in which the SDP offers and answers correctly meet the service policy. As they comply with the policy, they are not modified by the intermediate IMS nodes during session negotiation, leaving the results to the discretion of the two UEs.

The SDP messages are embedded in the SIP messages. Figure 8.58 shows an SDP offer sent by a UE that wants to initiate a session, which is called the *offerer* in IMS

```
C=IN IP4 aaa.bbb.ccc.ddd
b=AS:1065

m=audio 49152 RTP/AVP 105 97 98 99 100 101 102
b=AS:65
b=RS:0
b=RR:2500
a=rtpmap:105 EVS/16000/1
a=fmtp:105 br=5.9-48; bw=nb-fb
a=rtpmap:97 AMR-WB/16000/1
a=fmtp:97 octet-align=1
a=rtpmap:98 AMR-WB/16000/1
a=rtpmap:99 telephone-event/16000
a=fmtp:99 0-15
a=rtpmap:100 AMR/8000/1
a=fmtp:100 octet-align=1
a=rtpmap:101 AMR/8000/1
a=rtpmap:102 telephone-event/8000
a=fmtp:102 0-15
a=sendrecv
a=ptime:20
a=maxptime:120

m=video 49154 RTP/AVP 106 103 104 34
b=AS:1000
b=RS:0
b=RR:2500
a=rtpmap:106 H265/90000
a=fmtp:106 profile-id=1; level-id=93; \
   sprop-vps=QAEMAf//AWAAAAMAgAAAAwAAAwBdLAUg; \
   sprop-sps=QgEBAWAAAAMAgAAAAwAAAwBdoAKAgC0WUuS0i9AHcIBB; \
   sprop-pps=RAHAcYDZIA==
a=imageattr:106 send [x=1280,y=720] recv [x=1280,y=720]
a=rtpmap:103 H264/90000
a=fmtp:103 packetization-mode=1; profile-level-id=42c016; \
     sprop-parameter-sets=Z0LAFukDwKMg,aM4G4g==
a=imageattr:103 send [x=480,y=640] recv [x=480,y=640]
a=rtpmap:104 H264/90000
a=fmtp:104 packetization-mode=1; profile-level-id=42c00d; \
     sprop-parameter-sets=Z0LADekCg/I=,aM4G4g==
a=imageattr:104 send [x=320,y=240] recv [x=320,y=240]
a=rtpmap:34 H263/90000
a=fmtp:34 profile=0;level=10
a=imageattr:34 send [x=176,y=144] recv [x=176,y=144]
a=sendrecv
```

Fig. 8.58 SDP offer.

terminology. The offer includes three speech codecs and three video codecs, and the parameters to initialize the codecs and packetize their encoded media frames. In this session, the IPv4 is used, and the session level b = AS, the total bit-rate for both speech and video including the IP overhead but excluding the RTCP bandwidth, equals 1065 kbps. The Application Specific (AS) is understood to be the maximum bandwidth of the application. Several dynamic payload types, 97–105, are assigned to the speech and video codecs, and the telephone event representation of the Dual-Tone Multi-Frequency (DTMF) tones for narrowband and wideband.

Speech Session Description

For speech, a maximum bit-rate of 65 kbps is assigned, with EVS as the preferred codec located at the top of the list. AMR-WB or AMR can be selected if EVS is not available in the UE receiving this SDP offer. For EVS, all bit-rates from 5.9 to 48 kbps, and all bandwidths from narrowband to fullband are offered for a single audio channel (mono). In contrast, no constraints are made in the codec mode set of AMR-WB and AMR whose maximum bit-rates are within the transmission capability of the network and service policy. The grammar used in the description of media-specific configurations is defined for each codec.

The media level b = AS for speech is set to the maximum b = AS value of all payload types, each of which includes a speech codec. For EVS, the b = AS value is computed as follows. Each frame of EVS encoded at 48 kbps, excluding the SID, consists of 960 bits. Then the largest size of RTP payload header, two bytes when both CMR and ToC are present, is taken into account in the computation of b = AS. In EVS, either the CMR or the ToC can be excluded in the bit-stream if not necessary, unlike in AMR-WB or AMR where both fields should be always present. For example, when a single bit-rate is used in the session without media adaptation, they can be excluded. With the RTP/UDP/IPv4 headers, the size of the packet becomes 1296 bits, corresponding to 64.8 kbps. Then b = AS is computed as the integer value of the bit-rate in kbps, 65, after rounding up. b = AS can be considered as the maximum bit-rate that may be necessary during the session. The maximum bit-rate is reached in the core networks during voice activity. In practise, the ROHC and DTX reduce the average bit-rate significantly.

Figure 8.55 shows that the size of a packet including an AMR 23.85 kbps frame, after the RTP/UDP/IPv4 headers are attached, is 808 bits. Dividing it by 20 ms and rounding up the quotient to the nearest integer gives the b = AS value. Although the b = AS of AMR-WB in the octet-aligned format is in general slightly larger than that in the bandwidth-efficient format, in this case, they are both 41 kbps. As the b = AS of EVS 48 kbps is larger than that of AMR-WB 23.85 kbps, let alone that of AMR 12.2 kbps, the b = AS for speech is set to 65. The 2500 bps assigned for the RTCP Receiver Report (RR) is shared in the two directions, i.e., leaving 1250 bps for each direction. ptime is defined as the length of time in milliseconds represented by the media in a packet. maxptime is the maximum amount of media that can be encapsulated in each packet, expressed as the length of time in milliseconds. As with the packetization of AMR-WB and AMR, the octet-aligned format is used. Regardless of the audio bandwidth used in the session and the actual sampling rate, the clock rate of EVS is set to 16 kHz in the SDP.

Video Session Description

For video, a maximum bit-rate of 1000 kbps is assigned, with H.265 (HEVC) as the preferred codec. H.264 or H.263 can be selected if H.265 is not available in the UE receiving this SDP offer. If each RTP packet carries about 250 bytes of data, which is likely to be a fraction of an encoded video frame, by adding 40 bytes of RTP/UDP/IPv4 headers and two bytes of RTP payload header, b = AS of 1000 kbps corresponds to approximately 856.2 kbps of output bit-stream from the video encoder. The b = AS for video needs to be set also considering the fluctuation of the video bit-rate, depending on the contents and motion activity. For example, b = AS of a video bit-stream whose target bit-rate is 512 kbps can be set to 639 kbps when IPv4 is used, assuming a 20% margin for the bit-rate fluctuation.

The session conditions are specified by a = fmtp:98 and imageattr:98 confined to payload type 98, H.265. profile-id and level-id, which are integral representations, indicate the default profile and level of the bit-stream the receiver supports. In H.265 terminology, the profile represents the subset of coding tools that may have been used to generate the bit-stream or that the receiver supports, and the level means the maximum decoding capability. sprop-vps, sprop-sps, and sprop-pps contain optional information on the packetization and initialization of the video codec [3GPP (2017c)]. However, these codec-specific parameters are not considered in the session negotiation. imageattr indicates the resolution, i.e., the image size, to use in the session for the send and receive directions [Johansson and Jung (2011)]. Therefore, the level of H.265 needs to be high enough to handle video at a bit-rate of 856.2 kbps and resolution of 1280 × 720.

In the payload type for H.264, a single parameter specifying both profile and level of the codec, profile-level-id, is used, in addition to the codec-specific parameters specifying the information on the packetization and initialization [Wang *et al.* (2011)]. Note that the offered H.264 video codec may not be as capable as the H.265 as a smaller resolution, 640 × 480, is specified for H.264. According to the profile-level-id parameter, in a hexadecimal representation, the offerer proposes to use the constrained baseline profile of H.264 at level 1.2, which supports a bit-rate up to 384 kbps. H.263 is offered as the lowest-quality service, probably for the interworking with 3G-324M.

8.5.3 SDP Answer

Figure 8.59 shows an SDP answer received from the far-end UE, the *answerer*. The session level b = AS was reduced to 669 kbps, implying that the answerer considers the offered bit-rates for speech or video to be excessive. AMR-WB was selected but its bit-rates were limited to the three lowest values: 6.6, 8.85, and 12.65 kbps. Therefore all information on EVS and AMR was removed from the SDP answer. For video, the answerer selected H.264 at 640 × 480 for both directions, at a bit-rate of 639 kbps, which was reduced from the 1 Mbps offered with the larger resolution of 1280 × 720 in mind. Note that the user interface of VoLTE shown with the SDP answer does not necessarily match the negotiated QoS parameters. However, it is apparent that higher audiovisual quality can be achieved from higher bit-rate, resolution and audio bandwidth than those of 3G-324M, shown in Fig. 6.56.

8.5.4 Quality of Service Representation

The set of media parameters determined in the session negotiation is only a subset of the important quality factors, since additional aspects such as the maximum delay, or the allowed bit error or packet loss rate also need to be decided. The IMS reserves network resource for transmission by exchanging a set of information, among which

```
C=IN IP4 aaa.bbb.ccc.ddd
b=AS:669

m=audio 49152 RTP/AVP 97 99
b=AS:30
b=RS:0
b=RR:2500
a=rtpmap:97 AMR-WB/16000/1
a=fmtp:97 mode-set=0,1,2; octet-align=1
a=rtpmap:99 telephone-event/16000
a=fmtp:99 0-15
a=sendrecv
a=ptime:20
a=maxptime:120

m=video 49154 RTP/AVP 103
b=AS:639
b=RS:0
b=RR:2500
a=rtpmap:103 H264/90000
a=fmtp:103 packetization-mode=1;
  profile-level-id=42c016;
  sprop-parameter-sets=Z0LAFukDwKMg,aM4G4g==
a=imageattr:103 send [x=480,y=640] recv [x=480,y=640]
a=sendrecv
```

Fig. 8.59 SDP answer (image courtesy of NTT DOCOMO, Inc.).

the QoS Information Element (IE) specifies various quality criteria to meet in the session, with the network nodes in the transmission paths under its jurisdiction once the UEs have exchanged an SDP offer and an SDP answer, and reached a tentative agreement.

Figure 8.60 shows the structure of the QoS IE, which has a minimum length of 14 bytes and a maximum length of 18 bytes [3GPP (2017b)]. The level of QoS requested or granted is specified in bytes 3–5 and 6–14. For example, the three-bit delay class can have a value from one to four, which corresponds to the *conversational*, *streaming*, *interactive*, and *background* classes, respectively. The lower classes indicate that the packets should be handled with lower delay constraints. In VoLTE, speech, video, and RTCP packets are classified as conversational while the SIP messages exchanged in session negotiations are handled as interactive. Located in byte three, the three-bit reliability class can have a value from three to five, based on the partitioning of the ranges for the SDU error ratio and the residual bit error ratio, as illustrated in Fig. 8.61. Bytes 8–9, 12–13, and 15–18 are used to specify the maximum and guaranteed bit-rates for the uplink and downlink.

Suppose the offerer accepts the SDP answer in Fig. 8.59. The session level b = AS, originally defined for the Internet, is simply set to the sum of b = AS values for the payload types, i.e., speech and video, and is not taken into account in the resource reservation procedures. The maximum bit-rates for the uplink and downlink can be computed as follows. From b = AS:30 (kbps), b = RS:0 (bps), and b = RR:2500 (bps), the total maximum bit-rate for speech is 32.5 kbps for both links. Likewise, the total maximum bit-rate for video is 641.5 kbps. In addition to rounding these values up to the nearest

Octet	8		1
1	Quality of Service IEI		
	Length of QoS IE		
	Spare (0 0)	Delay Class	Reliability Class
	Peak Throughput	Spare (0)	Precedence Class
	Spare (0 0 0)	Mean Throughput	
6	Traffic Class	Delivery Order	Delivery of Erroneous SDU
	Maximum SDU Size		
	Maximum Bit-Rate for Uplink		
	Maximum Bit-Rate for Downlink		
	Residual BER	SDU Error Ratio	
11	Transfer Delay	Traffic Handling Priority	
	Guaranteed Bit-Rate for Uplink		
	Guaranteed Bit-Rate for Downlink		
	Spare (0 0 0)	SI	Source Statistics Descriptor
	Maximum Bit-Rate for Downlink (Extended)		
16	Guaranteed Bit-Rate for Downlink (Extended)		
	Maximum Bit-Rate for Uplink (Extended)		
	Guaranteed Bit-Rate for Uplink (Extended)		

Fig. 8.60 QoS information element.

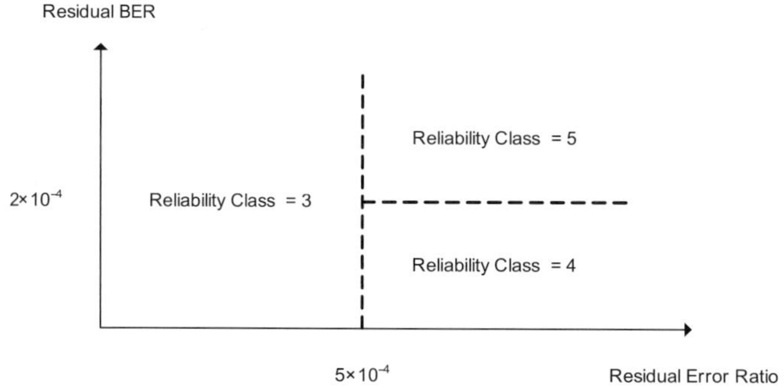

Fig. 8.61 Partitioning of reliability class.

integers, further rounding up, with nonlinear granularity, is necessary for the sum to be mapped to a byte in the QoS IE.

Table 8.12 compares the values of the QoS parameters for the speech and video. The development of these parameters and computational procedures for evaluating them can be found in [3GPP (2014)]. The total maximum bit-rates of the speech and video, the sum of b = AS (kbps) and RR (bps), are rounded up to 33 and 642 kbps, respectively, and set to the Maximum Bit-Rate (MBR) fields of the QoS IEs. It is assumed that the Guaranteed Bit-Rate (GBR) and MBR are identical, reducing the need for media adaptation. A separate QoS IE is constructed for each medium and the accompanying RTCP packet sharing the same QoS. It can be seen that the transparent mode of RLC is used, from the absence of delivery-order support but the SN of RTP payload header can be used instead for re-ordering.

Table 8.12 QoS profile for speech and video.

Parameter	Speech	Video
Delivery of erroneous SDU	No	No
Delivery order	No	No
Delay class	Conversational	Conversational
Maximum SDU size	130 bytes	500 bytes
GBR for uplink	33	642
GBR for downlink	33	642
MBR for uplink	33	642
MBR for downlink	33	642
Residual BER	10^{-5}	10^{-5}
SDU error ratio	7×10^{-3}	10^{-3}
Transfer delay	100 ms	250 ms

Bytes 8 and 9 of the QoS IE represent the bit-rate from 1 to 8640 kbps, in a fashion where the granularity is increased eight times at each step. First, 00000001 (1) to 00111111 (63) represent 1 to 63 kbps. In the next step, 01000000 (64) to 01111111 (127) represent 64 to 568 kbps in 8 kbps increments, which still cannot describe the maximum bit-rate of video, 642 kbps. 10000000 (128) to 11111110 (254) represent 576 to 8640 kbps in 64 kbps increments. 11111111 represents 0 kbps while 00000000 is reserved. As $642 = 576 + 1 \times 64 + 2$ kbps, the maximum bit-rate is mapped to $704 = 576 + 2\times$ 64 kbps, after rounding up to the nearest multiple of 64, 10000010. Note that no further information is necessary to signal the step or range used in the representation. It might appear that the resource is overly reserved, considering the original video bit-rate of 512 kbps. However, the radio access network minimizes the consumption by compressing the IP headers and scheduling only when there is data, and the transmission cost in the core network, where the headers are fully expanded, is much lower than in the access.

The maximum bit-rate of 8640 kbps, originally felt to be sufficient considering the limited capability of GSM and W-CDMA, can be a bottleneck in the resource management of mobile broadband networks such as LTE, where a bit-rate as high as 75 Mbps can be supported over a 10 MHz bandwidth. Bytes 15 and 17 are appended to extend the range of the maximum bit-rate. When the represented bit-rate is lower than 8640 kbps, these bytes are not present or set to zero. If byte 15 is included, then byte 16 will also be included. Likewise, if byte 17 is included, then byte 18 will also be included. When the maximum bit-rate for the uplink exceeds 8640 kbps, byte 8 is set to 11111110, and by setting byte 17, from 00000001 (1) to 01001010 (74), a range of 8700 to 16000 kbps can be represented in 100 kbps increments. In the next step, 01001011 (75) to 01001010 (186) represent 17 to 128 Mbps in 1 Mbps increments. Finally, 10111011 (187) to 11111010 (250) represent 130 to 256 Mbps in 2 Mbps increments. Figure 8.62 shows the maximum uplink and downlink bit-rates representable by the QoS IE. Apparently it is not possible to represent any bit-rate over the wide range with just two bytes and over-reservation of resource always happens, which would not result in a significant loss of spectral efficiency as the unused resource is rescheduled to other packet streams.

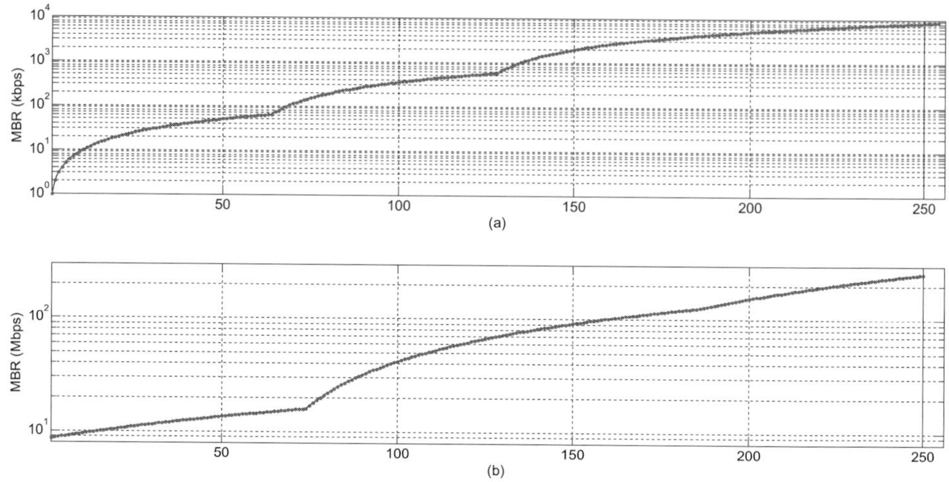

Fig. 8.62 Mapping binary representation to bit-rate.

MBR is not always supported by the network. For mission critical services that necessitate stable support of fixed bit-rates, GBR should be requested in the QoS IE. The representation of GBR follows the same grammar as that of MBR. If the MBR is greater than the GBR, the applications can take advantage of the additional capacity when it is available, while being flexible enough to react when such resources are no longer available. In typical operations of VoLTE, GBR and MBR are set identically. During the session negotiation between two UEs connected to networks operated by different service providers, it is necessary to use an identical GBR in both networks, as only GBR has well-defined quality criteria [3GPP (2017a)]. Located in byte 6, the 3-bit delivery of erroneous SDU field determines how the SDUs containing errors are handled by the intermediate network nodes. If the field is set to 010, the erroneous SDUs are delivered to the applications with an error mark while 011 indicates that the erroneous SDUs are dropped upon detection. As this field is typically set to 010 in VoLTE, the successfully received packets can be considered as error-free. However, as the corrupted packets are dropped, jitter is increased, as illustrated in Fig. 2.21(d).

Note that even if the network is serving a UE with GBR, it may not always be possible to meet the negotiated QoS as the eNodeB has finite resources such as the bandwidth or transmit power but the location or direction of the UE may physically limit the link quality. Then the media adaptation can be triggered by the UE even when the eNodeB and the UE communicate in GBR mode, asking the far-end UE to temporarily reduce the bit-rate lower than the negotiated GBR. Although the QoS IE contains extensive information on the required level of service quality, it is not efficient to control the behavior of the network nodes with such a complex parameter structure. The QoS Class Identifier (QCI) is a scalar used as a reference to a set of node-specific parameters that control packet forwarding treatment, e.g., scheduling weights, admission thresholds, queue management thresholds, and link layer protocol configurations. Table 8.13 summarizes a set of

Table 8.13 Characteristics of standard QCI values.

QCI	Type	Priority	PDB (ms)	PELR	Target services
1	GBR	2	100	10^{-2}	Conversational voice
2	GBR	4	150	10^{-3}	Conversational video
3	GBR	3	50	10^{-3}	Real-time gaming
4	GBR	5	300	10^{-6}	Buffered streaming
5	Non-GBR	1	100	10^{-6}	IMS signaling
6	Non-GBR	6	300	10^{-6}	TCP

standardized QCIs but custom QCIs can also be defined by a service provider to meet its specific needs for service quality.

The Packet Delay Budget (PDB) is an upper bound on the time that a packet may be delayed between the UE and the PCEF, which is used to support the configuration of scheduling and link layer functions, such as the setting of scheduling priority weights and HARQ target operating points (maximum number of re-transmissions). PDB should be interpreted as a maximum delay with a confidence level of 98%, which means that 98% of packets that have not been dropped due to congestion should not experience a delay exceeding the PDB. For a QCI value, the same PDB is applied in the uplink and downlink.

The Packet Error Loss Rate (PELR) is defined as an upper-bound on the rate of SDUs that have been processed by the sender of a link layer protocol, such as the RLC, but not successfully delivered by the corresponding receiver to the upper layer, such as the PDCP. Therefore the PELR can be considered as an upper-bound for the rate of packet loss not related to congestion, which enables appropriate configurations of the link layer protocol. For a QCI value, the same PELR is applied in the uplink and downlink.

The Priority is defined as the relative importance of a resource request for a service. It enables the eNodeB to decide whether a request for bearer establishment or modification can be accepted or rejected when the available resource is limited. The priority can also be used to decide which existing bearers to release in situations when blocking new sessions is not enough to manage the operation of the network. The PELR of conversational speech and video is set to match the target BLER of W-CDMA voice service and 3G-324M, 10^{-2} and 10^{-3}, respectively. In 3G-324M, there is no measure to apply separate delay values to speech and video as they are multiplexed into a single bit-stream. It is only possible to delay the play-out of speech frames, which become available before the larger video frames sampled at similar times are reconstructed, by a negotiated value of H223SkewIndication. In VoLTE, delay is a key resource that can be selectively enforceable to differentiate the QoS. The eNodeB can aggregate packet streams to and from a UE if they share a common QCI.

With the use of IMS and its various QoS parameters, the media and data transmission can be tailored to meet the detailed requirements of applications, reducing the needs for designing service-specific transport channels, such as the various types of voice channels for each bit-rate in circuit-switched networks. For example, one-way compression and transmission of media from drones or robots does not require the tight end-to-end delay of voice and video calls. By reducing the level of delay class but increasing the level of reliability class, from those of conversational services, the media stream would fulfill the needs of multimedia surveillance better. On the other hand, low bit-rate data channels with very low delay but high reliability can be configured for remotely controlling flying or moving objects.

8.5.5 Session Negotiation

Figure 8.63 illustrates the session negotiation procedures in which the media configurations are agreed upon between two UEs and resources for the wireless and wired links are reserved at the negotiated QoS. As the level of QoS in the first SDP answer is likely to be equal to or lower than that in the first SDP offer, the second SDP offer is likely to be identical to the first SDP answer, with some parameters related to the direction of transmission modified appropriately.

UE A and UE B are connected to eNodeB A and eNodeB B, respectively. UE A, which wants to have a multimedia conversation with UE B, sends an SIP Invite message including an SDP offer to UE B, which is connected to the network of another service provider. The message is routed via numerous IMS nodes to UE B, which inspects the contents of the SDP offer and selects a subset of the offered configurations to construct an SDP answer. The parameter values of the selected subset are adjusted to meet the media needs of UE B. The SDP answer is contained in an SIP 183 (Session Progress) message and sent to UE A, traversing the same paths in the opposite direction. The SIP 183 also includes a proposition, using the SDP precondition mechanism [Camarillo

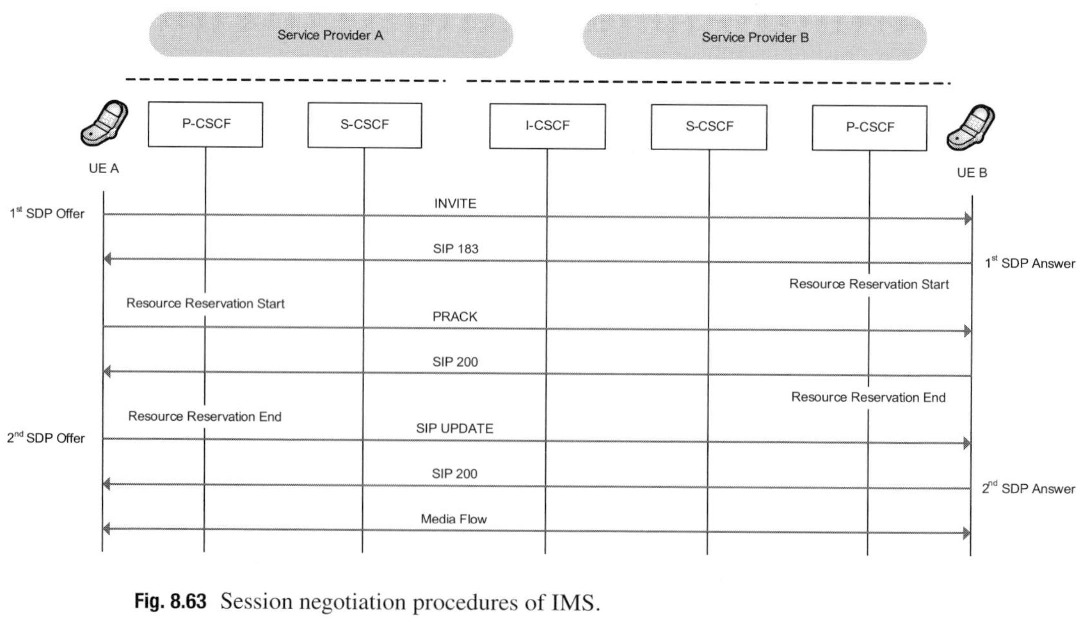

Fig. 8.63 Session negotiation procedures of IMS.

et al. (2002)], that UE B should wait without ringing for the confirmation from UE A that the resource has been successfully reserved at the far-end.

As both sides have already agreed upon a rough description of the services, the SIP 183 message triggers the reservation of resource in both networks while traveling through the various IMS nodes back to UE A. After receiving the SIP 183, the P-CSCF of service provider A passes the message in the direction of UE A while asking the PCRF to authorize the QoS described in the SDP answer. The QoS IE is constructed for each media. After receiving the message, UE A confirms reception with the Provisional Response Acknowledgement (PRACK) message, which does not include the SDP information as the QoS reservation may be still in progress. Meanwhile in the IMS, if the request by the P-CSCF is granted, the PCRF informs the P-CSCF and P-GW of the information necessary to enforce the granted QoS, such as the QCI, MBR, and GBR values.

PCEF, located in the P-GW, applies the QoS information to establish dedicated bearers for the media. The bearers in this context refer to the transport capability of all network nodes between the UE and the border of the network that meets the requested QoS. As soon as the bearers required for the services are established, UE A informs UE B by sending an SIP Update message, which includes a second SDP offer for which the resource is now ready. As soon as UE B receives the second SDP offer, it acknowledges with an SIP 200 message including the second SDP answer if the offer is acceptable, which is identical to the second SDP offer, and finally starts ringing to alert the user.

Note that although the P-CSCF on the side of UE B starts the resource reservation earlier, as it receives the SIP 183 message earlier, it does not always finish the establishment of bearers earlier than the P-CSCF of UE A. As the user of UE B pushes the answer button or icon, the encoded media starts flowing in both directions. The procedures typically take 2–3 seconds if the UEs understand the SDPs from their counterparts accurately. It would take the lowest delay if both UEs are in the connected mode of RRC, and additional delay would be required if the UEs have to transition from the idle mode of the protocol. As the resource reservation is triggered by the P-CSCF, this procedure is called the *network-initiated* resource reservation [Chakraborty *et al.* (2007)]. Although not widely used in VoLTE, it is also possible that a UE initiates the resource reservation itself, i.e., as the *mobile-initiated* resource reservation.

8.6 References

3GPP. 2014. TS 26.236 V12.0.0 Packet Switched Conversational Multimedia Applications; Transport Protocols. September.

3GPP. 2017a. TS 23.203 V14.3.0 Policy and Charging Control Architecture. March.

3GPP. 2017b. TS 24.008 V14.3.0 Mobile Radio Interface Layer 3 Specification; Core Network Protocols; Stage 3. March.

3GPP. 2017c. TS 26.114 V14.3.0 IP Multimedia Subsystem (IMS); Multimedia Telephony; Media Handling and Interaction. March.

3GPP. 2017d. TS 26.190 V14.0.0 Speech Codec Speech Processing Functions; Adaptive Multi-Rate – Wideband (AMR-WB) Speech Codec; Transcoding Functions. March.

3GPP. 2017e. TS 26.191 V14.0.0 Speech Codec Speech Processing Functions; Adaptive Multi-Rate – Wideband (AMR-WB) Speech Codec; Error Concealment of Erroneous or Lost Frames. March.

3GPP. 2017f. TS 26.192 V14.0.0 Speech Codec Speech Processing Functions; Adaptive Multi-Rate – Wideband (AMR-WB) Speech Codec; Comfort Noise Aspects. March.

3GPP. 2017g. TS 26.201 V14.0.0 Speech Codec Speech Processing Functions; Adaptive Multi-Rate – Wideband (AMR-WB) Speech Codec; Frame Structure. March.

3GPP. 2017h. TS 36.101 V14.3.0 Evolved Universal Terrestrial Radio Access (E-UTRA); User Equipment (UE) Radio Transmission and Reception. March.

3GPP. 2017i. TS 36.104 V14.3.0 Evolved Universal Terrestrial Radio Access (E-UTRA); Base Station (BS) Radio Transmission and Reception. March.

3GPP. 2017j. TS 36.201 V14.1.0 Evolved Universal Terrestrial Radio Access (E-UTRA); LTE Physical Layer; General Description. March.

3GPP. 2017k. TS 36.211 V14.2.0 Evolved Universal Terrestrial Radio Access (E-UTRA); Physical Channels and Modulation. March.

3GPP. 2017l. TS 36.212 V14.2.0 Evolved Universal Terrestrial Radio Access (E-UTRA); Multiplexing and Channel Coding. March.

3GPP. 2017m. TS 36.213 V14.2.0 Evolved Universal Terrestrial Radio Access (E-UTRA); Physical Layer Procedures. March.

3GPP. 2017n. TS 36.321 V14.2.0 Evolved Universal Terrestrial Radio Access (E-UTRA); Medium Access Control (MAC) Protocol Specification. March.

3GPP. 2017o. TS 36.322 V14.0.0 Evolved Universal Terrestrial Radio Access (E-UTRA); Radio Link Control (RLC) Protocol Specification. March.

3GPP. 2017p. TS 36.323 V14.2.0 Evolved Universal Terrestrial Radio Access (E-UTRA); Packet Data Convergence Protocol (PDCP) Specification. March.

3GPP. 2017q. TS 36.331 V14.2.0 Evolved Universal Terrestrial Radio Access (E-UTRA); Radio Resource Control (RRC); Protocol Specification. March.

Bormann, C., Burmeister, C., Degermark, M., *et al.* 2001. IETF RFC 3095 RObust Header Compression (ROHC): Framework and Four Profiles: RTP, UDP, ESP, and Uncompressed. July.

Camarillo, G., Marshall, W., and Rosenberg, J. 2002. IETF RFC 3312 Integration of Resource Management and Session Initiation Protocol (SIP). October.

Chakraborty, S., Frankkila, T., Peisa, J., and Synnergren, P. 2007. *IMS Multimedia Telephony over Cellular Systems: VoIP Evolution in a Converged Telecommunication World.* Wiley.

Deering, S., and Hinden, R. 1995. IETF RFC 1883 Internet Protocol: Version 6 (IPv6) Specification. December.

Fitzek, F. H., Rein, S., Seeling, P., and Reisslein, M. 2005. RObust Header Compression (ROHC) Performance for Multimedia Transmission over 3G/4G Wireless Networks. *Wireless Personal Communications: An International Journal*, **32**(January).

Halonen, T., Romero, J., and Melero, J. 2007. *GSM, GPRS, and EDGE Performance: Evolution Towards 3G/UMTS.* 2nd edn. Wiley.

Handley, M., Jacobson, V., and Perkins, C. 2006. IETF RFC 4566 SDP: Session Description Protocol. July.

Holma, H., and Toskala, A. 2006. *HSDPA/HSUPA for UMTS: High Speed Radio Access for Mobile Communications*. Wiley.

ISO/IEC. 1994. Open Systems Interconnection – Basic Reference Model: The Basic Model. November.

ITU-T. 2010. H.264 Advanced Video Coding for Generic Audiovisual Services. March.

Johansson, I., and Jung, K. 2011. IETF RFC 6236 Negotiation of Generic Image Attributes in SDP. May.

Olsson, M., Sultana, S., Rommer, S., Frid, L., and Mulligan, C. 2009. *SAE and the Evolved Packet Core: Driving the Mobile Broadband Revolution*. Academic Press.

Ozturk, O., Kapoor, R., Chande, V., Hou, J., and Mohanty, B. 2010. Circuit-Switched Voice Services over HSPA. *IEEE Vehicular Technology Conference*, May.

Poikselkä, M., and Mayer, G. 2009. *The IMS: IP Multimedia Concepts and Services*. 3rd edn. Wiley.

Postel, J. 1980. IETF RFC 768 UDP: User Datagram Protocol. August.

Postel, J. 1981a. IETF RFC 791 Internet Protocol. September.

Postel, J. 1981b. IETF RFC 793 TCP: Transmission Control Protocol. September.

Ramakrishnan, K., Floyd, S., and Black, D. 2001. IETF RFC 3168 The Addition of Explicit Congestion Notification (ECN) to IP. September.

Rein, S., Fitzek, F., and Reisslein, M. 2003. Voice Quality Evaluation for Wireless Transmission with ROHC (Extended Version and Software Code). *International Conference on Internet and Multimedia Systems and Applications*, August.

Schulzrinne, H., Casner, S., Frederick, R., and Jacobson, V. 2003. IETF RFC 3550 RTP: A Transport Protocol for Real-Time Applications. July.

Sjoberg, J., Westerlund, M., Lakaniemi, A., and Xie, Q. 2007. IETF RFC 4867 RTP Payload Format and File Storage Format for the Adaptive Multi-Rate (AMR) and Adaptive Multi-Rate Wideband (AMR-WB) Audio Codecs. April.

Taylor, D. E., Herkersdorf, A., Döring, A., and Dittmann, G. 2005. Robust Header Compression (ROHC) in Next-Generation Network Processors. *IEEE/ACM Transactions on Networking*, **13**(4).

Vanghi, V., Damnjanovic, A., and Vojcic, B. 2004. *The cdma2000 System for Mobile Communications: 3G Wireless Evolution*. Prentice Hall.

Wang, Y.-K., Even, R., Kristensen, T., and Jesup, R. 2011. IETF RFC 6184 RTP Payload Format for H.264 Video. May.

Westerlund, M., Johansson, I., Perkins, C., O'Hanlon, P., and Carlberg, K. 2012. IETF RFC 6679 Explicit Congestion Notification (ECN) for RTP over UDP. August.

9 Evolution of SC-FDMA/OFDMA Systems

The design, development, and deployment of LTE followed a different trajectory from those of the earlier circuit-switched mobile communications systems. First, the new system began its operation as the provider of non-real-time packet access services while the previous systems began by offering voice or video telephony from day one. Then VoLTE entered service but it was typically not possible to change the service provider by simply replacing the Universal Integrated Circuit Card (UICC). Moreover, interworking of two networks or global roaming was not automatically supported either, which had been taken for granted in the previous systems. These inconveniences or technical retrogressions were mainly due to the extensive use of IP and the IMS-related technical elements that make consistent implementations among manufacturers difficult.

Meanwhile some technical elements in the initial versions of VoLTE, originally designed for circuit-switched networks, showed limitations and necessitated enhancements in the signal processing chain. In this chapter, we discuss the improvements made to LTE for higher media quality and larger network capacity. Although LTE differs from the previous circuit-switched systems in many fundamental aspects, its improvements can still be classified into similar groups as in the enhancements of previous systems, including media compression, coordination of compression and transmission, session negotiation, and wireless transmission. Finally, the performance of LTE and its media and radio signal processing operations is discussed with experimental and field results.

9.1 Enhancements in Media Compression

Speech codecs discussed until AMR-WB were designed primarily for the operation of circuit-switched networks. During their development and standardization, the testing procedures focused on maximizing the speech quality at various bit error rate or frame error rate situations of interest. These errors were assumed to originate from thermal noise or fading, as illustrated in Figures 2.21(b) and 2.21(c). Later, the capabilities for jitter buffer management and error concealment were added to the codecs, but these external measures could only complement the operation of the speech decoder and did not exploit the opportunities in the entire signal processing chain of packet-switched networks.

Designing a new speech codec that is suited for a new operating environment will show higher performance than the one based on recycling technologies developed for different, more benign conditions. Another opportunity for improvement comes from the availability of additional computational complexity, as more than a decade has passed since the standardization of AMR-WB. The codec designers can take significantly faster digital signal processors (DSP) and larger memories into account when they create new speech compression algorithms.

H.264 has enjoyed tremendous success and has been used in applications ranging from mobile video telephony to Blu-ray disks. As the video resolution of broadcasting technologies evolved from HD to Ultra HD (UHD), H.264 displayed limitations in the compression of video at a resolution of UHD or 4K, i.e., 3840 × 2160 or higher, necessitating the development of a next-generation video codec capable of handling such large-scale video.

9.1.1 Enhanced Voice Services Speech Codec

Enhanced Voice Services (EVS) is a new speech codec designed to be the speech compression technology for LTE [3GPP (2017i)]. Although the available bit-rate for speech in LTE is expected to be higher than those of earlier mobile communications systems, it was recognized that higher complexity and bit-rate alone might not be sufficient to achieve the maximum improvement in quality and capacity. AMR-WB employs twice the sampling rate of AMR and correspondingly a larger source bandwidth to achieve higher quality, but EVS exploits this dimension even further, using a quadrupled (32 kHz) or an even higher sampling rate (48 kHz), which rivals those traditionally used to digitize the audio signals.

Figure 9.1 shows a series of audio waveforms which were constructed from four consecutive waveforms of the same length. The waveforms were generated by sampling the same input signal at four sampling rates, 8, 16, 32, and 48 kHz, and up-sampling the three waveforms to 48 kHz. The step-like spectrogram shows the gradual decrease of audio bandwidth from fullband to super-wideband, from super-wideband to wideband, and finally from wideband to narrowband. Note that the third and fourth waveforms and spectrograms are identical to those of Figures 8.48 and 2.1, respectively. Table 9.1 summarizes the parameters of EVS [3GPP (2011)], whose audio bandwidth is slightly wider than the conventional classification as illustrated in Fig. 8.49.

As with other speech codecs introduced so far, EVS operates with a frame size of 20 ms. In the fixed bit-rate mode, EVS operates at 7.2, 8, 9.6, 13.2, 16.4, 24.4, 32, 48, 64, 96, and 128 kbps. For NB and WB, EVS provides a Source-Controlled Variable Bit-Rate (SC-VBR) mode whose bit-rate is 5.9 kbps on average, fluctuating around the average value by ±5%. The SC-VBR mode consists of three component bit-rates: 8.0, 7.2, and 2.8 kbps, fewer than those of SMV or 4GV. Note that in the design of EVRC, it was taken into account that in QCELP, Rates 1 and $\frac{1}{8}$ constitute the majority of encoded frames while Rate $\frac{1}{4}$ is used less.

The accumulated rounding-ups of bit-rates to the nearest integers in the packetization and resource reservation procedures suggest that from the network's point of view, the same MBR corresponding to 8 kbps is requested when in the SDP for the EVS speech codec, br = 5.9, br = 5.9–7.2, br = 5.9–8, br = 7.2, br = 8, or br = 7.2–8. Here br is

Table 9.1 Key parameters of EVS.

	Audio bandwidth (Hz)	Bit-rate (kbps)
Narrowband (NB)	20–3500	5.9, 7.2, 8, 9.6, 13.2, 16.4, 24.4
Wideband (WB)	20–7000	5.9, 7.2, 8, 9.6, 13.2, 13.2 (channel-aware), 16.4, 24.4, 32, 48 64, 96, 128
Super-wideband (SWB)	20–16000	9.6, 13.2, 13.2 (channel-aware), 16.4, 24.4, 32, 48, 64, 96, 128
Fullband (FB)	20–20000	16.4, 24.4, 32, 48, 64, 96, 128

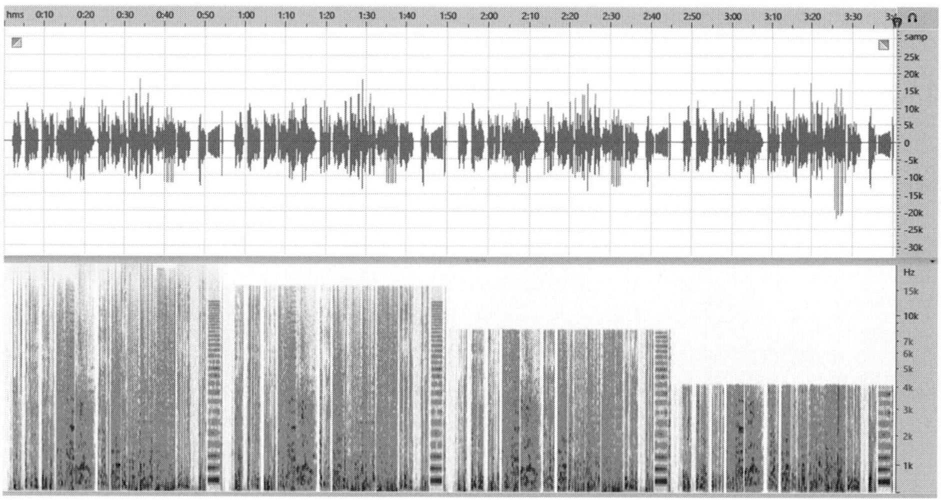

Fig. 9.1 Spectrogram of NB, WB, SWB, and FB speech waveforms.

the parameter specifying the range of bit-rate for the session. Considering the common operational practises of setting MBR equal to GBR, it would be advantageous to always use the highest bit-rate, 8 kbps, if the bit-rate needs to be in such a low range. Unlike in circuit-switched networks, the eNodeB would not obtain higher capacity, even if it uses 5.9 or 7.2 kbps, as similar transport block sizes (TBS) would be assigned to 7.2 and 8 kbps frames complemented with MAC control elements.

These EVS operations are collectively called the EVS Primary mode. EVS also includes a complete set of AMR-WB interoperable codec modes, called the EVS AMR-WB IO mode, whose quality is slightly higher than that of AMR-WB at each codec mode. The frames of the interoperable modes can be decoded by a conventional AMR-WB speech decoder. Table 9.2 summarizes the payload sizes defined for the EVS Primary mode.

To prepare for the aperiodic delivery or the loss of speech frames inherent in packet-switched networks, the EVS speech decoder has a built-in jitter buffer management (JBM) and a packet loss concealment (PLC), which were treated as post-processing performed after speech decoding in earlier codecs such as AMR-WB. With the large

Table 9.2 Payload sizes of EVS.

Bit-rate (kbps)	SID	2.8	7.2	8	9.6	13.2	16.4	24.4	32	48	64	96	128
Payload size (bits)	48	56	144	160	192	264	328	488	640	960	1280	1920	2560

Table 9.3 LTE transport block sizes and codec modes of AMR, AMR-WB, and bit-rates of EVS.

LTE transport block size	AMR	AMR-WB	EVS
208	5.9	6.6	7.2
224	7.4		8.0
256		8.85	9.6
328	12.2	12.65	13.2
392		15.85	16.4
552		23.85	24.4

number of bit-rates, a large variety of operations can be defined since EVS supports sampling rates of 8, 16, 32, and 48 kHz. To match the time-varying source characteristics and channel conditions, the EVS speech encoder can switch the bit-rate, bandwidth, or both during operation, which provides two-dimensional adaptation paths.

In circuit-switched networks, choice of the encoded speech bit-rate is limited by the channel structure and the in-band control signalling. In packet-switched networks, the same limitations are valid but it is also necessary to consider the overhead from the IP and radio protocols, and the out-of-band signalling. Table 9.3 shows that EVS matches the transport mechanisms of LTE better than the previous speech codecs, as the bit-rate and speech quality of EVS roughly scale up with the LTE transport block size (TBS).

For AMR-WB, the small transport block size, 224, is not used by any codec mode, and the speech quality does not improve significantly as the codec mode increases beyond 12.65 kbps [3GPP (2017c)]. In typical deployment scenarios, EVS is expected to operate in wideband until 9.6 kbps but switches to super-wideband at 13.2 kbps and above following a two-dimensional trajectory in the plane made up of the bit-rate and bandwidth axes. Thus in EVS, the tradeoff between speech quality and network capacity can be controlled in a more refined fashion than with the one-dimensional codec mode adaptation of GSM or W-CDMA.

After setting the target values for the bit-rate and audio bandwidth, the next step in the design of the speech codec is to define its complexity limitations. Table 9.4 outlines the complexity requirements for EVS, set with an algorithmic delay of less than or equal to 32 ms. The algorithmic delay here is defined as the frame size buffering delay plus any other delays inherent in the codec algorithms, e.g., look-ahead, sample-rate conversion, and decoder post-processing. The algorithmic delay excludes the processing delay at the speech encoder and the decoder, and the channel transmission delay. AMR and AMR-WB have smaller algorithmic delays, 20 and 25 ms, respectively.

Table 9.4 EVS design requirements for computational complexity.

	Computation (WMOPS)	RAM (kWords)	ROM (kWords)	PROM (compared with AMR-WB)
NB, WB, SWB (mono)	≤ 88	≤ 100	≤ 100	≤ 10×
Other modes (including stereo)	≤ 135	≤ 200	≤ 200	≤ 10×

Using AMR-WB as the reference, the maximum amount of computation, the maximum sizes of Random Access Memory (RAM), Read Only Memory (ROM), and Programmable ROM (PROM) of EVS are set to be equal to or less than 2, 16, 10, and 10 times those of AMR-WB, for bandwidths up to super-wideband. In the other modes, the limitations were set to be less than or equal to 3.1, 31, 20, and 10 times those of AMR-WB. However, these complexity requirements do not include that required for jitter buffer management, which has no precursors in AMR-WB or any other previous codecs. The computational complexity and PROM are measured using the methods specified in [ITU-T (2009)].

The complexity requirements for EVS show how the gain in computational resources can be systematically reflected in the design of new speech compression algorithms. The candidate codecs should compete in many aspects of speech quality under various channel conditions while meeting these requirements. In the completed EVS speech codec, the maximum complexity measured is 85.9 WMOPS, 56.7 WMOPS of which is spent for the encoder at 24.4 kbps for SWB with DTX on, and the remaining 29.2 WMOPS is spent for the decoder at 48 kbps for SWB with DTX off. The EVS speech codec uses 175 kWords of RAM, 157 kWords of ROM, and 116 kWords of PROM, without JBM. The standard JBM of EVS is observed to consume 18 WMOPS and 49 kWords of RAM [3GPP (2017b)].

Multi-Core Architecture

Figure 9.2 shows a simplified block diagram of the EVS speech encoder. It exhibits a significantly different architecture from those of the FBR or VBR speech encoders discussed so far. The input signal consists of 8000–48000 digitized speech samples per second, depending upon the base sampling frequency. If a lower sampling rate is used, the samples in each 20 ms frame are subsampled and filtered to match the sampling rate. The input signal is first notch filtered to remove the DC component, pre-processed, and fed into the core switching configure block that analyzes the spectral characteristics of the frame to generate signal classifier information. The information is sent to one of the coding cores for guiding the encoding process and is also sent to the bit-stream multiplex for in-band signaling. The network control can override the current encoding process and install a new process.

Compared with the earlier speech encoders, where the input signal is classified as either the speech or silence, the EVS speech encoder determines whether the input signal is voice, music, mixed content, or silence. Then the encoder applies different compression methods and bit-rates that best match the nature of the input signal. In the

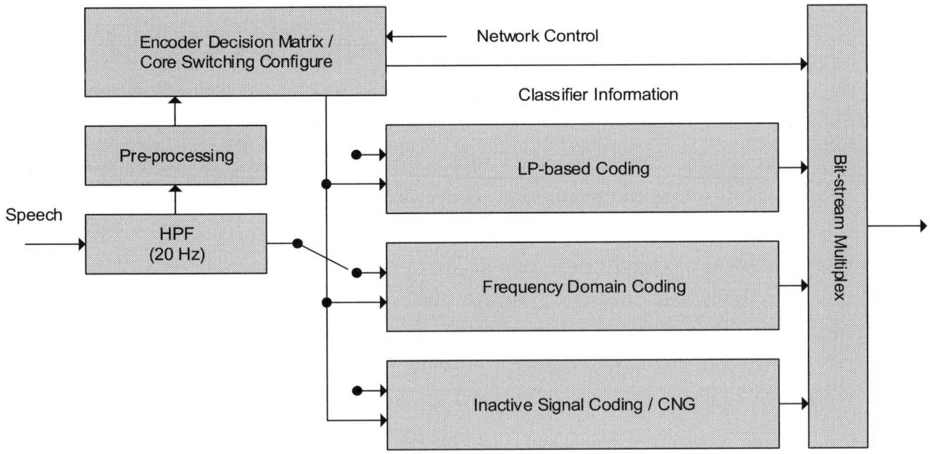

Fig. 9.2 Block diagram of EVS speech encoder.

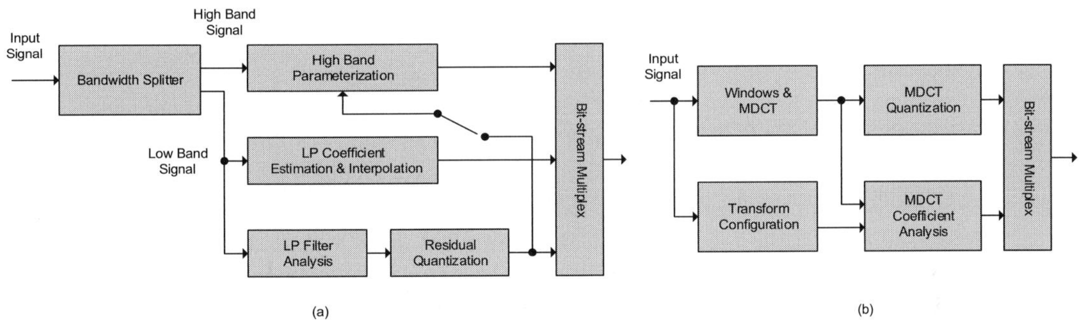

Fig. 9.3 EVS speech encoding cores. (a) LP-based encoder. (b) Frequency domain encoder.

SMV speech encoder, encoding modes with higher average bit-rates are applied if the background noise is detected to contain music or periodic signals. For each frame, based on the coding mode selected at the core switching configure block, either linear prediction (LP) based coding, the time-domain LPC-based speech compression algorithms, or frequency domain coding, the MDCT-based audio compression algorithms, are applied. If the input signal is not active, either a NO_DATA message or an SID is generated.

The LP-based and frequency-domain coding methods include encoding modes that are activated based on the type of input signal, available bandwidth, and the bit-rate. Figure 9.3 shows the architecture of two encoding cores used when the input signal is present. A third core is used to encode noise signals. The LP-based encoder has the architecture of a conventional speech encoder based on the ACELP principles; the frequency domain encoder has the architecture of a conventional audio encoder based on the Modified Discrete Cosine Transform (MDCT) principles. The MDCT used is the discrete cosine transform type IV, often used in video encoders, modified to reduce the memory storage and computational complexity. The definition of the L-point type IV DCT is

$$y(k) = \sum_{n=0}^{L-1} \tilde{x}(n) \cos\left[(n + \frac{1}{2})(k + \frac{1}{2})\frac{\pi}{L}\right], \tag{9.1}$$

where $\tilde{x}(n)$ is the windowed input signal of the current audio frame, and $y(k)$ is the kth DCT spectral component. There are also two hybrid schemes in which the ACELP and MDCT are combined, the Generic Signal Audio Coder (GSC) approach and the Transform Codec Excitation (TCX) approach.

In LP-based coding, a signal classification algorithm applies one of six encoding modes to capture the acoustic nature of the input signal: the Unvoiced Coding (UC) mode, Voiced Coding (VC) mode, Transition Coding (TC) mode, Audio Coding (AC) mode, Inactive Coding (IC) mode, and Generic Coding (GC) mode. Frequency-domain coding is selected when the input signal is not speech, or the bit-rate for encoding is high enough for the MDCT encoder. The selection of coder technology, from ACELP, High Quality (HQ) MDCT, GSC, or TCX, is determined by the bit-rate, bandwidth, signal classification, selected coding mode, and other parameters. Table 9.5 shows the allocation of coder technologies based on the bit-rate, bandwidth, and the content. Notice that beyond 64 kbps, only TCX is used.

Even at high bit-rates, ACELP can be employed as long as the input signal is identified as speech. On the other hand, application of GSC is limited to the bit-rates and bandwidths lower than or equal to 13.2 kbps and wideband. TCX and HQ MDCT cover wide ranges of the bit-rates and bandwidths. Note that the highest complexity measured is taken not at the highest or lowest bit-rate but at 24.4 kbps for the encoder, and at 48 kbps for the decoder, respectively.

The third coding core of EVS, for the Comfort Noise Generation (CNG), supports two schemes, LP-based CNG and frequency domain CNG, which are determined based on the acoustic nature of the input signal. The update rate of SID can be configured to a fixed number of frames or adaptively modified according to the characteristics of the background noise. In the former approach, the update rate can be set to once per 1–100 frames. In the adaptive approach, the SID can be adapted once per 8 or 50 frames, depending on the behavior of the noise. Note that the SID update rate may be adjusted via updating the firmware of UE, but not via network signaling such as the SDP.

The unique architecture of EVS having an LP-based core and a transform-based core to selectively handle the speech and audio signals does have a precedent in mobile communications, albeit for slightly different applications. The Extended AMR-WB (AMR-WB+) is an audio codec built by combining the ACELP core of AMR-WB with a TCX core, and allowing longer frame sizes up to 2048 samples and higher sampling rates up to 48 kHz [3GPP (2017h)]. Although the audio codec offers a lower-complexity, open-loop mode selection between the two cores, in addition to the exhaustive, closed-loop mode selection, real-time applications were not taken into account for its design objectives.

AMR-WB+ was designed with a focus on a narrow QoS range of 10–24 kbps, where the input signal consists mainly of speech mixed with music, assuming networks

Table 9.5 Allocation of coding methods for bit-rate, bandwidth, and content.

Bandwidth	Signal	Bit-rate				
		7.2	8	9.6	13.2	16.4
NB	Speech	ACELP	ACELP	ACELP	ACELP	
	Audio	HQ MDCT	HQ MDCT	TCX	TCX/HQ MDCT	
	Noise	GSC	GSC	TCX	GSC	
WB	Speech	ACELP	ACELP	ACELP	ACELP	ACELP
	Audio	GSC	GSC	TCX	TCX/HQ MDCT	TCX/HQ MDCT
	Noise	GSC	GSC	TCX	GSC	TCX
SWB	Speech				ACELP	ACELP
	Audio				GSC/TCX/HQ MDCT	TCX/HQ MDCT
	Noise				ACELP	TCX
FB	Speech					ACELP
	Audio					TCX/HQ MDCT
	Noise					TCX

Bandwidth	Signal	Bit-rate			
		24.4	32	48	64
NB	Speech				
	Audio				
	Noise				
WB	Speech	ACELP	ACELP	TCX	ACELP
	Audio	TCX	HQ MDCT	TCX	HQ MDCT
	Noise	TCX	ACELP	TCX	ACELP
SWB	Speech	ACELP	ACELP	TCX	ACELP
	Audio	TCX/HQ MDCT	TCX/HQ MDCT	TCX	HQ MDCT
	Noise	TCX	ACELP	TCX	ACELP
FB	Speech	ACELP	ACELP	TCX	ACELP
	Audio	TCX/HQ MDCT	TCX/HQ MDCT	TCX	HQ MDCT
	Noise	TCX	ACELP	TCX	ACELP

whose capability is equal to or lower than that of W-CDMA. In those limited conditions, AMR-WB+ can rival fully-fledged audio codecs such as the Enhanced Advanced Audio Coding Plus (Enhanced aacPlus) that use only transform coding. However, as the object-based coding of MPEG-4 [ISO (2001)] was not particularly successful in low bit-rate video communications, such as 3G-324M, AMR-WB+ could not establish itself well. One reason might be that HSPA and LTE increased the disposable bit-rate of each UE significantly, especially in such non-real-time services as audio streaming over the downlink. Thus, the envisioned opportunity for AMR-WB+ did not materialize as the typical bit-rates of audio exceeded 64 kbps by the time the new audio codec became ready for service. Similar technical attempts to apply separate compression methods for speech and audio were also made in the Unified Speech and Audio Coding (USAC) [MPEG (2012)].

Bandwidth Adaptation

Figure 9.4 shows the bandwidth selection process of the EVS speech encoder. The metrics used in the decision process, cnt_{nb}, cnt_{wb}, cnt_{swb}, and cnt_{fb}, are the counters that indicate the energy level of each audio bandwidth. Each plus (+) in the branch represents an increment of the corresponding counter by one. When the arrows fork, the highest bandwidth reached is used in the encoding. From the conditions, it is seen that the selection of audio bandwidth depends on both the values of the counters for the current frame and the bandwidth used for the last frame. The four bandwidth counters are computed from the per-band Complex-modulated Low Delay Filter Bank (CLDFB) energy values [3GPP (2017i)]. Since the values are computed directly from the sampled frame, without using any encoding parameters, the audio bandwidth of each frame is selected regardless of the bit-rate. Note that similar classification procedures are used to select the bit-rate for each frame in the speech codecs of CDMA systems based on the variable bit-rate (VBR) coding principles, as shown in Figures 5.15 and 5.17.

Figure 9.6 shows the audio bandwidth and coding core used for each frame when the synthetic multi-bandwidth audio waveforms shown in Fig. 9.1 are encoded by the EVS speech encoder at 24.4 kbps, with DTX turned off. The information on audio bandwidth and coding core is stored in a two-bit BW field and a one-bit CELP/MDCT core flag of each 488-bit frame, whose structure is shown in Fig. 9.5. The BW field is set to 0 (00), 1 (01), 2 (10), and 3 (11) when the audio bandwidth is diagnosed as NB, WB, SWB, and FB, respectively. The core flag is set to 0 if the CELP core is used, and to 1 if MDCT is used.

Considering the construction process of the series of audio waveforms, it is expected that the EVS speech encoder would reduce the audio bandwidth after each waveform, possibly after some analysis of the nature of the input signal. Starting with FB, the first switching occurs at the 2968th frame down to SWB, which is maintained until the 5728th frame. After the second switching, WB is maintained but ultimately switched down to NB at the 8638th frame. Therefore, as expected, the audio bandwidth shifts down in a stepwise fashion. Since each audio waveform is made up of 2760 frames, the switching occurs after the analysis of 200–400 frames belonging to the next waveform. This might be due to some hysteresis logic to avoid ping-pong effects between neighboring audio bandwidths while maintaining the same bit-rate. In GSM using AMR, such a logic is used to avoid too frequent switching of the codec mode while maintaining the same total bit-rate for the encoded speech and channel code, as shown in Fig. 3.12.

Figure 9.6(a) shows the difference in switching times between the source waveforms and the encoded frames. Although the stepwise switching of the BW field clearly illustrates the multi-bandwidth operating principles of EVS, audio waveforms that are strictly bandwidth-limited can only be artificially synthesized. With a fixed sampling rate, the ambient noise will be present up to a bandwidth limited by the sampling rate, and switching of the audio bandwidth will be an infrequent event. As the audio

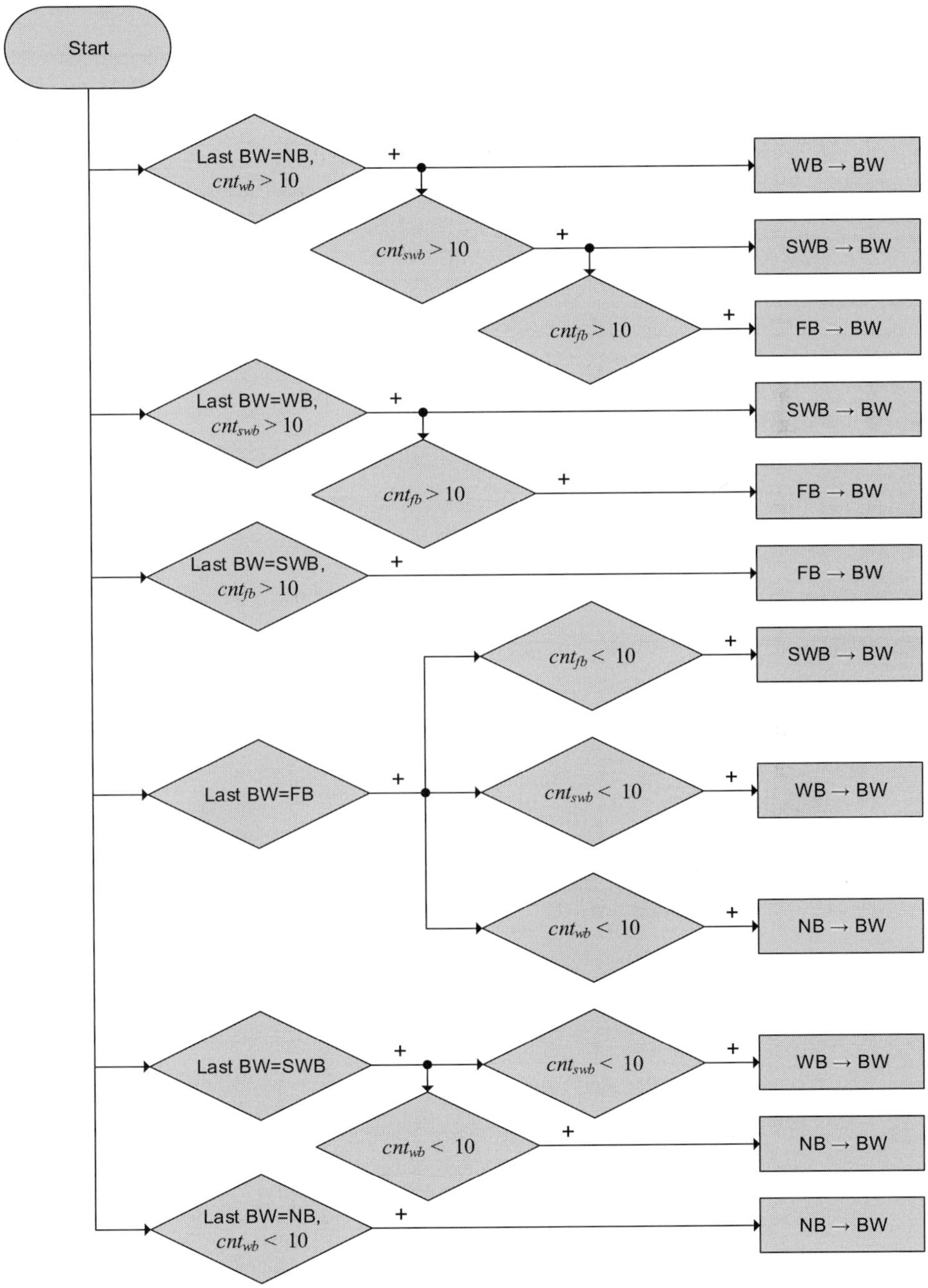

Fig. 9.4 Bandwidth selection algorithms.

Core	CELP	TCX	HQ-MDCT	CELP	TCX	CELP
Ext. Layer				SWB TBE	IGF	FB TBE
Number of Bits / Frame	488					
BW	2					
Reserved Flag	1					
CELP / MDCT Core Flag	1					
TCX / HQ-MDCT Core Flag		1	1		1	
CELP → HQ Core Switching Flag			1–2			
CT	2	2		2	2	2
Core Bits	482	481	481–482	422	481	423
SWB Ext. Layer Bits				57		55
FB Ext. Layer Bits						4
Padding Bits				3		

Fig. 9.5 Bit allocation for EVS frames at 24.4 kbps.

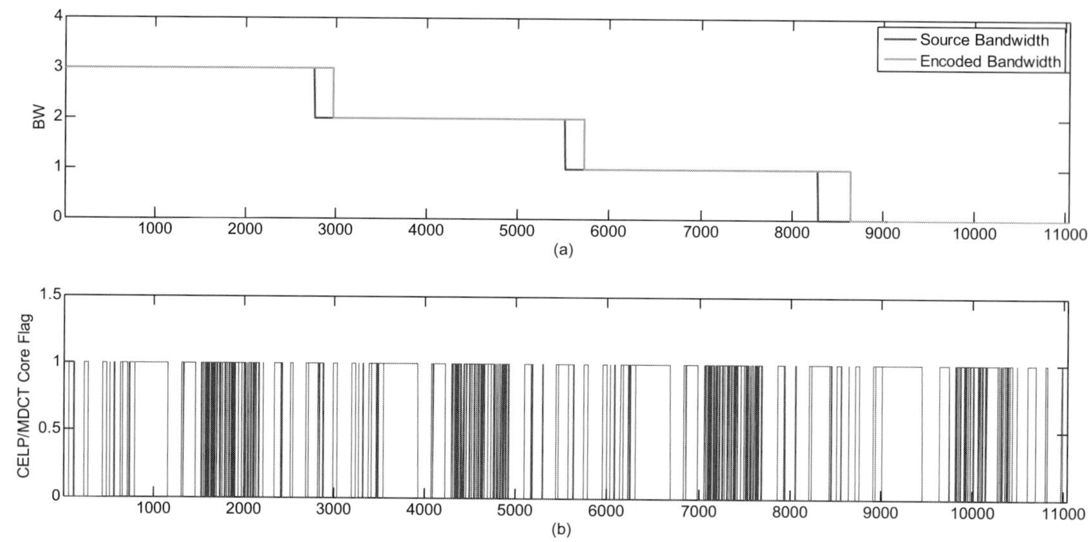

Fig. 9.6 Switching of parameters. (a) Audio bandwidth. (b) Coding core.

bandwidth is decided regardless of the bit-rate, bandwidth switching of different bit-rates would occur simultaneously, as long as the input signal does not change. As seen in Fig. 9.6(b), switching of coding cores depends on the bandwidth, even at the same bit-rate. Figure 9.7 compares the switching of bandwidth for 24.4, 13.2, and 9.6 kbps. Note that BW = 3 is not used in 13.2 and 9.6 kbps as FB is not supported at bit-rates below 16.4 kbps.

Bandwidth Extension

In higher-bandwidth operation, such as in super-wideband and fullband, further savings in the bit-rate can be achieved from an efficient representation of the upper spectrum. In

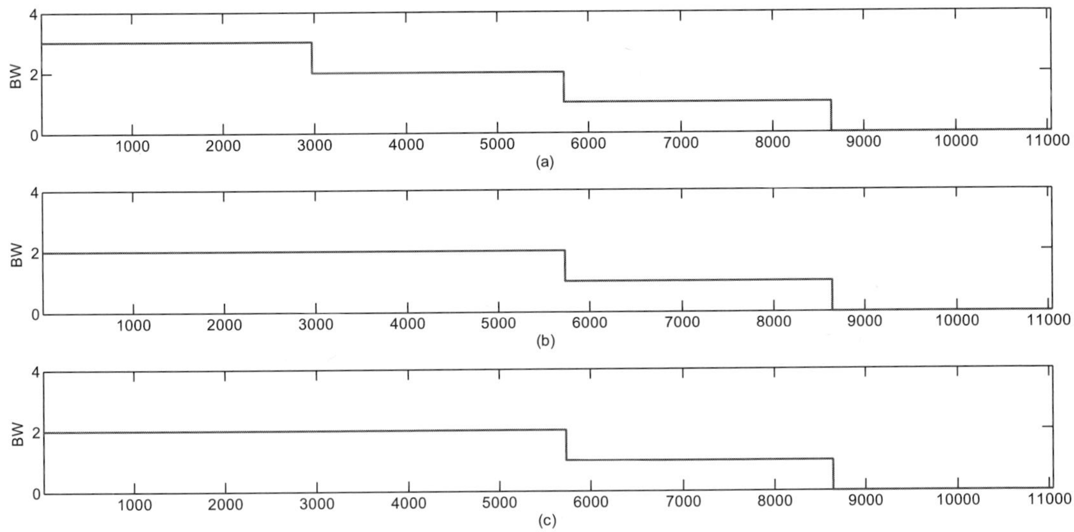

Fig. 9.7 Switching of bandwidth. (a) 24.4 kbps. (b) 13.2 kbps. (c) 9.6 kbps.

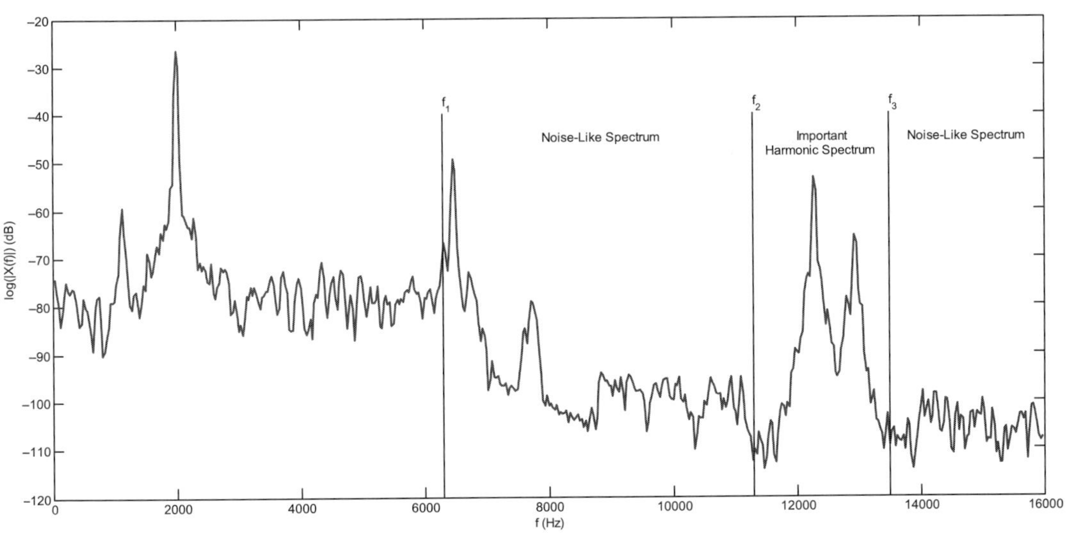

Fig. 9.8 Concept of bandwidth extension.

AMR-WB, the 6.4–7 kHz frequency band is modelled using parameters extracted from audio bandwidth up to 6.4 kHz. A more systematic approach, the Bandwidth Extension (BWE), is used in EVS so that the fine structure of the higher spectrum is represented partially using the information in the lower spectrum, saving bits for representing the higher spectrum.

Figure 9.8 illustrates the spectrum of a wideband signal where the higher band, $f \geq f_1$, consists of an acoustically important harmonic spectrum and also a noise-like spectrum.

If the important harmonic spectrum, $f_2 - f_3$, is related to the important harmonic spectrum in the lower band, it may be described to some extent by the harmonic spectrum in that band. As shown in Fig. 8.48, the spectrogram of a wideband speech waveform includes periodic elements. This is somewhat true for music and mixed contents too. The idea of representing a high frequency signal with the spectral information of a baseband signal was initially applied to the audio compression where fullband is basically considered. Spectral Band Replication (SBR) is one of such approaches used to improve the compression efficiency of audio codecs over a wide range of bit-rates [Dietz *et al.* (2002)].

The bandwidth extension of EVS exploits the inherent relationship between the lower and the higher spectra. When the fine signal structure in the higher spectrum is related to that in the lower spectrum in some way, the structure is not encoded in detail but extrapolated from the lower spectrum. With BWE, the higher spectrum is encoded in a less elaborate fashion, using fewer bits per unit spectrum, than the lower, baseband spectrum. For example, parameters containing information on the envelope of the higher spectrum, not the magnitude at each frequency, are computed and transmitted.

Figure 9.9 illustrates the concept of BWE used in LP-based coding, in which bandwidth higher than the core bandwidth is shifted to the baseband, flipped, and encoded. In the BWE of LP-based coding, two shapes, *steep onset* and *gentle onset*, are defined to describe the temporal envelope of the higher spectrum, and the higher spectrum is partitioned as specified in Table 9.6. The higher spectrum, after being shifted to the baseband, is flipped to better fit the psycho-acoustical model used. In the BWE of frequency domain coding, the higher spectrum is not flipped as the spectral flipping does not improve the alignment of the lower and higher bandwidths. The objective of manipulating the higher spectrum is not to reconstruct the original spectrum as faithfully as possible but to maximize the perceived quality.

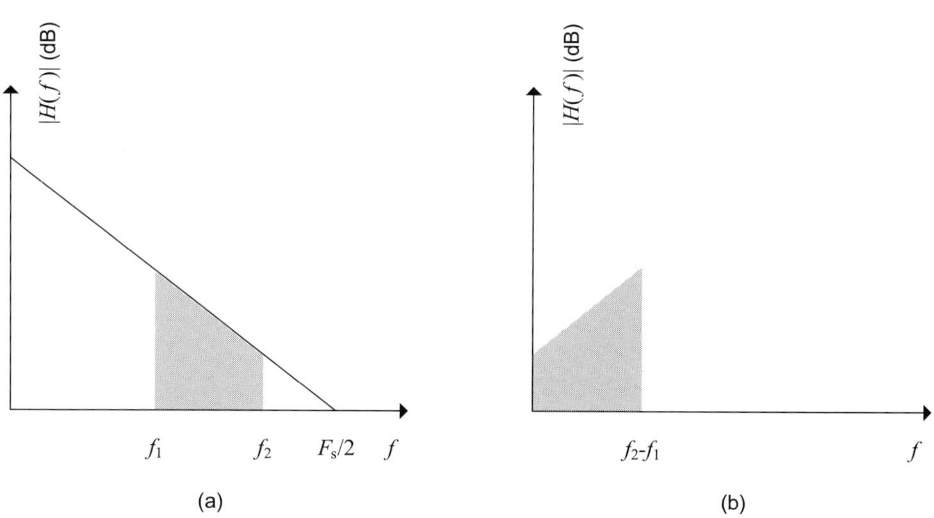

Fig. 9.9 Bandwidth extension in LP-based encoder. (a) Higher spectrum. (b) Shifted and flipped spectrum.

Table 9.6 Generation of higher spectrum for bandwidth extension.

Core bandwidth (kHz)	WB	SWB	
	6.4	6.4	8
f_1	6	6.4	8
f_2	8	14.4	16

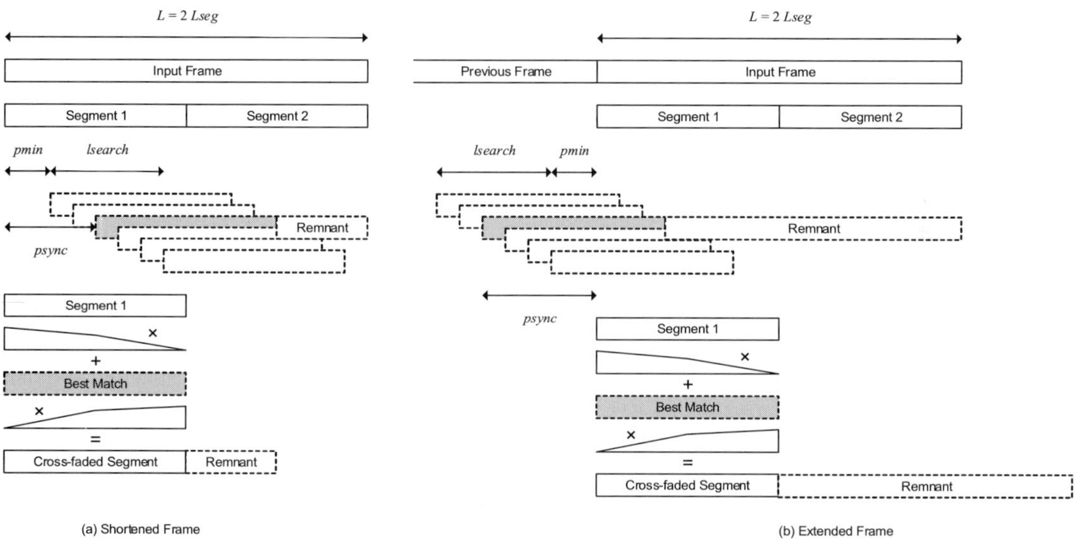

Fig. 9.10 Signal-based scaling. (a) Time-shrinking. (b) Time-stretching.

BWE can be applied as a kind of post-processing when the speech encoder does not process higher bandwidth or extend lower bandwidths. For example, the reconstructed spectrum of a narrowband signal may be duplicated and manipulated to make the signal wideband. Likewise, a wideband signal may be made super-wideband by the receiver. The quality of such a signal extended in the frequency domain may not be as high as that of the signal originally encoded for the higher bandwidth. Such a signal may also sound artificial as the information of the higher bandwidth is entirely fabricated. However, such a post-processing may be preferable in some situations, e.g., when the speech quality drops abruptly and significantly as EVS operating in SWB falls back to AMR, reducing the audio bandwidth to one-fourth of its previous value. This may happen when a UE goes out of LTE coverage and gets handed over to GSM networks via the Single Radio Voice Call Continuity (SRVCC) procedures.

Time Scaling

Figures 9.10(a) and 9.10(b) illustrate the shrinking and stretching procedures respectively, used in the jitter buffer management of EVS. In the shrinking, a 20 ms frame can be reduced to 10–17.5 ms, depending on the position of the speech segment that best

Table 9.7 Time-scale modification parameters.

		NB	WB	SWB	FB
L	Frame size	160	320	640	960
L_{seg}	Segment size	80	160	320	480
p_{min}	Minimum pitch	20	40	80	120
l_{search}	Search length	100	200	400	600
p_{max}	Maximum pitch	120	240	480	720

matches the first half of the input frame. The search starts from sample P_{min} for l_{search} samples. The best match is found to be a set of L_{seg} samples starting from sample p_{sync}.

The output frame is generated by *cross-fading* the first segment and the best matching candidate segment. The cross-fade is implemented by multiplying the two segments by a monotone-decreasing window and a monotone-increasing window, respectively, then overlapping them and adding, as illustrated in Fig. 9.10(a). The remnant samples after the chosen segment are appended to the cross-faded signal for continuity with the following frames. Therefore size of the shortened frame is $2L_{\text{seg}} - P_{\text{sync}} = L - P_{\text{sync}}$. In the stretching, a 20 ms frame can be extended to 22.5–35 ms, depending on the position of the speech segment that best matches the first segment. The search is made toward the previous frame. Following similar procedures, as illustrated in Fig. 9.10(b), the size of the stretched frame is $2L_{\text{seg}} + P_{\text{sync}} = L + P_{\text{sync}}$. Table 9.7 summarizes the parameters used in the time-scaling of EVS.

However, time-stretching is not a complete solution to the buffer underflow of JBM. To raise the buffer level, the media adaptation block shown in Fig. 8.56 may also request the far-end speech encoder to reduce the bit-rate, which may sound counterintuitive but indeed increases the likelihood of successful packet transmission, raising the level of the frame buffer, although at lower quality. This is because the scheduling algorithms used in the network nodes on the transmission paths often prioritize smaller packets. On the other hand, the buffer level may become temporarily too high, or buffer overflow may be expected when unusually many packets arrive within a short time window. The control block estimates the approximate play-out time of each frame in the frame buffer, and may drop some of them if the time exceeds the remaining time until the play-out of the following frames. When some of the stored frames are to be dropped, it is necessary to ensure that the effect on the reconstructed speech quality is minimized. To dilute the loss of speech quality, the impact of frame drop can be spread as widely as possible over the time domain, which will also benefit the operation of time scaling. Such a spreading strategy cannot be applied to the video directly, due to the strong dependence between consecutive frames. Moreover, dropping video frames usually initiates visually annoying error propagation to the following frames. Therefore, if some video frames have to be dropped, dropping predicted frames, rather than independently-encoded frames, may reduce the influence on the quality as an I-frame is often encoded when the scene changes significantly, and the motion estimation and compensation fail to find similar scenes in the previous frame.

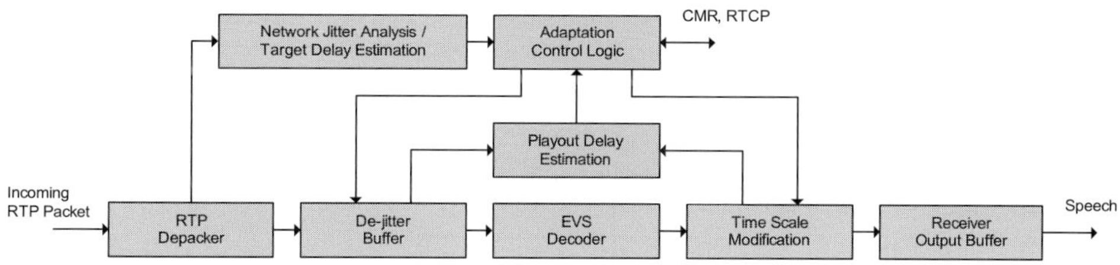

Fig. 9.11 Block diagram of JBM for EVS speech decoder.

The strategies for packet dropping can be extended to the operation of the scheduler at the eNodeB or router where the packets may temporarily stay before their next hop. For example, dropping video packets at the boundaries between GOBs, slices, or pictures can be recommended at the edge of networks. However, in practice, it may not be economical if the knowledge and capability to analyze the bit-stream of many standardized and proprietary video codecs have to be available at each eNodeB or packet router. A block diagram of JBM for the EVS decoder is shown in Fig. 9.11. Compared with the generic architecture of JBM shown in Fig. 8.56, the JBM of EVS has more clearly specified functions but is still based on similar principles to reduce the impact of jitter on speech quality. Although designed for EVS, the JBM of EVS is not tied to its decoder or control signal but can be used with other speech decoders.

9.1.2 High Efficiency Video Coding

Although officially called H.265, High Efficiency Video Coding (HEVC) is the successor of the H.264 video codec. As in the previous codec standards, HEVC is based on hybrid video coding principles. The video is compressed by creating a predicted signal using motion estimation and compensation. The difference between the original and the predicted signal is encoded using various data compression techniques. As in the transition from MPEG-4 to H.264, the focus of the design was on the compression efficiency. Figure 9.12 shows the encoder structure of HEVC [ITU-T (2013)], which is quite similar to that of H.264 shown in Fig. 8.51.

In contrast to H.264 where variable block sizes, from 4 × 4 to 16 × 16, were used in motion compensation, in HEVC, a new concept of the Coding Tree Unit (CTU) is introduced, which can vary from 4 × 4 to 64 × 64, enabling more flexible partitioning of the scenes and efficient allocation of bits. CTU is the basic processing unit in HEVC that specifies the decoding process, corresponding to the macroblock in previous video codecs. Each macroblock of H.264 is either inter- or intra-coded but each CTU of HEVC can have a combination of inter- and intra-coded blocks. The transforms used to convert the image to the frequency domain can also be adjusted from 4 × 4 to 32 × 32. Larger blocks and transforms of HEVC contribute to higher compression efficiency, especially at very high resolutions such as 3840 × 2160.

Spectral efficiency for intra-coding is also improved as HEVC provides 33 directional modes, compared to the eight directional modes of H.264. Like H.264, HEVC

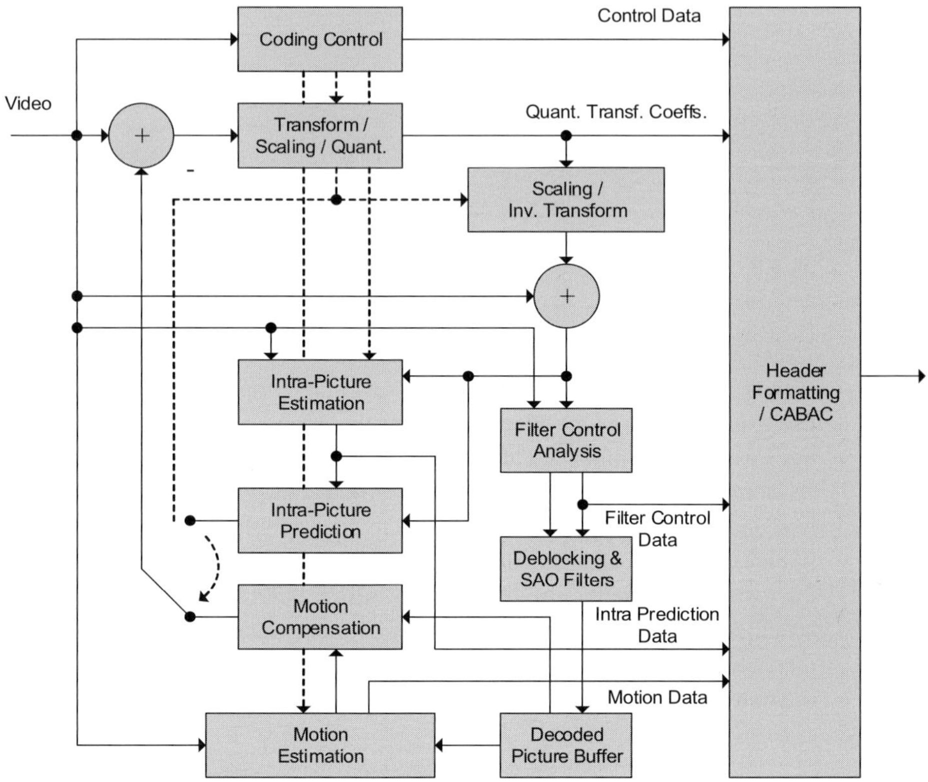

Fig. 9.12 Block diagram of HEVC video encoder.

uses the context adaptive binary arithmetic coding (CABAC) but the CABAC of HEVC compresses the syntax elements more efficiently while using lower complexity. HEVC employs a deblocking filter whose complexity is lower than that used in H.264, and a subsequent filter that adds a Sample Adaptive Offset (SAO) to each pixel during motion compensation. SAO classifies the reconstructed samples into different categories, based on the sample amplitudes or the edge characteristics, to improve the quality around sharp edges significantly.

Comparing the block diagrams of H.264 and HEVC, it is notable that in HEVC, the bit-stream includes the data for filter control and intra-prediction. For the 4 × 4 intra-transform, the Discrete Sine Transform (DST) is used, instead of the DCT. HEVC inherits the basic systems and transport interface designs, such as the codec parameter set and the network adaptation layer, from H.264. The key design objectives of HEVC were the handling of higher resolution video and architectural support for parallel processing.

9.1.3 Session Negotiation of Enhanced Media

Figure 9.13 shows an SDP answer, constructed from the offer shown in Fig. 8.58. EVS, the speech codec preferred by the offerer, was selected but its ranges of bit-rate and

```
C=IN IP4 aaa.bbb.ccc.ddd
b=AS:1042

m=audio 49152 RTP/AVP 105 99
b=AS:42
b=RS:0
b=RR:2500
a=rtpmap:105 EVS/16000/1
a=fmtp:105 br=13.2-24.4; bw=nb-swb
a=rtpmap:99 telephone-event/16000
a=fmtp:99 0-15
a=sendrecv
a=ptime:20
a=maxptime:120

m=video 49154 RTP/AVP 106
b=AS:1000
b=RS:0
b=RR:2500
a=rtpmap:106 H265/90000
a=fmtp:106 profile-id=1; level-id=93; \
   sprop-vps=QAEMAf//AWAAAAMAgAAAAwAAAwBdLAUg; \
   sprop-sps=QgEBAWAAAAMAgAAAAwAAAwBdoAKAgC0WUuS0i9AHcIBB; \
   sprop-pps=RAHAcYDZIA==
a=imageattr:106 send [x=1280,y=720] recv [x=1280,y=720]
a=sendrecv
```

Fig. 9.13 SDP answer.

audio bandwidth were reduced to from 13.2 to 24.4 kbps, and from narrowband to super-wideband, respectively. For video, the answerer selected H.265 at 1280 × 720 for both directions, at the offered bit-rate of 1 Mbps.

Figure 9.14 illustrates the session negotiation procedures when the EVS is offered for bit-rates from 5.9 to 48 kbps, and for audio bandwidths from narrowband to fullband, as shown in the SDP offer of Fig. 8.58. For the negotiation to converge, the range of bit-rates or bandwidths in the SDP answer is required to be equal to or a subset of that in the offer. Therefore, on the bit-rate vs. audio bandwidth plane, the rectangle made up of the two ranges gets smaller at each step of message exchanges between the UEs via IMS. In the case of bit-rate, the upper and lower bounds can be decreased and increased, respectively, when the SDP answer is constructed. In contrast, only the upper bound of audio bandwidth can be decreased as using only the upper bandwidths in a speech session is typically not possible. When the input signal can be bandlimited, such as in the fixed networks, it is allowed to configure a single audio bandwidth higher than narrowband.

9.2 Enhancements in Coordination of Compression and Transmission

9.2.1 Media Adaptation

In any mobile communications systems, it is important to avoid the effects of link degradation or cell overload. If such events happen to the VoLTE sessions, it is possible to

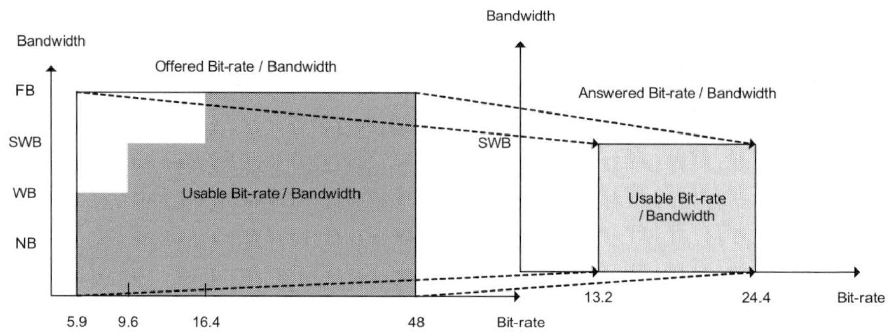

Fig. 9.14 Negotiation of bit-rate and audio bandwidth.

change the bit-rate by session re-negotiation, although this may not be fast enough to track the variations in transmission conditions. In 3G-324M or VoLTE, encoded speech and video frames can be transported over two or more wireless links whose channel characteristics are statistically independent, before they are decoded. However, the information on the total end-to-end link quality is not available to the network, since each NodeB or eNodeB can gather the link quality and cell loading information on just one wireless link. Because of this limitation, a conventional approach is to let the eNodeB control the tradeoff between media quality and network capacity for each cell with efficient scheduling algorithms.

However, scheduling of radio resources for a single cell cannot reflect the end-to-end transmission conditions completely. Moreover, developing scheduling algorithms that provide service quality close to that of circuit-switched networks under unfavorable conditions can be very difficult, because of the fundamental differences between dedicated circuit-switched and shared packet-switched networks. For example, in W-CDMA, a loss of speech frames typically occurs from temporary undershoot in power control but the link quality can be rapidly recovered by increasing transmit power at 1500 Hz. The open-loop power control can accelerate the recovery process. In LTE, a loss of speech frames typically occurs from the failure of HARQ in the link adaptation, or from losing the priority competition with other packet flows in the scheduling queue at the eNodeB. Thus, in packet-switched networks, it takes longer for the sender to learn the situation at the far-end and take appropriate measures, which makes the impact of packet losses on media quality more concentrated and perceivable.

Media adaptation is a class of techniques designed to address these situations. The idea behind media adaptation is that the UE not only has information on the transmission and reception status over the wireless channel but also has information on the overall status of communications with the far-end UE. For a UE, the RTCP Sender Report (SR) and the Receiver Report (RR) are the basic sources of such information. Lost RTP packets can be identified from the sequence number of the RTP header. RTCP RR includes such information as the fraction of transmitted packets lost, cumulative number of packets lost, highest sequence number received, inter-arrival jitter, reception time of the last SR, and the delay since the last RTCP SR. In VoLTE, additional messages can be used to enable more rapid and accurate control of the far-end speech and video encoders.

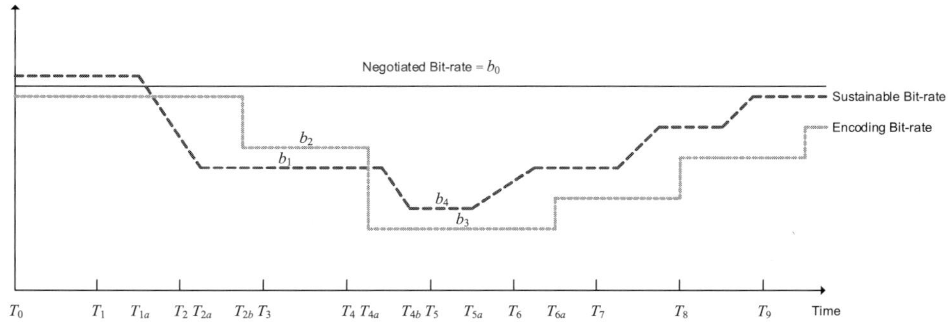

Fig. 9.15 Channel tracking in media adaptation.

Figure 9.15 illustrates the concept of media adaptation with a hypothetical channel-tracking example, in which the total bit-rate is controlled to match time-varying channel conditions. Here the total bit-rate is changed, in contrast to the link adaptation of GSM using AMR where a fixed total bit-rate is dynamically allocated between encoded speech and channel code. When the total bit-rate for packets, including encoded media and IP overhead, exceeds the bit-rate sustainable by the network, in other words, when the actual transport capability of the network falls below the negotiated QoS, the packets will be accumulated over the paths in the network, notably in front of the congested links such as the wireless channel. Unless the QoS is re-negotiated at a lower level or the sustainable bit-rate is increased, the network will eventually have to start dropping packets.

At time instant T_0, the bit-rate of the encoded media is set to the negotiated value, b_0. The network and the far-end UE participate in this negotiation. The channel conditions start to degenerate at T_{1a}, and the bit-rate sustainable by the network falls to b_1 at T_{2a}. This is recognized by the receiving UE based on several types of channel information, such as the reduced average bit-rate, detection of undelivered packets, or by the ECN-CE marking on the IP header, although the receiving UE overestimated the value to be b_2. Then the receiving UE asks the transmitting UE to reduce the bit-rate to b_2, which is accepted and executed at T_{2b}. However, the receiving UE recognizes that the first down-switching does not address the situation sufficiently, after observing the status of packet reception for a period. The receiving UE initiates another down-switch to b_3 at T_{4a}. Although the channel conditions deteriorate further and the sustainable bit-rate drops to b_4 at T_{4b}, the receiving UE does not experience further degradation of quality from packet loss as the transmitting UE is sending at a bit-rate lower than the sustainable bit-rate. Note that in this case, the degradation of quality from compression is larger but less noticeable than the degradation of quality from packet loss. Beginning at time T_{5a}, the channel conditions start to rebound, which is detected by the receiving UE. The bit-rate is increased in a stepwise fashion, by increasing the bit-rate by a small amount and waiting for the change of reception status for a period beginning at T_{6a}. The receiving UE increases the bit-rate in this conservative fashion until it returns to the negotiated bit-rate. Unlike in GSM using AMR where codec mode adaptation can track

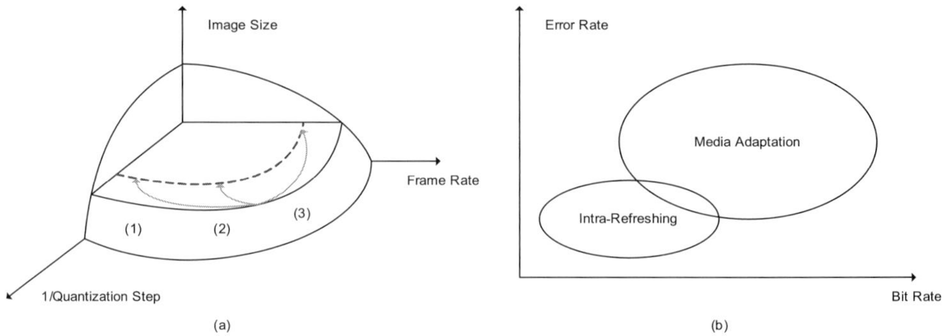

Fig. 9.16 (a) Spatiotemporal trajectory in media adaptation. (b) Favorable conditions for intra-refreshing and media adaptation.

the variations in channel quality with a time-lag less than several speech frames, tracking the fluctuations of sustainable bit-rate in the media adaptation of VoLTE has much more relaxed delay constraints. Therefore a recommended strategy for media adaptation is to reduce the bit-rate immediately once congestion is identified, but return to the original bit-rate reluctantly, step by step, after confirming the recovery of channel status [3GPP (2017f)].

From the network's point of view, the concept of media adaptation can be illustrated as above without specifying the nature of the media in detail. However, multi-dimensional signals such as video require more refined handling than one-dimensional signals such as speech, as the bit-rate should be controlled while maintaining a balance between the spatial and temporal quality, and the corresponding resolution. For example, with a fixed image size, reducing the bit-rate while keeping the same quantization step size or the same frame rate, which would conceptually correspond to trajectory (1) or (3) in the three-dimensional space illustrated in Fig. 9.16(a), will reduce the quality in either the temporal or spatial domain. Such an unbalanced encoding strategy can result in, for example, dynamic, but blurred, video quality or clear, but jerky, video quality.

Assuming a fixed bit-rate and a fixed image size, there can only be a finite number of frame rate and quantization step size combinations that can be used for typical video sequences. These can be projected as a set of coordinates on a plane defined with a fixed image size and two axes for the temporal and spatial parameters. The shape of the set has little reason to be a quadrant, as visualized in Fig. 9.16(a), but in any case the frame rate decreases monotonically as the reciprocal of the quantization step size increases, and vice versa. For a fixed image size, changing the bit-rate corresponds to relocating the operating point of the video codec from one coordinate to another in the three-dimensional space of Fig. 9.16(a), which may be closer to or farther from the origin. Depending on the location of the operating point, the perceived quality may vary significantly. Therefore caution is required in the media adaptation of video so that the trajectories match the nature of input sequences properly and maintain the balance of spatiotemporal quality.

Table 9.8 Video call QoS parameters.

Image Size	QCIF (176 × 144)	QVGA (320 × 240)	VGA (640 × 480)	720p (1280 × 720)
Bit rate (kbps)	50	250	500	800
Frame rate (fps)	7	15	20	30

Although the media adaptation is visualized as horizontal movement on the conceptual plane, vertical movement in the space, i.e., changing the resolution, can be considered if a more drastic change of bit-rate is required. For example, when the coverage of LTE comes to an end and the UE shifts the ongoing session to 3G-324M, through the SRVCC procedures, change of the video codec to MPEG-4 or H.263, and reduction of the bit-rate to 48–49 kbps and the resolution to QCIF (176 × 144) will be necessary. Even within the LTE coverage, changes in bit-rate and image size might be necessary if the system bandwidth supported by the network is abruptly reduced, e.g., from 20 to 10 MHz, when the UE moves into a less populated area. Such situations will require adaptation trajectories involving both vertical and horizontal movements. Table 9.8 outlines a set of typical bit-rate and frame rate combinations for each image size used in video sessions.

As the four image sizes partition the bit-rate from 0 to beyond 800 kbps, switching the image size up or down can be triggered once the bit-rate at a particular image size exceeds its limits. Unlike a temporary reduction of the bit-rate, it is necessary to update the session conditions and inform the billing systems if the bit-rate is changed permanently or if the image size is changed. Note that changing the resolution also requires changes in the video capture and display processes. Figure 9.17 shows an example of partitioning the bit-rate for the image sizes in Table 9.8, with some hysteresis bands set between two neighboring sizes. This partitioning suggests that unlike in one-dimensional transition with the codec mode adaptation of AMR operating in GSM, in media adaptation of video, the image size influences the bit-rate, and vice versa, limiting the possible trajectories for adaptation. Signals with fewer dimensions such as speech are likely to have smooth trajectories for media adaptation, as illustrated in Fig. 9.15.

It can be seen that certain bit-rates, located between those often used by two neighboring image sizes, should be avoided, since at these bit-rates the quality of a smaller size is saturated, i.e., does not improve noticeably even if the bit-rate is increased further, while the quality of a larger size is not acceptable. Such gray regions could be observed also in the partition of CQI indexes shown in Fig. 8.9. Note that wider hysteresis bands are used between larger image sizes as the bit-rate of video cannot be as finely controlled as that of speech, and the variance of video bit-rate is proportional to the target bit-rate. In contrast, in GSM, an identical hysteresis can be used for up to three rate boundaries. Note that such a two-dimensional adaptation is also necessary in the media adaptation of EVS as the change of bit-rate can necessitate a change of the audio bandwidth. For example, if the bit-rate is reduced from 13.2 to 8 kbps, the maximum audio bandwidth is also reduced from super-wideband to wideband.

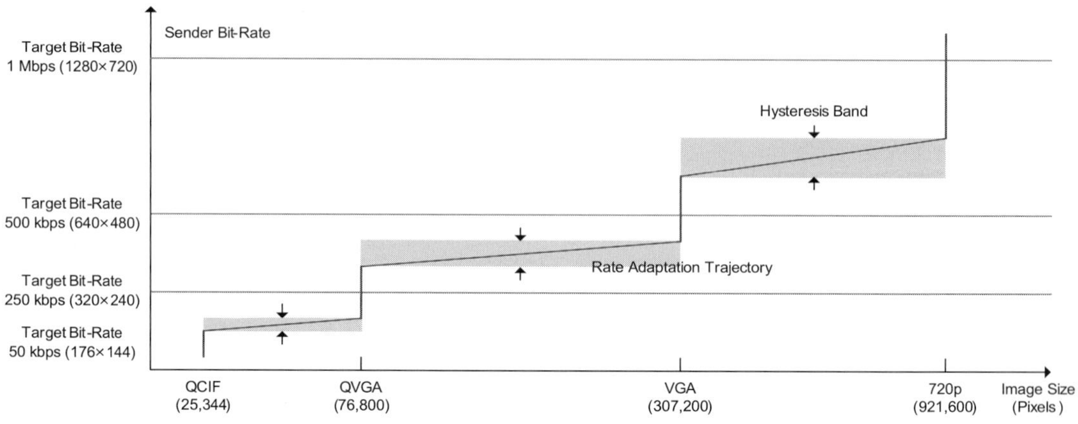

Fig. 9.17 Bit-rate and image-size adaptation.

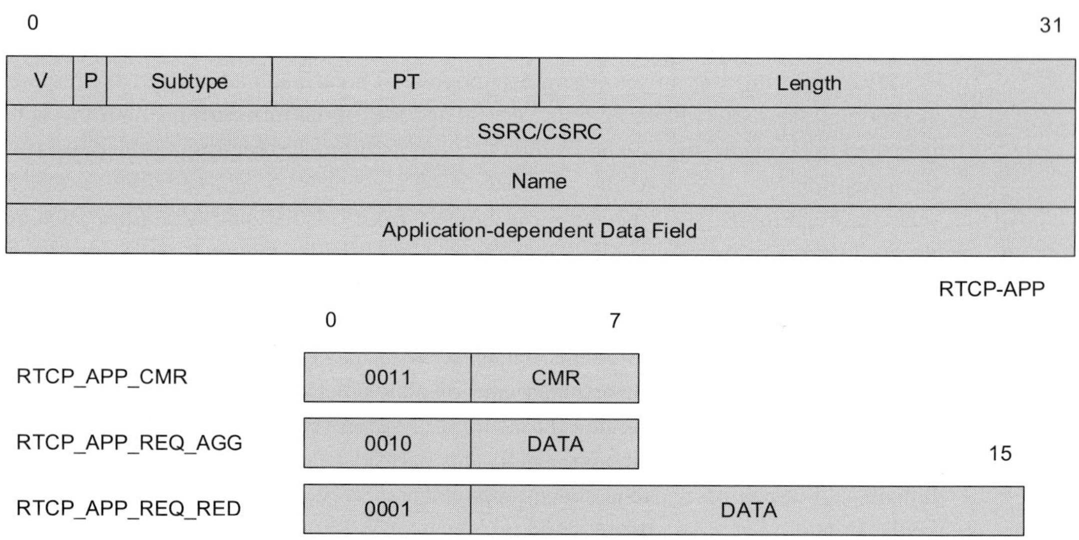

Fig. 9.18 RTCP-APP packet format.

Signaling Format

For AMR and AMR-WB, the RTCP-APP packet supports adaptation requests in three domains: bit-rate, packet-rate, and redundancy. As shown in Fig. 9.18, RTCP-APP is a control packet that carries one or more messages that request the far-end UE to change the compression or packetization methods. The basic message, RTCP_APP_CMR, requests a change in the codec mode. The value of the codec mode request (CMR) field, like that in the RTP payload header, is set to the frame type index of the corresponding speech bit-rate being requested. The four-bit frame type index can be 0–7 for AMR, and 0–8 for AMR-WB. A CMR value of 15 indicates that no mode request is present while the requested actions for reserved values are not defined [Sjoberg *et al.* (2007)]. The requested codec mode must be included in the negotiated set of codec modes. If both a CMR in the RTP payload header and an RTCP_APP_CMR message are received, and they are not equal, a recommended practice is to follow the request with a lower codec mode, regardless of its location. For EVS, RTCP_APP_REQ_EPRR requests a similar change in the bit-rate. Note that EVS needs another message type, RTCP_APP_REQ_EBWR, to control the maximum audio bandwidth.

Redundancy refers to the repetition of previously-transmitted payloads in the current packet. This is a robust transmission technique that can be used when the probability of packet loss is very high. For AMR, AMR-WB, and EVS, the RTCP_APP_REQ_RED message requests a change of the redundancy level and its offset, i.e., the number of redundant payloads and the time differences between the current payload and the redundant payloads. RTCP_APP_REQ_RED has a twelve-bit DATA field that designates the redundant payloads to be repeated. Up to three bits can be set in the DATA field, whose non-zero locations are matched to the pattern of the expected packet loss, e.g., random or bursty.

Frame aggregation refers to the concatenation of speech frames in an RTP packet to build *payload chunks*, which can be used when the opportunities for transmission or scheduling are limited, or when the transport channels are structured such that it is more efficient to send more frames less frequently, than to send less frames more frequently. The latter cases can be found in the networks originally designed for data transmission. For AMR, AMR-WB, and EVS, the RTCP_APP_REQ_AGG message requests a change of the frame aggregation method. RTCP_APP_REQ_AGG has a four-bit field where 0000, 0001, 0010, and 0011 represent payload chunks consisting of 1, 2, 3, and 4 speech frames, respectively. The limit for frame aggregation is set to four as aggregating more frames may delay transmission and reduce quality excessively.

The allowed bit fields of the RTCP_APP_CMR, RTCP_AP_REQ_RED, and RTCP _APP_REQ_AGG messages are restricted by the session description protocol (SDP) parameters, which are negotiated during session setup. The length of the application-dependent data, a properly combined form of three or fewer messages, should be an integer multiple of four bytes, and any unused bytes should be set to zero. Examples of applying redundancy and aggregation with the messages are shown in Fig. 9.19 where $N, N-1, \ldots, N-9$ denote the order of encoded speech frames, and P, $P-1$, and $P-2$ represent the sequence number in the RTP header. Table 9.9 shows the values of the RTCP_AP_REQ_RED and RTCP_APP_REQ_AGG combinations that generate packet formats (a)–(f). A NO_DATA frame represents that either no data for that frame has been produced by the speech encoder, or that no data for that frame is transmitted in the current payload.

With RTCP_APP_CMR, RTCP_APP_REQ_RED, and RTCP_APP_REQ_AGG, it is possible to finely control the bit-rate and packet rate during a speech session. This capability can be helpful in situations where the network clearly prefers certain configurations. For example, EDGE and Wi-Fi handle packets including multiple speech frames more efficiently than packets including one frame per packet because of their channel structures. There are additional messages designed for EVS, such as the one requesting a switch between the EVS Primary and EVS AMR-WB IO modes. However, it can be difficult, if not impossible, to identify the status of the entire network nodes that the

Table 9.9 Bit fields for redundancy and aggregation control.

	RTCP_APP_REQ_RED	RTCP_APP_REQ_AGG
(a)	000000000000	000
(b)	000000000001	000
(c)	000000000010	000
(d)	000000000000	001
(e)	000000000001	001
(f)	000000000010	001

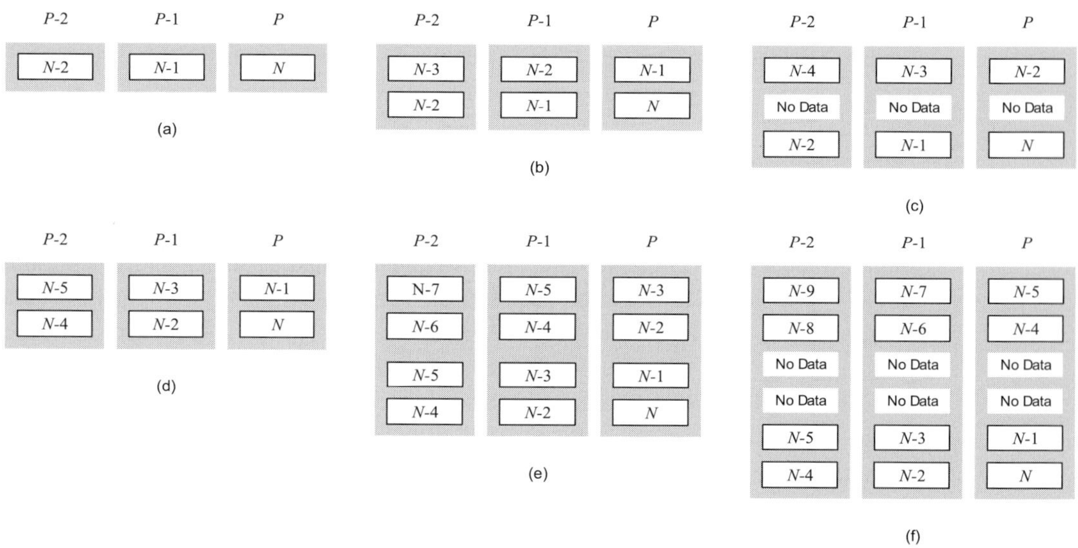

Fig. 9.19 Packetization control with RTCP-APP.

packets pass or the origin of problems that the packets encounter en route. Therefore, media adaptation techniques other than the CMR on the RTP payload header are rarely implemented. Redundancy and aggregation were originally designed to handle heavier congestion than codec mode adaptation was designed for, but the quality management of IMS and LTE is accurate and stable such that even the CMR is rarely transmitted. In operation over unreliable networks such as Wi-Fi on the public spectrum, these extreme measures can be considered when other measures fail.

For video, the Temporary Maximum Media Bit-rate Request (TMMBR) message can be used to request the far-end video encoder to control its maximum bit-rate [Wenger *et al.* (2008)]. The packet format of TMMBR is shown in Fig. 9.20(a), in which *Exp* and *Mantissa* are both unsigned integers. The maximum bit-rate is computed as $Mantissa \times 2^{Exp}$ bps. The measured overhead represents the measured average packet overhead, such as the RTP/UDP/IP headers, in bytes, whose maximum value is known to the UE from session negotiation. The measured overhead is present in the packet format as the TMMBR is originally not designed exclusively for media adaptation of conversational

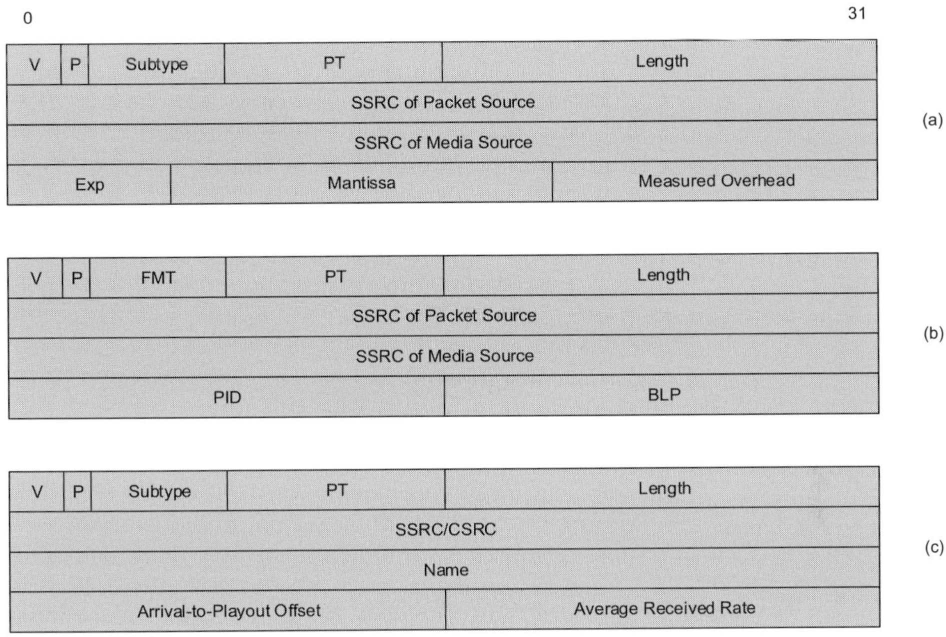

Fig. 9.20 Packet formats for video control. (a) TMMBR. (b) NACK. (c) APTO_ARR.

services. Therefore in VoLTE, the field does not request any action. Since the two fields of TMMBR cannot represent all bit-rates, and there can be more than one method to represent a bit-rate, some computation for determining the closest representation of a target bit-rate may be necessary if the negotiated or requested bit-rates are not predefined. Figure 9.21 illustrates a surface made up of two ranges of *Exp* and *Mantissa* when the TMMBR represents up to 10 Mbps. Note that the change of media bit-rate using CMR or TMMBR may necessitate the modification of radio access bearers, if the bit-rates stay as modified for a prolonged time. However, this complicates the quality and billings issues that limit the use of media bit-rate adaptation in services other than GSM with AMR.

Sender-Driven vs. Receiver-Driven Adaptation

For the codec mode adaptation of GSM and the media adaptation of VoLTE, the control logic that determines the media bit-rate is assumed to be at the receiver. For speech, where the available bit-rates are limited, there is little need to have the sender control the adaptation, as the receiver has the most accurate information about the end-to-end link quality in its direction from the sender. The sender cannot achieve more than a delayed decision even if the receiver reports the channel information to the sender using a higher bit-rate than the negligible bit-rate for the codec mode request. Therefore, *receiver-driven adaptation* is a natural approach for fixed bit-rate speech codecs such as AMR.

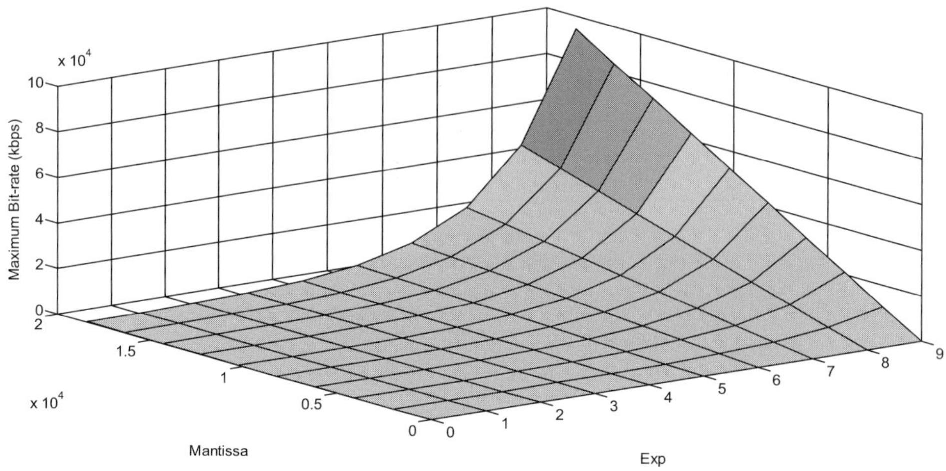

Fig. 9.21 Bit-rate representation with TMMBR.

For video, the *sender-driven adaptation* has some benefits if the information concerning the link quality can be efficiently compressed. However, that information needs to be sufficient to inform the sender about the ongoing situation at the receiver. Packet Switched Video Telephony (PSVT), a counterpart of VoLTE developed for cdma2000 1xRTT EV-DO, supports this type of adaptation for video, where the receiver gathers the information on the link quality and informs the sender by periodically transmitting the APTO_ARR messages, whose format is shown in Fig. 9.20(c). The channel information is contained in the two fields of APTO_ARR. The arrival-to-playout offset (ms) is defined as the difference between the arrival time of media packets and their properly scheduled play-out time. Negative values of the arrival-to-playout offset indicate that the packets are arriving earlier than their scheduled play-out time; positive values indicate that the packets are arriving later. The video frames in packets arrived too late can be discarded if the following frames arrive within time and can be decoded without the previous ones. The average received rate, expressed in units of 256 bps, is defined as the average rate of media received by the receiver.

Comparing the differences in the packet formats of the TMMBR and APTO_ARR messages, it can be understood that the transmission of feedback information can be aperiodic in the receiver-driven adaptation. The feedback information can be sent in an *event-driven* fashion, i.e., transmitted only when the receiver senses unusual situations and finds it necessary to modify the bit-rate of the far-end video encoder. This strategy can save the bit-rate for feedback information. In the sender-driven adaptation, the feedback information needs to be transmitted frequently enough for the sender to construct a statistical model for the link quality and follow its trajectory continuously.

Therefore, in the design of media adaptation algorithms, a tradeoff needs to be controlled between the bit-rate of the feedback channel and the accuracy of the channel estimation. Comparing the structures of the messages used in the speech and video adaptation reveals two different strategies. For speech, where the bit-rates or operating modes are limited, the control message has to be designed for the encoding modes or the bit-stream structure of each codec. In contrast, whether the control logic resides in the sender or the receiver, neither TMMBR nor APTO_ARR exploits any internal characteristics of the video codecs. There can be a large number of bit-rates or operating modes for video codecs, which are not designed with only mobile communications networks in mind, so it is more practical to feed back codec-independent information, such as the approximate bit-rates to encode video at or the status of packet reception.

Channel Information

When there are no explicit requests from the far-end UE concerning the source encoding bit-rates or packetization methods, or no explicit congestion notifications (ECN) from the intermediate network nodes, the sender may have to estimate the channel condition itself and take action preemptively. The Packet Loss Ratio (PLR) is a common metric for media adaptation in packet-switched networks. In the receive direction, the PLR can be estimated by tracking the sequence number of the received RTP packets. In the send direction, the value can also be estimated from the received RTCP RR. PLR can be manipulated to various forms for different types of channel conditions, e.g., by changing the time window of observation. This information can be gathered on the IP layer, whether or not the media client and the LTE modem are co-located.

In some cases, channel information on the physical layer, such as the SNR or CIR, may be available to the media client. Although more information on the channel condition is likely to improve the performance of media adaptation, information on the physical layer reflects mainly the local situation, i.e., the link quality between the UE and the eNodeB. Although their accuracy or frequency of reporting may exceed those for the information on the IP layer, the physical layer information cannot completely replace the IP layer information that represents the end-to-end situation. Because the two types of channel information, i.e., local and end-to-end, may conflict in some cases, an appropriate weighting may be necessary in the decision if both types are taken into account. The channel information on the radio protocol layers, such as the buffer status at the MAC, can also be considered with the information on the physical layer.

Channel-Aware Mode

EVS has two distinct operating modes at 13.2 kbps, the default mode and the channel-aware mode, as shown in Table 9.1. The channel-aware mode can be viewed as a limited form of redundancy supported only at this bit-rate for WB and SWB. In this mode, the speech encoding process is modified to save room for the key parameters of a previous frame that can be used to partially estimate the previous frame when that frame was not received. Such a reconstruction is enabled by the presence of a jitter buffer in the receiver that temporarily stores the received speech frames before decoding to absorb the jitter in the packet arrival time. The redundancy offset, the difference in the number of frames between the current and the previous frames, can be set to 2, 3, 5, or 7 during session negotiation or in the middle of a session with CMR or RTCP-APP.

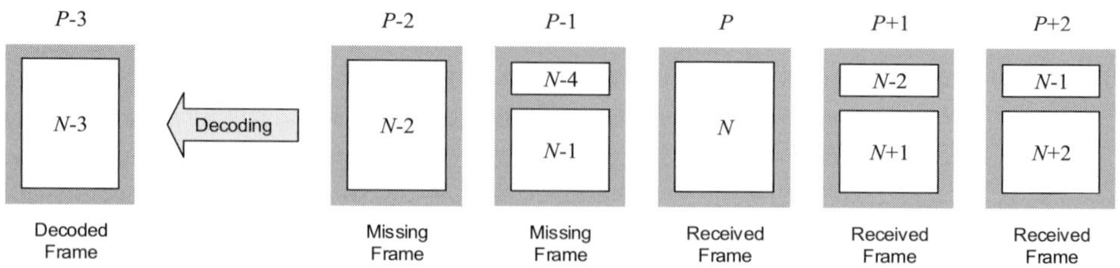

Fig. 9.22 Frame reconstruction with partial redundancy.

Figure 9.22 illustrates the operation of the channel-aware mode when the offset is 3. $N+2, N+1, \ldots, N-3$ denote the order of the encoded speech frames. $P+2, P+1, \ldots, P-3$ represent the sequence number of the RTP headers. Packets $P-3$, $P-2$, and P include only a speech frame while packets $P-1$, $P+1$, and $P+2$ include a speech frame, shown as the lower and larger rectangle inside each packet, and a partial redundancy element containing the key parameters of a frame encoded three frames earlier than the current one, shown as the upper and smaller rectangle. In Fig. 9.22, packets $P-2$ and $P-1$ have not been received when they need to be fed to the EVS speech decoder.

Fortunately, the far-end speech encoder is operating in the channel-aware mode, and some received frames include partial redundancy from the previous frames that are missing in the jitter buffer. The partial redundancy for frame $N-2$ is contained in packet $P+1$, which can be exploited by the speech decoder to reconstruct frame $N-2$, albeit at a lower quality. Likewise, the loss or delayed delivery of frame $N-1$ can be compensated to some extent with the partial redundancy in packet $P+2$. Note that the far-end EVS speech encoder is operating in the channel-aware mode from frame $N-1$, but frame N is encoded in the default 13.2 kbps mode, since frame N contains important acoustic information and its bit-rate cannot be reduced to host partial redundancy. Also note that in 3G-324M, an error-resilient transmission technique conceptually very similar to partial redundancy is used. In the four-byte header of H.223 MUX-PDU, as shown in Fig. 6.54(b), a one-byte header of the previous MUX-PDU in the format of level 0 follows a three-byte header of the current MUX-PDU in the format of level 2. Therefore, a previous MUX-PDU whose header is damaged but payload is intact can be salvaged from the redundant header in the current MUX-PDU.

Since the channel-aware mode sacrifices the quality of current frame to improve the quality of a previous frame, it can reduce the overall quality when the mode is not necessary. Therefore even if the channel-aware mode is activated, partial redundancy is transmitted only when certain conditions are met. The algorithms determine which frames have to be encoded at a reduced bit-rate, and also determine the amount of partial redundancy, based on the nature of the input signal formulated as a criticality measure. In the channel-aware mode, a speech frame is classified as *critical*, and therefore needs to be protected with partial redundancy, when the absence of the frame would reduce the speech quality significantly.

The parameters computed in the standard mode, such as the coder type classification information or the subframe pitch lag, are used to measure the criticality of a frame. If declared critical, the frame is re-encoded and its partial redundancy is packetized later with another frame. The criticality can also depend on whether or not previous frames were lost. An acoustically non-critical frame can be converted to a critical one if its previous frames are lost. The threshold used to determine criticality is a configurable parameter, which can be set to reflect the channel conditions. If a frame error rate (FER) of lower than 5% is estimated, an internal parameter of the EVS speech encoder, the frame erasure rate indicator, is set to *LO* (low), and to *HI* (high) otherwise. The frame erasure rate indicator controls the thresholds that determine whether a speech frame is critical or not, thereby controlling the frequency of partial redundancy transmission. Figures 9.23(a) and 9.23(b) compare the bit allocations of the standard and channel-aware modes. As in the case of redundancy or aggregation in media adaptation, the advantage of channel aware may not be significant in packet-switched networks where the QoS is well-managed, such as in LTE.

9.2.2 Selective Intra-Refreshing

In addition to TMMBR, the Negative Acknowledgement (NACK), Picture Loss Indication (PLI), and Full Intra-Request (FIR) messages are provided in VoLTE [Ott *et al.* (2006)] to provide quality recovery mechanisms similar to the intra-refreshing of 3G-324M with the videoFastUpdatePicture message. These messages are used to ask the

Core	CELP	HQ-MDCT	TCX	CELP			TCX
Ext. Layer				WB TBE	WB FD	SWB TBE SWB FD	IGF
Number of Bits / Frame	264						
BW, CT, RF	5						
TCX / HQ-MDCT Core Flag		1	1				1
TCX CT			2				2
TD / FD Ext. Layer Flag				1	1	1	
Core Bits	259	258	256	238	252	227	256
WB / SWB Ext. Layer Bits				20	6	31	

(a)

Core	CELP		TCX
Ext. Layer	WB TBE	SWB TBE	IGF
Number of Bits / Frame	264		
BW, CT, RF	5		
Core Bits (Primary)	183–248	171–236	189–254
WB / SWB Ext. Layer Bits (Primary)	6	18	
Core Bits (Partial RF)	0–60	0–60	0–65
WB / SWB Ext. Layer Bits (Partial RF)	0–5	0–5	
RF Offset	2		
RF Frame Type	3		

(b)

Fig. 9.23 Bit allocation for EVS frames at 13.2 kbps. (a) Standard mode. (b) Channel aware mode.

far-end video encoder to generate independently encoded frames at its earliest opportunity, to stop the error propagation initiated by corrupt or lost video data. Figure 9.20(b) shows the packet format of NACK, where the Packet ID (PID) field represents the RTP sequence number of the lost packet. The Bitmask of following Lost Packets (BLP) field can be used to report the losses of any of the 16 RTP packets immediately following the RTP packet indicated by the PID.

Historically, intra-refreshing was conceived to recover video quality in networks with low bit-error rate, for example, about 10^{-6} for PSTN or 10^{-8} or lower for ISDN, for which the H.32x series of multimedia communications systems and the H.245 call control protocol were designed. Under such benign channel conditions, a preferred strategy to encode video is to generate the independently encoded frames as infrequently as possible. This strategy maximizes the frame rate without reducing the image clarity. Occasional errors, such as those originating from thermal noise, can be accommodated by intra-refreshing signaled by the receiver. In contrast, in 3G-324M, the W-CDMA networks are typically configured with a target block error rate (BLER) of $10^{-3} - 10^{-2}$. Reducing the BLER further, e.g., by allowing higher transmit power, can drain the network capacity significantly. Since the deployment of 3G-324M, inefficiency of this classic error-control technique for video has been frequently observed. In Fig. 9.24, two such cases are illustrated.

In Fig. 9.24(a), the adaptation layer of 3G-324M in UE B detects a loss of video data at a P-frame, P_l, and sends a videoFastUpdatePicture message, shown as VFU, to the far-end UE A. However, it takes more than a two-way delay for the requested I-frame to arrive at UE B. The gradation beginning at P_l illustrates the deterioration of the perceived video quality as the error propagates to the following P-frames. Although there is no further loss of video data after P_l, the corruption of visual information proceeds as the influence of the error propagates to the following P-frames. At P_m, another loss of video data occurs. As the gradation almost reaches black, the visual information is

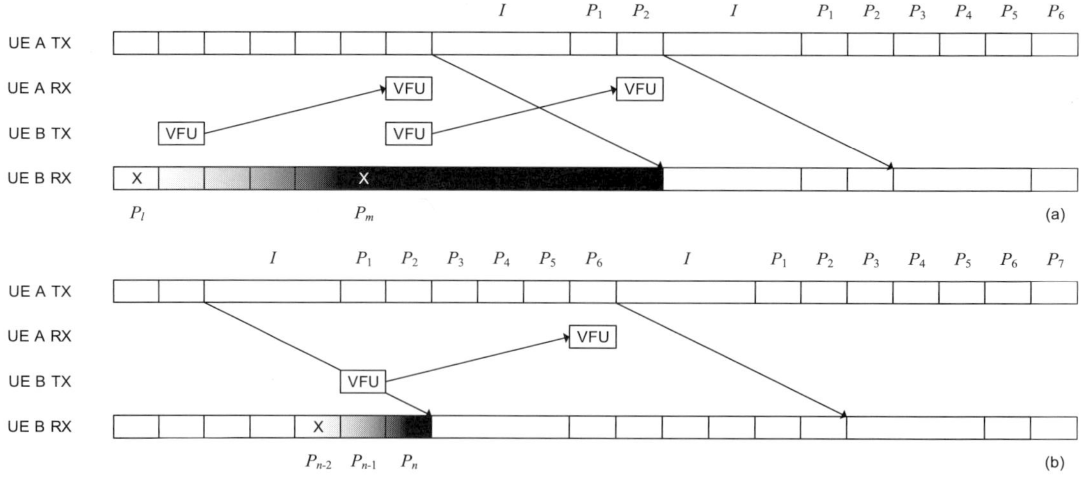

Fig. 9.24 Redundant intra-refreshing situations.

already irrecoverably damaged. The adaptation layer of UE B sends another videoFastUpdatePicture message, which is not necessary since the I-frame requested by the first VFU should arrive earlier. Unless additional errors occur between the two I-frames, the second I-frame would waste the bit-rate since several P-frames, which are of quality comparable to that of an I-frame, could have been packed into that period instead of an I-frame. Bursty errors, a series of errors occurring within a short time period, can cause this type of tricky situation.

Another possibility for redundant intra-refreshing is shown in Fig. 9.24(b). After a loss of video data is detected at P_{n-2}, a videoFastUpdatePicture message is sent to the far-end. However, a periodically-encoded I-frame arrives earlier than the requested I-frame. It can be seen that the gradation implying quality degradation has reached only dark gray when the first I-frame arrives. If no error occurs between the two I-frames, the I-frame generated by the videoFastUpdatePicture message only reduces the average frame rate without contributing to the image clarity. These observations show that intra-refreshing is conceptually equivalent to replacing the distortion from transmission with that from compression, which is, in general, considered to be visually more acceptable when the channel condition is not favorable. Therefore, unnecessary intra-refreshing can reduce video quality. In 3G-324M where bit-rates as low as 48–49 kbps are typically assigned to the video, it is a conservative but commonly used strategy to encode an I-frame every 2–3 seconds and ignore any received videoFastUpdatePicture messages.

It is possible to intra-refresh at levels lower than a frame or a picture, as H.245 also provides the videoFastUpdateGOB and videoFastUpdateMB messages, which refresh a group of blocks and a macro block, respectively. As these messages were designed with much more benign channels such as PSTN and ISDN in mind, they are not considered to be capable of handling the challenges of 3G-324M channels. Such messages for sub-frame intra-refreshing are also tied to the bit-stream structure of specific codecs. Therefore, the Slice Loss Indication (SLI) message provided by RTCP is not used in VoLTE. The intra-refreshing strategy for VoLTE can be considered with the observations in 3G-324M. If poorly-timed intra-refreshing reduces the quality, a straightforward approach will be to use it only when it is absolutely necessary. Such a decision can be made at either the sender or the receiver. In the case when the decision logic is located at the sender, the receiver sends an intra-refreshing request whenever a loss of video data occurs. However, the sender decides whether or not to accept the request. When the decision logic resides at the receiver, the receiver decides whether or not to send an intra-refreshing request when video data is lost. On the other hand, the sender encodes an I-frame whenever an intra-refreshing request is received.

The advantage of locating the decision logic at the sender is that accurate information on the encoder status, such as when the last I-frame was encoded and transmitted, is available there. If the decision logic is located at the receiver, such information has to be estimated by inspecting the received video bit-stream. Another advantage of locating the decision logic at the sender is the higher resilience against the loss of an intra-refreshing request. The impact on the visual quality from a loss of intra-refreshing request, transmitted by the receiver in a controlled way, will in general be larger than in the case when the receiver automatically transmits a request upon the loss of video data. As a system for executing the *selective intra-refreshing* strategy, it is recommended that the sender consider whether the requested intra-refreshing, if executed, can improve the situation at the far-end better than by ignoring the received request and maintaining the current video encoding process. In other words, the sender should take action only when it is deemed necessary. For example, an intra-refreshing request received just after transmitting an I-frame can be safely ignored.

Figure 9.16(b) illustrates the areas on the plane defined by the bit-rate and error-rate axes where intra-refreshing and media adaptation are relatively effective. When the majority of frames are predicted frames and the errors occur rarely, intra-refreshing can be an efficient measure to limit error propagation while maximizing the quality. As the operating environments for 3G-324M, W-CDMA can be characterized as low bit-rate and low error-rate, intra-refreshing fits the situation illustrated as a small ellipse in the low and left area of the plane. In contrast, media adaptation fits the more general transmission conditions of wide bit-rate and error-rate ranges. The transmission conditions of HSPA or LTE may be represented as the low and right area of the plane, i.e., high bit-rate and low error-rate while the conditions of non-QoS networks such as Wi-Fi are represented as the high and right area, high bit-rate and high error-rate.

9.2.3 Coordination of Video Orientation

In addition to the classical sources of media distortion, such as compression, transmission errors, or delays, the quality of real-time video transmitted over mobile communications networks can be degraded by factors not applicable to speech, because of the fact that video is a multi-dimensional signal. When the camera capturing a rectangular image is rotated, so will be the decoded and displayed video, making the image difficult to recognize at some angles. The Coordination of Video Orientation (CVO) addresses this problem by transmitting information on the angle of the camera, shown in Fig. 9.25 as θ, with respect to the direction of gravity. This information is included in a two-byte extended header of RTP, at a maximum resolution of $\frac{360}{64}$ degrees, with which the receiver can compensate for the rotation.

CVO can be considered as a generalization of the automatic screen switching between the portrait and horizontal modes, as seen in Fig. 9.26, which rotates the screen by 90 degrees and adjusts the display layout correspondingly, depending on the direction of the reference axis of the UE. Assuming a rectangular display, two angles are defined with respect to a vertical axis pointing towards the center of the earth and a plane tangential to the ground at the location of the UE. The two angles, ρ and θ, can be measured with an accelerometer aligned to the display, which limits the location and direction of the camera so that the orientation of the camera is aligned with that of the display. Another angle, ϕ, is defined with respect to an axis pointing the magnetic pole, and can be measured with a magnetometer.

To derive the angle for CVO, θ, as a function of the accelerometer output, a three-dimensional Cartesian coordinate system can be defined with the X, Y, and Z axes, as illustrated in Fig. 9.27. The rectangular image sensor of the camera lies in the plane defined by the X and Y axes. The gravitational acceleration on the image sensor can

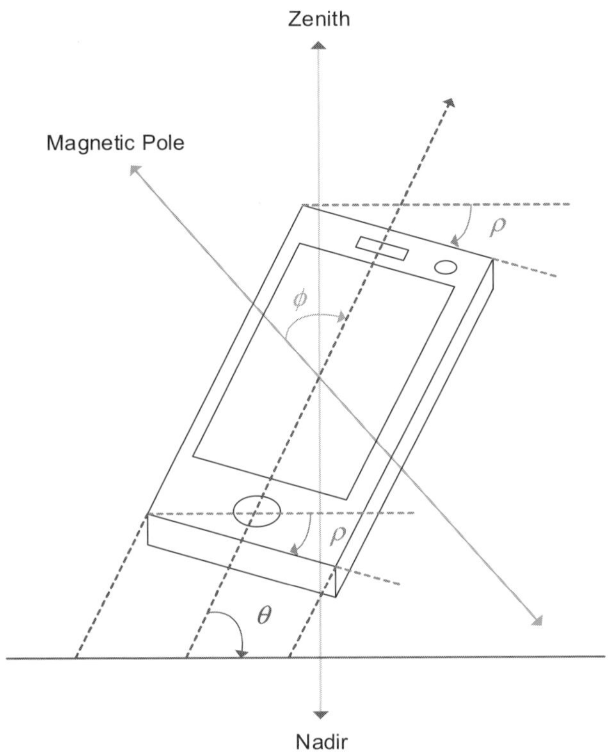

Fig. 9.25 Coordinate system for video orientation control.

(a)

(b)

Fig. 9.26 Video call user-interfaces. (a) Portrait mode. (b) Horizontal mode (courtesy of Samsung Electronics Co., Ltd.).

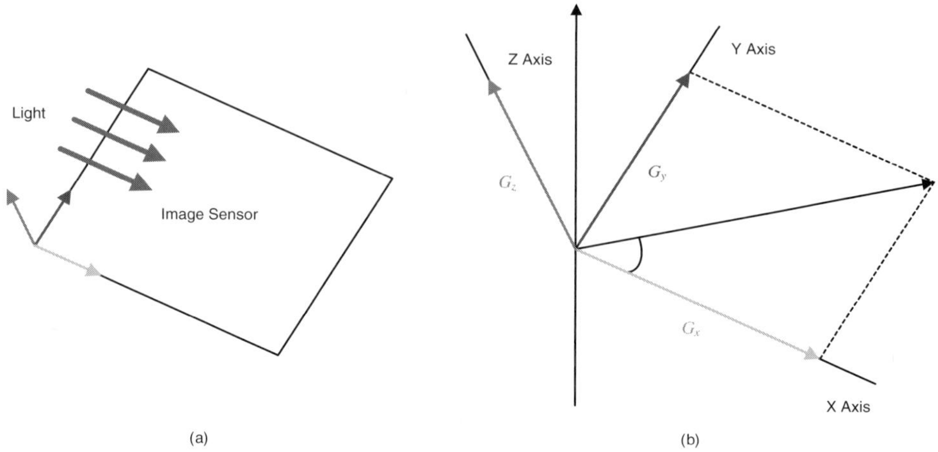

Fig. 9.27 (a) CVO coordinate system. (b) Measuring angle from gravitational acceleration.

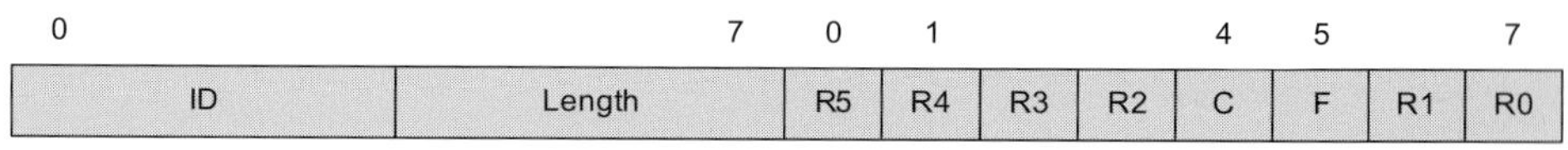

Fig. 9.28 Extended RTP header for CVO.

be expressed as a sum of three linear accelerations, G_x, G_y, and G_z, measured in the directions of the X, Y, and Z axes. When the UE is positioned on the surface of the earth so that the Z axis points to the center of the earth, G_z is close to the standard gravitational acceleration, $g = 9.8\ m/s^2$, while G_x and G_y are close to zero.

Then θ is defined as

$$\theta = \begin{cases} 90 - \tan^{-1}\left(\frac{G_y}{G_x}\right) & \text{if } G_x > 0, \\ 270 - \tan^{-1}\left(\frac{G_y}{G_x}\right) & \text{if } G_x < 0, \end{cases} \tag{9.2}$$

where $-90 < \tan^{-1}\left(\frac{G_y}{G_x}\right) < 90$. Table 9.10 and Fig. 9.28 show the format of the extended RTP header [Singer and Desineni (2008)] that transports the quantized θ values. The extended header can be added to the RTP header periodically or in an event-driven fashion. In the SDP offer or answer, a 4-bit ID, which can have a value from 1–14, identifies the RTP packet stream whose header is extended. The 4-bit Length is defined as the number of data bytes of this header extension following the one-byte header minus one. Therefore Length = 0 indicates that one byte of data follows, as in Fig. 9.28, and up to 16 bytes of data can follow.

In typical UEs, when the display is activated, θ is continuously monitored to control the screen orientation, even when the video call is not used. For example, when θ approaches 90 or 270 degrees, the screen layout is switched from the portrait to horizontal mode. As long as the camera and display are co-planar, i.e., located on the same

Table 9.10 Rotation signaling for CVO.

R1	R0	R5	R4	R3	R2	Rotation (degrees counter-clockwise)
0	0	0	0	0	0	0
0	0	0	0	0	1	$\frac{360}{64}$
0	0	0	0	1	0	$\frac{2\times360}{64}$
...	...	...	...	...	...	...
0	1	0	0	0	0	90
0	1	0	0	0	1	$90+\frac{91\times360}{64}$
...	...	...	...	...	...	...
1	0	0	0	0	0	180
1	0	0	0	0	1	$180+\frac{360}{64}$
...	...	...	...	...	...	...
1	1	0	0	0	0	270
1	1	0	0	0	1	$270+\frac{360}{64}$
...	...	...	...	...	...	...

plane, and share the same coordinate system, the θ measured for display orientation can be used for CVO.

In the low-resolution mode of CVO, only R0 and R1 bits are used to signal the change of orientation, by an integer multiple of 90 degrees. In the high-resolution mode, the four bits, R2–R5, are also used to represent θ in 64 levels. Whether to use CVO or not, and whether to use the low-resolution or high-resolution mode is determined in the session negotiation. In the default camera orientation, i.e., when $G_y = g$ and $G_x = G_z = 0$, θ is theoretically zero. However, the noise level can be the highest at this orientation, as the computation of θ in this condition involves division by zero. Similar instability can be observed when the Y axis points to the nadir. To avoid the ping-pong effects, abrupt or frequent changes of angle around $\theta = 0$ can be reduced by representing the ranges of quantized angle, $[\frac{127\times360}{2\times64}, \frac{360}{2\times64})$, $[\frac{360}{2\times64}, \frac{3\times360}{2\times64})$, …, $[\frac{125\times360}{2\times64}, \frac{127\times360}{2\times64})$, with R1R0R5R4R3R2 = 000000, 000001, …, 111111, respectively. The mapping effectively transforms the quantization scheme from Uniform Mid-rise to Uniform Mid-zero, increasing the probability of setting R1R0R5R4R3R2 to 000000 at the default position.

The C bit signals the change of camera, e.g., from the secondary to the primary or vice versa, but the measured θ value may not be valid unless all cameras are co-planar with the display and use the same axes. The F bit signals that the image captured is mirrored horizontally. Information about the accelerometer principles can be found in [Huyghe *et al.* (2009), Freescale (2012)]. Note that some of the assumptions made in the development of CVO may not be valid in recent display technologies or UE designs, as the display may not be flat or rectangular, and the display may be physically separated from the camera. As the video resolution gets higher, the necessity for CVO operation diminishes as larger resolutions, e.g., from 720p (1280 $\times$ 720), are mainly for the horizontal view.

9.3 Enhancements in Session Negotiation

9.3.1 Reduction of Resizing-Induced Spectral and Computational Inefficiency

As media oriented negotiation acceleration (MONA) reduced the call setup time of 3G-324M, a new type of enhancement in the media negotiation was applied to VoLTE to reduce the bit-rate and power consumption, and increase the image clarity. Observing the quality improvement of 3G-324M or conventional IP-based multimedia communications systems while increasing the bit-rate or decreasing the delay and error-rate, or employing advanced video codecs reveals the existence of an unexpected limitation such favorable conditions for compression and transmission cannot remove completely. Figure 9.29 illustrates the end-to-end media compression and transmission paths in which the pre-processing of the media before source encoding and the post-processing of the media after source decoding are indicated.

For example, initially the camera digitizes input video, after optical infra-red (IR) filtering, and feeds the captured raw video to an image signal processor (ISP), which filters and subsamples the video data to generate the required image size and frame rate. It then outputs the pre-processed video, in the YCbCr or RGB565 format, to the video encoder. The distortion, i.e., the loss of media quality, can be associated with where it originates. First, the distortion from compression, D_1, occurs from those operations between the pre-processing and compression of the video. The Rate-Distortion (R-D) optimization theory deals with the minimization of this classical source of distortion. Then there is distortion from transmission, D_2, which occurs from transmission over one or more channels with errors. A third type of distortion is introduced during reconstruction, which includes decoding, resizing, filtering, and analog modulation of the

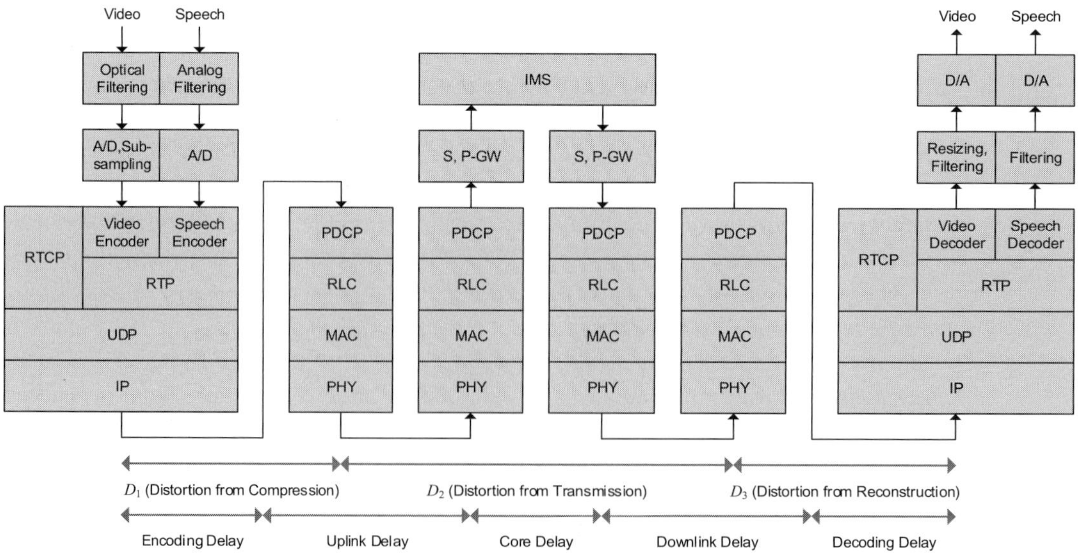

Fig. 9.29 End-to-end distortion and delay partitioning of VoLTE.

pixels of the display. The problem of maximizing video quality, in principle, follows the framework of joint source-channel coding formulated in [Ortega and Ramchandran (1998)],

$$\min_{\text{(source parameters,channel parameters)}} E(D) \tag{9.3}$$

where the expected end-to-end distortion $E(D)$ is minimized based on some criterion, with a proper selection of source and channel parameters while meeting the total bit-rate constraints of R_{budget}. The problem of maximizing speech quality can be formulated in a similar way.

Let R_{source} and R_{channel} be the bit-rates for the source and channel codes respectively. The channel code in this context includes the bits for error resilience, including the CRC, error correction code, and other redundancies added to the encoded source information before transmission. GSM operating with AMR involves cases where $R_{\text{source}} + R_{\text{channel}} = R_{\text{budget}}$ while for VoLTE, $R_{\text{source}} + R_{\text{channel}} \leq R_{\text{budget}}$. In 3G-324M, both R_{source} and R_{channel} are fixed. Note that the above simplified framework does not take some practical factors, such as the delay, or the interactions with the network or the far-end UE, into account. Although D is the total distortion reflecting the combined effects of D_1, D_2, and D_3, it would be very difficult to model D as the distortions are likely to be interrelated in a complex fashion. However, it is possible to remove the influence of one or two sources in some cases. For example, if the channel is error-free, then D_2 will be equal to zero. The contribution of D_3 will be insignificant in the case of speech where the frequencies of the A/D and D/A conversions are finite and known to the sender and the receiver precisely. In the case of video, if the media-handling capabilities of both sides are similar, and the image sizes used in the video encoder and the decoder are comparable, D_3 will not have noticeable effects either. When the resolutions of the encoded video and the finally displayed video are different, the magnitude of D_3 becomes significant as the distortion from resizing can be considerable.

In 3G-324M, as illustrated in Fig. 9.30, the decoded image is slightly upscaled but rarely pruned or clipped as typical UEs capable of 3G-324M are equipped with a display whose resolution is larger than QCIF. Distortion from resizing can often be experienced, however, for example in the playback of video encoded in SD (704×576) format displayed on televisions with Full HD (1920×1080) resolution. In Fig. 9.30, $X_1 \times Y_1$ and $X_2 \times Y_2$ are the resolutions of the encoded video and the decoded, resized, and displayed

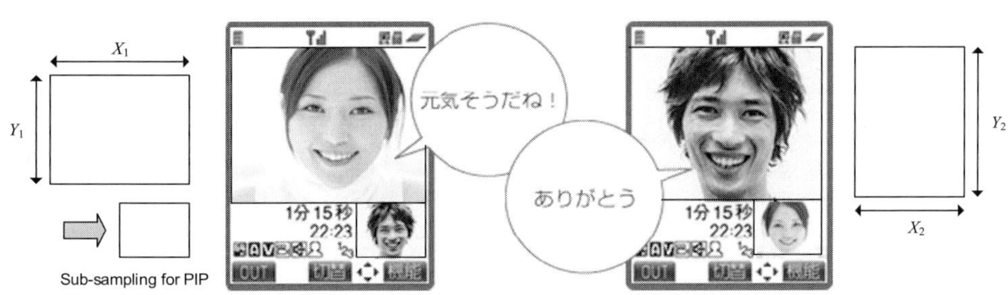

Fig. 9.30 Resolutions of encoded and finally displayed video (courtesy of NTT DOCOMO, Inc.).

video, respectively. Therefore $X_1 \times Y_1$ refers to the resolution at which the video of the man displayed in the small PIP of the left display will be encoded.

Therefore, the end-to-end distortion D can be re-formulated as

$$D(R_s, P_p, \frac{X_1}{X_2}, \frac{Y_1}{Y_2}), \tag{9.4}$$

where R_s and P_p are the video encoding bit-rate and the packet loss rate, respectively. Figure 9.29 shows two major sources of quality loss, distortion and delay, where the end-to-end distortion is partitioned into D_1, D_2, and D_3, and the end-to-end delay is partitioned into (media) encoding delay, uplink delay, (core) network delay, downlink delay, and (media) decoding delay. In VoLTE, the distortion can be further simplified,

$$D(R_s, P_p, \frac{X_1}{X_2}, \frac{Y_1}{Y_2}) \approx D(\frac{X_1}{X_2}, \frac{Y_1}{Y_2}), \tag{9.5}$$

since R_s and P_p are managed as QoS by the network. In other words, R_s and P_p can be assumed to be fixed.

For delay, although it is not possible to partition the end-to-end delay over the end-to-end transmission paths as precisely as in circuit-switched networks, as shown in Tables 2.10 and 2.11 for GSM, a rough estimate may be made for VoLTE as follows: 35 ms for the encoding delay, 40 ms for the uplink delay, 10 ms for the core network delay, 40 ms for the downlink delay, and 35 ms for the decoding delay. The encoding delay is the sum of delay elements from the A/D, pre-processing, media encoding, and the RTP/UDP/IP packetization procedures. Likewise, the decoding delay is the sum of delay elements from the RTP/UDP/IP de-packetization, media decoding, de-jitter buffering, post-processing, and finally the D/A procedures.

If a hands-free or a headset is connected to the UE via Bluetooth [SIG (2016)], an additional delay of about 30 ms is added, which consists of speech encoding and decoding delays with Continuously Variable Slope Delta Modulation (CVSD) for narrowband or modified Sub-Band Codec (SBC) for wideband, and transmission delay. The uplink delay of VoLTE is dominated by HARQ, and the downlink delay is caused mainly by the scheduling. Since the variances of the other delay elements are negligible, the variation in the end-to-end delay, jitter, is dominated by the uplink and downlink delays. It is often observed in commercially operational LTE networks that the variations of encoding delay, core network delay, and decoding delay are not significant. Therefore, the combined delay contribution from these elements can be considered as fixed. On the other hand, the uplink and downlink delays are key parameters controllable by the network to manage the tradeoff between media quality and network capacity.

We do not attempt to manipulate D further but it can be expected that the end-to-end distortion will increase as $\frac{X_1}{X_2}$ or $\frac{Y_1}{Y_2}$ deviates from unity. Note that in Fig. 9.29, the filtering before source encoding refers to analog signal processing while the signals after source decoding are filtered in the digital domain. However, additional analog or digital signal processing can be applied before the source encoder or after the source decoder to enhance the media quality.

Conventional session negotiation procedures based on SIP/SDP do not explicitly exchange information on the supported video resolutions. Instead, the largest size

Fig. 9.31 Resizing distortion. (a) Original video (CIF, aspect ratio 1.23). (b) Pruned video (36.7% of pixels removed). (c) Reshaped video (aspect ratio 0.86).

allowed at the negotiated codec level is commonly encoded. Therefore the available image sizes are limited and the sender cannot be sure of proper presentation at the far-end. These limitations necessitate a computationally-expensive resizing process in which the decoded video is distorted to fit display areas that are different from the encoded sizes or areas with different aspect ratios. Figure 9.31 shows examples of video distorted from resizing where the reconstructed video is pruned or reshaped. It is not easy to quantify the loss and decide which distortion type reduces the subjective quality most but it would be possible for the sender to improve the video quality as much as possible if it knows the resolution preferred by the receiver. With the use of the imageattr parameter in session negotiation, X_1 and Y_1 can be identical to X_2 and Y_2, respectively.

Considering the wide variety of display resolutions and pixel sizes (dot pitch) of contemporary UEs, ranging from two inch QCIF+ (220×176) to 5.5 inch Ultra HD (UHD) (3840×2160) displays, flexible negotiation methods that enable the bit-rate for video to be spread spatially and temporally in a controlled fashion are required, which is provided by the imageattr parameter. Multiple resolutions for the send and receive directions can be offered in a single line of the parameter, and a pair that best matches the needs of the answerer can be selected and returned to the offerer [Johansson and Jung (2011)]. By switching the width and height values of imageattr and sending the modified SDP in an SIP Update message, similar effects to those of CVO can be achieved in a simpler implementation.

On the other hand, the inefficiency of session negotiation based on the conventional SIP/SDP reflects the optimistic design philosophy of IP where it is assumed that the computers have enough capability and flexibility to encode and decode video of any bit-rate and resolution, and that the displays have enough space for presenting the video. In 3G-324M, these assumptions are generally not valid, as the bit-rate can be either doubled or halved by controlling the spreading factor of W-CDMA. In LTE, more refined control over the bit-rate is possible through scheduling, which can be exploited for the pixel-level control of the far-end video.

9.3.2 Asymmetric Media Configuration

Needs for asymmetric media configuration may arise in sessions between UEs with different computational complexity and media handling capabilities, or in sessions between UEs with different service policies and quality levels. As the UEs become more and more powerful, the needs of the second type may become more significant. If the media is not generated or presented in the UE, but in subordinate devices connected to the UE via Wi-Fi or Bluetooth, such as a video camera, it may be necessary to configure the media asymmetrically or for one-way transmission.

In addition to the reduction of distortion from resizing, another benefit of imageattr is that this SDP parameter enables a simple configuration of asymmetrical video, in which different resolutions can be set for the send and receive directions. However, the efficiency of asymmetric video configuration using this parameter decreases as the difference between the two resolutions increases, e.g., in the number of pixels or aspect ratios, unless the bit-rates are also adjusted separately to compensate for the differences. As a new opportunity to exploit for higher spectral efficiency, VoLTE supports asymmetric configurations for speech and video with the help of SDP parameters defined for recent media codecs.

Figure 9.32 shows an SDP answer in which EVS and H.264 are configured asymmetrically. The bit-rates of EVS for the send and receive directions are set to 16.4 and 24.4 kbps, respectively. In the send direction, from the answerer to the offerer, two audio channels are transported, as signaled by ch-send = 2. The channels parameter, in EVS/16000/2, is set to two for sessions between mono and stereo UEs. Therefore the bit-rate for the send direction is higher than b = AS, which represents the maximum bit-rate for the receive direction if different bit-rates are used for the send and receive directions. The maximum bandwidth is set to super-wideband for both directions, but it is possible to set different maximum audio bandwidths for the two directions with the bw-send and bw-recv parameters.

Suppose that in the SDP offer, the b = AS, profile-level-id, level-asymmetry-allowed, and max-recv-level parameters of H.264 are set to 2000, 42e00d, 1, and e01f, respectively. From the value of profile-level-id, the default level offered is 1.2, which can handle the bit-rate at up to 384 kbps. From level-asymmetry-allowed = 1, the offerer informs the answerer that it can handle the bit-streams of different bit-rates for the send and receive directions. From the value of max-recv-level, the offerer declares that it can receive H.264 bit-stream encoded at up to 14 Mbps but the bit-rate is limited by setting b = AS to 2 Mbps. By repeating level-asymmetry-allowed = 1, the answerer agreed to use asymmetric video if possible, but reduced the default level to 1.1, which allows handling of video at up to 192 kbps. For the receive direction, the answerer can accept H.264 bit-stream at up to 384 kbps, by setting max-recv-level to level 1.2. The network, if unaware of the usage of the level-asymmetry-allowed parameter, would reserve resource to the b = AS value in the SDP answer, 416 kbps for both directions. Those capable of asymmetric video can assign 384 kbps for VGA video to the direction from the offerer to the answerer, and 192 kbps for QVGA video to the other direction. Assuming the IP overhead plus 384 kbps, the b = AS of video is set to 416 kbps. Note

```
C=IN IP4 aaa.bbb.ccc.ddd
b=AS:458

m=audio 49152 RTP/AVP 105 99
b=AS:42
b=RS:0
b=RR:2500
a=rtpmap:105 EVS/16000/2
a=fmtp:105 br-send=16.4; br-recv=24.4; bw=nb-swb; ch-send=2; ch-recv=1;
a=rtpmap:99 telephone-event/16000
a=fmtp:99 0-15
a=sendrecv
a=ptime:20
a=maxptime:120

m=video 49154 RTP/AVP 106
b=AS:416
b=RS:0
b=RR:2500
a=rtpmap:106 H264/90000
a=fmtp:106 packetization-mode=0; profile-level-id=42e00b; \
     sprop-parameter-sets=J0LgDJWgUH6Af1A=,KM46gA==;
     level-asymmetry-allowed=1; \
     max-recv-level=e00c
a=imageattr:99 send [x=320,y=240] recv [x=640,y=480]
a=sendrecv
```

Fig. 9.32 SDP answer for asymmetric media configuration.

that setting a higher level to max-recv-level than the level signaled by profile-level-id does not incur excessive complexity problems as the computational complexity of the video decoder is, in general, much lower than that of the video encoder. The speech and video codecs before EVS and H.264 do not provide the SDP parameters and negotiation mechanisms for asymmetric encoding of media.

9.4 Enhancements in Wireless Transmission

The radio signal processing of LTE was enhanced using methods that exploited the unique characteristics of LTE as well as conventional approaches whose effects had been proven in previous wireless generations.

9.4.1 Spectrum Usage Analysis

Figure 9.33 shows an actual frequency allocation for three service providers, A, B, and C, in an area where multiple generations of mobile communications systems are being operated. The detailed guard bands are not shown for brevity.

The earliest generation still operational is cdma2000, in the 800 and 1800 MHz (downlink) bands, run by A and C. The originators of A and B were a non-wireline

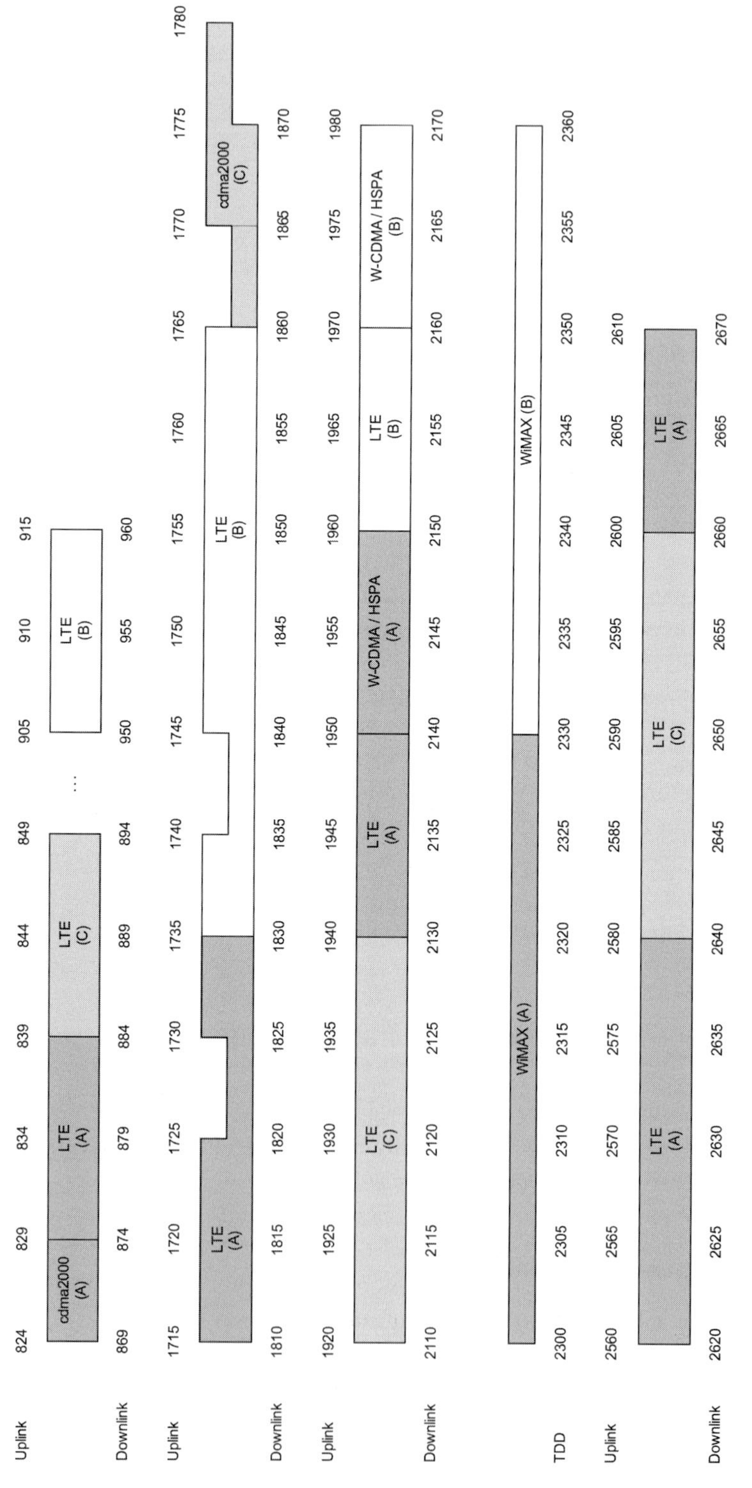

Fig. 9.33 Frequency allocation for multiple generations of mobile communications systems.

operator and a wireline operator, following the practices used in Fig. 1.4, but B has already discontinued its cdma2000 system. C joined the cellular business with PCS, a high-frequency version of IS-95 using QCELP-13. Note that some carriers of cdma2000 exclusively use the EV-DO mode, although simultaneous support of circuit-switched voice and packet-switched data as in W-CDMA/HSPA became feasible in later versions of cdma2000. In the 2100 MHz band, A and B operate two carriers of W-CDMA/HSPA respectively, which provide the voice, 3G-324M, and other data services. These networks are used also for SRVCC when the UE moves out of LTE coverage. The 1800 MHz cdma2000 band of C was initially paired with LTE in a completely independent fashion, but the co-operation halted as running two modems drained the battery rapidly, and the voice coverage became provided solely by LTE.

In the 2300 MHz band, A and B operate the mobile version of WiMAX respectively, which is facing sunset and used for lower-priority services such as the backhaul for Wi-Fi networks. Note that WiMAX is operated in the TDD mode, using a 30 MHz bandwidth for both the uplink and downlink. Finally, each operates 3–4 bands of LTE but the largest, e.g., nationwide, coverage is typically offered in the lower bands that host VoLTE.

As was shown in Table 6.3, the coverage of cellular systems decreases rapidly as the frequency increases, requiring a greater infrastructure investment. However, the cost of the frequency spectrum decreases noticeably. For example, a 10+10 MHz band at 1800 MHz, for the uplink and downlink, costs more than twice as much as a band with the same width at 2100 MHz. Thus, it can be a good strategy to balance the network load if higher bit-rate traffic, such as video, is carried on the higher bands while lower bit-rate traffic, such as speech, is handled on the lower ones. Note that the cost of spectrum is determined also by other factors, such as availability in wide geographic areas or countries, which might be as important as favorable propagation characteristics. For example, the bands at 1800 MHz, globally used for LTE roaming, are not necessarily cheaper than the 700 MHz bands, still used in many countries for broadcasting.

9.4.2 Carrier Aggregation

There have been alternative approaches to increase the overall transmission rate or throughput when there is little margin to improve the efficiency of radio signal processing operations. An early idea was to combine multiple carriers of a fixed channel-width system, e.g., using three adjacent carriers simultaneously in the forward link or a single carrier whose width is three times that of 1xRTT in the reverse link, as in cdma2000 3xRTT. This approach has limitations when applied in the wider spectrum as the chip rate and the receiver complexity need to be excessively increased. When there are few available chunks of bandwidth in the spectrum with good propagation characteristics, the TDD mode of LTE is a good candidate that can exploit larger, lower-cost spectrum but its operation is not as flexible or efficient as that of FDD, especially in two-way communications.

Carrier Aggregation (CA) is an advanced operating mode of LTE-Advanced, an evolved version of LTE, that can meet these situations that are commonly encountered in typical networks such as the 800, 1800, 2100, and 2600 MHz bands of service provider A shown in Fig. 9.33, which employ multiple adjacent or separate bands in the frequency spectrum. With CA, the receiver complexity increases linearly with the number or width of simultaneously supported bands. Control of the QoS can be facilitated by taking the channel condition and loading level of each band into account. When multiple bands are supported by an eNodeB, the higher ones would have smaller coverage.

CA is flexible enough to optimize the combined use of scattered bandwidths, called *component carriers* in LTE terminology. Each carrier can have a different width, and the uplink and downlink can have different numbers of carriers. It is not required that only a single cell receives or transmits the aggregated carriers. There are a number of important requirements for the CA operation, however. The number of uplink carriers should be equal to or less than the number of downlink carriers, and there is a maximum of five carriers. Therefore up to 100 MHz can be aggregated in either link. In the modem protocols, only the MAC and lower layers are aware of the CA operation. Note that PDCP and RLC are not influenced from CA but RRC is heavily involved in its operation. The basic relationship between the PCCs in the uplink and downlink is signaled via SIB2 of the RRC messages. It is not permitted for a re-transmission to be attempted in another HARQ process, let alone in another carrier.

When there are a number of cells transmitting or receiving their component carriers, one of the cells is appointed to be the Primary Cell (PCell), and the other cells become the Secondary Cells (SCell). A component carrier of PCell becomes the Primary Component Carrier (PCC) that is used to manage the RRC connection of all carriers of the PCell and the SCells. Carriers that are not the PCC are the Secondary Component Carriers (SCC). Therefore determination of PCC depends on the UE, not the eNodeB, and UEs can have their PCCs on different bands although they are connected to the same network. Once a PCC is assigned to a UE, it can only be changed in handover. When the cells share component carriers, the terms PCell and SCell are interchangeably used with PCC and SCC respectively. In the idle RRC mode, the UE receives system information from the downlink PCC. Only the PCell can transport the PUCCH, including the scheduling request, power headroom report, and other control information reporting the status of UE. The one PUCCH and one or more PUSCHs can be power controlled independently.

The transmission timing needs to be set for each carrier independently, although initial versions of CA did not support this. Note that a component carrier is not necessarily a band but an aggregation of subcarriers. Therefore, a band can have multiple component carriers that may or may not be contiguous. The former case is called intraband contiguous, and the latter case is called intraband non-contiguous. The intraband CA is easier to implement in the uplink as it does not require separate chains of radio signal processing. On the other hand, in the interband CA where carriers in different bands are aggregated, separate transmit chains are necessary. With the many advantages of carrier aggregation, the RF requirements, especially of UEs, become very challenging.

Fig. 9.34 3-channel carrier aggregation (courtesy of Accuver Co., Ltd.).

To facilitate the implementation and deployment of CA, pre-selected combinations of bands are provided.

The set of plots and text windows included in Fig. 9.34 shows the configuration and transmission status of an LTE system operating in the FDD mode, with three aggregated component carriers. From the values of E-UTRAN ARFCN (EARFCN), it can be seen that the carriers are in the 900, 1800, and 2100 MHz bands respectively, which matches the frequency allocation of service provider B shown in Fig. 9.33. It is shown that the PCell uses a carrier at 1800 MHz. Therefore, this band is likely to be the primary band that serves VoLTE or key services over the widest coverage. The plots also show the status of the components carriers used in the neighboring cells. The Reference Signal Received Power (RSRP) and Reference Signal Received Quality (RSRQ) are metrics that indicate the health of each carrier, corresponding to the RXLEV and RXQUAL values of GSM.

The plots included in Fig. 9.35 show the link quality and throughput of each carrier for the downlink, configured as in Fig. 9.34 for data services. In the second plot, it is shown that the channel condition of an SCell in 900 MHz, expressed in RSRP, is higher, perhaps due to its lower loading level or favorable propagation characteristics. In the third plot, it is shown that the BLERs of three carriers are maintained in similar ranges

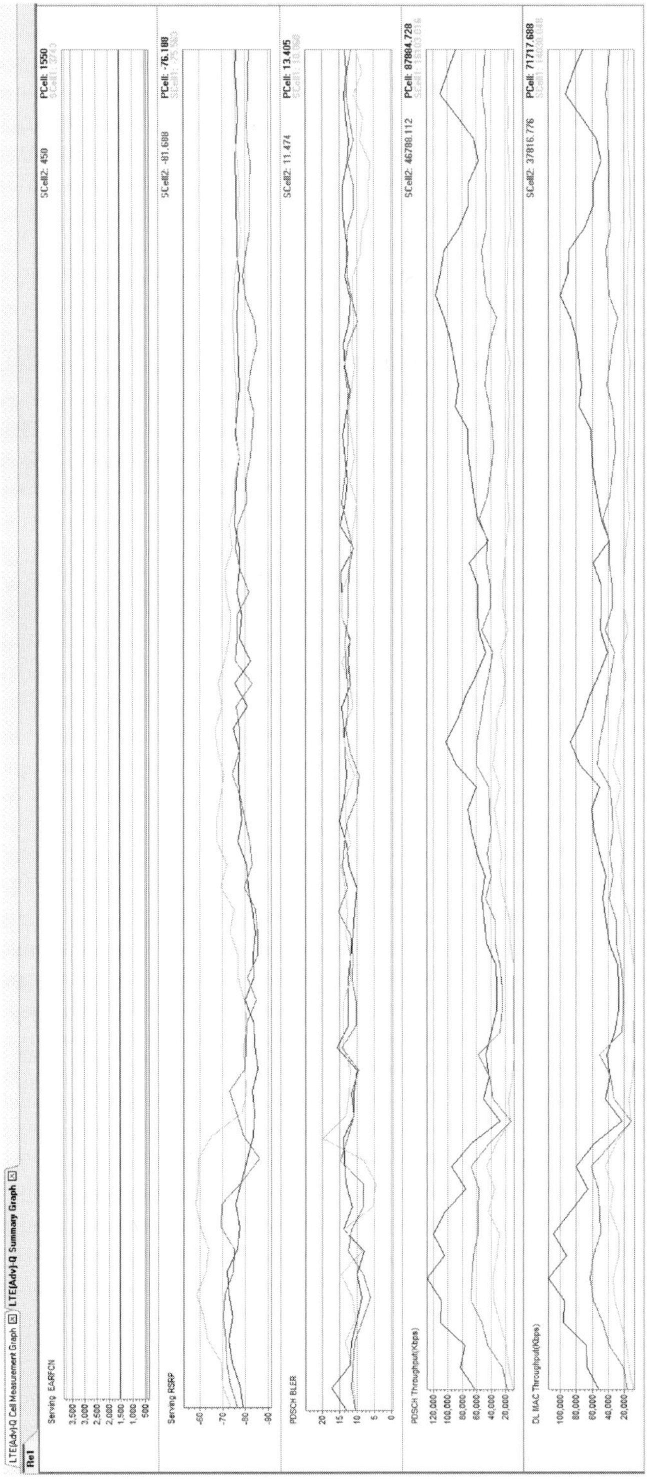

Fig. 9.35 CA performance analysis (courtesy of Accuver Co., Ltd.).

using various schemes for link adaptation. The fourth and fifth plots show the throughput of each carrier measured in the L2 and the physical layer, respectively. It can be seen that the PCell is the major workhorse in the transport of packets, followed by another SCell in the 2100 MHz band. The results are understandable as the PCell would carry the most traffic considering its wider bandwidth, and the SCell at 2100 MHz would outperform the other SCell in the 900 MHz band, as voice traffic, which reduces the scheduling efficiency due to its real-time constraint, is likely to be accommodated in the lower frequencies.

9.4.3 Recommendation of Media Bit-Rates

The application of H.324, which was originally designed for almost error-free fixed networks, to harsher mobile environments after some structural reinforcements, revealed an unexpected source of quality degradation, extended call setup delay, that necessitated the introduction of media oriented negotiation acceleration (MONA). Likewise, the introduction of ECN, which was designed as a crude measure of quality control for the Internet, to mobile environments, where precise control of radio resources is essential, showed the limitations of this technique. The ambiguity on the required actions, when packets marked with ECN-CE are received, was a key limitation that could not easily be fixed as the UE cannot be sure of where and by whom the ECN-CE was marked.

The eNodeB does not possess the authoritative measures of circuit-switched networks, such as the CMC or RATE_REDUC messages, because of architectural limitations. It cannot intervene in the session negotiation procedures managed by the IMS. So a new approach was taken to send control information that was clearer for the UE to understand, although not necessarily easier to follow, using signaling mechanisms whose origin could not be misunderstood. In this approach, the eNodeB can transmit media bit-rates recommended to each UE via the MAC control elements. This approach is more refined and personalized than the ECN, or the statistical control of the SSAC whose messages are broadcast and applicable to all UEs in the cell. However, with this approach, the level of enforcement is still no more than a recommendation, and whether or not to apply the recommended bit-rates via the SIP Invite or Update messages, or the RTCP packets is left to the discretion of the implementation or service policy.

There are other means to convey the information on recommended media bit-rates to the UEs, such as using the PDCP application messages, i.e., by *injecting* the TMMBR or RTCP-APP packets between packets carrying media frames at this protocol layer. However, the use of such approaches needs to be explicitly negotiated using the IMS, whose results are not visible to the eNodeB. On the other hand, use of the MAC control elements implies that the eNodeB cannot influence the actions of UE before the session begins. Another limitation of this approach is that the eNodeB is typically not intelligent enough to assess the impact on subjective quality from the change of media bit-rates it makes, as it is endowed only with the QoS information from the IMS.

9.5 Remote Management of Operation

In circuit-switched networks, the MS informs the BSC, to which it is indirectly attached via the BTS, of the list of all speech codecs and their bit-rates it supports, using standardized message formats. Then the BSC selects the codec it prefers to use, which must be supported by the local TRAU, and configures the bit-rates or operating modes of the chosen codec for both directions. In tandem free operation (TFO) or transcoder free operation (TrFO) of the UMTS, similar negotiation principles are used. The call-originating UE indicates the list of speech codecs it supports, which is conveyed to the terminating MSC that, after asking the far-end UE to select a preferred codec, conveys the information on the selected codec back to the originating MSC. Thus, introduction of a new speech codec requires the update of the MSCs, TRAUs, and the end-to-end signaling procedures. Figure 9.36 shows an example of the supported codec list for GSM and UMTS [3GPP (2017e)], which includes AMR and AMR-WB.

In Fig. 9.36, the Organization Identifier (OID) and the Codec Identifier (CoID) precede the configuration information of each codec. The list specifies the Supported Codec-mode Set (SCS) and Active Codec-mode Set (ACS) of AMR and AMR-WB. For AMR, one byte is enough to represent any set of codec modes to use. By setting the bit corresponding to each codec mode, the SCS and ACS can be configured. For AMR-WB, which includes nine codec modes, four bits are used to specify a pre-defined set of codec modes, Config-WB-Code, for SCS and ACS. For example, Config-WB-Code = 0 refers to the commonly-used low bit-rate set of 6.6, 8.85, and 12.65 kbps. Therefore, an arbitrary set of codec modes of AMR-WB can be used in VoLTE but not in W-CDMA. To signal the support of 3G-324M, a Multi-Media (MuMe) dummy codec can be included in the supported codec list. The MuMe dummy codec includes only a one-byte Bandwidth Multiplier (BWM), which represents the bit-rate in multiples of

	8							1
1	Codec List							
	Length Indication							
	Compatibility Information							
	Single Codec							
	Length Indication							
6	Compatibility Information							
	ETSI OID							
	UMTS_AMR2_CoID							
ACS	12.2	10.2	7.95	7.40	6.70	5.90	5.15	4.75
SCS	12.2	10.2	7.95	7.40	6.70	5.90	5.15	4.75
11	Spare	Spare	Spare	Spare	OM	MACS		
	Single Codec							
	Length Indication							
	Compatibility Information							
	ETSI OID							
16	UMTS_AMR-WB_CoID							
	Spare	Spare	Spare	Spare	Config-WB-Code			

Fig. 9.36 Supported codec list for UMTS.

64 kbps. As 3G-324M consumes 64 kbps for multiplexed speech and video, the BWM is set to 00000001.

These signaling methods are sufficient as each generation of circuit-switched networks supports only a few speech codecs whose signal processing procedures are well-matched to the radio signal processing procedures of the access networks. In 3G-324M, none of the NodeB, RNC, or MSC is involved in configuration of the media, which is initiated after the 64 kbps bearers are established between two UEs for the two directions. No inspection by the network is applied to the information on media capability exchanged. It may be concluded that circuit-switched networks manage the media configuration in a tight, but inflexible fashion, as all media bit-rates to use have to be precisely defined and registered to the network before service.

9.5.1 Session Negotiation Management

In VoLTE, reporting the codecs supported by the UE via a single message, as in UMTS, is not efficient, as a single codec can be configured or packetized in many ways. Since the contents of SDP become more complex, the session negotiations are more easily delayed and the results are less predictable. Furthermore, different configurations may have to be used, depending on the situations. For example, only basic codecs at low bit-rates may be allowed for roaming UEs while advanced codecs at high bit-rates or wide bandwidths, or multiple channels may be reserved when the UEs are attached to the home network. It may also be necessary to increase the bit-rate of existing codecs when the service plan is upgraded or the system bandwidth is extended, for example, from 10 to 20 MHz.

These new requirements necessitate the mechanisms to manage session negotiation (1) by selecting an SDP offer from several candidates or constructing an SDP answer based on some priority or decision logic so that the highest quality is achieved for the given situation, and (2) by remotely modifying the SDPs or decision logic stored in the UEs. Note that installing a new codec after the UEs have been manufactured and activated is not feasible in most cases even if typical speech codecs can be implemented in software and are downloadable over-the-air (OTA), as the conformation to terminal acoustics requirements cannot easily be confirmed without calibration. In the case of video, supporting a new video codec, which includes a significant amount of hardware, is not feasible.

While the two design requirements can be met using several approaches, in VoLTE, one of the simplest methods to update only the key parameters of the SDPs is provided as an option. To meet the first requirement above, suppose that the UE is equipped with SDPs designed for several situations. The UE may select an SDP, depending on, for example, the radio access network to which it is connected, information on the service provider broadcast by the eNodeB, date and time, in addition to the service policy. For example, when the UE is attached to GPRS, support of high-quality video may not be possible, considering the capability of the network.

In VoLTE, the key parameters of SDP can be updated using the Device Management (DM) protocol [OMA (2007)], in which the servers remotely manage key session parameters organized in a tree-like data structure, the Management Object (MO), stored

Fig. 9.37 Management object for session negotiation.

inside the UE. Each SDP parameter is mapped from a leaf of the tree structure. Figure 9.37 shows the structure of the Multimedia Telephony Service for IMS Network Preference (MTSINP) MO [3GPP (2017f)], which includes three media categories: speech, video, and text.

A format that matches the nature of the parameters it represents is assigned to each leaf. ID (integer) and TAG (character) are used to identify a set of session parameters.

Multiple sets of session parameters may be necessary for an SDP offer since multiple media configurations are typically included, as in Fig. 8.58. The IPver (character) node is defined to represent the version of the IP protocols that are used. For example, when packetized in the octet-aligned format for ptime = 20, the b = AS values of AMR-WB 23.85 kbps in IPv4 and AMR-WB 15.85 kbps in IPv6 are both 41. In this case, when different configurations result in the same b = AS value, IPver can be used to determine the codec mode.

The parameters related to the bit-rate, b = AS (integer), RS (integer), and RR (integer), are grouped under the Bandwidth node. RateSet (char) represents the set of bit-rates to be used for the speech codec specified in the Codec (char) node. In AMR and AMR-WB, the bit-rates contained in RateSet are translated to the codec modes included in the mode-set parameters of the SDP offer or answer. If the Codec and RateSet nodes are set to "AMR-WB" and "0,1, 2", respectively, only the first three codec modes of AMR-WB, 6.6, 8.85, and 12.65 kbps, can be used in the session. Note that these standardized parameters enable only basic configurations of AMR and AMR-WB, excluding such configurations as packetizing at a ptime other than 20. The parameters related to more sophisticated session negotiations can be located under the Ext node. The configuration of the EVS speech codec, or video and text can be managed in a similar fashion. The DM protocol and the structure of MTSINP illustrated in Fig. 9.37 are not the only way to manage session negotiation. With the flexibility of contemporary mobile operating systems and computational resources available at the UE, not just the parameters of SDP but the complete session negotiation logic may be updated.

9.5.2 Media Adaptation Management

In some sense, media adaptation of VoLTE is conceptually close to the link adaptation of the GSM operating AMR, as both schemes enable the MSs or UEs to influence the far-end source encoding to match time-varying channel conditions. The control messages for media adaptation, RTCP-APP for speech or TMMBR for video, roughly correspond to CMC/CMR for the codec mode adaptation. In GSM, the signaling methods for BTS to control the MS are defined in detail and enforced as mandatory to all MSs supporting AMR. Moreover, control messages are provided for network management, such as RATSCCH, THRESH_REQ, or AMR_CONFIG_REQ, which can be used to modify the behavior of the adaptation. With the messages, it is possible to change the codec modes, thresholds, and even the hysteresis values, when it is necessary to reconfigure the link adaptation algorithms. When the MS is handed over to the BTS of another operator, or the traffic pattern is changed, the link adaptation algorithms can be reconfigured to meet the new need for quality and capacity.

In VoLTE, RTCP-APP and TMMBR are provided for similar objectives. However, the format and usage of these messages are more complex than those in GSM, which consist of just a few bits. The link adaptation of GSM is based on a normalized, one-dimensional metric for channel quality, the quality indicator. In contrast, the media adaptation of VoLTE needs to be controlled based on the end-to-end transmission results and, therefore, requires more parameters to describe the adaptation algorithms. However, the media adaptation algorithms are not specified in the VoLTE standards to facilitate the introduction of new channel estimation and media adaptation techniques.

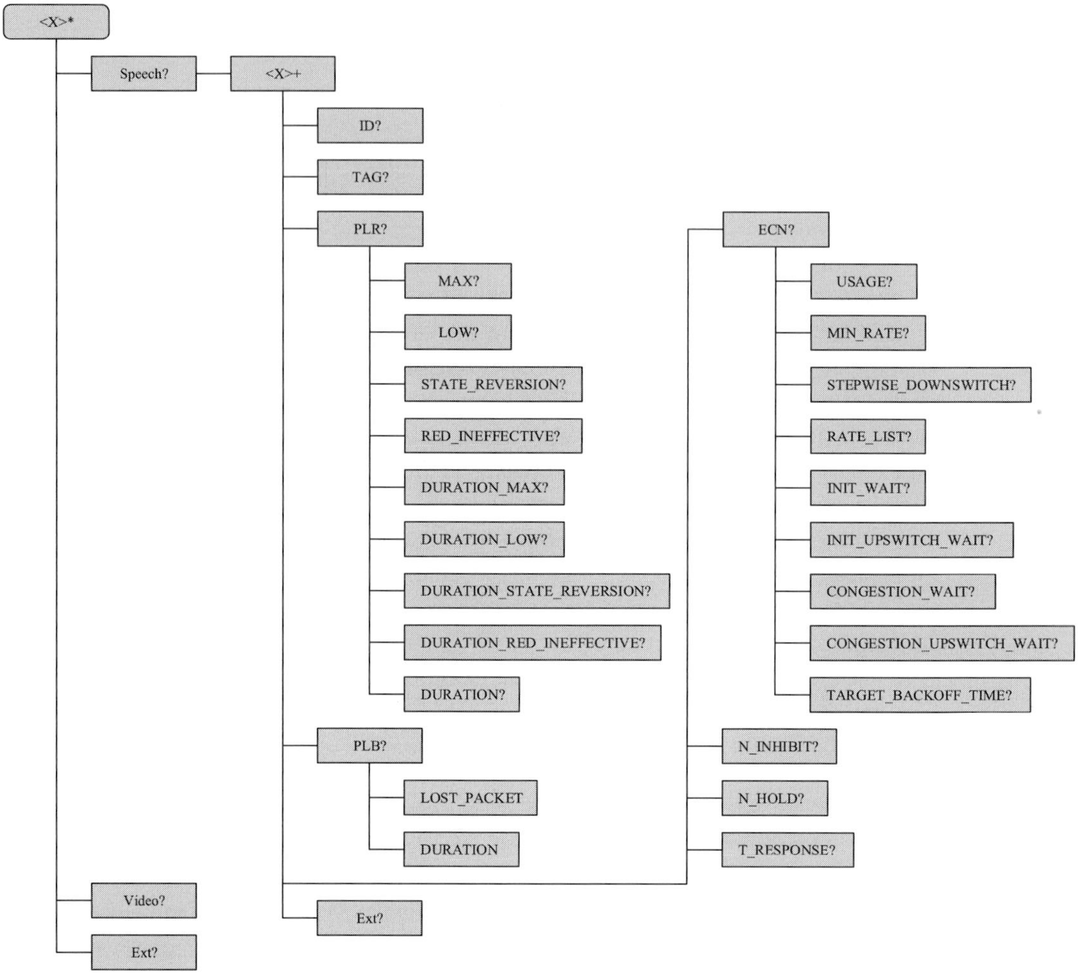

Fig. 9.38 Management object for media adaptation.

The media adaptation algorithms for speech and video might consist of parameters with different origins, e.g., parameters determined in session negotiation, parameters related to the status of transmission and reception, parameters measured during session, and others depending on the implementations. If these parameters are standardized, flexibility of implementation or the room for improvement would be limited. Instead, VoLTE only introduces a set of parameters for each media, which can be taken into account when constructing the media adaptation algorithms. Such a set may include parameters denoting packet loss rate and the period over which the rate is computed, parameters describing the conditions of certain events, target quality values, or the limits adaptation algorithms should avoid exceeding or falling short of, and parameters for the bit fields of RTCP-APP or TMMBR. As in the management of session negotiation, these parameters are mapped to the nodes of a tree-like structure, the Multimedia Telephony Service for IMS Media Adaptation (MTSIMA) MO [3GPP (2017f)], which can be managed with the DM protocol. As shown in Fig. 9.38, the media adaptation algorithms can be built with some off-the-shelf parameters provided in the MO, and other implementation-specific parameters located under the Ext node.

Under this architecture for management, media adaptation algorithms may not be altered completely once the UE is manufactured and activated. Instead, the conditions for state transitions or arguments of messages can be updated. Note that updating the MO is a time-consuming procedure that cannot be used to handle urgent situations, unlike the GSM messages for link adaptation whose effects can be reflected within a few speech frames. Therefore, in VoLTE, it is only possible for the network to control the media bit-rate in an indirect and non-real-time fashion. In contrast to the master and slave relationship between a BTS and an MS, which is clear from the notations of the *commands* transmitted by the BTS and the *requests* transmitted by the MS, there is no such vertical relationship between two VoLTE UEs. Therefore, received messages for media adaptation can be turned down if the local conditions do not allow their execution. For example, when the received RTCP-APP or TMMBR message asks a UE to increase the bit-rate, the UE may not accept the request if more data than can be transmitted is left in the transmitter buffer. Tables 9.11 and 9.12 summarize the adaptation parameters for speech and video, shown as the nodes of MTSIMA. PLR and PLB represent the packet loss rate and packet loss burst respectively.

9.6 Performance Evaluation

In the typical deployments of W-CDMA and cdma2000, a few channels in the lower frequency bands are assigned for voice services only, leaving the other higher frequency bands equipped with HSPA and EV-DO, for packet data services. In these situations, the objective of network operation is to maximize voice capacity while meeting the minimum quality requirements with efficient power control and call admission control algorithms. Although the quality at certain bit-rates, such as AMR 12.2 kbps or EVRC Operating Mode 0, is considered to be of the toll quality for narrowband voice services, measures are taken to reduce the maximum or average bit-rate when the cell loading level approaches the maximum capacity or certain MSs at the cell edge require excessive transmit power, with control messages such as RATE_REDUC of cdma2000 or CMC of W-CDMA.

As the number of MSs or UEs connected to these circuit-switched networks decreases, their channels are gradually reduced although they are not re-used for the time being by new networks, to control the operating cost. The remaining networks are also re-configured to maximize the capacity. Therefore some of the networks whose performance was analyzed in previous chapters were no longer operational, or used with different configurations for their media codecs, by the time advanced versions of LTE became operational. New media codecs are often deployed at low bit-rates, such as the SC-VBR mode of EVS over W-CDMA, to squeeze the capacity out of old networks as much as possible.

In LTE, as the bandwidth is shared not only by voice services but also by data services with QoS requirements that are different from those of speech or video calls, estimating the maximum voice capacity assuming that all UEs in the cell are using only voice services makes less sense. Nevertheless, estimation of the capacity may provide a sense of

Table 9.11 Speech adaptation parameters.

Name	Format	Definition
PLR/MAX	Float (%)	Maximum PLR tolerated when redundancy is not used (before receiver signals sender to attempt adaptation or operate at bit-rates more robust to packet loss)
PLR/LOW	Float (%)	Minimum PLR tolerated before receiver signals sender for higher bit-rate, reduce redundancy, or perform other procedures exploiting favorable conditions
PLR/STATE_REVERSION	Float (%)	Maximum PLR tolerated after media adaptation is activated (once threshold is exceeded, receiver decides adaptation was not effective)
PLR/RED_INEFFECTIVE	Float (%)	Maximum PLR tolerated after redundancy is increased (once threshold is exceeded, receiver decides situation was not improved but degraded)
PLR/DURATION_MAX	Integer (ms)	Duration of sliding window over which PLR is computed and compared against MAX
PLR/DURATION_LOW	Integer (ms)	Duration of sliding window over which PLR is computed and compared against LOW
PLR/DURATION_STATE_REVERSION	Integer (ms)	Duration of sliding window over which PLR is computed and compared against STATE_REVERSION
PLR/DURATION_RED_INEFFECTIVE	Integer (ms)	Duration of sliding window over which PLR is computed and compared against RED_INEFFECTIVE
PLB/LOST_PACKET	Integer	Number of packets lost during a period of PLB/DURATION
PLB/DURATION	Integer (ms)	Period of PLB event for which LOST_PACKET is counted
ECN/USAGE	Boolean	Switch to turn on or off ECN-based adaptation
ECN/MIN_RATE	Integer (bps)	Minimum bit-rate allowed when ECN is used (excluding IP overhead)
ECN/STEPWISE_DOWNSWITCH	Boolean	Switch between direct or step-wise bit-rate reduction
ECN/RATE_LIST	Character	List of bit-rates speech encoder should use during ECN-based stepwise rate reduction
ECN/INIT_WAIT	Integer (ms)	Time sender should wait when ECN is used (before initial up-switch is attempted)
ECN/INIT_UPSWITCH_WAIT	Integer (ms)	Time sender should wait when ECN is used (before each up-switch is attempted)
ECN/CONGESTION_WAIT	Integer (ms)	Minimum interval receiver should wait (between ECN-CE detection and up-switch)
ECN/CONGESTION_UPSWITCH_WAIT	Integer (ms)	Time sender should wait (before up-switch from each step is attempted)
ICM/INITIAL_CODEC_RATE	Integer (bps)	Bit-rate speech encoder should use when session begins (excluding IP overhead)
ICM/INIT_WAIT	Integer (ms)	Time sender should wait before bit-rate is increased (after session begins) if no information on rate control is available or reception quality feedback is received
ICM/INIT_UPSWITCH_WAIT	Integer	Time sender should wait before up-switch is attempted (after session begins)
N_INHIBIT	Integer	Period (number of speech frames) for which adaptation is disabled to avoid ping-pong effects
N_HOLD	Integer	Integer used to compute an integer multiple of other period
T_RESPONSE	Integer (ms)	Maximum period receiver should wait before concluding transmitted request is not granted by far-end

Table 9.12 Video adaptation parameters.

Name	Format	Definition
PLR/MAX	Float (%)	Maximum PLR tolerated before receiver signals sender to reduce bit-rate
PLR/LOW	Float (%)	Minimum PLR before receiver decides situations have improved and signals sender for higher bit-rate
PLR/DURATION_MAX	Integer (ms)	Duration of sliding window over which PLR is computed and compared against MAX
PLR/DURATION_LOW	Integer (ms)	Duration of sliding window over which PLR is computed and compared against LOW
PLB/LOST_PACKET	Integer	Number of RTP packets lost during a period of PLB/DURATION
PLB/DURATION	Integer (ms)	Period of PLB event for which LOST_PACKET is counted
MIN_QUALITY/BIT_RATE/ABSOLUTE	Float (kbps)	Minimum bit-rate video encoder should use
MIN_QUALITY/BIT_RATE/RELATIVE	Float (kbps)	Minimum bit-rate video encoder should use (as a proportion of negotiated b = AS)
MIN_QUALITY/FRAME_RATE/ABSOLUTE	Float (fps)	Minimum frame-rate video encoder should use
MIN_QUALITY/FRAME_RATE/RELATIVE	Float (fps)	Minimum frame-rate video encoder should use (as a proportion of maximum rate negotiated codec level limits)
MIN_QUALITY/QP/H263	Integer	Maximum luminance quantization parameter (QUANT) video encoder should use if H.263 is negotiated
MIN_QUALITY/QP/MPEG4	Integer	Maximum luminance quantization parameter (quantiser_scale) video encoder should use if MPEG-4 is negotiated
MIN_QUALITY/QP/H264	Integer	Maximum luminance quantization parameter (QP_Y) video encoder should use if H.264 is negotiated
ECN/MIN_RATE_ABSOLUTE	Float (kbps)	Minimum bit-rate video encoder should use when ECN is used (excluding IP overhead)
ECN/MIN_RATE_RELATIVE	Float (kbps)	Minimum bit-rate video encoder should use when ECN is used (as a proportion of b = AS, excluding IP overhead)
ECN/STEP_UP	Integer (%)	Proportion of current video encoding rate (used to ask for an increase by this value)
ECN/STEP_DOWN	Integer (%)	Proportion of current video encoding rate (used to ask for a decrease by this value)
ECN/INIT_WAIT	Integer (ms)	Time sender should wait when ECN is used (before initial up-switch is attempted)
ECN/INIT_UPSWITCH_WAIT	Integer (ms)	Time sender should wait when ECN is used (before each up-switch is attempted)
ECN/CONGESTION_WAIT	Integer (ms)	Minimum interval receiver should wait (between ECN-CE detection and up-switch)
ECN/CONGESTION_UPSWITCH_WAIT	Integer (ms)	Time sender should wait (before up-switch from each step is attempted)

Table 9.12 *(cont.)*

RTP_GAP	Integer (ms)	Maximum period between RTP packets tolerated (before measures are taken to reduce)
INC_FBACK_MIN_INTERVAL	Integer (ms)	Minimum interval at which TMMBR should be sent for bit-rate increase
DEC_FBACK_MIN_INTERVAL	Integer (ms)	Minimum interval at which TMMBR should be sent for bit-rate decrease
TP_DURATION_HIGH	Integer (ms)	Maximum duration of sliding window over which interval between packet arrival and play-out is observed
TP_DURATION_MIN	Integer (ms)	Minimum duration of sliding window over which interval between packet arrival and play-out is observed
TARGET_PLAYOUT_MARGIN_HI	Integer (ms)	Maximum interval between packet arrival and scheduled play-out
TARGET_PLAYOUT_MARGIN_MIN	Integer (ms)	Minimum interval between packet arrival and scheduled play-out
RAMP_UP_RATE	Float (kbps)	Bit-rate by which video encoder increases bit-rate
RAMP_DOWN_RATE	Float (kbps)	Bit-rate by which video encoder decreases bit-rate
DECONGEST_TIME	Integer (ms)	Time receiver commands sender to spend in de-congesting transmission path
HOLD_DROP_END	0, 1, 2	Switch determining actions when video quality does not meet requirements
INITIAL_CODEC_RATE	Float (kbps)	Bit-rate (as a proportion of negotiated b = AS) video encoder should use when session begins
X_PERCENTILE	Float	Percentile point of packet arrival distribution (used with TARGET_PLAYOUT_MARGIN parameters)

the effectiveness of the radio signal processing procedures of LTE. Moreover, as multiple bands can be simultaneously supported in the evolved versions of LTE such as the LTE-Advanced (LTE-A) Pro networks, it is likely that most voice traffic will be transported over the lowest bands, because of the advantages in coverage and handover at lower frequencies, following a similar pattern of bandwidth usage as in circuit-switched networks.

The following sections evaluate the quality and capacity of VoLTE. The estimates are based on simulations and are close to the results observed in commercial deployments. In any estimation, the downlink capacity is larger than the uplink capacity, because of the intrinsic advantage of the downlink in packet-switched networks. In addition to the information on channel conditions reported by the UE, the eNodeB can also exploit the downlink buffer status of each UE in its scheduling. In the uplink, the buffer status needs to be reported to the eNodeB in any scheduling request, requiring additional signaling over the air not required in the downlink. In circuit-switched networks, the capacity of the uplink tends to be larger as the power amplifier of the BTS usually reaches its limit when the MSs in the cell can still increase their transmit power.

The results from live VoLTE calls, measured from commercially operational networks, are presented to visualize the dynamic operation of their media and radio signal processing procedures. It can be seen that the voice sessions using AMR-WB were run on the early versions of LTE networks where the carriers were not aggregated. It is also shown that video sessions and voice sessions using EVS were run on the more advanced LTE networks with 3 CA operation. However, a proper QoS as negotiated is maintained in each session.

9.6.1 Speech Compression and Transmission Performance

In this section, the compression efficiency and error resilience of the speech codecs introduced to VoLTE are compared in diverse channel conditions and input signal types. With the multi-bitrate and multi-bandwidth nature of EVS, its performance is evaluated across a wide range of bit-rates and audio bandwidths. In the testing for error-prone channel conditions, the bandwidth is limited to wideband. In the tests, the term Direct Source (DS) refers to the quality of an uncompressed and error-free input signal. Notice that in the codec configurations for EVS-NB, EVS-WB, and EVS-SWB, only the maximum audio bandwidth is set to NB, WB, and SWB, respectively. Therefore it is not impossible that the EVS-SWB speech encoder outputs frames encoded in NB or WB, depending on the input signal. The performance of codecs was subjectively determined using methods specified in [ITU-T (1996)]. DTX was activated in the following tests. Detailed analysis of the tests can be found in [3GPP (2017b)].

Clean Speech under Clean Channel Conditions

In this experiment, the performance of AMR, AMR-WB, and EVS is evaluated for clean speech in clean channel conditions. Therefore, noise is present neither in the input nor in the channel. The test is conducted in North American English. For EVS, the performance is measured for NB, WB, and SWB. Figure 9.39 shows the performance of EVS

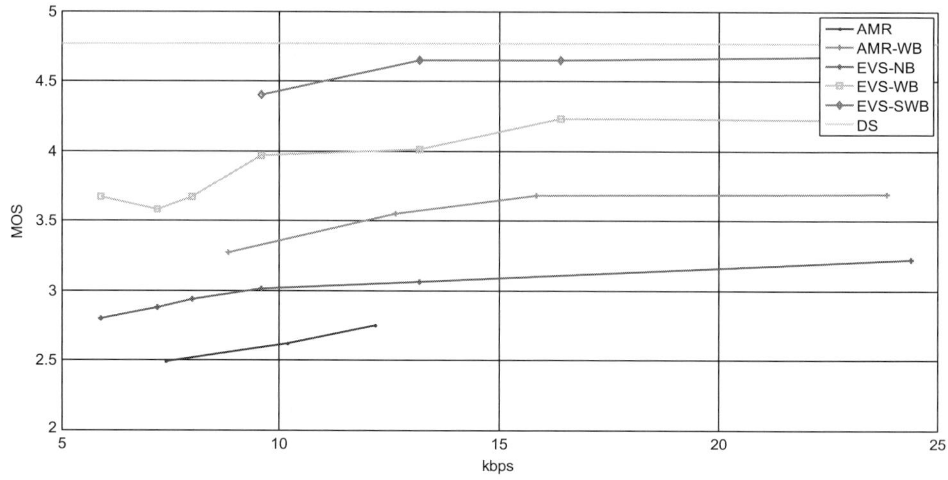

Fig. 9.39 Test results of AMR, AMR-WB, and EVS (for clean English in clean channel conditions).

from 5.9 to 24.4 kbps, for the three audio bandwidths. AMR and AMR-WB are tested from 7.4 to 12.2 kbps, and from 8.85 to 23.85 kbps, for NB and WB respectively.

It can be seen that EVS in NB performs significantly better than AMR over the range of bit-rates compared. Even at the lowest bit-rate, average 5.9 kbps of SC-VBR, EVS achieves subjective quality comparable to that of AMR at its highest bit-rate, 12.2 kbps. EVS operating in NB at its highest bit-rate tested, 24.4 kbps, approaches the quality of AMR-WB at its lowest bit-rate tested, 8.85 kbps, which clearly shows the limitation in quality improvement when the audio bandwidth is limited. In WB, EVS outperforms AMR-WB by a large margin at each bit-rate. SC-VBR exceeds the quality of AMR-WB at 8.85 kbps, approaching the quality of AMR-WB at 12.65 kbps. Note that SC-VBR exceeds the quality of EVS in WB at 7.2 kbps, and matches the quality of EVS in WB at 8 kbps, showing that the VBR can be effective at low bit-rates, with a clean input and a clean channel.

The quality of EVS in WB at 9.6 kbps is significantly higher than that of AMR-WB at its highest bit-rate. It can be seen that the quality of EVS in WB increases in a steady fashion from 7.2 to 16.4 kbps, and enters a state of quality saturation beyond this bit-rate. Finally, in SWB, EVS exhibits quality significantly higher than that of either EVS in WB or AMR-WB. EVS in SWB at 9.6 kbps exceeds the quality of both AMR-WB 23.85 kbps and EVS in WB at 24.4 kbps. At 13.2 kbps, EVS in SWB approaches the quality of DS and negligible improvement is observed when the bit-rate is increased further. It can be concluded that AMR-WB achieves much higher compression efficiency than AMR, and EVS in WB or SWB outperforms AMR-WB significantly over the entire bit-rate range.

Noisy Speech under Clean Channel Conditions

In this experiment, the performance of AMR, AMR-WB, and EVS is evaluated for noisy speech in clean channel conditions. The test is conducted in Finnish, and the background

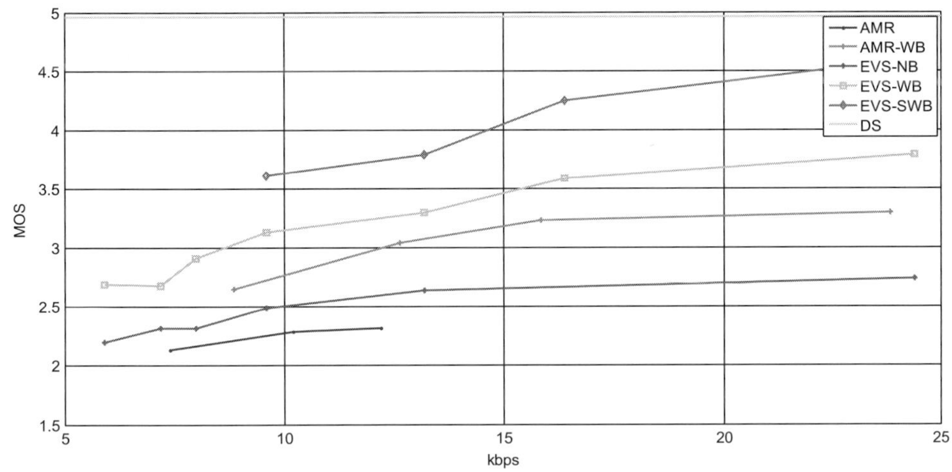

Fig. 9.40 Test results of AMR, AMR-WB, and EVS (for noisy Finnish in clean channel conditions).

noise is car noise at an SNR of 20 dB. It can be seen in Fig. 9.40 that EVS in NB performs significantly better than AMR over the range of bit-rates compared. At 7.2 kbps, EVS in NB achieves subjective quality comparable to that of AMR at its highest bit-rate, 12.2 kbps. It is seen that the quality of EVS increases in a steady fashion from 5.9 to 24.4 kbps.

In WB, EVS outperforms AMR-WB significantly over the entire bit-rate range. SC-VBR in WB achieves subjective quality comparable to that of AMR-WB at 8.85 kbps. EVS in WB at 13.2 kbps matches the quality of AMR-WB at the highest bit-rate. It is seen that the quality of EVS in WB increases in a steady fashion from 7.2 to 24.4 kbps but enters a state of quality saturation beyond this bit-rate. Finally, in SWB, EVS exhibits quality beyond that of either EVS in WB or AMR-WB for any bit-rate. For example, EVS in SWB at 13.2 kbps exceeds the quality of AMR-WB at 23.85 kbps. The quality of EVS in SWB increases gradually from 9.6 to 24.4 kbps, approaching the quality of DS. It can be seen that higher audio bandwidth increases not only the quality but also the resilience against acoustic noise, as the spectrum spreading of CDMA increases the resilience against multi-path fading.

Music and Mixed Content under Clean Channel Conditions

In this experiment, the performance of AMR, AMR-WB, and EVS is evaluated for music/mixed contents in clean channel conditions. The test is conducted in German. Although SC-VBR is designed to achieve an average bit-rate of 5.9 kbps for active speech, the mode may result in different average bit-rates depending on the acoustic nature of input signals. For NB and WB, the measured average bit-rates of SC-VBR are 7.11 and 7.68 kbps, respectively, as the music detector inside the EVS speech encoder assigned more 8 kbps frames for this type of input signal.

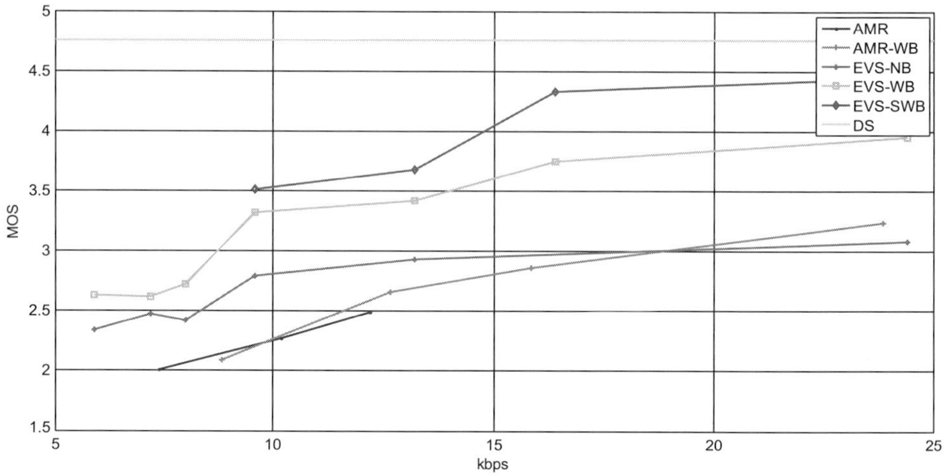

Fig. 9.41 Test results of AMR, AMR-WB, and EVS (for music and mixed contents in clean channel conditions).

As shown in Fig. 9.41, in NB, EVS performs significantly better than AMR over a wide range of bit-rates. The quality of EVS in NB at 7.2 kbps exceeds that of AMR at 7.4 kbps. From 9.6 kbps, EVS in NB outperforms AMR at any bit-rate compared. Moreover, up to 13.2 kbps, EVS in NB performs equally to or better than even AMR-WB at comparable bit-rates. The results show that depending on the contents, an NB speech codec with advanced compression algorithms can outperform even a WB codec.

In WB, the quality of EVS is superior to that of AMR-WB at any comparable bit-rates. EVS in WB at 13.2 kbps achieves subjective quality not matched by any bit-rate of AMR-WB. In fact, EVS in WB at 9.6 kbps achieves higher subjective quality than AMR-WB at any bit-rate. In SWB, the gaps between EVS and AMR-WB or EVS in lower audio bandwidths are expanded. The subjective quality of EVS in SWB at 13.2 kbps is significantly higher than that of AMR-WB at 23.85 kbps. EVS in SWB at 16.4 kbps achieves quality beyond that of AMR-WB at any bit-rates or EVS in WB at up to 24.4 kbps. The quality of EVS increases in a steady fashion from 9.6 to 24.4 kbps, approaching the quality of DS.

Noisy Speech under Error-prone Channel Conditions

In this experiment, the performance of AMR-WB and EVS in WB is evaluated for clean speech under error-prone channel conditions. The test is conducted in Spanish, and the background noise is office noise at an SNR of 20 dB. As shown in Fig. 9.42, EVS in WB outperforms AMR-WB at frame erasure rates (FER) of 3% and 6%, at each bit-rate. In particular, EVS in WB at 6% FER exhibits comparable performance to AMR-WB at 3% FER. It can be seen that the built-in JBM and PLC of EVS increase its error resilience significantly.

To evaluate the performance of JBM, EVS in WB at 13.2 kbps and AMR-WB at 15.85 kbps are tested for six delay and error profiles whose characteristics are outlined in

Table 9.13 Delay and error profiles.

Profile	Characteristics	Packet loss rate (%)
1	Low-amplitude, static jitter characteristics, 1 frame/packet	0
2	Hi-amplitude, semi-static jitter characteristics, 1 frame/packet	0.24
3	Low/high/low amplitude, changing jitter, 1 frame/packet	0.51
4	Low/high/low/high amplitude, changing jitter, 1 frame/packet	2.4
5	Moderate jitter with occasional delay spikes, 2 frames/packet	5.9
6	Moderate jitter with severe delay spikes, 1 frame/packet	0.1

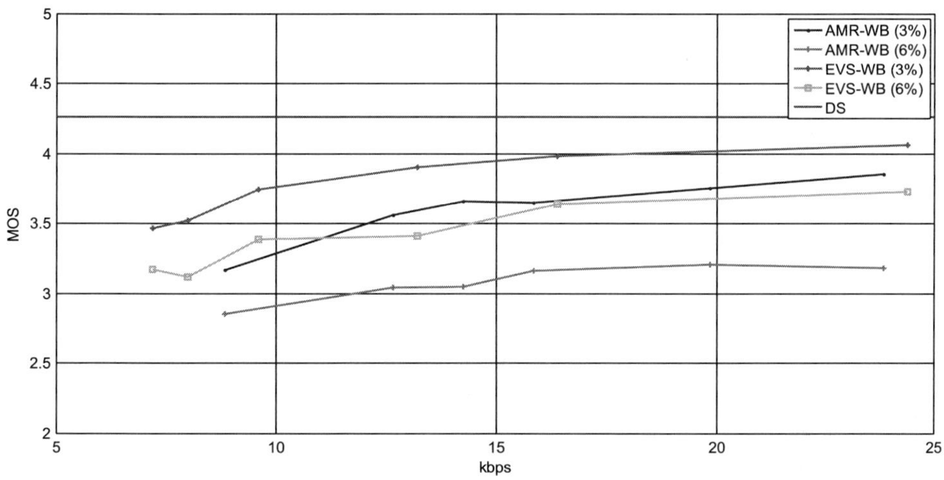

Fig. 9.42 Test results of AMR-WB and EVS (for clean Spanish in error-prone channel conditions).

Table 9.13. Profiles 4 and 5 correspond to abnormal operating conditions of LTE where the QoS is not properly managed, but the conditions are not so excessive in applications such as the Mission Critical Push To Talk (MCPTT) [3GPP (2017g)]. In the sidelink [3GPP (2017j)] operation of MCPTT, in which the networks are absent or not functional, speech frames are transmitted between two or more nearby UEs, not traversing any network node.

Figure 9.43 shows that EVS in WB at 13.2 kbps consistently achieves higher quality than AMR-WB at 15.85 kbps. The gap in quality increases if higher bandwidths of EVS are used. Under unfavorable conditions, i.e., when noisy speech, or music and mixed contents are considered, EVS in WB always exhibits superior performance [3GPP (2017b)]. The higher error-resilience of EVS can be exploited for reduced probability of triggering SRVCC, by setting lower SNR thresholds for switching networks, as the handover to previous networks where codecs with narrower audio bandwidth are used would reduce the quality more noticeably.

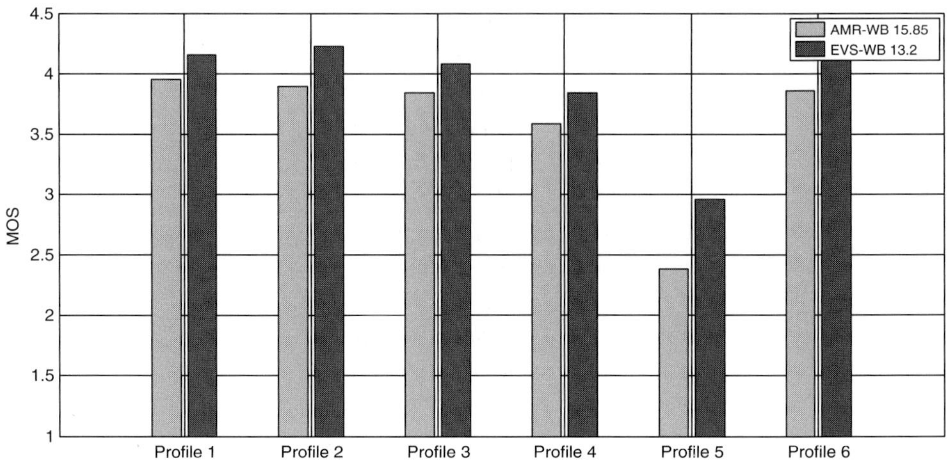

Fig. 9.43 JBM performance of AMR-WB 15.85 kbps and EVS 13.2 kbps.

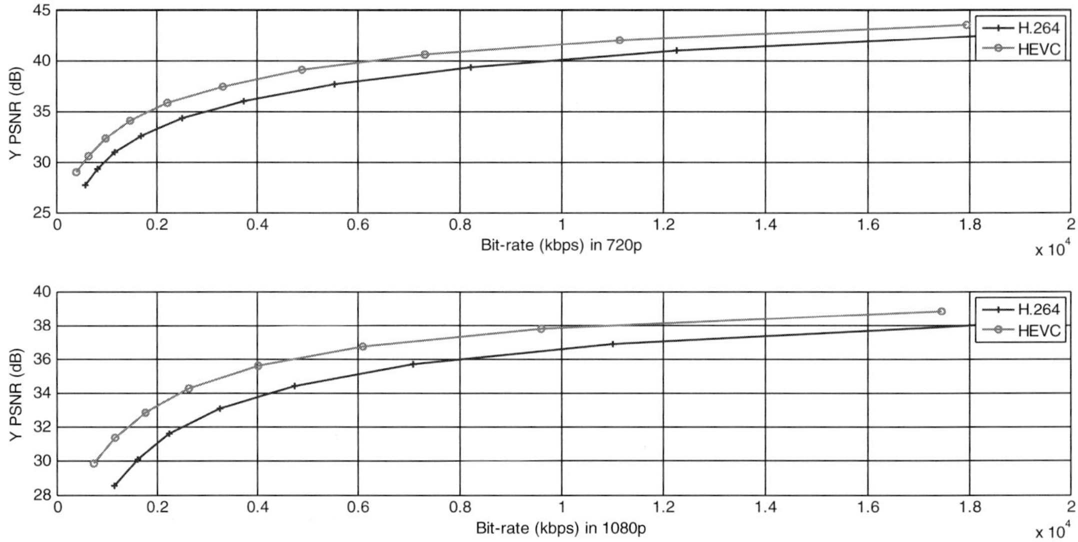

Fig. 9.44 Comparison of (luminance) PSNR.

9.6.2 Video Compression and Transmission Performance

Figure 9.44 compares plots of PSNR versus bit-rate for a typical sequence, *Basketball Drive*, generated with the H.264 and HEVC video codecs for resolutions of 720p (1280 × 720) and 1080p (1920 × 1080), over a wide range of bit-rates. The test conditions are explained in [3GPP (2017a)], where the quantizer parameter (QP) values are fixed and the previous two pictures in the decoding order are used for inter-frame prediction to reduce the uncertainty from applying different rate-control algorithms. The frame rate is set to 50 fps for both 720p and 1080p. After the first frame is encoded as an IDR picture, the following frames are encoded as P pictures, assuming clean channel.

It can be seen that HEVC achieves quality comparable to H.264, albeit using on average 40–45% lower bit-rates. However, since the gain from using HEVC for bit-rates up to 1–2 Mbps range is not significant, 1–2 dB at most, replacing H.264 with HEVC does not automatically result in enhanced quality or capacity if the same low bit-rate is maintained. Therefore H.264 is typically employed for services whose bit-rate is 300–500 kbps and resolution is 640 × 480 or lower, to compress typical head-and-shoulder video sequences. HEVC exhibits its superior compression efficiency when the bit-rate is higher than the range and the resolution is beyond 720p. Note that in the experiments, performances for error-prone channels are not considered, as the packet loss rate of LTE can be controlled, and the bit-rate of interest is high enough such that an IDR can be encoded in less than a Round Trip Time (RTT), preventing the effects of packet loss from propagating.

In [Ohm *et al.* (2012)], the compression performances of H.263, MPEG-4, H.264, and HEVC are compared for wide bit-rate ranges and diverse input types. It is shown that on average HEVC can reduce the bit-rate by 67.9%, 72.3%, and 40.3% for *interactive* applications, compared with H.263, MPEG-4, and H.264, respectively. With the interactive applications considered in the test, it was assumed that all pictures were coded in the order of display, as in video telephony or conference. Therefore only the first picture is encoded as an IDR and all subsequent pictures are P pictures. In VoLTE, an IDR picture is typically encoded every 2–3 seconds to limit error propagation. Note that the compression efficiency of video codecs needs to be taken with caution. In Fig. 6.40, it was shown that the PSNR values can also depend on the complexity of video encoder even when the same contents are compressed using the same video codec standard, which is not valid in the case of speech.

9.6.3 Live Session Analysis

Figure 9.45, consisting of four plots, shows the time-varying parameters of a live VoLTE speech session using AMR-WB 23.85 kbps. The first and second plots show the BLER values of the downlink and uplink measured at the physical layer, when the cell is not loaded and the link quality is stable. It can be seen that the average BLER of the downlink is about 0.5% while the average BLER of the uplink is close to 10%. Therefore, more frequent re-transmissions are required in the uplink.

It can also be seen that the target BLER of the uplink is set to a higher value so that the HARQ is used several times for each speech frame, to reduce transmit power and suppress the interference. In the downlink, a lower BLER can be achieved by transmitting packets when the link quality is favorable. As one of the essential criteria for the negotiated QoS, the target BLER is maintained by the eNodeB within acceptable ranges using packet scheduling and power control. The cell used in the experiments is loaded with a small number of UEs, and the FER of the downlink is maintained at a negligible level.

The next two plots visualize the fluctuations of throughput, i.e., the bit-rate, measured at each radio protocol layer for the downlink and uplink. From the rise and fall of the throughput values, it can be seen that the speech frames are periodically generated and

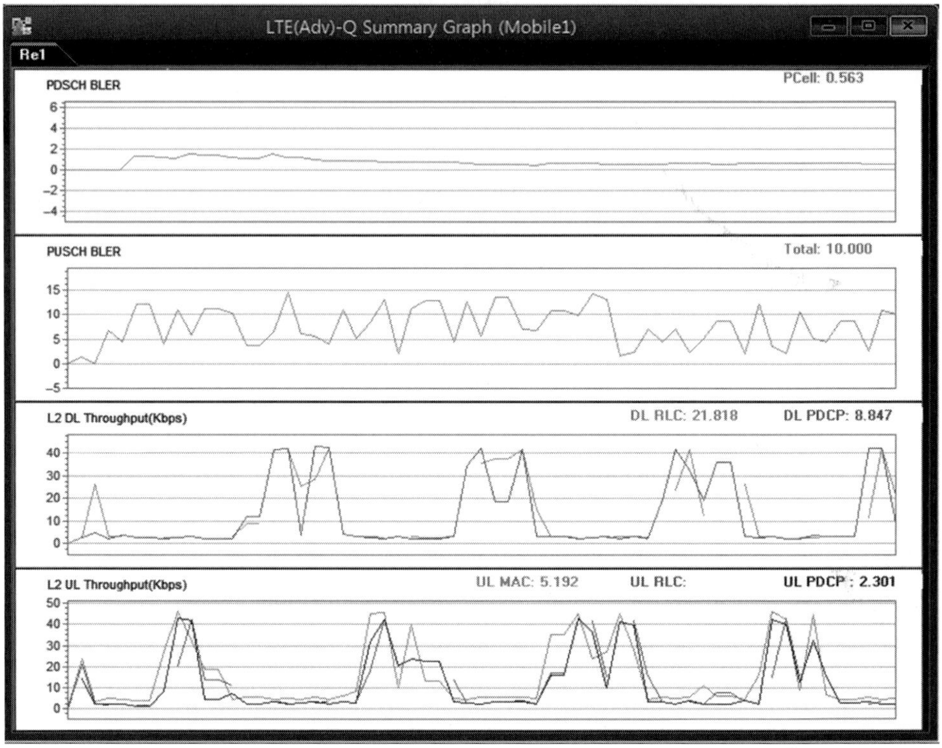

Fig. 9.45 VoLTE voice session analysis for L2 (courtesy of Accuver Co., Ltd.).

transmitted. The peak throughput of over 40 kbps can be considered as the b = AS for AMR-WB 23.85 kbps, which is reduced by the robust header compression (ROHC) at the PDCP but increased by the headers of PDCP, RLC, MAC, and control information, such as the scheduling request and buffer status report.

Figure 9.46 shows the time-varying parameters in similar conditions to Fig. 9.45. The first two plots show the BLER values of the downlink and uplink measured at the physical layer. The next two plots visualize the fluctuations of the throughput measured at the layer. In Fig. 9.45, the throughput curves measured at the protocol layer are quite similar, although the BLER of the uplink is much higher. This similarity can be expected as there is no re-transmission in the radio protocol layers, with the RLC set to the unacknowledged mode. The shape of the curve is dominated by the bit-rate of the speech encoder. On the other hand, in Fig. 9.46, the throughput curves are noticeably different, as the throughput in the uplink is higher and non-zero even when there are no speech frames to transmit. The re-transmissions from HARQ, which can be measured only in the physical layer, seem to have increased the throughput of the uplink. The throughput curve of the downlink measured at the physical layer is similar to those measured at the radio protocol layers.

From the expensive lessons learned from the initial deployments of 3G-324M, whose combined call set-up delay often exceeded 10 seconds and thereby necessitated the

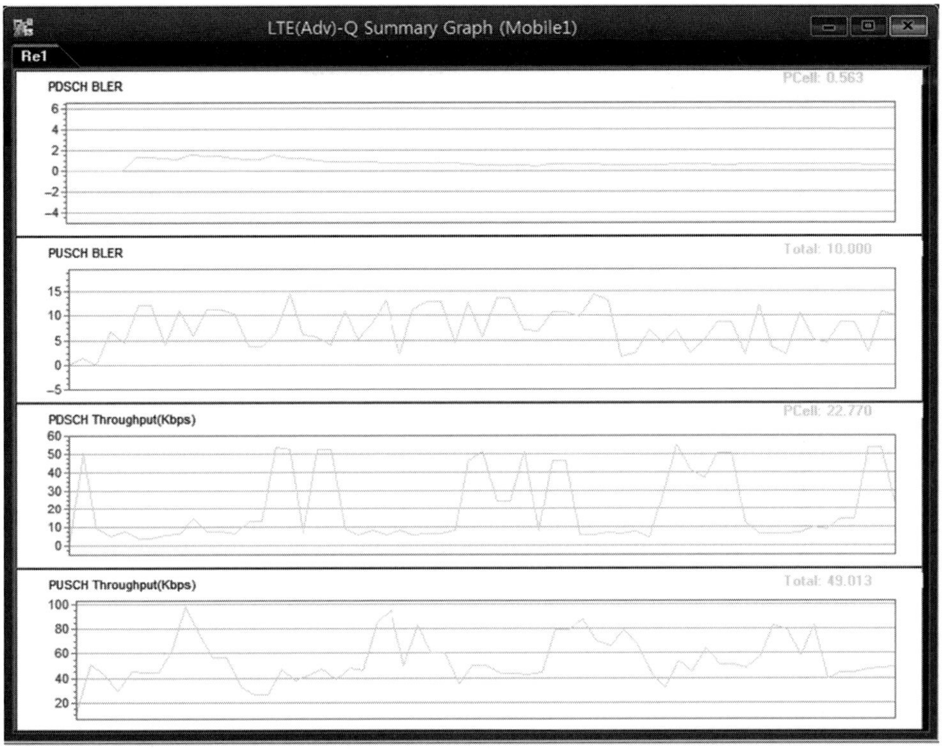

Fig. 9.46 VoLTE voice session analysis for physical layer (courtesy of Accuver Co., Ltd.).

band-aid of MONA, minimization of session negotiation delay has been one of the key objectives in the design of VoLTE. As the IMS nodes start reserving the resource for wireless and wires links as the SIP 183 message including the first SDP answer passes, media configurations and resource reservations are almost simultaneously initiated, reducing the delay to 3–4 seconds, which is also faster than the voice call set-up of previous networks. In addition, there is no difference between the negotiation delays of voice and video sessions.

Figures 9.47 and 9.48, each of which consists of five plots, show the time-varying parameters of a live VoLTE video session configured as the SDP answer shown in Fig. 8.59, for the uplink and downlink, respectively. The H.264 video codec is operating at 512 kbps for 640 × 480 frames, but its b = AS value is set to 639 kbps, after taking the IP overhead and the fluctuation of the bit-rate into account. The plots were captured at the beginning of a session, showing the bit-rate increase when the assigned bearers are getting filled with encoded video.

The first plot in each figure shows the reference signal received power (RSRP) values of the data and control channels, both of which are well under control staying in narrow ranges. The second plot in each figure shows the block error rate (BLER). The downlink has in general lower values since the eNodeB can benefit from more flexible frequency-domain scheduling, by choosing resource blocks with favorable channel conditions that

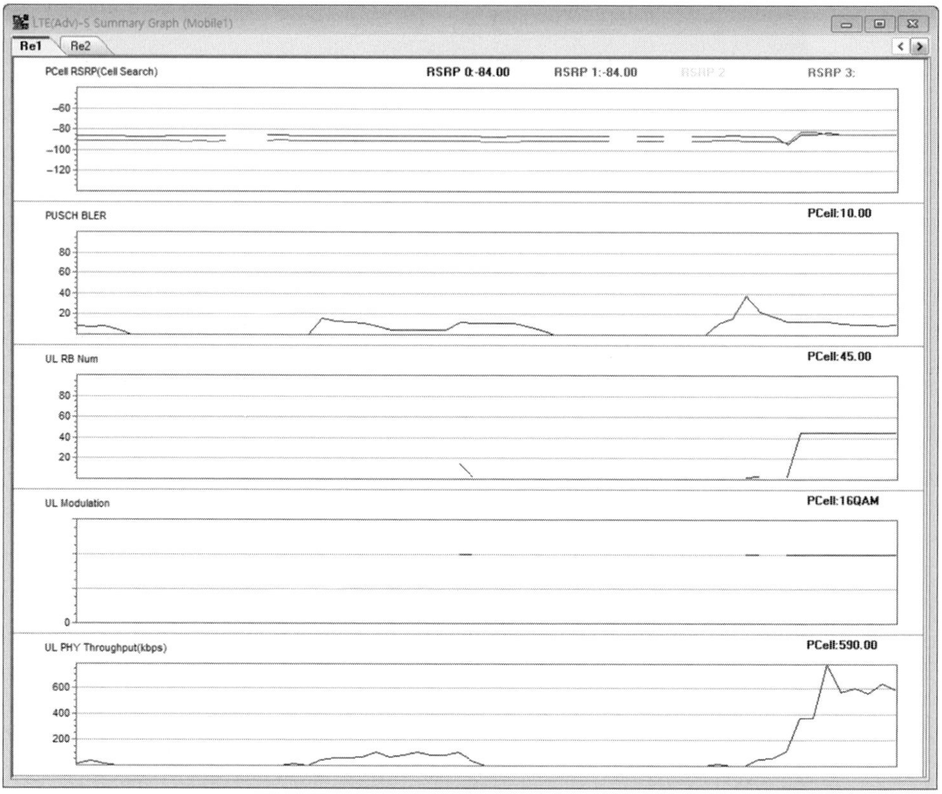

Fig. 9.47 VoLTE video session analysis for uplink (courtesy of Accuver Co., Ltd.).

may or may not be contiguous. However, the uplink has a higher BLER because the use of HARQ is preferred for managing link quality, e.g., in the scheduling or power control. The third plot shows the number of resource blocks used. Notice that the uplink uses more blocks when the actual transmission of video begins, as its modulation, shown in the fourth plot, is fixed at 16QAM while the downlink is able to use higher-order modulation schemes such as 64QAM, each symbol of which represents 6 bits, when the channel condition allows. With less limitation on its transmit power and more flexibility in its scheduling, it is easier to exploit higher-order modulation schemes in the downlink. The fifth plot shows the throughput of each link measured in the physical layer, which matches the values expected from the negotiated session conditions.

Note that using a new video codec does not necessarily increase the compression efficiency as the algorithms can vary significantly depending on the implementation, and a sophisticated codec can often outperform more recent ones without enough refinement. However, with the codec, bit-rate, and the link quality vastly improved from those of W-CDMA, the differences between the uncompressed and compressed video, and between the compressed, and the compressed and transmitted video, as shown in Fig. 6.41 for 3G-324M, would be smaller in LTE. To improve the quality noticeably using

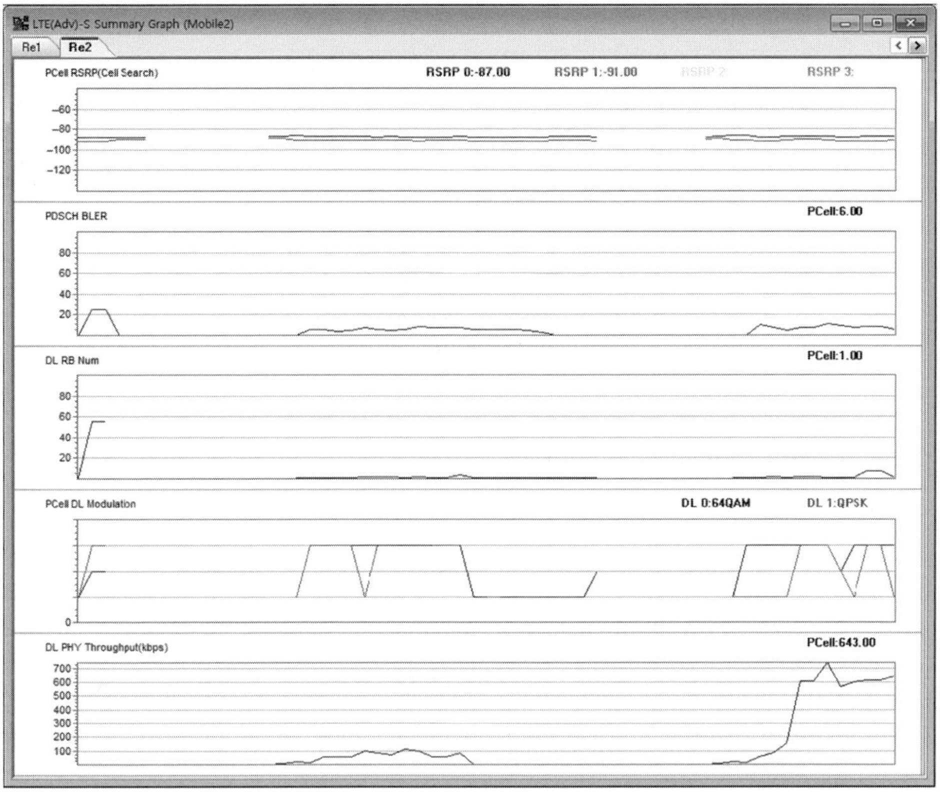

Fig. 9.48 VoLTE video session analysis for downlink (courtesy of Accuver Co., Ltd.).

a minimum amount of resource, a configuration deployed was the HEVC video codec operating at 800 kbps for 1280 × 720 frames, whose b = AS value is set to 820 and 840 for IPv4 and IPv6, respectively.

Figure 9.49, consisting of four plots, shows the time-varying parameters of a live VoLTE speech session that uses EVS 24.4 kbps, whose pattern should be similar to those of AMR-WB operating at 23.85 kbps. To show the aspects of speech transmission that are different from those shown in Figures 9.45 and 9.46, the parameters measured in the IP layer are presented. Although both links are established, the uplink is not active, i.e., it is not transporting any packets. In the second plot, the throughput of the RTP payload is shown. Notice that in the compact format of the EVS Primary mode where no RTP payload header is attached, there can be only three types of output from the EVS speech encoder, a 488-bit frame, a 48-bit SID, and silence. Therefore the values shown in the second plot, other than the three types, are generated from the throughput estimation process. The packet loss rate, shown in the first plot, is maintained at negligible levels. The third plot shows the delay from the expected arrival of packets. Notice that the delay can be negative, showing that sometimes packets arrive earlier when the actions of HARQ in the uplink or scheduling in the downlink take less time than expected.

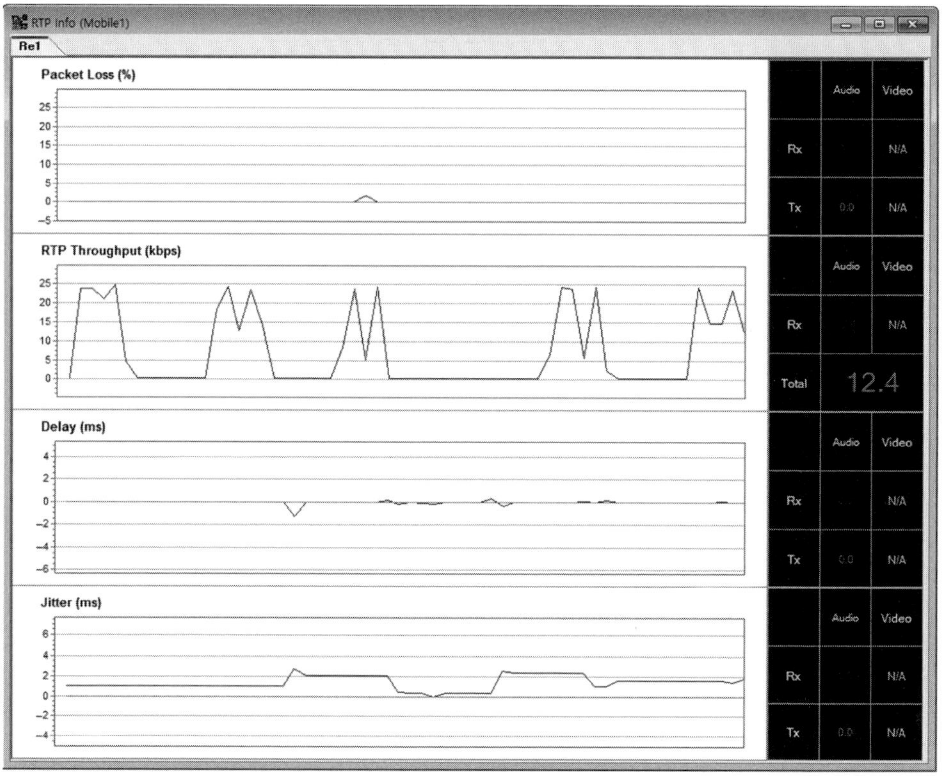

Fig. 9.49 VoLTE voice session analysis for EVS (courtesy of Accuver Co., Ltd.).

The jitter, shown in the fourth plot, was defined as an average of the absolute values of delay shown in the above plot. As shown in Fig. 9.6, the audio bandwidth is likely to vary even though a single bit-rate is used in this session.

Unlike video, there is little difference in the quality of speech that depends on the codec complexity or implementation, as both the encoder and decoder algorithms are strictly specified. In practical conversations, the gap in quality as envisioned in the controlled testing environments for evaluating codecs is perceived less significant, and inexperienced listeners would not be able to differentiate it. In addition, pre- or post-processing of speech such as noise suppression or echo cancellation tends to equalize the quality, as some of the emotional or personal aspects of audio information, typically in the higher bandwidth, are lost during the signal processing operations.

9.6.4 Voice Capacity

Table 9.14 shows the estimated voice capacity based on simulations assuming AMR 12.2 kbps for 5 MHz of system bandwidth [Holma and Toskala (2011)]. Using lower codec modes might be considered to increase network capacity but in commercial LTE deployments, AMR-WB 12.65 kbps is typically the lowest codec mode used, leaving

Table 9.14 Voice capacity of VoLTE.

	Downlink	Uplink
Dynamic scheduling	210	210
Dynamic scheduling (bundling)	370	
Semi-persistent scheduling	320	240

Table 9.15 Controlling uplink capacity with delay.

Maximum delay (ms)	50	60	70
Uplink capacity	144	155	163

AMR 12.2 kbps for interworking with W-CDMA. Note that typically bandwidths larger than 5 MHz are assigned for VoLTE, as the spectral efficiency of LTE would not be noticeably higher than those of cdma2000 or W-CDMA. Although LTE uses more advanced signal processing, the overhead from layers of IP and modem protocols dilutes the gain from sharing channels with scheduling. Therefore the return on investment for the network migration might not be sufficient at such a narrow bandwidth.

With only dynamic scheduling, the downlink capacity does not increase, even if the codec mode is reduced to 7.95 and 5.9 kbps, as the shortage of downlink control channels limits the capacity. In the simulations, a simplified quality criterion is used such that the quality is considered unsatisfactory if more than 2% of the packets are lost. The capacity is analyzed as operational techniques to boost capacity such as packet bundling and semi-persistent scheduling are applied. It is seen that the downlink capacity is, in general, larger than that of the uplink, as envisioned from the advantages of the downlink where the eNodeB can coordinate the allocation of radio resource and transmit power to each UE in a tight fashion.

Packet bundling in the downlink does increase the capacity, albeit at a lower quality. A similar increase, which is compensated by losses in other aspects, can be achieved by increasing the maximum delay each packet spends in the uplink. However, unless both the uplink and downlink belong to the same network, such an operating strategy of adaptively allocating the end-to-end delay cannot be utilized. Table 9.15 compares the capacity for different delay values simulated over 5 MHz of system bandwidth [Holma and Toskala (2011)]. As expected, larger delay results in larger capacity, albeit at lower quality. As a rule of thumb, it is often considered that the voice capacity of LTE is slightly less than three times that of W-CDMA at similar media configurations and radio conditions.

9.6.5 Derivation of LTE Voice Capacity

As in CDMA, a rough bound for voice capacity can be derived for the uplink of LTE. We follow a system model and simplifying assumptions used in [Wang *et al.* (2007)], which do not take the radio parameters such as the SNR or the use of control signaling messages into account.

Let R_U and R_T be the numbers of used resource blocks and the total number of resource blocks assigned for voice traffic for 20 ms by the eNodeB. R_T is defined to be (Number of TTIs/20 ms)(Number of RBs/TTI). In typical operations of VoLTE, a 1 ms TTI is used for each speech frame but re-transmissions from the HARQ require additional TTIs. R_U can be represented as

$$R_U = \sum_{k=1}^{K} N_{\text{speech},k}(1 + R_{\text{HARQ\_}k}\varepsilon_k) + \sum_{l=1}^{L} N_{\text{SID,l}}(1 + R_{\text{HARQ\_}l}\varepsilon_l), \tag{9.6}$$

where K and L are the numbers of active and inactive UEs respectively. Here $N_{\text{speech},k}$ and $N_{\text{SID,l}}$ represent the average numbers of required resource blocks for a speech frame and an SID transmitted from UEs k and l, respectively, $R_{\text{HARQ\_}k}$ and $R_{\text{HARQ\_}l}$ are the average numbers of re-transmissions for a speech frame and an SID, and ε_k and ε_l represent the ratios of resource blocks for re-transmissions to those of initial transmissions, which is unity if chase combining is used.

As R_U is smaller than or equal to R_T,

$$\frac{R_U}{R_T} \leq 1. \tag{9.7}$$

The expression of R_U can be simplified further if we assume $\varepsilon_k = \varepsilon_l = 1$, $N_{\text{speech},k} = N_{\text{SID,l}} = N_{\text{ave}}$ and $R_{\text{HARQ\_}k} = R_{\text{HARQ\_}l} = R_{\text{HARQ}}$. Let ν and p be the voice activity factor and the probability that an SID is generated during a silence period. If the speech encoder is configured to transmit an SID each $20 \times M$ ms, then p can be set to $\frac{1}{M}$. For example, in AMR, as a new SID is transmitted if eight consecutive frames are marked with VAD flags of 0, i.e., diagnosed as non-speech [3GPP (2017d)], p is set to $\frac{1}{8}$.

Then R_U can be approximated as

$$R_U = (1 + R_{\text{HARQ}})N_{\text{ave}}N_{\text{sup}}(\nu + p(1 - \nu)), \tag{9.8}$$

where N_{sup} is the number of UEs in voice sessions that can be supported in the cell, which can be upper-bounded as

$$N_{\text{sup}} \leq \frac{R_T}{(1 + R_{\text{ave\_}re})N_{\text{ave}}((1 - p)\nu + p)}. \tag{9.9}$$

As it is unlikely that a speech frame, which consumes a low bit-rate, will be transmitted over RBs of the same TTI but separated in the frequency spectrum, the above bound can also be used for the downlink. Additional factors that would influence the network capacity include the interference from neighboring cells, and the total bandwidth applicable to voice services as the relative overhead for control channel is likely to decrease as the total bandwidth increases, thereby increasing the number of simultaneously supportable UEs per unit bandwidth.

9.7 References

3GPP. 2011. S4-110710 EVS Permanent Document 4 (EVS-4): EVS Design Constraints. August.

3GPP. 2017a. TR 26.906 V14.0.0 Evaluation of High Efficiency Video Coding (HEVC) for 3GPP Services. March.

3GPP. 2017b. TR 26.952 V14.0.0 Codec for Enhanced Voice Services (EVS); Performance Characterization. March.

3GPP. 2017c. TR 26.976 V14.0.0 Performance Characterization of the Adaptive Multi-Rate Wideband (AMR-WB) Speech Codec. March.

3GPP. 2017d. TS 26.093 V14.0.0 Mandatory Speech Codec Speech Processing Functions; Adaptive Multi-Rate (AMR) Speech Codec; Source Controlled Rate Operation. March.

3GPP. 2017e. TS 26.103 V14.0.0 Speech Codec List for GSM and UMTS. March.

3GPP. 2017f. TS 26.114 V14.3.0 IP Multimedia Subsystem (IMS); Multimedia Telephony; Media Handling and Interaction. March.

3GPP. 2017g. TS 26.179 V14.0.0 Mission Critical Push To Talk (MCPTT); Codecs and Media Handling. March.

3GPP. 2017h. TS 26.290 V14.0.0 Audio Codec Processing Functions; Extended Adaptive Multi-Rate – Wideband (AMR-WB+) Codec; Transcoding Functions. March.

3GPP. 2017i. TS 26.445 V14.0.0 Codec for Enhanced Voice Services (EVS); Detailed Algorithmic Description. March.

3GPP. 2017j. TS 36.331 V14.2.0 Evolved Universal Terrestrial Radio Access (E-UTRA); Radio Resource Control (RRC); Protocol Specification. March.

Dietz, M., Liljeryd, L., Kjörling, K., and Kunz, O. 2002. Spectral Band Replication, a Novel Approach in Audio Coding. *Audio Engineering Society 112th Convention*, May.

Freescale. 2012. Semiconductor Application Note: Tile Sensing Using a Three-Axis Accelerometer. *Document Number: AN3461*, February.

Holma, H., and Toskala, A. 2011. *LTE for UMTS: Evolution to LTE-Advanced.* 2nd edn. Wiley.

Huyghe, B., Doutreloigne, J., and Vanfleteren, J. 2009. 3D Orientation Tracking Based on Unscented Kalman Filtering of Accelerometer and Magnetometer Data. *IEEE Sensors Applications Symposium*, February.

ISO. 2001. MPEG-4 Information Technology – Coding of Audio-Visual Objects – Part 2: Visual. December.

ITU-T. 1996. P.800 Methods for Subjective Determination of Transmission Quality. August.

ITU-T. 2009. Recommendation G.191 STL-2009 Manual. November.

ITU-T. 2013. H.265 High Efficiency Video Coding. April.

Johansson, I., and Jung, K. 2011. IETF RFC 6236 Negotiation of Generic Image Attributes in SDP. May.

MPEG. 2012. ISO/IEC 23003-3:2012 Information Technology – MPEG Audio Technologies – Part 3: Unified Speech and Audio Coding. April.

Ohm, J.-R., Sullivan, G. J., Schwarz, H., Tan, T. K., and Wiegand, T. 2012. Comparison of the Coding Efficiency of Video Coding Standards – Including High Efficiency Video Coding (HEVC). *IEEE Transactions on Circuits and Systems for Video Technology*, **22**(12).

OMA. 2007. Enabler Release Definition for OMA Device Management, Approved Version 1.2. February.

Ortega, A., and Ramchandran, K. 1998. Rate-Distortion Methods for Image and Video Compression. *IEEE Signal Processing Magazine*, **15**(November).

Ott, J., Wenger, S., Sato, N., Burmeister, C., and Rey, J. 2006. IETF RFC 4585 Extended RTP Profile for Real-time Transport Control Protocol (RTCP)-Based Feedback (RTP/AVPF). July.

SIG. 2016. Bluetooth Core Specification V5.0. December.

Singer, D., and Desineni, H. 2008. IETF RFC 5285 A General Mechanism for RTP Header Extensions. July.

Sjoberg, J., Westerlund, M., Lakaniemi, A., and Xie, Q. 2007. IETF RFC 4867 RTP Payload Format and File Storage Format for the Adaptive Multi-Rate (AMR) and Adaptive Multi-Rate Wideband (AMR-WB) Audio Codecs. April.

Wang, H., Jiang, D., and Tuomaala, E. 2007. Uplink Capacity of VoIP on LTE System. *Asia-Pacific Conference on Communications*, October.

Wenger, S., Chandra, U., Westerlund, M., and Burman, B. 2008. IETF RFC 5104 Codec Control Messages in the RTP Audio-Visual Profile with Feedback (AVPF). February.

10 Signal Processing in 5G Systems

10.1 Technical Background

Long Term Evolution (LTE) was successful in finally accomplishing High Speed Packet Access's purpose, a transition of mobile communications services to an All-IP system. Like GSM or IS-95, LTE was initially designed with a significantly different set of system parameters than previous systems. However, its scalable bandwidths and features facilitated the migration from circuit-switched networks of service providers running various types of networks with low deployment risks. In the early stages of LTE development, competition between LTE and Ultra Mobile Broadband (UMB) [3GPP2 (2008)], a descendant of cdma2000 EV-DO, was joined by a mobile version of Worldwide Interoperability for Microwave Access (WiMAX) [IEEE (2005)], a new system originating from Wi-Fi that provides cellular-class coverage and mobility.

In their radio signal processing aspects, the three systems, all of which are based on the OFDMA principles, could meet the performance requirements of 4G, the IMT-2010: 100 Mbps at vehicular speed and 1 Gbps in a fixed position [ITU-R (2016a)]. Note that the evolved versions of LTE met the requirements by using multiple bandwidths, using carrier aggregation. However, unlike in the early days of 2G when GSM, IS-54, and IS-95 competed as commercially operational systems, the three-way standoff for the dominance of 4G did not last long, as the complexity and cost of 4G systems prevented minor or new systems from being deployed or even implemented although they were technically competitive, leaving only LTE by the time the limitations of this dominant system began to be recognized, but could not be solved because of its fundamental design constraints.

In addition to the absence of competition between systems, there are other unique characteristics of fifth generation mobile communications systems. However, most of the element technologies required for the media and radio signal processing have already matured. Holma and Toskala have shown that the transmission techniques of LTE, whose performance is estimated in bps/Hz/cell, come to within 0.3–0.4 dB of the Shannon theoretical limit if applied for AWGN [Holma and Toskala (2011)]. This gap can be reduced by simply adding more MCS steps. Further increases in the network capacity can be achieved more easily by using additional bandwidths than by further refining the radio signal processing algorithms. In spite of these limitations, the performance requirements of 5G, the IMT-2020, were set ambitiously: a user experienced bit-rate of 100 Mbps and a peak bit-rate of 20 Gbps in a fixed position [ITU-R (2016b)], for which there might not be enough bandwidth left in the frequency spectrum with favorable propagation characteristics. Note that in the spectrum shown in Fig. 9.33, the bands of various 2–4G systems are located below 2600 MHz.

The compression efficiencies of speech (EVS) and video (HEVC) codecs, as shown in Figures 9.39 and 9.44, have also matured to the point where improving conventional multimedia communications using mobile handsets does not result in recognizable quality improvements. Moreover, new features of handsets, such as the desire that they be waterproof, mean that the speakers and microphones have to be sealed. In addition the desire for an ever-decreasing form factor for the handsets, especially their depth, compromises any improvement from advanced speech compression algorithms. Likewise, increasing the video resolution and bit-rate while keeping display sizes of 5–6 inches diagonal returns limited gain in visual quality, since the density of the pixels in high-end UE displays already exceeds 800 pixels per inch (PPI).

All of this means that increasing the speech bit-rate or bandwidth, number of audio channels, or the video resolution, e.g., from the QoS levels shown in the SDP answer of Fig. 8.59, will not necessarily provide higher quality if the media is still captured, transmitted, and rendered in typical UEs, such as those shown in Fig. 9.26. However, there still remain seizable opportunities for higher network capacity, using the ever-increasing complexity and larger spectrum previously not considered. This all suggests that it is the improvement from a larger system scale rather than a higher system efficiency that can be expected to be the major source of innovations for this generation. However, new types of services that cannot be delivered by conventional UEs, regardless of their physical shapes or processing capabilities, would require the transport capability of new radio access technologies and also necessitate redefinitions of media quality.

In this generation, additional types of performance requirements are set to accommodate new services or services currently provided by other networks. Such requirements include the area traffic capacity, measured in Mbps/m^2, and the connection density, measured in the number of devices per square kilometer. The latter was contrived to represent the capability of a network supporting a large number of UEs that do not require any mobility or high bit-rate, or the capability for conversation but have to operate for a prolonged time without recharging, which are typical of the Internet of Things (IOT) devices.

In this chapter, we outline the architectures and features that are envisioned for 5G radio access and its core network technologies. The massive bit-rate requirements are expected to be met by exploiting large amounts of lower-cost, higher-frequency spectrum, and by replacing key elements unchanged during the previous transition to 4G, such as turbo codes, with new ones that use the available complexity for higher performance. Compared with the performance of radio access technologies in previous generations, transmission of 5G at the highest bit-rates is likely to be faster and cheaper, but not always more maneuverable, as such bit-rates might require not just fixed positions but also LOS links or even alignments of antennas and UEs. The generic framework of LTE will be maintained and refined, but compatibility will not be supported. We also discuss the characteristics of immersive media services considered for this generation, which can provide an elevated level of audiovisual quality, using fundamentally different signal processing operations for media capturing, compression, transmission, and presentation. Finally, we outline the issues of controlling the tradeoff between media quality and network capacity when new types of audiovisual front-ends for immersive services, such as omnidirectional cameras and head-mounted displays (HMD), are connected to or integrated into the handsets.

10.2 Network Architecture

The network architecture of fifth generation mobile communications systems is not significantly different from its precursors, consisting of the UE, the gNB that replaces the eNodeB, and the 5G core network that replaces the EPC. Figure 10.1 shows the network architecture of 5G systems with the interfaces between the UE and the gNB, and between the gNB and the core network nodes. The range of media and radio signal processing is similar to that of LTE shown in Fig. 8.3. In the 5G core network, the User Plane Function (UPF) and the Data Network (DN) are network nodes on the user plane (UP), with similar roles as the S- and P-GWs of EPC, and the IP backbone respectively [3GPP (2017c)].

A new concept introduced in this architecture is the *master node*, in which a particular gNB or eNodeB can be given higher authority in the management of network operation than other neighboring nodes with identical or similar positions in the architecture. This means that a gNB appointed as master node in an area can control other gNBs or eNodeBs to some extent. The introduction of a master node is a compromise between the hierarchical architecture of UMTS and the flat architecture of LTE. After the radical transition to the latter architecture, it was soon recognized that the absence of network nodes that assume the responsibility for controlling quality and capacity had

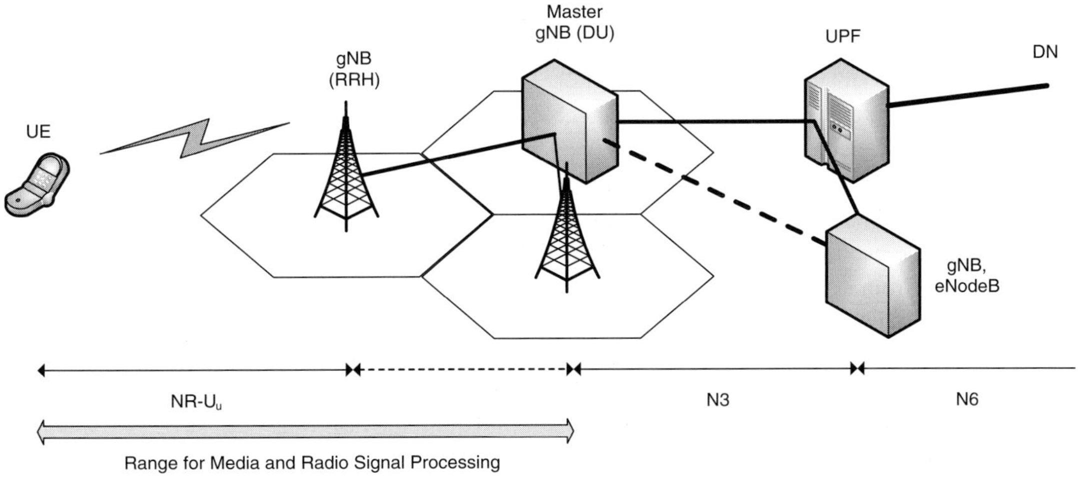

Fig. 10.1 Network architecture of 5G systems.

to be compensated, which resulted in the eNodeB functionality that marked ECN-CE on the packets or recommended media bit-rates when the available capacity was not sufficient for demand.

Either a gNB or an eNodeB can be connected to the 5G core network, and either can serve as the master node in a group of gNBs and eNodeBs. Note that a gNB does not necessarily become a master node even if it is surrounded by eNodeBs, and vice versa. Since the IMS can serve a diversity of access networks including fixed and Wi-Fi networks, the 5G core network can be used to interconnect the new radio access NR and Wi-Fi. A master node can manage the uplink and downlink transmission of neighboring nodes, by coordinating data flow across groups of nodes, to increase the throughput or reduce the interference. With the functional separation of the eNodeB into a remote radio head and a digital unit, the operational practice of running digital signal processing software on general purpose, off-the-shelf servers instead of dedicated hardware is expected to expand into more functionality of the core network and the IMS. This will facilitate the support of multiple networks or service types on the same, cloud-based infrastructure, thereby realizing Network Function Virtualization (NFV).

10.3 New Radio Access

New Radio (NR) is the tentative name of 5G radio access technologies that replace LTE [3GPP (2017a), 3GPP (2017b)]. The NR modem has a similar set of radio protocols as LTE, which include PDCP, RLC, MAC, PHY, and RRC, whose technical responsibilities and methods of operation differ slightly. Table 10.1 shows that although they are incompatible with each other, the physical layers of the NR and LTE share many system parameters. Compared with LTE, whose parameters are outlined in Table 8.6, the lowest-order modulation, BPSK, is no longer supported, and 256QAM, which LTE began providing in its evolved versions, is included as one of the basic modulation schemes. However, 256QAM requires a very high SNR to be activated by the adaptive modulation and coding (AMC). This modulation scheme is used only when LOS links over short distances are available.

A key characteristic of the New Radio is its scalable subcarrier width, 15×2^n kHz, where n is a non-negative integer. This approach is a natural feature as the maximum channel bandwidth has to be significantly increased to meet the target bit-rate. Serving a wide bandwidth range with a fixed subcarrier width would not always result in optimal performance. At very high carrier frequencies, such as 28 GHz, the subcarrier spacing needs to be increased to, e.g., wider than 100 kHz, to facilitate the receiver operation. The maximum channel bandwidth is 400 MHz, and up to 16 carriers can be aggregated. Even with the scaled-up subcarrier widths, covering the larger channel bandwidths over more component carriers than LTE would require the NR modem to have the capability of simultaneously handling thousands of subcarriers.

OFDMA is used for the downlink modulation, and is also used in the uplink in addition to SC-FDMA. This is because the need to limit the variation of transmit power to

Table 10.1 System parameters of NR.

Multiple access	SC-FDMA/OFDMA (uplink), OFDMA (downlink)
Modulation	QPSK, 16QAM, 64QAM, 256QAM
Maximum channel bandwidth (MHz)	400
Maximum number of component carriers	16
Subcarrier width (kHz)	15–480

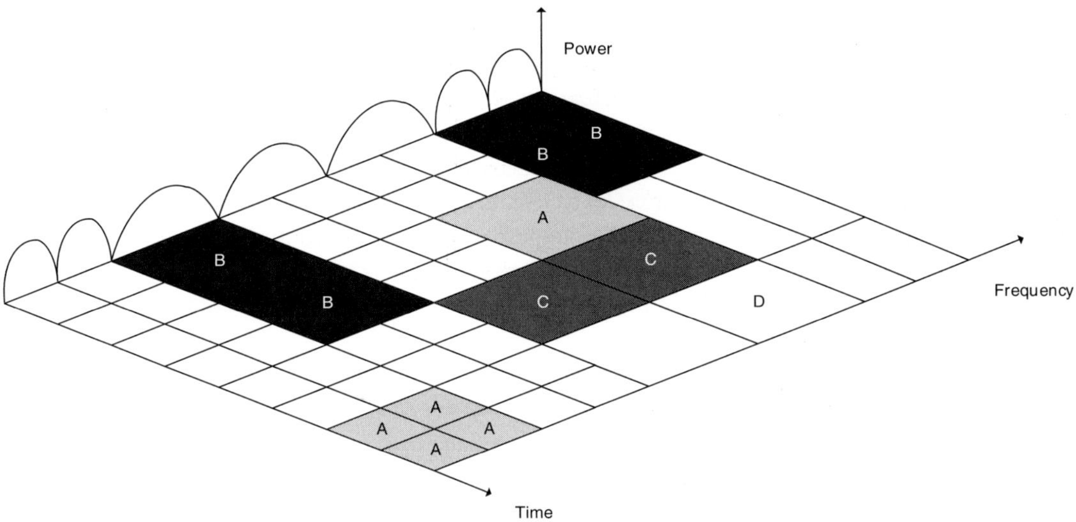

Fig. 10.2 Frequency-, time-, and power-domain representation of physical layer.

the linear operating range of the power amplifier at the UE is no longer considered to be urgent, as the efficiency of the RF components has improved significantly. The transmission time interval (TTI) is also expected to be scalable. Shorter TTIs are necessary to enable lower latencies, and to pack a number of transmissions into a time period that is shorter than a subframe.

Figure 10.2 illustrates the time–frequency structure of the New Radio for the uplink and downlink, where the resource blocks are dynamically assigned to four UEs, A–D. Notice that in contrast to the time–frequency structure of the LTE, shown in Fig. 8.18, either the subcarrier width or the TTI, or both, can change to match the channel condition of each UE and the cell loading level. These previously fixed channel parameters are now adjustable for more flexible allocations of radio resources. The two channel codes used in W-CDMA and LTE, convolutional code and turbo code, are likely to be replaced by polar code [Arıkan (2009)] and Low Density Parity Check (LDPC) code [Gallager (1962)].

Polar code is a type of linear block code. It is constructed by multiple recursive concatenations of a basic code, which corresponds to transforming the channel into multiple *virtual channels*. As the number of recursions increases, the virtual channels approach one of two channel types, one with low reliability and another with high reliability, thereby effectively *polarizing* the channels. During the transformation process, the channel capacity is conserved but re-distributed unevenly among the virtual channels. As the number of virtual channels increases, the proportion of channels with high reliability rapidly increases, with a combined channel capacity that closely approaches the theoretical channel capacity of many binary channel types. One benefit of the polar code is its modest complexity for the encoder and decoder implementations. This might be the reason that the code is being considered even though it was introduced only relatively recently and actual incorporations in products and services are few.

LDPC code is also a linear block code, whose parity check matrix has a small number of non-zero elements. The concept of LDPC code was introduced as early as in the 1960s but its practical decoding algorithms were not found until the 1990s. With *Belief Propagation* [MacKay (1999)], a type of iterative message-passing algorithms, the decoding time of LDPC code can be controlled. It is roughly proportional to the length of codewords. Like turbo or polar code, LDPC code can closely approach the Shannon limit, with other favorable features that help the code meet the radio signal processing needs of 5G. It is expected that in the New Radio, LDPC code will protect the user plane data of large block sizes, while polar code is used with control plane signaling and for data with small block sizes.

In earlier packet-switched networks, MIMO provided higher throughput, serving the surging need for data and media services. In NR, a spatially or vertically extended version of MIMO, *massive MIMO*, is being considered to exploit the larger bandwidths in the higher frequency spectrum, using a large number, e.g., 32–64, of small antennas typically integrated onto a cylindrical shape on top of the gNB. As with previous MIMO techniques, resource blocks belonging to the same time–frequency grid can be assigned to multiple UEs. However, a notable difference is that the beam has to be formed, taking three-dimensional directions into account. In the MIMO systems, time division duplex (TDD) typically outperforms frequency division duplex (FDD) as the gNB can estimate the channel condition not only from the pilot elements such as the demodulation reference signal, but also from the uplink data streams. A key benefit of massive MIMO is that the throughput can be theoretically scaled up with the increased number of antennas and the higher spatial resolution, which would be realized to some extent. In addition, the energy efficiency of the gNB, defined as the amount of data transported using a unit of energy measured in MB/Joule, is higher than conventional MIMO systems using fewer, e.g., 2–8, antennas.

Figure 10.3 illustrates the operation of massive MIMO using a gNB with 64 antennas, where the UEs share the same resource blocks transmitted in spatially separated or overlapped directions. Each antenna is a half wavelength by a half wavelength in size, distanced a half wavelength apart circumferentially, and distanced a half wavelength apart vertically [Marzetta (2015)]. Note that the capacity for real-time traffic would not be scaled up with the number of antennas. With tight delay constraints, it would be more difficult to improve than the throughput. It was shown in the VAMOS of GSM that the MS falls back to TCH/AFS or TCH/AHS when the CIR required to differentiate the subchannels sharing the same frequency and time slot is not met. If the gNB in massive MIMO can maintain the separation of resource blocks belonging to the same

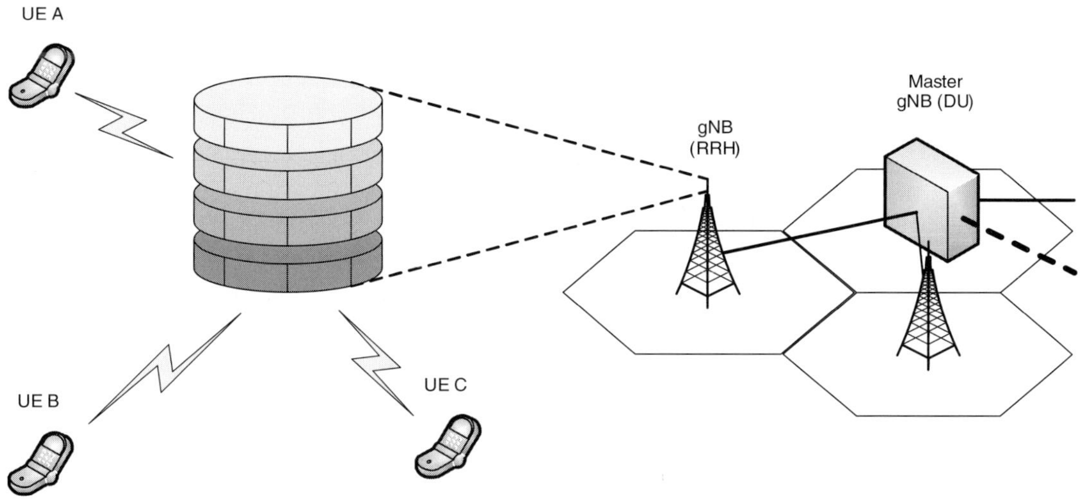

Fig. 10.3 Massive MIMO operation.

time-frequency grid, enough to meet the negotiated QoS of each session, which conventional MIMO systems with a smaller number of antennas could not do well, the transport of real-time traffic can be considered using this technique.

10.4 Immersive Media Service

10.4.1 Virtual Reality

Each generation of mobile communications systems has a unique service that characterizes the combined capability of the state-of-the-art handsets and networks. Such services, digital speech coding in 2G, 3G-324M in 3G, and VoLTE in 4G, are likely to be real-time and conversational, since services with relaxed delay constraints or from third-party providers, i.e., the *over-the-top* (OTT) services, would not have a coordinated care from the network such as quality of service indicators 1 and 2. Services requiring higher bit-rates than were supported by earlier systems are also preferred as these clearly differentiate the need for new systems.

Virtual Reality (VR), which provides a higher level of *immersive quality*, is one such service often envisioned for 5G systems. The usage of VR is not confined to entertainment or conversation. The bit-rate requirements for VR media at UHD and 60 fps would be as high as 15 Mbps, which would not be supported by a 5 MHz channel operating W-CDMA or HSPA, even if a UE is given all resources of the cell. Such bit-rates would be available to less than 5 UEs in a cell operating 10 MHz for LTE. VR can serve as an interface for presenting the audiovisual information of a wide area in a compact but lively fashion, e.g., enabling tasks such as remote control of drones or robots. Its wide-angle video can also be used by driver assistance systems or self-driving vehicles.

Conventional approaches to achieve immersive quality, in which the viewer feels that she or he is actually involved in a remote or artificial environment described with

Fig. 10.4 Virtual reality presentation (courtesy of Samsung Electronics Co., Ltd.).

the presented media, have increased the size of the display, and the number of audio channels and loud speakers [NHK (2011)], which limit their applications. In contrast, VR brings displays and speakers with a wearable form factor close to the eyes and ears using a head mounted display (HMD). Figure 10.4 illustrates the presentation of VR media where the scenes and sounds corresponding to the viewing angle of the user change as the viewer moves her/his head. In Fig. 9.27, it was shown that the orientation of a display can be estimated for the coordination of video orientation (CVO) operation using an accelerometer and a magnetometer. HMD uses a similar set of sensors to track the orientation of the head, and take that information into account in the presentation of media.

10.4.2 Ambisonic Audio Signal Processing

In Fig. 9.32, the EVS speech codec of a UE is configured to send two audio channels but receive only one, which represents the hypothetical situation of an omnidirectional camera built into or connected to a UE transmitting VR media. Although stereo audio is transmitted, unless the microphones on the camera or the camera itself move according to the orientation of the head at the far-end, the audio will not match the video, as stereo audio cannot cover all directions. Thus, with conventional *channel-based* audio signal processing, either a large number of audio channels covering all directions must be simultaneously transmitted or the audio channels must be updated based on the orientation of the far-end head, using some kind of beam-forming techniques and feedback channels. With the former approach, the camera may not be aware of the layout of speakers at the far-end, or the speakers may not be aligned with the orientation of the microphones. In the latter approach, the audio may trail the video with a noticeable latency when the head moves. Therefore it would be beneficial if the layout issues of microphones and speakers can be separated without these problems.

Ambisonic, often called *scene-based*, audio signal processing uses a radically different approach to represent audio signals and avoid these shortcomings of channel-based audio [Daniel (2001)]. Consider a spherical coordinate system where the location of an arbitrary region of space is represented by distance r from the origin, elevation θ from the horizontal plane, and azimuth ϕ from the reference axis. Then the sound pressure field in the region at time t can be approximated using spherical harmonic basis functions as

$$\begin{aligned} p(r,\theta,\phi,\omega,t) &= \left[\sum_{n=0}^{\infty} j_n\Big(\frac{\omega r}{c}\Big) \sum_{m=-n}^{n} a_n^m(\omega,t)Y_n^m(\theta,\phi)\right]e^{j\omega t} \\ &\approx \left[\sum_{n=0}^{N} j_n\Big(\frac{\omega r}{c}\Big) \sum_{m=-n}^{n} a_n^m(\omega,t)Y_n^m(\theta,\phi)\right]e^{j\omega t}, \end{aligned} \tag{10.1}$$

where c is the speed of sound, and ω is the angular frequency. $j_n(\frac{\omega r}{c})$ and $Y_n^m(\theta,\phi)$ are the spherical Bessel function of degree n and the spherical harmonic function of order n and degree m, respectively. Then the spherical harmonic coefficient, $a_n^m(\omega,t)$, which can be estimated from the values of the acoustic pressure at sampled locations around the region, contains all the information about the sound field. In ambisonic audio signal processing, it is the set of spherical harmonic coefficients that is compressed and transmitted. For each order $n \geq 0$, $2n+1$ new spherical harmonics are introduced. The set of coefficients for an order n ambisonic system includes those for all lower-order systems. Thus the higher the order, the higher the spatial resolution of the soundfield.

Figure 10.5 visualizes the squared magnitude of the spherical harmonic functions for $0 \leq n \leq 3$ and $0 \leq m \leq n$. The first row shows Y_0^0, and the fourth row shows Y_3^0, Y_3^1, Y_3^2, and Y_3^3. As the order increases, the shape of the harmonic functions becomes more complicated, which is necessary to describe a soundfield with more complex structures with a higher level of spatial detail. The spherical harmonic functions with negative indexes have identical magnitude to their counterparts with positive indices but differ in polarity.

Since the information on the whole structure of the soundfield around the camera is transmitted, instead of the audio samples for some directions, audio signals of arbitrary orientations can be estimated at the receiver by reconstructing the soundfield and capturing the signals for proper orientations of the speakers. Therefore it is not required to align the numbers or orientations of microphones and speakers. Another benefit of ambisonic audio is that the rotation or shaping of the soundfield can be implemented by mathematical operations on the spherical harmonic coefficients. Notice that ambisonic audio is not a panacea for all VR applications. For applications in a controlled environment such as concerts in a studio or a hall, channel-based audio is very efficient. Ambisonic audio is particularly good at handling the ambient noise and sound in an outdoor environment.

Figure 10.6 shows the ambisonic audio signal processing operations of a VR system, assuming a similar protocol architecture and an end-to-end transmission path such as that shown in Fig. 8.4. A notable difference from 3G-324M or VoLTE is that the media client of VR is likely to be located in a separate device from the UE, because of its

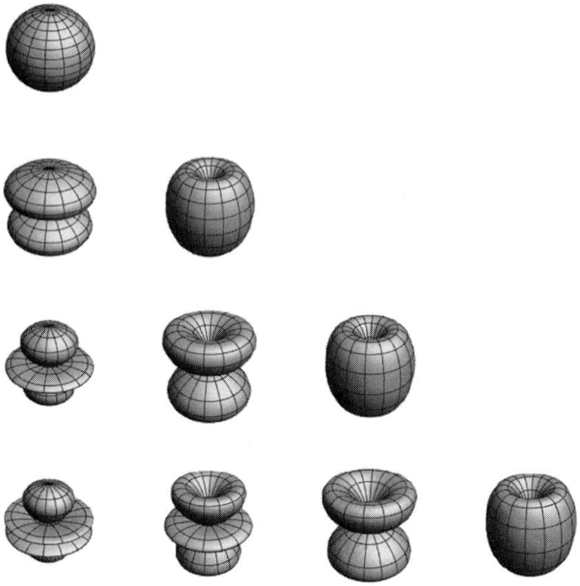

Fig. 10.5 Spherical harmonic functions of orders 0–3.

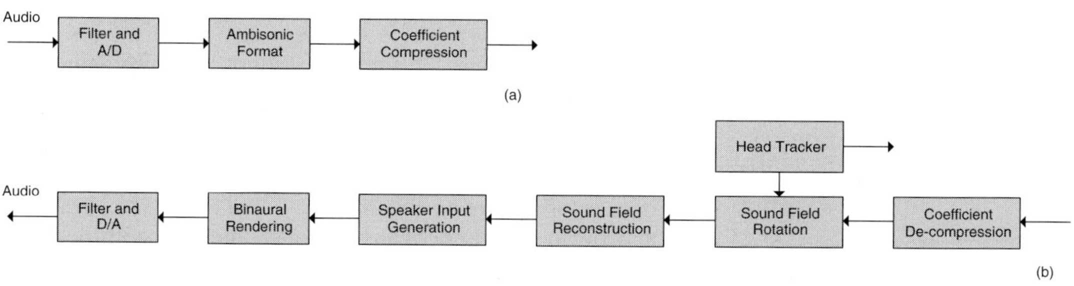

Fig. 10.6 Ambisonic audio signal processing operations of VR. (a) Transmitter side. (b) Receiver side.

strong optical and acoustic requirements for the capture and presentation of media that might not be realized with the limited form factor of conventional UE types, as shown in Fig. 9.26.

At the transmitter side, for each 20 ms frame, audio signals spanning 360 degrees horizontally and vertically are first converted into an nth-order ambisonic format, such as the AMBIX [Nachbar *et al.* (2011)], in which the number of coefficients for each sampling instant is equal to $(n+1)^2$. For order n, $(n+1)^2$ audio channels are required to transport the coefficients. The number of channels can be reduced by exploiting the correlation between coefficients. Thus, by varying n, the tradeoff between audio quality and network capacity can be controlled to some extent. Then, the coefficients for the 20 ms frame are compressed and loaded onto an RTP packet, which is followed by the RTP/UDP/IP packetization and radio signal processing procedures of LTE or NR.

If the delay requirements are relaxed, e.g., to reduce the cost for bit-rate or increase the integrity of data, a longer frame size, a streaming protocol instead of RTP, the AM of RLC, or more HARQ re-transmissions at PHY can be considered. A 64 kbps channel is typically used to transport audio signals sampled at 48 kHz. Therefore the bit-rates of VR required for audio, whether it is channel-based or scene-based, are not significant even when as many as eight channels are used, compared with those for video.

At the receiver, the received spherical harmonic coefficients are decoded after the protocol headers are removed. Then, simple matrix operations are applied to the coefficients to rotate the soundfield to the orientation of the head. The soundfield is reconstructed and the audio signals corresponding to the directions of the speakers are extracted. When a headphone is used for the presentation of audio, the stereo audio signals can be *binaural* rendered to match the anatomical characteristics of the ears, the distance and direction between the ears and the sound sources, and the reverberation characteristics of the rooms.

Head Related Transfer Function (HRTF) [Cheng and Wakefield (2001)] is a complex function defined for binaural rendering. Let the impulse response for the left and right ears be $h_L(t)$ and $h_R(t)$, respectively. Then the pressure at the ears, $y_L(t)$ and $y_R(t)$, can be described as the convolutions of the sound source, $x(t)$, and the impulse responses. HRTF is the Fourier transform of the impulse response. The measurement of HRTF is a time-consuming process requiring a rotating loop of loudspeakers around a listener [Gardner and Martin (1995)]. In the measurement, a known audio signal is transmitted by a loudspeaker placed at a direction and a distance from the head. The signal is measured at the entrance of the ear canals. The procedure is repeated for different signals and directions. HRTF is influenced by the height of the listener, the number and location of sound sources, and the shapes of the head and torso. As a method to accelerate the procedures, approaches to synthesize HRTF based on three-dimensional modeling of the head and torso, and the internal shapes of ears have been proposed, as the measurement does not require any subjective evaluations and therefore can be estimated by simulating the propagation and reflection of sound waves.

10.4.3 Omnidirectional Video Signal Processing

Figure 10.4 shows a user viewing scenes covering 96 degrees horizontally, but the viewing angle, the Field of View (FOV), can vary depending on the implementations. With a FOV less than 360 degrees, video data wider than the FOV needs to be transmitted to the HMD because the scenes may change immediately following the movement of the head. If the available bit-rate is sufficient, which is unlikely in mobile communications regardless of the generation, scenes covering 360 degrees in both the horizontal and vertical directions can be transmitted in a uniform quality. Such a wide-angle video can be generated by using an omnidirectional camera with multiple lenses and sensors, whose captured images are *stitched*, i.e., interwoven onto the surface of a sphere, and projected onto a single rectangular image to be encoded by the video encoder, as illustrated in Fig. 10.7.

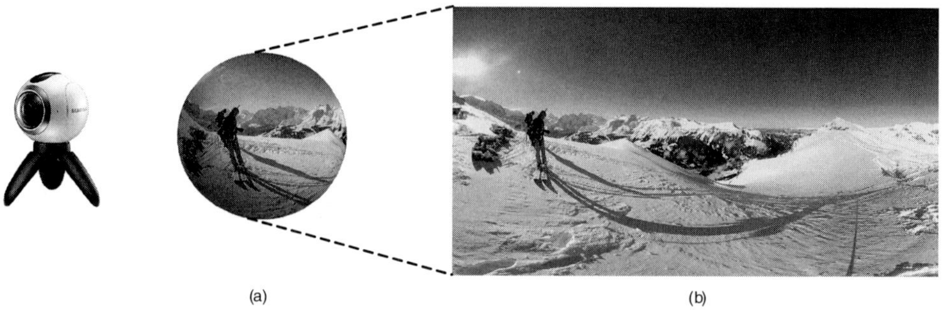

Fig. 10.7 Signal processing prior to compressing omnidirectional video. (a) Spherical stitching. (b) 2D projection (courtesy of Samsung Electronics Co., Ltd.).

This must take the geometry of the image sensors into account to reduce the distortion of the reconstructed image. The more lenses used in the capture, the less distortion in the image area covered by each lens, but the higher the complexity when stitching images whose contents are dynamically changing and overlapped in the edges. Note that differences in the compression efficiency between H.264 and HEVC, which were insignificant in the relatively low QoS for conventional video telephony, as shown in Fig. 9.44, become noticeable as the bit-rate and resolution of video increase to the higher levels used for immersive media.

Several approaches can be considered to reduce the complexity and bit-rate requirements, e.g., 15 Mbps for 360-degree video encoded in UHD at 50 fps. Note that not only the spatial quality but also the temporal quality has to be improved for higher immersive quality. One approach is to encode the scenes for 360 degrees but quantize the areas outside the FOV more crudely. In this approach, the scenes being presented have higher clarity than the other areas, which can be realized by applying smaller quantization steps to the blocks being watched. The strategy can also be implemented by sending a high resolution bit-stream for the FOV and a low resolution bit-stream for the background areas. Another approach is to encode the scenes for less than 360 degrees, probably as negotiated via the IMS, but still include enough pixels outside the FOV area to serve as margins. Then the user would not perceive an abrupt quality reduction immediately after head movement, and the presented scenes would be updated within a short time.

In either approach, it is necessary to use head tracking information that is fed back to the omnidirectional camera from the HMD. The second approach has the advantage of higher quality when the head is stationary, as only the areas within FOV are encoded in the maximum bit-rate and resolution, but it can be vulnerable to the loss of feedback information and a rapid change of head direction, as is the case with the receiver-driven media adaptation of VoLTE. However, if head tracking information is lost during transmission, only the image clarity is reduced in the first approach but there can be insufficient video data for the current direction in the second approach. The direction of the head, represented in elevation and azimuth if a spherical coordinate system is used, can be encoded and transmitted to the far-end to inform the camera of the current direction.

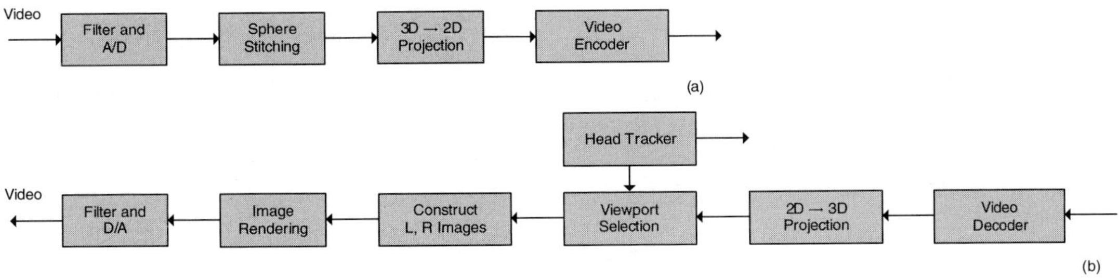

Fig. 10.8 Omnidirectional video signal processing operations of VR. (a) Transmitter side. (b) Receiver side.

At the HMD, the video is decoded and the areas corresponding to the current FOV are presented to the displays in front of the left and right eyes, after appropriate adjustment for stereoscopic effects. Figure 10.8 shows the video signal processing operations of a VR system. If the delay requirements are relaxed, e.g., from real-time, application of a streaming protocol instead of RTP, the AM of RLC, or more HARQ re-transmissions at PHY can be considered as in the case of audio. Keeping the TM or UM of RLC but replacing RTP/UDP with TCP is another option that can be employed when the delay budget is enough.

10.4.4 Controlling Quality–Capacity Tradeoff of Immersive Media

Since the ambisonic order affects audio quality and network capacity, it can be expected that the order is considered in the session negotiation, or to handle network congestion or link failure, feedback signaling from the receiver to the transmitter would carry the order as well as the bit-rate or audio bandwidth requested to use in the next encoding opportunity. Note that channel-based audio, such as the 5.1 or 7.1 speaker layouts, cannot be controlled for the tradeoff between quality and capacity as gracefully as scene-based audio. For the video, the FOV is a key parameter to be considered in the session negotiated, or included in the feedback information, accompanying control information that directs the bit-rate or resolution. Finally, the head tracking information can help the transmitter handle audio or video efficiently. Figure 10.4 illustrates a case when a cylindrical coordinate system is used to represent the orientation of head, and only the azimuth ϕ needs to be fed back.

10.5 References

3GPP. 2017a. TR 38.802 V14.0.0 Study on New Radio Access Technology; Physical Layer Aspects. March.

3GPP. 2017b. TR 38.804 V14.0.0 Study on New Radio (NR) Access Technology; Radio Interface Protocol Aspects. March.

3GPP. 2017c. TS 23.501 V1.0.0 System Architecture for the 5G System. December.

3GPP2. 2008. C.S0084-000-A V1.0 Overview for Ultra Mobile Broadband (UMB) Air Interface Specification. December.

Arıkan, E. 2009. Channel Polarization: A Method for Constructing Capacity-Achieving Codes for Symmetric Binary-Input Memoryless Channels. *IEEE Transactions on Information Theory*, **55**(7).

Cheng, C. I., and Wakefield, G. H. 2001. Introduction to Head-Related Transfer Functions (HRTFs): Representations of HRTFs in Time, Frequency, and Space. *The Journal of the Audio Engineering Society*, **49**(4).

Daniel, J. 2001. Représentation de Champs Acoustiques, Application à la Transmission et à la Reproduction de Scènes Sonores Complexes dans un Contexte Multimédia. *Thèse de Doctorat de l'Université Paris 6*, Juillet.

Gallager, R. G. 1962. Low-Density Parity-Check Codes. *IRE Transactions Information Theory*, **IT-8**(1).

Gardner, W. G., and Martin, K. D. 1995. HRTF Measurements of a KEMAR. *The Journal of the Acoustical Society of America*, **97**(6).

Holma, H., and Toskala, A. 2011. *LTE for UMTS: Evolution to LTE-Advanced*. 2nd edn. Wiley.

IEEE. 2005. Standard 802.16e-2005. Part 16: Air Interface for Fixed and Mobile Broadband Wireless Systems – Amendment for Physical and Medium Access Control Layers for Combined Fixed and Mobile Operation in Licensed Band. December.

ITU-R. 2016a. M.2012 Detailed Specifications of the Terrestrial Radio Interfaces of International Mobile Telecommunications-Advanced (IMT ADVANCED). March.

ITU-R. 2016b. Submission, Evaluation Process and Consensus Building for IMT-2020. June.

MacKay, D. J. C. 1999. Good Error-Correcting Codes based on Very Sparse Matrices. *IEEE Transactions on Information Theory*, **45**(2).

Marzetta, T. L. 2015. Massive MIMO: An Introduction. *Bell Labs Technical Journal*, **20**(March).

Nachbar, C., Zotter, F., Deleflie, E., and Sontacchi, A. 2011. AMBIX – A Suggested Ambisonics Format. *Ambisonics Symposium*, June.

NHK. 2011. 22.2 Multichannel Audio Format Standardization Activity. *STRL Broadcast Technology*.

Index